Study Guide
to Accompany Chang: **CHEMISTRY**

Study Guide to Accompany Chang: CHEMISTRY

Fifth Edition

Kenneth W. Watkins
Colorado State University

McGRAW-HILL, INC.
New York St. Louis San Francisco Auckland Bogotá Caracas Lisbon
London Madrid Mexico City Milan Montreal New Delhi
San Juan Singapore Sydney Tokyo Toronto

Study Guide to Accompany Chang: CHEMISTRY

Copyright ©1994, 1991 by McGraw-Hill, Inc. All rights reserved. Previously published under the title of *Study Guide for Chang's Chemistry*. Copyright ©1988, 1984, 1981 by McGraw-Hill, Inc. All rights reserved. Printed in the United States of America. Except as permitted under the United States Copyright Act of 1976, no part of this publication may be reproduced or distributed in any form or by any means, or stored in a data base or retrieval system, without the prior written permission of the publisher.

 This book is printed on recycled paper containing a minimum of 50% total recycled fiber with 10% postconsumer de-inked fiber.

3 4 5 6 7 8 9 0 SEM SEM 9 0 9 8 7 6 5

ISBN 0-07-011005-0

This book was set in Times Roman by Science Typographers, Inc.
The editors were Maggie Lanzillo, Jennifer Speer, and Jack Maisel;
the production supervisor was Janelle S. Travers.
Semline, Inc., was printer and binder.

CONTENTS

Preface vii
1 Tools of Chemistry **1**
2 Atoms, Molecules, and Ions **21**
3 Chemical Reactions I: Chemical Equations and Reactions in Aqueous Solution **45**
4 Chemical Reactions II: Mass Relationships **65**
5 The Gaseous State **85**
6 Thermochemistry **106**
7 Quantum Theory and the Electronic Structure of Atoms **127**
8 Periodic Relationships among the Elements **145**
9 Chemical Bonding I: Basic Concepts **163**
10 Chemical Bonding II: Molecular Geometry and Molecular Orbitals **185**
11 Intermolecular Forces and Liquids and Solids **207**
12 Physical Properties of Solutions **228**
13 Chemical Kinetics **244**
14 Chemical Equilibrium **268**
15 Acids and Bases: General Properties **288**
16 Acid-Base Equilibria **305**
17 Solubility Equilibria **328**
18 Chemistry in the Atmosphere **345**
19 Entropy, Free Energy, and Equilibrium **355**
20 Electrochemistry **368**
21 Metallurgy and the Chemistry of Metals **390**
22 Nonmetallic Elements and Their Compounds **405**
23 Transition Metal Chemistry and Coordination Compounds **418**
24 Nuclear Chemistry **433**
25 Organic Chemistry **448**
26 Organic Polymers: Synthetic and Natural **460**

PREFACE

This *Study Guide* is intended for use with the fifth edition of Raymond Chang's *Chemistry*. Its purpose is to help you learn chemistry. For this guide to be of maximum assistance, it should be incorporated into a plan for studying general chemistry. The *Study Guide* contains material to help you organize your studying, practice your problem-solving skills, and test yourself. Each chapter in the guide corresponds to one in the text. Each chapter is organized into four or five sections for easier presentation and comprehension. Every chapter contains the following main features.

Study Objectives. Each chapter section begins with a set of performance objectives. These objectives state in a straightforward manner what you must be able to do with the topics discussed in the section. One problem students frequently experience during a lecture is knowing what is important enough to put in notes. Previewing the objectives on specific topics before attending a lecture will help you focus on important material, listen better, and retain more from your first exposure to the material.

Reviews. The reviews provide a summary of essential material. They differ slightly in approach from the text. They are meant to reinforce and enhance your understanding of the material presented in the text and in lectures. The thorough descriptions and explanations provided by the reviews should help you solve the accompanying problems.

Example Problems. After each summary there are example problems with detailed solutions. These solutions often contain further explanation of important concepts. They emphasize how to initiate the solution. The steps toward the solution are outlined first and are then worked through. The detailed solutions should help you develop problem-solving skills. Almost every example contains Method of Solution and Calculation sections. The general outline followed is:

EXAMPLE

Statement of the problem.

METHOD OF SOLUTION

This section outlines the method of solution and relates the solution of the problem to the principles involved. It provides information on how to initiate the solution and what equations, if any, are applicable.

CALCULATION

This section shows how to set up equations and substitute into them to achieve a numerical answer.

COMMENT

Some example problems also contain a section in which helpful hints or relationships to previous material are pointed out.

True-False Questions. These questions provide a quick test of your familiarity with the terminology and principles in the chapter. Much of your success in chemistry will depend on your learning the meanings and applications of chemical and scientific terms. The answers to these questions are provided at the end of each chapter along with explanations of what is false about the false statements.

Self-Tests. Each chapter contains questions to test your knowledge of the material. The self-tests are organized into Self-Tests A, B1, and B2. Many of the questions in Self-Test A have been taken from old examinations that the author has written or collected from other chemistry instructors. Some self-tests contain a few general problems that require integration of concepts from previous material. Self-Tests B1 and B2 are designed to complement each other. The questions in B2 are the "reverse" of those in B1. Students sometimes feel "tricked" on an exam if a question is asked "backwards" from the way it was presented in lecture or quiz section. Usually no trick is involved. The instructor wants to know if you can think rather than memorize. Thus a question in section B1 may give information about X and asks you to find Y. The complementary question in B2 gives information about Y and asks you to find X. Chapters 18 and 21 to 26 contain descriptive material and do not emphasize problem solving. These chapters contain only Self-Test A.

Answers. Answers to all the problems are conveniently found at the end of each chapter. If you cannot work a significant number of the problems in the self-tests, you should go back and study the material on that subject in the textbook. Work through the example problems by *writing down* the main steps of the solutions. Don't look at the answers first. Act as if you are taking an exam. Use the answers only to check your own work.

Hints. The responsibility for organizing your study so that you learn the required material, of course, rests with you. No one will check to see whether you are taking good notes and are working the problems assigned in class. You must organize your study and test yourself in order to be certain that you know the material. A helpful rule is "Don't get behind." You will find that your college chemistry course moves much faster than your high school course did. Keep in mind that the content of the course builds on concepts and principles developed earlier in the course. General chemistry will be much more rewarding if you set aside time each day to learn the material as it is presented in class. For most chemistry courses, it is necessary to study 2 to 3 hours outside of class for every hour in class. Study could include reading the text, working lots of problems, and memorizing facts and vocabulary. Reading in science takes longer than reading in other subject areas. You cannot expect to "speed-read" a chemistry text. It is important that what you read makes sense. To this end, time must be set aside daily so that the concepts and principles have time to sink in. Cramming is not a proven or successful way to learn chemistry or any other science.

Hopefully you will establish a plan for studying chemistry that includes the suggestions mentioned above. Some of you may wish to read through a chapter in the textbook before using the *Study Guide*. Others may wish to read the *Study Guide* reviews before attending a lecture on that subject. Be sure to work the problems at the end of each chapter of the text as soon as they are assigned. Finally, work the true-false questions and self-tests to check your knowledge.

Learning chemistry, like learning anything else, takes time, effort, and concentration. In my teaching experience I have never had anyone say "I *plan to fail* your course." Yet, there are some who *fail to plan* and end up with the same unfortunate result. My hope is that you will be a successful learner and chemistry student.

ACKNOWLEDGMENTS

I gratefully acknowledge the assistance of Professor Raymond Chang, who made numerous suggestions to better correlate the *Study Guide* with his text. I also thank those who reviewed the manuscript for their insights and helpful suggestions.

I thank Jack Maisel, the editing supervisor, who has done an outstanding job of overseeing the editing and production of this text. In addition, it is a pleasure to acknowledge the enthusiastic support and assistance given to me by the members of the McGraw-Hill staff, especially Denise Schanck, Maggie Lanzillo, Jennifer Speer, and Beatrice Ruberto.

Kenneth W. Watkins

Chapter One
TOOLS OF CHEMISTRY

- Introduction
- Matter and Its Properties
- Units of Measure
- Math Review
- Factor-Label Method

INTRODUCTION

About Chemistry. Chemistry is the study of the properties of matter. *Matter* is defined as anything that has mass and occupies space. The three physical states of matter are solid, liquid, and gas. All matter exists in one or another of these three states, depending on the temperature and pressure of the surrounding environment. *Chemistry* is the branch of science that deals with the nature of matter and energy. Chemists are concerned with developing the tools used to study matter and the concepts useful in describing the properties of matter. They direct their efforts toward the purposeful changing of given forms of matter into new and different substances and to the discovery of the properties and uses of these new materials. Chemists usually observe matter and the changes it undergoes in the *macroscopic world*. This refers to the objects we can see and touch and deal with everyday. However, our interpretations of matter involve atoms and molecules and their properties. Because atoms and molecules are so extremely small, we refer to them as belonging to the *microscopic world*.

Chemists, as do other scientists, make use of the scientific method, which can be broken down into four parts: observation and experiment, hypothesis, laws, and theory. The first step is to define the problem clearly. That is, you must know what you are trying to find out about. During *experiments* observations are made and information is collected about the system being studied. After a large amount of data related to a certain phenomenon have been collected, sometimes the information can be summarized in a simple verbal or mathematical statement called a *law*. At this point a *hypothesis* may be formulated to provide a tentative explanation of the facts. A hypothesis is only temporary and is meant to be a working model or explanation. It is often adjusted as new information is discovered. As a hypothesis grows and successfully survives many experimental tests, it develops into a theory. A *theory* is a unifying principle that explains a large body of facts and laws. Besides providing explanations for the laws of science, its role is to aid in making predictions that lead to new knowledge. In fact theories are often tested by carefully devised experiments that confirm or disprove their predictions.

MATTER AND ITS PROPERTIES

STUDY OBJECTIVES

You should be able to:
1. Distinguish between elements, compounds, and solutions.
2. Distinguish between homogeneous and heterogeneous mixtures.
3. Distinguish between physical and chemical properties of a substance.

Elements and Compounds. A *pure substance* is a form of matter that has a definite composition and distinct properties. Examples are water, table salt, and iron. Just as each individual person has a set of characteristics, such as fingerprints and color of eyes and hair, each pure substance has characteristic properties. There are two types of pure substance: elements and compounds. A *chemical element* is a pure substance that cannot be decomposed into simpler substances by ordinary chemical reactions. Elements are the building blocks of which all compounds are composed. Nitrogen, oxygen, and iron are examples of pure substances that are elements. A number of common elements and their symbols are listed in the textbook in Table 1.1.

Compounds are pure substances that are composed of two or more elements. Compounds can be broken down into the elements of which they are composed by chemical means. Water and table salt, mentioned earlier, are compounds. The action of an electric current, called electrolysis, is one method that can be used to decompose both water and molten table salt into their constituent elements. Pure water consists of 89% oxygen and 11% hydrogen by mass. Pure salt contains 30% sodium and 61% chlorine. Because a compound has a constant composition, it exhibits distinct and characteristic properties, regardless of its source.

Mixtures. Pure substances can be brought together to form mixtures. *Mixtures* are combinations of two or more substances. They can be homogeneous or heterogeneous depending on the state of subdivision of the components. Salt water is a uniform mixture of table salt (NaCl) and water. On the ordinary scale of observation we cannot detect any chemical or physical differences between adjacent regions of the mixture. The original crystals of salt have dissolved and are dispersed evenly. The particles of salt are too small to observe. A mixture that has the same composition throughout is said to be a *homogeneous mixture*. Mixtures of gases are homogeneous. Dry, dust-free air, for example, is a homogeneous mixture of which the components are mostly nitrogen, oxygen, and argon along with a large number of trace gases. The mixture is essentially of constant composition anywhere on the Earth's surface. Polluted cites are excluded, of course. The properties of a homogeneous mixture vary since they depend on the percent composition. For example, the hardness of steel, a solid mixture of iron and carbon, depends on the percentage of carbon that is added to iron. Homogeneous mixtures are also called *solutions*.

Heterogeneous mixtures are not uniform in composition. And indeed the individual particles of their components can often be seen by the unaided eye. For example, when preparing home-made ice cream, you use a mixture of ice and rock salt. This is a heterogeneous mixture. The individual chunks (particles) of ice and salt are clearly visible. Also, because the particles are so large, they are not evenly dispersed. The composition of this mixture varies from place to place within the mixture itself. Natural air is actually a heterogeneous mixture. In addition to the gases mentioned above, air contains solid particles of pollen and dust. Dust particles, called *particulate matter*, can be blown up into the atmosphere by winds and injected by large smokestacks and by volcanic eruptions and meteors. Any mixture, whether it be homogeneous or heterogeneous, can be separated into its pure components by physical means.

Properties of Matter. *Physical properties* are those properties that can be measured and observed without changing the identity or composition of the substance. Physical properties include color, hardness, solubility, density, specific heat, melting point, and boiling point. *Physical changes* are those that take place with no change in chemical composition. Changes of a substance from one state of matter to another do not change its chemical composition and are examples of physical changes. The three forms of water we call ice, liquid water, and steam are all the same substances, just different physical states.

Chemical properties most often are descriptions of reactions that a substance undergoes when brought in contact with other substances. In a chemical reaction the original substance or substances are changed into new substances. When sodium metal and chlorine gas are heated together, a white solid known as table salt or sodium chloride is formed. That this is a chemical change is evident by the observation that sodium, a shiny metal, and chlorine, a pale yellow-green gas, have disappeared and in their place is a substance with a completely new set of properties. Table salt is a white solid that melts at extremely high temperatures.

All properties of matter are either extensive or intensive properties. *Extensive properties* depend on the amount of matter being considered. Volume and mass are examples. In contrast, temperature and density are two properties that do not depend on the amount of mass present. Thus, they are *intensive properties*.

EXAMPLE 1.1 Chemical and Physical Properties

The following are properties of the element silicon; classify them as physical or chemical properties:
a. Melting point, 1410°C
b. Reacts with fluorine to form silicon tetrafluoride
c. Gray
d. Not affected by most acids

METHOD OF SOLUTION

Physical properties can be observed without a change in composition, while chemical properties describe reactions with other substances.
 a. Melting involves a change in physical state but no chemical change. *Answer:* The melting point is a physical property.
 b. This statement describes the change of silicon into another substance on reaction with fluorine. *Answer:* The reaction is a chemical property.
 c. *Answer:* The color of a substance is a physical property. No change in composition occurs while observing the color.
 d. *Answer:* The lack of reactivity with another substance or class of substances such as acids is a chemical property.

EXAMPLE 1.2 Homogeneous or Heterogeneous Mixture

Classify each of the following as a homogeneous or a heterogeneous mixture:
a. The beverage tea
b. Oil and water
c. Cow's milk
d. Wine

METHOD OF SOLUTION

Recall that homogeneous mixtures are uniform throughout, while heterogeneous mixtures have components which can be physically observed to be separate.
 a. The mixture called tea is uniform throughout and no particles of tea can be observed (if there are no tea leaves). *Answer:* A homogeneous mixture.
 b. Since oil floats on water, the oil and water components can be observed to be separate. Samples from the top part of the mixture are different from samples taken from the bottom. *Answer:* A heterogeneous mixture.
 c. Cow's milk contains fats and solids suspended in water. On standing, the cream (fat) will rise to the top and the solids will settle. *Answer:* A heterogeneous mixture.
 d. Wine contains ethyl alcohol, water, flavor components, dye molecules, and many other substances. The composition is uniform throughout. No solid particles are visible in the liquid. *Answer:* A homogeneous mixture.

EXAMPLE 1.3 Classification of Matter

Classify each of the following as an element, a compound, or a mixture:
a. Iodine
b. Dry ice
c. Gasoline

4 / *Tools of Chemistry*

METHOD OF SOLUTION
a. Iodine cannot be broken down into simpler substances. *Answer:* It is an *element* whose symbol I can be found in the periodic table of the elements in Group 7A.
b. *Answer:* Dry ice is a *compound* made from carbon and oxygen. Its chemical name is carbon dioxide. The percentages of carbon and oxygen in dry ice are constant.
c. The properties of gasoline vary. Regular, unleaded, and premium grades of gasoline have different octane ratings, a real clue that gasoline is a mixture. *Answer:* Gasoline is a *mixture* of liquid hydrocarbons and antiknock compounds.

EXAMPLE 1.4 Chemical and Physical Changes

Classify the following changes as physical or chemical:
a. Purification of water by distillation
b. Gasoline burning in air

METHOD OF SOLUTION
a. Distillation involves phase changes such as vaporization and its reverse process, condensation. These physical changes are used to separate water from dissolved components such as salt. Water is made more pure by distillation, but it is still water. *Answer:* A physical change.
b. Liquid gasoline undergoes combustion with oxygen (burning) to product carbon dioxide and water. This process is accompanied by the evolution of heat. *Answer:* A chemical change.

UNITS OF MEASURE

STUDY OBJECTIVES

You should be able to:
1. Express base units and derived units in terms of the SI.
2. Use prefixes in the SI.
3. Convert temperatures between the Celsius and Fahrenheit scales.

The International System of Units. Since 1960 a coherent system of units, known as the SI, has been in effect and is gaining acceptance among scientists and engineers. SI is the abbreviation for *Le Système International d'Unités*. Most of the units used in the textbook will be from the SI. A few units from outside the SI are still used because some fields of science have developed specialized units for their convenience.

The SI has as its *base units* the kilogram (kg) for mass, the meter (m) for length, the second (s) for time, the kelvin (K) for temperature, and the mole (mol) for amount of substance. Combinations of base units produce *derived units*. For example, velocity is distance traveled per unit of time. Therefore, velocity has units of meters per second (m/s) and is a derived unit rather than a base unit. Examples of derived units for selected physical quantities commonly used in chemistry are given in Table 1.1.

Note that unit names are sometimes derived from the proper names of scientists. In these cases the unit *symbols* are capitalized, but the unit *names* are not. For example, the newton has the symbol N.

Examples of non-SI units in current use are the angstrom (Å, a unit of length); the atmosphere (atm, a unit of pressure); millimeters of mercury (mm Hg, another unit of pressure); the calorie (cal, a unit of energy); and the liter (L, a unit of volume).

Prefixes. A major advantage of the SI is that it uses the decimal system. A complete list of the SI prefixes is given in Table 1.2. Be sure to memorize each prefix, its symbol, and its meaning, as these will be used throughout your study of chemistry. Note that the prefixes are used with all the base units. This

Table 1.1 Some Derived Units

Quantity	Definition of Unit	Name of Unit	Symbol
Force	kg m/s^2	newton	N
Pressure	N/m^2	pascal	Pa
Energy	kg m^2/s^2	joule	J
Frequency	cycles/s or s^{-1}	hertz	Hz

is a major difference between the current U.S. system and the SI system. In the U.S. system, a pound can be divided into 16 ounces, a gallon into 4 quarts, and a mile into 5280 feet. In the SI, all base and derived units are made into larger or smaller units that are based on the decimal system, rather than on the random divisions of the U.S. system. In the SI the same prefix can be applied to any base unit or derived unit. Thus the prefix *milli* can be used to describe a unit that is $\frac{1}{1000}$ of a gram, or $\frac{1}{1000}$ of a meter, or $\frac{1}{1000}$ of any SI unit:

1 milligram = $\frac{1}{1000}$ gram

1 millimeter = $\frac{1}{1000}$ meter

1 millisecond = $\frac{1}{1000}$ second

The prefixes in Table 1.2 apply to any unit in the SI.

Table 1.2 Prefixes Used in the SI

Prefix	Symbol	Meaning	
tera	T	1,000,000,000,000	or 10^{12}
giga	G	1,000,000,000	or 10^{9}
mega	M	1,000,000	or 10^{6}
kilo	k	1,000	or 10^{3}
deci	d	0.1	or 10^{-1}
centi	c	0.01	or 10^{-2}
milli	m	0.001	or 10^{-3}
micro	μ	0.000 001	or 10^{-6}
nano	n	0.00 000 001	or 10^{-9}
pico	p	0.000 000 000 001	or 10^{-12}

Temperature Scales. Chemists use two temperature scales, the Kelvin scale (K) and the Celsius scale (°C). A third scale, the Fahrenheit scale (°F), is commonly used in the United States but not in the rest of the world. To make conversions between the Fahrenheit and Celsius scales requires a scale shift and a zero shift. The Celsius scale has 100 degrees between the boiling point and the freezing point of water. In the Fahrenheit scale there are 180 degrees between the boiling and freezing points of water. This means that one degree Fahrenheit is smaller than one degree Celsius by a factor of $\frac{100}{180}$, or $\frac{5}{9}$. This is the shift that adjusts for the differences in scale. Thus, positive Celsius temperatures are *always* smaller than the corresponding Fahrenheit temperatures. However, you cannot simply multiply degrees Fahrenheit by $\frac{5}{9}$ to get degrees Celsius. This is because 0° on each scale corresponds to a different temperature—0°F is much lower than 0°C. To correct for this shift in starting points (the zero shift), you must subtract 32 from the Fahrenheit reading. After doing this, you can multiply by $\frac{5}{9}$ to get the Celsius temperature.

The equation for converting from degrees Fahrenheit to degrees Celsius is

$$°C = (°F - 32°F) \times \frac{5°C}{9°F}$$

To convert from degrees Celsius to degrees Fahrenheit, rearrange the equation. See Example 1.6. The Kelvin scale is discussed in Chapter 5.

EXAMPLE 1.5 SI Prefixes

Express the following amounts using prefixes from the SI:
a. Convert 10,000 m to kilometers.
b. Convert 0.02 g to centigrams.
c. Convert 0.000005 L to microliters.
d. Convert 10^6 J to megajoules.

METHOD OF SOLUTION

a. 10,000 m is 10×10^3 m; substitute k (kilo) for 10^3. *Answer:* 10 km.
b. 0.02 g is 2×10^{-2} g; substitute c (centi) for 10^{-2}. *Answer:* 2 cg.
c. 0.000005 L is 5×10^{-6} L; substitute μ (micro) for 10^{-6}. *Answer:* 5 μL.
d. 10^6 J is the same as 1×10^6 J. Substitute M (mega) for 10^6. *Answer:* 1 MJ.

EXAMPLE 1.6 Temperature Scale Conversion

The freezing point of a 50-50 mixture of antifreeze (ethylene glycol) and water is $-36.5°C$. Convert this temperature to degrees Fahrenheit.

METHOD OF SOLUTION

First write the equation for temperature conversions:

$$°C = (°F - 32°F) \times \frac{5°C}{9°F}$$

Next substitute $-36.5°C$ and solve for degrees Fahrenheit.

CALCULATION

$$-36.5°C = (°F - 32°F) \times \frac{5°C}{9°F}$$

$$-65.7°C = (°F - 32°F)$$

$$°F = -33.7°F$$

MATH REVIEW

STUDY OBJECTIVES

You should be able to:
1. Express numbers and perform calculations in exponential notation.
2. Determine the number of significant figures in a given number and in a number that is the result of a mathematical operation.

Exponential Notation. Many of the quantities (numbers) encountered in chemistry are more easily manipulated when they are written in a form known as scientific or exponential notation. Very large or very small numbers are expressed as $N \times 10^n$, where N is a number between 1 and 10 and n corresponds to the exponents $1, 2, 3, \ldots$. Recall that the exponent tells you how many times to multiply the number 10 by itself. Thus,

$$10^3 = 10 \times 10 \times 10 = 1000$$
$$10^2 = 10 \times 10 = 100$$
$$10^1 = 10 = 10$$
$$10^0 = 1$$

To write the number 5000 in exponential notation, we could rewrite it as 5×1000. But 1000 can be written $10 \times 10 \times 10$, or just 10^3. Therefore, 5000 can be written as 5×10^3.

More simply, to find the exponent of 10, just count the number of places that the decimal point must be moved to the left to give the number N. For the number 5000, moving the decimal point three places to the left gives 5×10^3.

Fractional numbers can also be expressed in scientific notation, but in this case the decimal point will be moved to the right and the exponent will be a negative number ($-n$). For the number 0.05, the decimal point must be moved two places to the right to give N (a number between 1 and 10). In exponential notation 0.05 becomes 5×10^{-2}.

Multiplication and division of numbers that are expressed in exponential notation are accomplished by operating on the coefficients (N) and exponentials (n) separately. Recall when multiplying two exponential numbers that the exponents are added. See Example 1.8. When dividing exponential numbers, we subtract the exponent in the denominator from the exponent in the numerator, as shown in Example 1.9.

In some calculations you must take the square or cube root of an exponential number. The general rule is

$$(X^a)^n = X^{an}$$

where n is $\frac{1}{2}$ for the square root, $\frac{1}{3}$ for the cube root, and so on. Thus,

$$\sqrt{9.0 \times 10^{-16}} = (9.0 \times 10^{-16})^{1/2} = (9.0)^{1/2} \times 10^{-16/2}$$

Taking the square root of the coefficient and exponent separately, we get

$$\sqrt{9.0 \times 10^{-16}} = 3.0 \times 10^{-8}$$

Significant Figures. Measurements, and calculations based on measurements, must be reported so as to convey information about the number of meaningful digits. The *significant digits* in a number are those that give reasonably reliable information. If we measure the length of a desk top with a meter stick calibrated in millimeters and find the length to be 972.5 mm, then the digits 9, 7, and 2 mean there are 9 hundreds, 7 tens, and 2 ones, and because of the calibrations, these digits are certain. Five-tenths, however, is an estimate because it falls between the finest calibrations. Someone else might read it as 0.6 or 0.4. To express the result with more digits, as in 972.523, would be to include two meaningless digits. This practice should be avoided. (Note that although the 5 is uncertain, it is still meaningful. With the 5 being uncertain, the 2 and 3 must be meaningless.)

The result could be expressed with error limits as 972.5 ± 0.1 mm. This measurement provides three certain digits and one uncertain digit. All of these provide useful information, and we say the number 972.5 has four significant digits or figures.

Always take care that the numbers you write reflect the proper number of meaningful digits. The process of determining the correct number of significant figures after a calculation depends on the type of calculation. First let us review the rules for expressing significant figures.

Guidelines for Writing Significant Figures. To determine the number of significant digits that are present in a *written number*, use the following rules:

- Any nonzero digit is significant.
- Zeros between nonzero digits are significant. The digits in the number 506 mean there are 5 hundreds, 0 tens, and 6 ones. The zero digit is significant because it tells us there are no tens. All three digits are significant.

- Leading zeros are those to the left of the first nonzero digit, as in 0.02. They are not significant figures. These zeros serve to mark the position of the decimal point. The rule is that zeros used only to place the decimal point are *not* significant. Thus 0.6 g and 0.0006 g both have only one significant figure.
- Zeros written to the right of nonzero digits are called trailing zeros. They are considered to be significant only if the number contains a *decimal point*. Thus the number 1.0 mg has two significant figures, and 0.500 mg has three significant figures. The following figure further exemplifies these rules:

nonsignificant zeros → 0.00260740 (six significant figures), with significant zeros indicated

234,600 (four significant figures)

- Trailing zeros in numbers that do not contain decimal points may or may not be significant. The number 450 g has two significant figures if the number is 450 ± 10 and three significant figures if the number is 450 ± 1. The number of significant figures can be indicated unambiguously with scientific notation. The number 450 can be written as 4.5×10^2 if it has two significant figures and as 4.50×10^2 if it has three significant figures.
- Exact numbers are those obtained from a definition. Thus, when we say 1 ft = 12 in, both the 1 and the 12 are exact numbers, and they are said to have an infinite number of significant figures. Thus $1.00\ldots$ ft = $12.00\ldots$ in. On the other hand, conversion factors between unit systems usually are not exact numbers. For example, the conversion from quarts to liters is 1.06 qt = 1 L. Here the 1 is one exact liter, but the 1.06 has three significant figures.

Significant Figures and Calculations. In most cases the numbers we measure are used to calculate other quantities. Care must be exercised to report the proper number of significant figures in the calculated result. This is extremely important when electronic calculators are used because they give answers with 8 and 10 digits. The rule used is that the accuracy of the result is limited by the least accurate measurement. In *multiplication* and *division*, the number of significant figures in a calculated result is determined by the original measurement that has the *fewest* number of significant digits. Thus, if a number with three significant figures is multiplied by a number with four significant figures, the computed result will have only three significant figures. Even if a calculator computes eight digits, most of them will be meaningless in terms of their physical significance.

For example, the density of an object is its mass divided by its volume. The density of an object with a mass of 2.157 g that occupies 1.66 cm³ can be known to only three significant figures:

$$\frac{2.157 \text{ g}}{1.66 \text{ cm}^3} = 1.29939759 \text{ g/cm}^3 \quad \leftarrow \text{calculator result}$$

$$= 1.30 \text{ g/cm}^3 \quad \leftarrow \text{rounded off}$$

Rounding Off Numbers. A number is rounded off to the desired number of significant figures by dropping one or more digits to the right. The following rules should be observed when rounding off numbers:

1. When the first digit dropped is less than 5, the last digit retained remains unchanged. Thus, 4.2643 rounded to three significant figures becomes 4.26.
2. When the first digit dropped is equal to or greater than 5, the last digit retained is increased by 1. Thus, 2.38 and 2.35 rounded to two significant figures both become 2.4.
3. For calculations involving more than one step, carry out the calculations without rounding off the intermediate results. Round off the final answer to the correct number of significant figures.

Addition and Subtraction. In *addition* and *subtraction*, the number of significant figures in the answer depends on the original number in the calculation that has the *fewest digits to the right of the decimal point*.

For example, when adding,

```
  102.226
    2.51
  736.      ← fewest digits to the right
  ──────      of the decimal point
  840.736
       ‿‿‿
        └─ nonsignificant digits

  841       ← rounded off
```

and when subtracting,

```
  102.25
 − 99.3     ← fewest digits to the right
  ─────       of the decimal point
    2.95
     ‿
     └─ nonsignificant digit

    3.0     ← rounded off
```

It is important to remember, when adding or subtracting, that the number of significant figures in the answer is not determined by the quantity having the fewest significant figures, which is true for multiplication and division.

EXAMPLE 1.7 Exponential Notation

Write the following numbers using exponential notation:
a. 7,620,000
b. 0.000495

METHOD OF SOLUTION
a. 7,620,000—Count the number of places that the decimal point must be moved to the left to give a number between 1 and 10, in this case, 7.62. Since this requires six places, the exponent of 10 must be 6. *Answer:* 7.62×10^6.
b. 0.000495—Fractional numbers like this have negative exponents in scientific notation. Count the number of places that the decimal point must be moved to the right to give 4.95. Since this requires a move of four places, the exponent must be -4. *Answer:* 4.95×10^{-4}.

EXAMPLE 1.8 Math Operations with Exponential Notation

Multiply 4.0×10^4 by 3.5×10^{-6}.

METHOD OF SOLUTION

$$(4.0 \times 10^4)(3.5 \times 10^{-6})$$

Regroup so that the coefficients are separated from the exponentials, and add the exponents algebraically.

CALCULATION

$$(4.0 \times 3.5) \times 10^{4+(-6)} = 14 \times 10^{-2} = 1.4 \times 10^{-1}$$

COMMENT

In these days of inexpensive calculators, most people use an electronic calculator for these types of calculations. On most calculators, you would carry out the above multiplication problem in the following way:

$$(4.0 \times 10^{4})(3.5 \times 10^{-6})$$

When entering the factors, the exponents 4 and −6 are entered as [EXP] then 4 and [EXP] then [+/−] 6. Thus, 4.0×10^{4} is entered as 4.0 [EXP] 4, and 3.5×10^{-6} is entered as 3.5 [EXP] [+/−] 6. The calculation, if written out, would look like the following:

4.0 [EXP] 4 [×] 3.5 [EXP] [+/−] 6 [=]

The calculator will display the answer either as just 0.14 or in exponential form as 1.4^{-01} or as 1.4 −01 depending on the brand of calculator.

EXAMPLE 1.9 Math Operations with Exponential Notation

Divide 4.2×10^{-7} by 5.0×10^{-5}.

METHOD OF SOLUTION

$$\frac{4.2 \times 10^{-7}}{5.0 \times 10^{-5}}$$

Divide the coefficients, and subtract the exponent in the denominator from the exponent in the numerator.

CALCULATION

$$= 0.84 \times 10^{-7-(-5)}$$

$$= 0.84 \times 10^{-7+5}$$

$$= 0.84 \times 10^{-2}$$

Because 0.84 is less than 1, the usual practice is to move the decimal point to give a coefficient between 1 and 10. Move the decimal point one place to the right and add −1 to the exponent.

$$= 8.4 \times 10^{-3}$$

COMMENT

On most calculators, you would carry out the above division problem,

$$\frac{4.2 \times 10^{-7}}{5.0 \times 10^{-5}}$$

by entering the numerator first, then pushing the [÷] key, and then entering the denominator followed by the [=] key. The exponents −7 and −5 are entered as [EXP] then [+/−] 7 and [EXP] then [+/−] 5. The calculation, if written out, would look like this:

4.2 [EXP] [+/−] 7 [÷] 5.0 [EXP] [+/−] 5 [=]

The calculator will display the answer as 8.4^{-03} or 8.4 −03; both stand for 8.4×10^{-3}.

EXAMPLE 1.10 Significant Figures

Determine the number of significant figures in each of the following numbers:
a. 6.02
b. 0.012
c. 1.23×10^7
d. 1.5400

METHOD OF SOLUTION

Refer to the guidelines for writing significant figures.
a. Recall that zeros between nonzero digits are significant. *Answer:* three significant figures.
b. Zeros to the left of the first nonzero digit are not significant. *Answer:* two significant figures.
c. The use of scientific notation implies that only significant figures are shown. *Answer:* three significant figures.
d. When a number contains a decimal point and has trailing zeros, these zeros are significant. *Answer:* five significant figures.

EXAMPLE 1.11 Significant Figures

Carry out the following operations, rounding off the answer to the correct number of significant digits:
a. Add:

    6.356
    29.
    115.91
    ‾

b. Subtract:

    287.12
    −95.333
    ‾

c. $7.25 \times 10^{12} \div 92$

METHOD OF SOLUTION
a. The sum is 151.266, but it is limited to three significant figures. None of the digits to the right of the decimal point are significant. *Answer:* 151.
b. *Answer:* 191.787; round off to 191.79.
c. *Answer:* 7.9×10^{10}; two significant figures.

FACTOR-LABEL METHOD

STUDY OBJECTIVES

You should be able to:
1. Construct conversion factors from given equalities.
2. Solve problems using the factor-label method.
3. Solve problems involving density.

Conversion Factors. The factor-label, or dimensional analysis, method is a simple but powerful technique that can be applied to a great variety of calculational problems. In this method the units (labels) of the quantities involved are used as a guide in setting up a calculation. The method is based on the use of conversion factors and on the idea that units or dimensions of various quantities can be handled algebraically in the same way as numbers are handled.

Conversion factors are constructed from equalities such as the following:

$$2.54 \text{ cm} = 1 \text{ in}$$

$$1000 \text{ g} = 1 \text{ kg}$$

$$24 \text{ h} = 1 \text{ day}$$

From the first one we can also write "2.54 cm per inch." Likewise, there are 1000 grams per kilogram and 24 h per day. Since the "per" can mean "divided by," we can write the following conversion factors:

$$\frac{2.54 \text{ cm}}{1 \text{ in}}, \quad \frac{1000 \text{ g}}{1 \text{ kg}}, \quad \text{and} \quad \frac{24 \text{ h}}{1 \text{ day}}$$

These conversion factors are examples of *unit factors*. Each unit factor is equal to 1 (unity) because the numerator equals the denominator.

The Factor-Label Method. If we treat units in the same fashion as numbers are treated in algebra, then multiplying inch by inch yields inches squared, or square inches, or in^2. One hour divided by 1 h is equal to 1, and we say that the units cancel:

$$\frac{h}{h} = 1$$

These two operations can be put together in a problem: How many hours are there in 5 days? First write down in algebraic form what is being asked:

$$? \text{ h} = 5 \text{ days}$$

Hours and days are related by the equality

$$24 \text{ h} = 1 \text{ day}$$

This equality can be made into two unit factors:

$$\frac{24 \text{ h}}{1 \text{ day}} \quad 24 \text{ h per day}$$

and

$$\frac{1 \text{ day}}{24 \text{ h}} \quad 1 \text{ day per 24 h}$$

Inserting the first unit factor into the algebraic equation, we obtain

$$? \text{ h} = 5 \text{ days} \times \frac{24 \text{ h}}{1 \text{ day}}$$

Now treating the units algebraically, the days "cancel":

$$? \text{ h} = 5 \text{ days} \times \frac{24 \text{ h}}{1 \text{ day}}$$

and taking $5 \times 24 = 120$, we obtain

$$? \text{ h} = 120 \text{ h}$$

With a little practice, the correct conversion factor can be chosen every time. In this case, we wanted hours for the units of the answer; therefore, we put hours in the *numerator* of the conversion factor. Also, we wanted the unit days to cancel; therefore, we put days in the denominator. Writing "? h" first serves as a guide to remind us of what units we want in the result. Note that we *might* have picked the second of the two conversion factors. In this case we would have obtained

$$? \text{ h} = 5 \text{ days} \times \frac{1 \text{ day}}{24 \text{ h}} = 0.208 \frac{\text{day}^2}{\text{h}}$$

The units in this answer are incorrect and tell us we have chosen the wrong conversion factor. As you can see, the units serve as a good check of your answer.

When several steps are necessary to carry out the conversion, a "road map" is very helpful. A football field has a length of 100.00 yd. What is its length in meters? We know that 2.540 cm = 1 in. First we convert the yards to inches, then the inches to centimeters, and then the centimeters to meters. Our road map, or plan, is

$$\text{yards} \rightarrow \text{inches} \rightarrow \text{centimeters} \rightarrow \text{meters}$$

Each arrow requires a conversion factor. The first conversion is within the U.S. system:

$$1 \text{ yd} = 36 \text{ in}$$

The second conversion is a U.S.-to-SI conversion (1 in = 2.540 cm), and the third is a conversion within the SI:

$$10^{-2} \text{ m} = 1 \text{ cm}$$

Writing the problem in algebraic form, we get

$$? \text{ m} = 100.00 \text{ yd}$$

Now we write the first unit factor with yards in the denominator so that the units of yards cancel:

$$? \text{ m} = 100.00 \text{ yd} \times \frac{36 \text{ in}}{1 \text{ yd}}$$

Write the second unit factor with inches in the denominator so that the units of inches cancel and units of centimeters are left:

$$? \text{ m} = 100.00 \text{ yd} \times \frac{36 \text{ in}}{1 \text{ yd}} \times \frac{2.540 \text{ cm}}{1 \text{ in}}$$

Then we multiply by the third conversion factor so that the centimeter units cancel and the meter units remain:

$$? \text{ m} = 100.00 \text{ yd} \times \frac{36 \text{ in}}{1 \text{ yd}} \times \frac{2.540 \text{ cm}}{1 \text{ in}} \times \frac{10^{-2} \text{ m}}{1 \text{ cm}}$$

Carrying out the indicated operations yields

$$? \text{ m} = 91.44 \text{ m}$$

How many significant figures can we show in the answer? Remember to look for the multiplier with the least number of significant figures. Let us say that the measured length of the football field was accurate to $\frac{1}{100}$th of a yard. That is 100.00 ± 0.01 yd, which corresponds to five significant figures. In the conversion factor the exact number of inches in a yard is 36. The number 36 has an infinite number of significant figures. Similarly, the conversion factor 2.540 cm per inch, *unlike* many conversion factors between unit systems, happens to have an infinite number of significant figures! This is exceptional because the inch is now defined as exactly 2.540... cm. (Yes, the U.S. inch is defined in terms of the metric system.) Also, the definition of *centi* as 0.01 is exact. Therefore, the number in the calculation with the fewest significant figures is 100.00, which limits the number of significant figures in our answer to five. The result should be expressed as 91.440 m.

EXAMPLE 1.12 The Factor-Label Method

How long will it take to fly from Denver to New York, a distance of 1631 miles, at a speed of 815 km/h?

METHOD OF SOLUTION

First write the question

$$? \text{ h} = 1631 \text{ mi}$$

In this case, the speed provides the conversion factor converting distance to time. The statement "815 km per hour" means 815 km is equivalent to 1 hour of time:

$$\frac{815 \text{ km}}{1 \text{ h}} \quad \text{or} \quad \frac{1 \text{ h}}{815 \text{ km}}$$

However, we must first convert 1631 mi to kilometers. Our plan is

$$\text{miles} \rightarrow \text{kilometers} \rightarrow \text{hours}$$

Here we will shorten the calculation by looking up the following equivalence in a handbook:

$$1.61 \text{ km} = 1 \text{ mi}$$

CALCULATION

Write the unit factor relating miles to kilometers so that the units of miles cancel. Then insert the conversion factor relating kilometers to time in such a way that the units of kilometers cancel:

$$? \text{ h} = 1631 \text{ mi} \times \frac{1.61 \text{ km}}{1 \text{ mi}} \times \frac{1 \text{ h}}{815 \text{ km}}$$

$$? \text{ h} = 3.2219754 \text{ h}$$

Round to three significant figures:

$$? \text{ h} = 3.22 \text{ h}$$

COMMENT

Two factors in the above calculation have three significant figures, and one factor has four. Therefore, the answer cannot have more than three significant figures.

EXAMPLE 1.13 Using Density

The density of plutonium metal (Pu) is 19.8 g/cm^3. What is the volume in cubic centimeters of 21 lb Pu?

METHOD OF SOLUTION

Write the problem in algebraic form:

$$? \text{ cm}^3 = 21 \text{ lb Pu}$$

Since we are given the density in grams per cubic centimeters, the pounds of plutonium must first be converted to grams. One pound is 453.6 g:

$$453.6 \text{ g} = 1 \text{ lb}$$

The road map is

$$\text{pounds} \rightarrow \text{grams} \rightarrow \text{cm}^3$$

CALCULATION

Two unit factors are required:

$$?\ \text{cm}^3 = 21\ \text{lb Pu} \times \frac{453.6\ \text{g}}{1\ \text{lb}} \times \frac{1\ \text{cm}^3}{19.8\ \text{g Pu}}$$

$$= 481.0909\ \text{cm}^3$$

Round to two significant figures:

$$?\ \text{cm}^3 = 4.8 \times 10^2\ \text{cm}^3$$

COMMENT

Note that the second conversion factor is the reciprocal of the density. If there are 19.8 g Pu per cm^3, then there is 1 cm^3 per 19.8 g Pu. In the case of Pu, 1 cm^3 of Pu is equivalent to 19.8 g Pu.

EXAMPLE 1.14 Using Density

What is the mass in grams of a lead (Pb) brick that measures 8.1 in × 3.9 in × 2.0 in? The density of Pb is 11.4 g/cm^3.

METHOD OF SOLUTION

Density is defined by the formula

$$\text{density} = \frac{\text{mass}}{\text{volume}}$$

If you know any two of the variables in the equation, you can solve for the remaining unknown. Upon rearranging, we get

$$\text{mass} = \text{density} \times \text{volume}$$

This equation tells us that the mass of an object can be determined if we know its density and volume. In this problem we are given the dimensions of a Pb brick, which we can use to calculate the volume in cubic inches since volume = length × width × height. However, the volume units appearing in the density are cubic centimeters. We must first convert the volume in cubic inches to cubic centimeters:

$$?\ \text{cm}^3 = (8.1\ \text{in} \times 3.9\ \text{in} \times 2.0\ \text{in}) = 63.2\ \text{in}^3$$

We will carry one extra significant figure and drop it at the end. Each inch could be converted to centimeters separately, but we can simplify the notation by writing

$$\frac{2.54\ \text{cm}}{1\ \text{in}} \times \frac{2.54\ \text{cm}}{1\ \text{in}} \times \frac{2.54\ \text{cm}}{1\ \text{in}} = \frac{(2.54\ \text{cm})^3}{1\ \text{in}^3} = 16.4\ \text{cm}^3/\text{in}^3$$

$$?\ \text{cm}^3 = 63.2\ \text{in}^3 \times \frac{16.4\ \text{cm}^3}{1\ \text{in}^3} = 1040\ \text{cm}^3$$

CALCULATION

$$\text{mass} = \text{density} \times \text{volume} = \frac{11.4\ \text{g}}{1\ \text{cm}^3} \times 1040\ \text{cm}^3 = 1190\ \text{g Pb}$$

Our answer to two significant figures is

$$\text{mass} = 1.2 \times 10^4 \text{ g Pb}$$

EXAMPLE 1.15 The Factor-Label Method

The *Voyager 2* mission to the outer planets of our solar system transmitted by radio signals many spectacular photographs of Neptune. Radio waves, like light waves, travel at a speed of 3.00×10^8 m/s. If Neptune was 2.75 billion miles from Earth during these transmissions, how many hours were required for radio signals to travel from Neptune to Earth?

METHOD OF SOLUTION

First state the problem:

$$? \text{ h} = 2.75 \times 10^9 \text{ mi}$$

The speed of light can be written as a conversion factor because 1 s of time is equivalent to 3.00×10^8 m of distance:

$$\frac{3.0 \times 10^8 \text{ m}}{1 \text{ s}} \quad \text{or} \quad \frac{1 \text{ s}}{3.00 \times 10^8 \text{ m}}$$

This will convert units of distance into units of time in seconds, but first the distance must be expressed in meters rather than miles. To convert from the U.S. system to the SI, we can use 1.61 km = 1.0 m. In general, our plan will be

distance → time

miles → km → m → s → min → h

Notice that the time in seconds must be converted to hours in order to answer the question.

CALCULATION

Insert one conversion factor per arrow in the road map:

$$? \text{ h} = 2.75 \times 10^9 \text{ mi} \times \frac{1.61 \text{ km}}{1.0 \text{ mi}} \times \frac{10^3 \text{ m}}{1.0 \text{ km}} \times \frac{1 \text{ s}}{3.00 \times 10^8 \text{ m}} \times \frac{1 \text{ min}}{60 \text{ s}} \times \frac{1 \text{ h}}{60 \text{ min}}$$

$$= 4.099537 \text{ h}$$

Following the rules for significant figures, we must round off this answer to three significant figures. Therefore

$$? \text{ h} = 4.10 \text{ h}$$

TRUE-FALSE QUESTIONS

1. A pure compound has variable properties that depend on its source.

2. The color, melting point, and density of a substance are examples of physical properties.

3. Heterogeneous mixtures have uniform composition throughout.

4. The melting of ice involves a change in chemical composition, and water is formed.

5. During combustion, substances such as wood or gasoline are converted into new substances.

6. 10,000,000 g is greater than 1 Mg.

7. The unit of energy in the SI is a derived unit called the joule (J), where $1 \text{ J} = 1 \text{ kg m}^2/\text{s}^2$.

8. Density is the mass of an object divided by its volume.

9. The density of an apple on Earth has a certain value. When the same apple is in orbit around the Earth, its density is zero.

10. 0°F is colder than 0°C.

11. The number 0.0375 has five significant figures.

12. The number 3.2200×10^4 has five significant figures.

13. The conversion factor 1.06 qt per liter has an infinite number of significant figures.

14. The number 18.502 rounded off to two significant figures is 18.

SELF-TEST A

1. Which of the following describe physical properties and which describe chemical properties?
   a. Color
   b. Decomposition upon heating
   c. Boiling point
   d. Hardness
   e. Density
   f. A change of color on exposure to air

2. Which of the following statements describe physical changes and which describe chemical changes?
   a. Fresh-cut apples turning brown
   b. Baking bread
   c. Cutting a soft metal with a knife
   d. Drying clothes
   e. Fermentation of sugar

3. Classify each of the following as a pure substance, a heterogeneous mixture, or a homogeneous mixture:
   a. Raw milk
   b. Tap water
   c. Gold
   d. Table sugar
   e. Concrete

4. Classify each of the following substances as an element, a compound, or a mixture:
   a. Hot tea
   b. Dry ice
   c. Sulfur
   d. Pure aspirin
   e. Natural gas
   f. Beer

5. Convert:
   a. 9.8 cm to millimeters
   b. 170 mm to centimeters
   c. 325 mL to liters
   d. 50 mi/h to m/s
   e. 20 GJ to kilojoules

6. The melting point of iron is 1540°C. What is this temperature in degrees Fahrenheit?

7. To determine the density of a piece of metal, a student finds its mass to be 13.62 g. The metal is placed into a flask that can hold exactly 25.00 mL. The student finds that 19.26 g of water is required to fill the flask. Given that the density of water is 0.9970 g/mL, what is the density of the metal?

8. The density of lead is 11.4 g/cm^3. What is the mass of a lead brick with dimensions 2.0 in by 4.0 in by 8.0 in?

9. 1.00 lb gold occupies what volume in cm^3? The density of gold is 19.3 g/cm^3 at 20°C.

10. If gasoline costs $1.21 per gallon, compare the expense of driving exactly 10,000 miles in an old "dinosaur" that gets 12 mi per gallon to that of a new compact car that gets 12 km/L. Recall that 1.06 qt = 1 L.

11. On a certain day the concentration of carbon monoxide in the air over Denver reached 1.8×10^{-5} g/L. Convert the concentration to mg/m^3.

12. In order to walk 3.5 mi, the average adult consumes 270 kcal. A can of beer contains 150 kcal. If you plan to walk a distance of 65 mi, how much beer should you take along as fuel?

13. The radius of an aluminum (Al) atom is 0.125 nm. How many Al atoms would have to be lined up in a row to form a line 1 cm in length?

14. Seawater is 4.0% salt, and the density of seawater is 62.4 lb/ft^3. How many pounds of salt are there in 1.00×10^2 gal of seawater? One cubic foot is 7.5 gal.

15. Soft solder contains 70.0% tin and 30.0% lead by mass. If you had 10.0 g of tin, how many grams of solder could you make?

16. Copper (Cu) is a trace element that is essential for nutrition. Newborn infants require 80 μg of Cu per kilogram of body mass per day. The Cu content of a popular baby formula is 0.48 μg of Cu per milliliter. How many milliliters of formula should a 7.0-lb baby consume per day to obtain the minimum daily Cu requirement?

17. It takes 31,000 J of energy to make 1.0 g of aluminum (Al) metal from the ore bauxite. Burning 10.0 g of gasoline releases 4.20×10^5 J. If burning gasoline is used to supply the energy to produce Al from bauxite, then how many grams of gasoline are needed to make enough Al for one beverage can containing approximately 15 g Al?

SELF-TEST B1

1. Express the following numbers in scientific notation:
   a. 24,000   b. 0.00014   c. 740,000,000   d. 0.0906

2. Express the following amounts using prefixes from the SI:
   a. 7×10^{-6} g   b. 8.0×10^{-9} m   c. 1.4×10^5 L   d. 1.0×10^3 s   e. 2.1×10^{-4} g

3. Fill in the blank with the correct unit and prefix:
   a. 1.6×10^{-2} m = 1.6 _____
   b. 2.7×10^{-9} m = 2.7 _____
   c. 7.2×10^{-3} g = 7.2 _____
   d. 3.50×10^{-1} mL = 350 _____
   e. 52 Mg = 5.2×10^7 _____

4. Determine the number of significant figures expressed in the following numbers:
   a. 0.609   b. 1.0×10^3 c. 0.000222   d. 238.0   e. 1.30×10^{-2}

5. Determine the number of significant figures to be expressed as a result of the following arithmetical operations:
   a. 12×2143.1
   b. $3.09 \div 7$
   c. $(2.2 \times 10^{-3})(1.40 \times 10^6)$
   d. $12.70 + 1.222$
   e. $6.271 - 24.33 + 595.2$

6. The density of mercury (Hg) is 13.6 g/cm^3. What is the volume occupied by 2.00 g Hg?

7. The melting point of cesium metal (Cs) is 83°F. Convert this to degrees Celsius.

8. Convert 350 in^3, the displacement volume of a certain automobile engine, into liters.

9. The world record for the 100-yd dash is 9.1 s. What is the average speed in m/s?

10. What mass of carbon monoxide (CO) in kilograms is present in the lower 500 m of air above a section of a city that measures 10 km × 10 km if the concentration of CO is 10 mg/m^3?

SELF-TEST B2

1. Write out the following exponential numbers by moving the decimal point in the proper direction:
   a. 2.4×10^4 b. 1.4×10^{-4} c. 7.4×10^8 d. 9.06×10^{-2}

2. Express the following amount in exponential notation in terms of the base unit with no prefix:
   a. 7 µg   b. 8.0 nm   c. 0.14 ML   d. 1.0 ks   e. 0.21 mg

3. Fill in the blank with the correct unit:
   a. 1.6 cm = 1.6×10^{-2}_____
   b. 2.7 mm = 2.7×10^{-6}_____
   c. 7.2 mg = 0.0072_____
   d. 350 µL = 3.50×10^{-1}_____
   e. 5.2×10^7 g = 52_____

4. Round off the following numbers to the number of significant figures requested:
   a. 0.60945 to three significant figures
   b. 1.012×10^3 to two significant figures
   c. 0.00022174 to three significant figures
   d. 237.95 to four significant figures
   e. 1.303 to two significant figures

5. Carry out the following operations and express the result to the correct number of significant figures:
   a. 12 × 2143.1
   b. 3.09 ÷ 7
   c. $(2.2 \times 10^{-3})(1.40 \times 10^6)$
   d. 12.70 + 1.222
   e. 6.271 − 24.33 + 595.2

6. Calculate the density of mercury given that a spherical droplet of mercury with a radius of 0.328 cm has a mass of 2.00 g.

7. The melting point of cesium metal is 28.4°C. What is the melting point in degrees Fahrenheit?

8. A certain automobile engine has a displacement of 5.74 L. Convert this volume to cubic inches.

9. How many seconds are required to run 1.00×10^2 yd at an average speed of 10.0 m/s.

10. Calculate the concentration (in mg/m^3) of CO in urban air if the air in the first 500 m of altitude above 100 km^2 of the city contains 500,000 kg of CO.

ANSWERS

TRUE-FALSE QUESTIONS

1. False. Pure substances have characteristic and reproducible properties.
2. True.
3. False. The description fits a homogeneous mixture.

4. False. Melting is a physical change.
5. True.
6. True. 10,000,000 g is 10 Mg.
7. True.
8. True.
9. False. In orbit the apple is weightless, but it has the same mass and hence the same density.
10. True. 0°F is about −18°C.
11. False. It has three significant figures.
12. True.
13. False. It has three significant figures.
14. False. It is 19, not 18.

SELF-TEST A

1. a. physical b. chemical c. physical d. physical e. physical f. chemical
2. a. chemical b. chemical c. physical d. physical e. chemical
3. a. heterogeneous b. homogeneous c. pure substance d. pure substance e. homogeneous
4. a. mixture b. compound c. element d. compound e. mixture f. mixture
5. a. 98 mm b. 17.0 cm c. 0.325 L d. 22 m/s e. 20,000,000 kJ
6. 2804°F
7. 2.37 g/mL
8. 12 kg
9. 23.5 cm^3
10. Dinosaur $1008 per year; subcompact $430 per year
11. 18 mg/m^3
12. 33 cans
13. 8.0×10^8 atoms (800 million)
14. 33 lb of salt
15. 14.3 g solder
16. 530 mL
17. 11 g gasoline

SELF-TEST B1

1. a. 2.4×10^4 b. 1.4×10^{-4} c. 7.4×10^8 d. 9.06×10^{-2}
2. a. 7 μg b. 8.0 nm c. 0.14 ML d. 1.0 ks e. 0.21 mg
3. a. cm b. nm c. mg d. μL e. g
4. a. 3 b. 2 c. 3 d. 4 e. 3
5. a. 2 b. 1 c. 2 d. 4 e. 4
6. 0.147 cm^3
7. 28°C
8. 5.74 L
9. 10.0 m/s
10. 5.0×10^5 kg

SELF-TEST B2

1. a. 24,000 b. 0.00014 c. 740,000,000 d. 0.0906
2. a. 7×10^{-6} g b. 8.0×120^{-9} m c. 1.4×10^5 L d. 1.0×10^3 s e. 2.1×10^{-4} g
3. a. m b. mm c. g d. mL e. Mg
4. a. 0.609 b. 1.0×10^3 c. 0.000222 d. 238.0 e. 1.3
5. a. 2.6×10^4 b. 0.4 c. 3.1×10^3 d. 13.92 e. 577.1
6. 13.5 g/cm^3
7. 83°F
8. 350 in^3
9. 9.17 s
10. 10 mg/m^3

Chapter Two
ATOMS, MOLECULES, AND IONS

- Atoms and Subatomic Particles
- Atomic and Molecular Masses
- The Mole Concept
- Chemical Formulas and Percent Composition
- Laws of Chemical Combination
- Naming Inorganic Compounds

ATOMS AND SUBATOMIC PARTICLES

STUDY OBJECTIVES

You should be able to:
1. Name the three subatomic particles of most importance to chemistry and list the values of their charge and mass.
2. Write the symbol of an isotope having been given its mass number and element name.
3. Determine the number of protons, neutrons, and electrons in an atom given its isotopic symbol.

Atoms. According to Dalton's atomic theory, developed in 1808, elements are composed of extremely small particles called *atoms*. Atoms of the same element all have the same mass, and atoms of different elements have different masses. Dalton assumed atoms to be indivisible. His theory explained two laws that were known at that time, the law of conservation of mass and the law of definite proportions. These laws will be discussed later in this chapter.

In the early 1900s, scientists learned that all atoms are constructed from the same three subatomic particles: the electron, proton, and neutron. The electron has a single unit of negative charge; the proton has a single unit of positive charge; and the neutron has no charge. The neutron and proton have essentially the same mass, 1.67×10^{-24} g. This is equivalent to 1.00 amu (atomic mass unit). The electron mass is much smaller: 9.11×10^{-28} g, or only 0.00055 amu.

Rutherford's experiments on alpha particle scattering in 1910 led the way to the nuclear model of the atom by showing that all the positive charge is localized in an extremely small region of the atom. A number of years later researchers discovered that both neutrons and protons occupy the *nucleus*, which is in the center of the atom. The electrons occupy the outer part of an atom. Atoms such as those in pure elements are neutral. They do not exhibit an electrical charge because each atom contains equal numbers of protons and electrons. The number of neutrons in atoms of a specific element varies somewhat from atom to atom, as we will discuss presently.

An atom of one element is distinguished from an atom of another element by its number of protons. The atomic number Z of an element is the number of protons in the atomic nucleus. For instance, the atomic number of oxygen is 8; therefore, all oxygen atoms contain eight protons and eight electrons.

The total mass of an atom is determined almost entirely by the number of protons and neutrons. In many cases the mass of the electrons can be discounted because it is so much smaller. The mass number A is the same as the total number of neutrons and protons present in the nucleus of an atom. The number of neutrons in an atom is $A - Z$.

Isotopes.

Contrary to one of Dalton's ideas, atoms of the same element do not always possess the same mass but often exhibit several masses that differ by increments of 1.0 amu. For example, there are two atoms of the element lithium, one with a mass of 6 amu and another with 7 amu. Atoms of the same element (that is, having the same atomic number) but with different mass numbers are called *isotopes*. The different mass numbers of isotopes are determined by the different numbers of neutrons per atom. Isotopes can be referred to by their mass numbers, as in lithium-6 and lithium-7. Also symbols are used to designate a specific isotope.

The general symbol for an isotope is

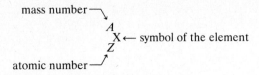

The symbols for the isotopes of lithium are

$$^{6}_{3}\text{Li} \quad \text{and} \quad ^{7}_{3}\text{Li}$$

An atom of the isotope lithium-6 contains three protons, three electrons, and three neutrons, whereas an atom of lithium-7 contains three protons, three electrons, and four neutrons. Both atoms contain three protons, and so both are of the element lithium. In nature, elements are found as mixtures of isotopes.

EXAMPLE 2.1 Isotopic Symbols

The three isotopes of oxygen found in nature are oxygen-16, -17, and -18. Write their isotopic symbols.

METHOD OF SOLUTION

The general form of the symbols is $^{A}_{Z}\text{O}$, where A is the mass number and Z is the atomic number. The atomic number of oxygen is 8, and so all oxygen atoms contain eight protons. The mass numbers are 16, 17, and 18, respectively. *Answer:* The isotopic symbols are

$$^{16}_{8}\text{O} \quad ^{17}_{8}\text{O} \quad ^{18}_{8}\text{O}$$

EXAMPLE 2.2 Isotopic Symbols

How many neutrons are present in the nucleus of each of the oxygen isotopes in the previous example?

METHOD OF SOLUTION

The number of neutrons is given by $A - Z$, the mass number minus the proton number. *Answer:*

For ^{16}O, there are $16 - 8 = 8$ neutrons.

For ^{17}O, there are $17 - 8 = 9$ neutrons.

For ^{18}O, there are $18 - 8 = 10$ neutrons.

ATOMIC AND MOLECULAR MASSES

STUDY OBJECTIVES

You should be able to:
1. Describe the atomic mass unit scale and its use of an arbitrary standard.
2. Determine the average atomic mass of an element from a knowledge of the masses and abundances of its isotopes.
3. Determine the molecular mass of a compound given its molecular formula.

The Atomic Mass Unit Scale. The atomic mass unit scale works like this. The masses of individual atoms cannot be measured with a balance; but *relative* masses of the atoms of different elements can be measured. For instance, it is possible to determine that an atom of ^4_2He is very close to $\frac{1}{3}$ the mass of an atom of $^{12}_6\text{C}$. Next we assign a certain value for the mass of a $^{12}_6\text{C}$ atom. By international agreement an atom of carbon-12 has a mass of exactly 12 atomic mass units amu. Carbon-12 is the standard of the amu scale. On this scale a helium-4 atom has a mass of 4.00 amu, which is $\frac{1}{3}$ of 12.... Atoms heavier than carbon-12 have masses greater than 12 amu. Fluorine atoms, for instance, are 1.583 times heavier than carbon-12 atoms. Thus the mass of a fluorine atom is 19.00 amu. Other experiments showed that the lightest element is hydrogen, H atoms having about $\frac{1}{12}$ the mass of carbon-12 atoms. In this way, a scale of relative masses of atoms has been established.

Modern Atomic Masses. The atomic masses that appear in the present-day periodic table are determined by taking into account the fact that atoms of the elements exist as isotopes. First the mass of each isotope of an element is measured with a mass spectrometer relative to the atomic mass of $^{12}_6\text{C}$. Then the percent abundance of each isotope is measured. This information permits calculation of the *average* atomic mass.

For instance, the element lithium has two isotopes that occur in nature: ^6_3Li with 7.5% abundance and ^7_3Li with 92.5% abundance. The atomic mass of ^6_3Li is 6.01513 amu, and that of ^7_3Li is 7.01601 amu. The average mass of such a mixture of Li atoms is given by

$$\text{average atomic mass} = (\text{fraction of isotope } X)(\text{mass isotope } X)$$
$$+ (\text{fraction of isotope } Y)(\text{mass isotope } Y)$$
$$= (0.075)(6.01513 \text{ amu}) + (0.925)(7.0161 \text{ amu})$$
$$= 0.45 \text{ amu} + 6.49 \text{ amu}$$
$$= 6.94 \text{ amu}$$

Note that neither ^6_3Li nor ^7_3Li has an atomic mass of 6.94 amu. This is the average value for the mixture of the two Li isotopes.

Molecular Masses. Compounds result when atoms of different elements bond to each other. There are two kinds of compounds: molecular and ionic. *Molecules* are tiny particles that are formed when

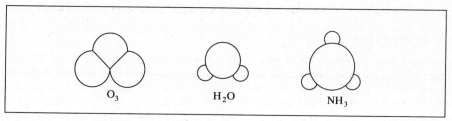

Figure 2.1 Molecules of ozone, water, and ammonia.

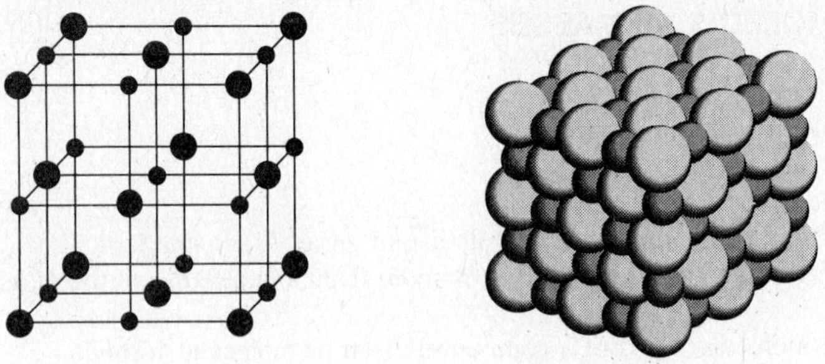

Figure 2.2 Two ways of depicting the structure of solid sodium chloride. There are no identifiable NaCl molecules in solid NaCl. The larger spheres are chloride ions and the smaller ones are sodium ions. Each chloride ion makes contact with six sodium ions, and each sodium ion makes contact with six chloride ions.

atoms combine and are held together by chemical forces. A molecule of a compound is formed by the combination of atoms of different elements. The *molecular formula* is written in terms of the symbols of the elements of which the compound consists and shows the exact numbers and kinds of atoms in a molecule of the compound. Figure 2.1 shows some common compounds and their molecular formulas.

The mass corresponding to a molecular formula is the sum of the atomic masses of the constituent atoms. For example, the molecular masses of the nitrogen oxides NO_2 and N_2O_5 are as follows:

molecular mass of NO_2 = atomic mass of N + 2 (atomic mass of O)

$$= 14.01 \text{ amu} + 2(16.00 \text{ amu})$$

$$= 46.01 \text{ amu}$$

atomic mass of N_2O_5 = 2(atomic mass of N) + 5(atomic mass of O)

$$= 2(14.01 \text{ amu}) + 5(16.00 \text{ amu})$$

$$= 108.02 \text{ amu}$$

Many compounds do not exist as molecules. The *ionic compounds*, such as sodium chloride (NaCl), for example, exist as individual sodium and chloride ions. The formula NaCl just gives the ratio of Na^+ ions to Cl^- ions in any sample of sodium chloride. Thus, in sodium chloride the ratio of Na^+ ions to Cl^- ions is 1:1. (See Figure 2.2.) For ionic compounds, the sum of the atomic masses in the formula is called the *formula mass* rather than molecular mass. The formula mass of sodium chloride is 58.45 amu, the sum of the atomic mass units for Na and Cl, 23.0 and 35.45 amu, respectively.

EXAMPLE 2.3 Average Atomic Mass

The element boron (B) consists of two stable isotopes with atomic masses of 10.012938 amu and 11.009304 amu. The average atomic mass of B is 10.81 amu. Which isotope is more abundant?

METHOD OF SOLUTION

The atomic mass is an average of the masses of the naturally occurring isotopes of an element. This means that each isotopic mass is multiplied by its percent abundance. If both isotopes in this example had a natural abundance of 50%, then the atomic mass would be just the average of the two masses, or approximately 10.5 amu. *Answer:* Because the average mass of B (10.81 amu) is between 10.5 and 11.0 amu, the abundance of $^{11}_{5}B$ must be greater than that of the $^{10}_{5}B$ atom. The actual abundances are $^{10}_{5}B$, 19.6%; and $^{11}_{5}B$, 80.4%.

EXAMPLE 2.4 Molecular Mass

Calculate the molecular mass of carbon tetrachloride (CCl_4).

METHOD OF SOLUTION

The molecular mass is the sum of the atomic masses of all the atoms in the molecule:

$$\text{molecular mass } CCl_4 = (12.01 \text{ amu}) + 4(35.45 \text{ amu})$$

$$= 153.81 \text{ amu}$$

EXAMPLE 2.5 Formula Mass

Determine the formula mass of $NaNO_3$.

METHOD OF SOLUTION

The formula mass is the sum of the atomic masses of all the atoms in the formula:

$$\text{formula mass } NaNO_3 = 23.0 \text{ amu} + 14.0 \text{ amu} + 3(16.0 \text{ amu})$$

$$= 85.0 \text{ amu}$$

THE MOLE CONCEPT

STUDY OBJECTIVES

You should be able to:
1. Define the mole as a base unit of the SI.
2. Calculate the number of moles in a given amount of an element of a compound.
3. Calculate the number of atoms, molecules, or formula units in a given amount of a substance.

The Mole. In the laboratory, amounts of elements and compounds are measured in units of grams. In order to measure out equal numbers of atoms of two elements, say carbon and silicon, we must measure a gram ratio that is the same as the mass ratio of one Si atom to one C atom. The atomic masses of Si and C are 28.1 amu and 12.0 amu, respectively. Thus, one Si atom has a mass 2.3 times greater than that of a C atom. It follows that any amounts of Si and C that have a mass ratio of Si to C of 2.3:1.0 will have equal numbers of Si and C atoms.

The *molar mass* of an element is a mass equal to the atomic mass expressed in grams. A molar mass of any element will contain the same number of atoms as a molar mass of any other element. Therefore, 28.1 g Si contains the same number of atoms as 12.0 g C. The number of atoms in a molar mass, called *Avogadro's number*, is equal to 6.022×10^{23} atoms. The quantity that contains Avogadro's number of atoms or other entities is called a *mole*. The fact that 1 mole is a number of items is often emphasized by comparing it to the term *dozen*:

$$1 \text{ dozen} = 12 \text{ objects}$$

$$1 \text{ mole} = 6.022 \times 10^{23} \text{ atoms}$$

Examples include the following:

One mole of carbon contains 6.022×10^{23} C atoms and has a mass of 12.01 g (the molar mass).

One mole of silicon contains 6.022×10^{23} Si atoms and has a mass of 28.1 g (the molar mass).

For most problems involving Avogadro's number, it is sufficient to use only three significant figures, 6.02×10^{23}.

The term *mole* can be used in relation to any kind of particle, such as atoms, ions, or molecules. For clarity the particle must always be specified. We say 1 mole of O_3 (ozone), or 1 mole of O_2 (diatomic oxygen), or 1 mole of Na^+ (sodium ions).

Other examples of molar amounts are

$$1 \text{ mole } Na^+ \text{ ions} = 6.02 \times 10^{23} \text{ } Na^+ \text{ ions} = 23.0 \text{ g } Na^+$$

$$1 \text{ mole } O_2 \text{ molecules} = 6.02 \times 10^{23} \text{ } O_2 \text{ molecules} = 32.0 \text{ g } O_2$$

$$1 \text{ mole } O_3 \text{ molecules} = 6.02 \times 10^{23} \text{ } O_3 \text{ molecules} = 48.0 \text{ g } O_3$$

The mole is an SI unit that is defined in relation to the mass of the carbon-12 isotope: *the mole is the amount of substance of a system that contains as many elementary entities as there are atoms in 0.012 kg of carbon-12*. In 0.012 kg of carbon-12 there are 6.022×10^{23} carbon-12 atoms. The symbol for mole is *mol*. The mole concept allows us to count atoms by weighing large numbers of them on a balance.

Let us also take a look at the relationship between the atomic mass units and grams. We will use hydrogen, which has an atomic mass of 1.008 amu, as an example. We can write

$$\text{mass of 1 H atom} \times 6.022 \times 10^{23} \text{ H atoms} = 1.008 \text{ g}$$

Then we can substitute the mass of an H atom:

$$1.008 \text{ amu/H atom} \times 6.022 \times 10^{23} \text{ H atoms} = 1.008 \text{ g}$$

$$6.022 \times 10^{23} \text{ amu} = 1.008 \text{ g}$$

$$1 \text{ amu} = 1.67 \times 10^{-24} \text{ g}$$

Compounds. The molar mass of a compound is the molecular mass expressed in grams. The molar masses of NO_2 and N_2O_5 are 46.0 g and 108 g, respectively. Since 46.0 g of NO_2 contains 6.02×10^{23} molecules of NO_2, this amount is 1 mol of NO_2. On the other hand, 1 mol of N_2O_5 has a mass of 108.0 g:

$$1 \text{ mol } NO_2 = 6.02 \times 10^{23} \text{ } NO_2 \text{ molecules} = 46.0 \text{ g } NO_2$$

$$1 \text{ mol } N_2O_5 = 6.02 \times 10^{23} \text{ } N_2O_5 \text{ molecules} = 108 \text{ } N_2O_5$$

Also note that 1 mol of NO_2 consists of 1 mol of N atoms and 2 mol of O atoms:

$$1 \text{ mol } NO_2 \text{ contains 1 mol N atoms} = 14.0 \text{ g N}$$

$$1 \text{ mol } NO_2 \text{ contains 2 mol O atoms} = 32.0 \text{ g O}$$

$$1 \text{ mol } NO_2 = 46.0 \text{ g } NO_2$$

Just as the molecular mass is the sum of the atomic masses, the molar mass of a molecular compound is the sum of the molar masses of the atoms in the molecule.

EXAMPLE 2.6 The Mole Concept

Complete the following sentences:

a. One mole of ^6Li contains _____ ^6Li atoms (isotopic mass = 6.015 amu) and will have a mass of _____ g.
b. One mole of Li atoms consists of _____ ^6Li atoms and _____ ^7Li atoms and will have a mass of _____ g.

METHOD OF SOLUTION

a. One mole of ^{6}Li contains 6.02×10^{23} ^{6}Li atoms and will have a mass of 6.015 g.
b. One mole of Li atoms consists of both stable isotopes of lithium: ^{6}Li (7.4%) and ^{7}Li (92.5%). See the preceding section for percent abundances. One mole of Li atoms contains $(0.074)(6.02 \times 10^{23})$ ^{6}Li atoms, which is equal to 4.45×10^{22} ^{6}Li atoms, and $(0.925)(6.02 \times 10^{23})$ ^{7}Li atoms, which is 5.57×10^{23} ^{7}Li atoms. Note that the total number of Li atoms is 6.02×10^{23} and thus will have a mass of 6.94 g.

EXAMPLE 2.7 Relative Atomic Masses

Compare the number of neon atoms in 20.1 g Ne to the number of helium atoms in 4.0 g He.

METHOD OF SOLUTION

Both amounts correspond to the molar masses of the respective elements. *Answer:* Each contains 6.02×10^{23} atoms. Also, note that both masses contain the same number of atoms because a Ne atom is on average five times heavier than a He atom (see atomic masses). Thus, any amounts of Ne and He that show a 5:1 mass ratio will contain equal numbers of Ne and He atoms.

EXAMPLE 2.8 Mass of a Given Number of Moles

What is the mass of 2.50 moles of zinc (Zn)?

METHOD OF SOLUTION

To convert moles of Zn to grams of Zn, first write the relationship between moles, atoms, and mass:

$$1 \text{ mol Zn} = 6.02 \times 10^{23} \text{ Zn atoms} = 65.39 \text{ g Zn}$$

The problem asks

$$? \text{ g Zn} = 2.50 \text{ mol Zn}$$

Since the atomic mass of Zn is 65.39 amu, then 1 mol of Zn has a mass of 65.39 g. Setting up the unit conversion factor,

$$\frac{65.39 \text{ g Zn}}{1 \text{ mol Zn}} = 1$$

CALCULATION

Since 1 mol Zn has a mass of 65.39 g, then 2.5 mol Zn must have a mass of

$$? \text{ g Zn} = 2.50 \text{ mol Zn} \times \frac{65.39 \text{ g}}{1 \text{ mol Zn}} = 163 \text{ g Zn}$$

EXAMPLE 2.9 Number of Atoms in a Given Mass

How many zinc atoms are present in 20.0 g Zn?

METHOD OF SOLUTION

We know that 1 mol of Zn contains 6.02×10^{23} atoms. First find how many moles of Zn atoms are in 20.0 g Zn and then multiply by Avogadro's number. Finding the number of moles of Zn is the key to

28 / Atoms, Molecules, and Ions

finding the number of atoms. The relationship between moles, atoms, and grams is

$$1 \text{ mol Zn} = 6.02 \times 10^{23} \text{ Zn atoms} = 65.39 \text{ g Zn}$$

The problem asks

$$? \text{ Zn atoms} = 20.0 \text{ g Zn}$$

Our road map is

$$\text{g Zn} \rightarrow \text{mol Zn} \rightarrow \text{number of Zn atoms}$$

CALCULATION

One mole has a mass of 65.39 g. Therefore, 20.0 g Zn corresponds to

$$? \text{ Zn mol} = 20.0 \text{ g Zn} \times \frac{1 \text{ mol Zn}}{65.39 \text{ g Zn}} = 0.3058 \text{ mol Zn}$$

Here we will carry one more digit than the correct number of significant figures. To convert moles of Zn to atoms of Zn, we use Avogadro's number, the number of atoms per mole, as the unit factor:

$$\frac{6.02 \times 10^{23} \text{ Zn atoms}}{1 \text{ mol Zn}} = 1$$

$$? \text{ Zn atoms} = 0.3058 \text{ mol Zn} \times \frac{6.02 \times 10^{23} \text{ Zn atoms}}{1 \text{ mol Zn}} = 1.84 \times 10^{23} \text{ Zn atoms}$$

COMMENT

Instead of calculating the number of moles separately, we could have strung the conversion factors together into one calculation:

$$? \text{ Zn atoms} = 20.0 \text{ g Zn} \times \frac{1 \text{ mol Zn}}{65.39 \text{ g Zn}} \times \frac{6.02 \times 10^{23} \text{ Zn atoms}}{1 \text{ mol Zn}}$$

$$= 1.84 \times 10^{23} \text{ Zn atoms}$$

EXAMPLE 2.10 Moles of a Molecular Compound

a. How many molecules of ethane (C_2H_6) are present in 50.3 g of ethane?
b. How many atoms each of H and C are in this sample?

METHOD OF SOLUTION FOR a

The problem asks

$$? \text{ } C_2H_6 \text{ molecules} = 50.3 \text{ g } C_2H_6$$

To convert grams to molecules, you need the molar mass. The molecular mass of C_2H_6 is $(2 \times 12.01) + (6 \times 1.008) = 30.1$ amu. Therefore, we can write

$$1 \text{ mol } C_2H_6 = 6.02 \times 10^{23} \text{ } C_2H_6 \text{ molecules} = 30.1 \text{ g } C_2H_6$$

Finding the number of moles of C_2H_6 is the key to finding the number of molecules:

$$\text{g } C_2H_6 \rightarrow \text{mol } C_2H_6 \rightarrow \text{number of } C_2H_6 \text{ molecules}$$

$$? \text{ mol } C_2H_6 = 50.3 \text{ g } C_2H_6$$

Since 1 mol of C_2H_6 molecules is 30.1 g, 50.3 g C_2H_6 must be

$$? \text{ mol } C_2H_6 = 50.3 \text{ g } C_2H_6 \times \frac{1 \text{ mol } C_2H_6}{30.1 \text{ C}_2H_6} = 1.67 \text{ mol } C_2H_6$$

Since 1 mol of C_2H_6 contains 6.02×10^{23} molecules, then 1.67 mol must contain

$$? \; C_2H_6 \text{ molecules} = 1.67 \text{ mol } C_2H_6 \times \frac{6.02 \times 10^{23} \text{ molecules}}{1 \text{ mol } C_2H_6} = 1.01 \times 10^{24} \; C_2H_6 \text{ molecules}$$

CALCULATION FOR a

$$? \; C_2H_6 \text{ molecules} = 50.3 \text{ g } C_2H_6 \times \frac{6.02 \times 10^{23} \text{ molecules}}{30.1 \text{ g } C_2H_6}$$

$$= 1.01 \times 10^{24} \text{ molecules}$$

METHOD OF SOLUTION FOR b

The molecular formula shows the number and kinds of atoms in a molecule:

$$\frac{6 \text{ H atoms}}{C_2H_6 \text{ molecule}} \quad \text{and} \quad \frac{2 \text{ C atoms}}{C_2H_6 \text{ molecule}}$$

CALCULATION FOR b

Multiplying the number of C_2H_6 molecules by the number of atoms of each kind per molecule gives the number of each kind of atom present:

$$1.01 \times 10^{24} \text{ molecules} \times \frac{6 \text{ H atoms}}{\text{molecule}} = 6.06 \times 10^{24} \text{ H atoms}$$

$$1.01 \times 10^{24} \text{ molecules} \times \frac{2 \text{ C atoms}}{\text{molecule}} = 2.02 \times 10^{24} \text{ C atoms}$$

CHEMICAL FORMULAS AND PERCENT COMPOSITION

STUDY OBJECTIVES

You should be able to:
1. Determine the percent composition of a compound with a known formula.
2. Determine the empirical formula of a compound given the percent composition and also find the molecular formula when the molar mass is given.

Percent Composition. The percent composition of a compound is the percentage by mass of each element in the compound. The percent composition can be determined by chemical analysis or it can be calculated directly if the formula of the compound is known. The formula is

$$\text{percent of element} = \frac{\text{mass of element per mole of compound}}{\text{molar mass of compound}} \times 100\%$$

Let us take sodium chloride (NaCl) as an example. The molar mass, 58.5 g, is the sum of the mass of

1 mol of Na, 23.0 g, and the mass of 1 mol of Cl, 35.5 g. The percentage of Na by mass is

$$\%\text{Na} = \frac{23.0 \text{ g Na}}{58.5 \text{ g NaCl}} \times 100\% = 39.3\%$$

The percentage of Cl by mass is

$$\%\text{Cl} = \frac{35.5 \text{ g Cl}}{58.5 \text{ g NaCl}} \times 100\% = 60.7\%$$

Empirical and Molecular Formulas. There are two types of formulas that give information about compounds. The *empirical formula* of a compound gives the simplest whole-number ratios between the numbers of atoms of the elements making up the compound. In sulfur dioxide (SO_2) there are twice as many O atoms in the compound as S atoms. The *molecular formula* indicates the numbers of atoms of each element in a molecule of a substance. The molecular formula and the empirical formula may be the same for some substances, as in the case of SO_2, where one molecule contains two atoms of oxygen and one atom of sulfur. On the other hand, the molecular formula of benzene is C_6H_6, but the empirical formula is CH because the smallest or simplest ratio of C atoms to H atoms is 1:1.

Chemical formulas of new compounds are determined from a knowledge of the percent composition of the elements. For example, potassium carbonate is a solid with a percent composition of 56.6% K, 8.68% C, and 34.7% O. Its empirical formula will give the ratios of K atoms to C atoms to O atoms. A convenient starting point is to assume you have 100 g of potassium carbonate. Therefore, 100 g of compound will contain 56.6 g K, 8.69 g C, and 34.7 g O.

The number of moles of each element present is

K: $56.6 \text{ g K} \times \frac{1 \text{ mol K}}{39.1 \text{ g K}} = 1.45 \text{ mol K}$

C: $8.69 \text{ g C} \times \frac{1 \text{ mol C}}{12.0 \text{ g C}} = 0.724 \text{ mol C}$

O: $34.7 \text{ g O} \times \frac{1 \text{ mol O}}{16.0 \text{ g O}} = 2.17 \text{ mol O}$

Dividing each number of moles by the smallest of the three molar amounts yields the following mole ratios:

$$\frac{1.45 \text{ mol K}}{0.724 \text{ mol C}} = 2.01 \text{ K to } 1.00 \text{ C}$$

$$\frac{2.17 \text{ mol O}}{0.724 \text{ mol C}} = 3.00 \text{ O to } 1.00 \text{ C}$$

The simplest whole-number ratio of K:C:O is 2:1:3, and so the empirical formula of potassium carbonate is K_2CO_3.

When the molar mass is known, the molecular formula can be deduced from the empirical formula. For example, in the case of benzene, which has an empirical formula of CH, the molar mass is known to be 78.11 g. Since the empirical mass is 13.01 g, there are 78.11/13.01 or 6.00 CH units per molecule, and so the molecular formula is C_6H_6.

EXAMPLE 2.11 Percent Composition

Calculate the percent composition of glucose ($C_6H_{12}O_6$).

METHOD OF SOLUTION

First the molar mass of glucose must be found from the atomic mass of each element:

$$\text{molar mass of } C_6H_{12}O_6 = 6 \text{ mol C}\left(\frac{12.01 \text{ g}}{1 \text{ mol C}}\right) + 12 \text{ mol H}\left(\frac{1.008 \text{ g}}{1 \text{ mol H}}\right) + 6 \text{ mol O}\left(\frac{16.00 \text{ g}}{1 \text{ mol O}}\right)$$

$$= 72.06 \text{ g} + 12.10 \text{ g} + 96.00 \text{ g} = 180.16 \text{ g}$$

Then the percentage of each element present is found by dividing the mass of each element in 1 mol of glucose by the total mass of 1 mol of glucose.

CALCULATION

$$\text{mass percent of C} = \frac{\text{mass of C in 1 mol } C_6H_{12}O_6}{\text{mass of 1 mol } C_6H_{12}O_6} \times 100\%$$

$$\%C = \frac{6 \text{ mol C} \times (12.01 \text{ g/mol C})}{180.16 \text{ g } C_6H_{12}O_6} \times 100\% = \frac{72.06 \text{ g}}{180.16 \text{ g}} \times 100\% = 40.0\% \text{ C}$$

After practicing the calculation several times, you can simplify the notation:

$$\%H = \frac{12 \times 1.008 \text{ g}}{180.16 \text{ g}} \times 100\% = \frac{12.1 \text{ g}}{180.16 \text{ g}} \times 100\% = 6.72\% \text{ H}$$

$$\%O = \frac{6 \times 16.00 \text{ g}}{180.16 \text{ g}} \times 100\% = \frac{96.0 \text{ g}}{180.16 \text{ g}} \times 100\% = 53.3\% \text{ O}$$

Thus glucose ($C_6H_{12}O_6$) contains 40.0% C, 6.72% H, and 53.3% O by mass.

COMMENT

A good way to check the results is to sum the percentages of the elements, because their total must add to 100%. Here we get 100.02%.

EXAMPLE 2.12 Using Percent Composition

Calculate the mass of carbon in exactly 10 g of glucose ($C_6H_{12}O_6$).

METHOD OF SOLUTION

In Example 2.11 we calculated that glucose contains 40.0% carbon. Then

$$\text{mass C} = 40.0\% \text{ of } 10 \text{ g } C_6H_{12}O_6 = 4.0 \text{ g C}$$

CALCULATION

This calculation would need to be done from scratch if we did not already know the percentage of carbon. Our road map is

$$\text{mass of glucose} \rightarrow \text{mol glucose} \rightarrow \text{mol carbon} \rightarrow \text{g carbon}$$

$$? \text{ mass C} = 10 \text{ g } C_6H_{12}O_6 \times \frac{1 \text{ mol } C_6H_{12}O_6}{180.16 \text{ g } C_6H_{12}O_6} \times \frac{6 \text{ mol C}}{1 \text{ mol } C_6H_{12}O_6} \times \frac{12.01 \text{ g C}}{1 \text{ mol C}} = 4.0 \text{ g C}$$

EXAMPLE 2.13 Empirical Formula

An elemental analysis of a new compound reveals its percent composition to be 50.7% C, 4.23% H, and 45.1% O. Determine the empirical formula.

METHOD OF SOLUTION

The empirical formula simply gives the relative numbers of atoms of each element in a formula unit. To start, assume you have exactly 100 g of compound, and then determine the number of moles of each element present. Then find the ratios of the number of moles.

32 / Atoms, Molecules, and Ions

CALCULATION

$$50.7 \text{ g C} \times \frac{1 \text{ mol C}}{12.01 \text{ g C}} = 4.22 \text{ mol C}$$

$$4.23 \text{ g H} \times \frac{1 \text{ mol H}}{1.008 \text{ g H}} = 4.20 \text{ mol H}$$

$$45.1 \text{ g O} \times \frac{1 \text{ mol O}}{16.00 \text{ g O}} = 2.82 \text{ mol O}$$

This gives the mole ratio for C : H : O of 4.22 : 4.20 : 2.82.
   Dividing by the smallest of the molar amounts gives

$$\frac{4.22 \text{ mol C}}{2.82 \text{ mol O}} = 1.50 \text{ C to } 1.00 \text{ O} \quad \text{and} \quad \frac{4.20 \text{ mol H}}{2.82 \text{ mol O}} = 1.50 \text{ H to } 1.00 \text{ O}$$

Therefore, the empirical formula is

$$C_{1.5}H_{1.5}O_{1.0}$$

which of course is not an acceptable formula because it does not contain only whole numbers. In this case, it is necessary to convert these fractions to whole numbers. We multiply each subscript by the same number, in this case 2. The empirical formula therefore is

$$C_3H_3O_2$$

EXAMPLE 2.14 Molecular Formula

Mass spectrometer experiments on the compound in Example 2.13 show its molecular mass to be about 140 amu. What is the molecular formula?

METHOD OF SOLUTION

First find how many empirical formula units there are in 1 mol of compound which is 140 g. The molar mass of the empirical formula $C_3H_3O_2$ is 71 g.

CALCULATION

Dividing 140 g by 71 g shows that there are two empirical units per molecule. Therefore, the molecular formula must show twice as many atoms as the empirical formula:

$$\text{molecular formula} = C_6H_6O_4$$

EXAMPLE 2.15 Carbon-Hydrogen Analysis

When a 0.761-g sample of a compound of carbon and hydrogen is burned in a C-H combustion apparatus (see Figure 2.11 in the text), 2.23 g CO_2 and 1.37 g H_2O are produced. Determine the percent composition of the compound.

METHOD OF SOLUTION

Here we want the masses of carbon and hydrogen in the original compound. All the C is now present in 2.23 g of CO_2. All the H is now present in 1.37 g of H_2O. How many grams of C are there in 2.23 g of CO_2 and how many grams of H are there in 1.37 g of H_2O? We know there is 1 mol of C atoms (12.01 g) in every mole of CO_2, or 44.01 g. Our road map is

$$g\ CO_2 \rightarrow mol\ CO_2 \rightarrow g\ C$$

CALCULATION

$$?\ g\ C = 2.23\ g\ CO_2 \times \frac{1\ mol\ CO_2}{44.01\ g\ CO_2} \times \frac{12.01\ g\ C}{1\ mol\ CO_2} = 0.608\ g\ C$$

$$?\ g\ H = 1.37\ g\ H_2O \times \frac{1\ mol\ H_2O}{18.0\ g\ H_2O} \times \frac{2.01\ g\ H}{1\ mol\ H_2O} = 0.153\ g\ H$$

The percent composition is

$$\%H = \frac{0.153\ g\ H}{0.761\ g\ compound} \times 100\% = 20.1\%\ H$$

$$\%C = \frac{0.608\ g\ C}{0.761\ g\ compound} \times 100\% = 79.9\%\ C$$

LAWS OF CHEMICAL COMBINATION

STUDY OBJECTIVES

You should be able to:
1. State the law of definite proportions and the law of multiple proportions.
2. Describe how Dalton's atomic theory explains these laws.

The *law of definite proportions* (or constant composition) refers to the elemental constituents of compounds and states: *The composition of a compound is invariant; the elements in a compound are combined in fixed proportions by mass.* Water, for example, can be decomposed by electrolysis into hydrogen and oxygen. When 9.0 g of water is decomposed, 8.0 g of oxygen and 1.0 g of hydrogen are obtained. If 90 g is decomposed, 80 g of oxygen and 10 g of hydrogen are obtained, and so on. Repeated analyses show that the ratio of the mass of oxygen to that of hydrogen is always 8.0 to 1.

This law was extremely important to Dalton because his theory was designed to show why the law of definite proportions holds true. If water molecules always contain two H atoms, each with a mass of 1.01 amu, and one O atom, with a mass of 16.0 amu, then the mass ratio of oxygen to hydrogen in water is always

$$\frac{mass\ O}{mass\ H} = \frac{16.0\ amu}{2 \times 1.0\ amu} = \frac{8.0}{1}$$

While working on the atomic theory, Dalton discovered the *law of multiple proportions*: *When two elements form more than one compound, the various masses of one element, combining with a fixed mass of another element, are related by small whole-number ratios.* Thus, for N_2O, NO, and NO_2, he found the masses of oxygen per 1.0 g of nitrogen to be 0.65 g, 1.3 g, and 2.6 g, respectively. Dividing each mass by the smallest, 0.65 g, Dalton obtained the whole-number ratios 1 : 2 : 4.

EXAMPLE 2.16 Law of Definite Proportions

Common table salt is sodium chloride (NaCl). If a given sample of NaCl contains 2.0 mg of Na, how many milligrams of Cl would it contain?

METHOD OF SOLUTION

NaCl is a compound with a constant composition. From the atomic masses, we know that in NaCl the ratio of the mass of Cl to Na is

35.5 g Cl per 23.0 g Na

CALCULATION

Using this ratio as a conversion factor, we get

$$? \text{ mg Cl} = 2.0 \text{ mg Na} \times \frac{35.5 \text{ g Cl}}{23.0 \text{ g Na}} = 3.08 \text{ mg Cl}$$

EXAMPLE 2.17 Law of Multiple Proportions

Sulfur forms two compounds with fluorine. In one of them 0.800 g of sulfur combines with 1.90 g of fluorine, and in the second compound 0.800 g of sulfur combines with 2.85 g of fluorine. Show that these data illustrate the law of multiple proportions.

METHOD OF SOLUTION

According to this law, if two elements form more than one compound, when one of the elements is kept at a fixed mass, then the masses of the other element combining are present in ratios of small whole numbers. In the first compound 0.800 g S combine with 1.90 g F, and in the second compound 0.800 g S combine with 2.85 g F.

CALCULATION

Thus, the ratio of the masses of fluorine combining with 0.800 g sulfur is

$$\frac{2.85 \text{ g}}{1.90 \text{ g}} = 1.5$$

This is the same ratio as 3:2, a ratio of small whole numbers.

COMMENT

The formulas of the two sulfur compounds are SF_4 and SF_6.

NAMING INORGANIC COMPOUNDS

STUDY OBJECTIVES

You should be able to:
1. Write the names of simple inorganic compounds given their formulas.
2. Write the formulas of inorganic compounds from their names.

Chemical Names. *Nomenclature* refers to the naming of chemical substances. Chemical nomenclature and formulas form the basis of the language of chemistry. Chemists communicate with each other using chemical names, but chemical formulas are more useful when doing calculations. In the textbook, inorganic compounds are divided into four categories: *ionic compounds, molecular compounds, acids and bases*, and *hydrates. Binary compounds* are those with only two elements and are the easiest to name.

How do you go about learning a new language? First, you build up a vocabulary of words and, second, then you learn rules for putting words together meaningfully (syntax). To learn a vocabulary requires strict memorization, and thus the names of cations and anions should be treated as you would words in a new language. Get a package of 3 × 5 cards, and write the name of an ion on one side of each card and its chemical formula on the other side. Table 2.3 in the textbook has 20 cations and 26 anions. If you learn their names and formulas, you will be able to name 572 possible chemical compounds!

Ionic Compounds. In binary ionic compounds, the metal ion is positively charged (a cation) and is named first. The name of a simple cation is the same as the name of the element. The negative ion (anion) name appears second and is derived from the name of the nonmetallic element in which an *-ide* ending replaces the usual ending of the element name. Following are the formulas and names of some binary ionic compounds:

NaI	sodium iodide
CaF_2	calcium fluoride
SnO	tin oxide

Some metals can form more than one kind of cation. This is true of tin, for example where Sn^{2+} and Sn^{4+} are known. These are called tin(II) and tin(IV), respectively. In the preceding list SnO should more correctly be named tin(II) oxide. This distinguishes it from SnO_2, tin(IV) oxide, which is a different compound. The transition metals commonly exhibit a variety of ionic charges. The system in which a Roman numeral is used to designate the ionic charge of the cation was developed by Alfred Stock and is called the Stock system:

MnO_2	manganese(IV) oxide
Mn_2O_3	manganese(III) oxide

In *ternary ionic compounds*, those consisting of three elements, the metal ion is named first followed by the name of the *polyatomic anion*. Some examples are

Na_2SO_4	sodium sulfate
$Fe(NO_2)_2$	iron(II) nitrite
$Mg(NO_3)_2$	magnesium nitrate

These polyatomic ions are examples of *oxoanions*. They contain a central atom, such as sulfur or nitrogen, that is bonded to several oxygen atoms. Some elements form several oxoanions that differ only in the number of oxygen atoms. In the case of an element that can have two oxoanions, the name of the anion with more oxygen atoms ends in *-ate*, and the name of the anion with fewer oxygen atoms ends in *-ite*. See nit*rate* and nit*rite* in the preceding examples.

Molecular Compounds. In binary molecular compounds, the more metallic element is named first, followed by the less metallic elements, whose name ends in *-ide*. When a pair of elements can form several compounds, questions about which compound is being referred to are removed by adding prefixes to denote the number of atoms of each element. The prefixes taken from Greek are given in Table 2.4 of the textbook. Some examples are

SF_4	sulfur tetrafluoride
SF_6	sulfur hexafluoride
PCl_3	phosphorus trichloride
P_2Cl_4	diphosphorus tetrachloride

Acids and Bases. Binary acids are compounds of hydrogen and a nonmetal. Many hydrogen compounds have completely different properties when they are in the pure gaseous or liquid state than when they are dissolved in water. In the gaseous state, HCl and H_2S, for instance, are molecular compounds called hydrogen chloride and hydrogen sulfide, respectively, whereas water solutions of these compounds contain hydrogen ions (H^+) and anions and have slightly different names. *Acids* are substances that release H^+ ions when they dissolve in water.

In naming binary acids, the word *hydrogen* is replaced by *hydro-*, and the *-ide* ending of the anion name is replaced with *-ic acid*. For example,

HCl(*g*)	hydrogen chloride
HCl(*aq*)	hydrochloric acid
H_2S(*g*)	hydrogen sulfide
H_2S(*aq*)	hydrosulfuric acid

The *oxoacids* are compounds containing hydrogen and oxoanions. The acids of nitrate and sulfate are

HNO_3	nitric acid
H_2SO_4	sulfuric acid

Note that the *-ate* ending of the anion name has been changed to *-ic acid*. For oxoanions with fewer oxygen atoms, the *-ite* ending of the anion name is changed to *-ous acid* for the oxyacid:

HNO_2	nitrous acid
H_2SO_3	sulfurous acid

Bases are substances that release hydroxide ions (OH^-) in aqueous solution. The metal hydroxides are named as if they were binary compounds and employ the *-ide* ending.

LiOH	lithium hydroxide
$Ca(OH)_2$	calcium hydroxide

Formulas of Ionic Compounds. The names and formulas of the cations and anions in Table 2.3 (textbook) can be used to write formulas of compounds. To write the correct formula for a compound, you must know the ionic charges of the cations and anions. Keep in mind that all compounds must be electrically neutral. The sum of the positive charges of cations and the negative charges of anions must be equal to zero. Consider the following examples:

1. *Potassium bromide.* The ions are K^+ and Br^-. Adding the charges of these ions, we get

$$(+1) + (-1) = 0$$

   The ratio of K^+ to Br^- must be 1:1. The formula is KBr.

2. *Potassium sulfide.* The ions are K^+ and S^{2-}. When the charge on the positive ion is less than the charge on the negative ion, more positive ions will be required to balance the negative charge. If there are two K^+ ions for each S^{2-} ion, the compound will be neutral:

$$2(+1) + (-2) = 0$$

   The formula is K_2S.

3. *Magnesium sulfate.* The ions are Mg^{2+} and SO_4^{2-}. A compound with one Mg^{2+} ion and one SO_4^{2-} ion will be neutral:

$$+2 + (-2) = 0$$

   The formula is $MgSO_4$.

4. *Aluminum oxide*. The ions are Al^{3+} and O^{2-}. Two cations have a charge of $+6$, and three anions have a charge of -6:

$$2(+3) + 3(-2) = 0$$

The formula is Al_2O_3.

EXAMPLE 2.18 Naming Compounds

Name the following compounds according to the Stock system:
a. Hg_2I_2
b. HgI_2

METHOD OF SOLUTION

In the Stock system the symbol of the metallic element in a compound is followed by a Roman numeral derived from the charge of the cation.
a. Assuming iodide to be -1 as in the I^- ion, Hg must be $+1$ in order to have a neutral compound. Therefore, Hg_2I_2 is named mercury(I) iodide.
b. In this case mercury must be $+2$ and the compound is mercury(II) iodide.

TRUE-FALSE QUESTIONS

1. The atomic number of an element is the number of protons in the atomic nucleus.

2. The mass number of an atom is the number of neutrons in the atom.

3. The electron has more mass than either the proton or the neutron.

4. The symbols ^3H, ^2H, and ^1H represent isotopes of hydrogen.

5. All atoms of fluorine contain nine protons.

6. On the average, Br atoms have approximately twice the mass of K atoms.

7. All carbon atoms are assigned the mass of exactly 12 amu as the current standard of the atomic mass unit scale.

8. All chlorine atoms have a mass of 35.45 amu.

9. A crystal of sodium chloride consists of NaCl molecules.

10. The molecular mass of CO_2 is 44 g.

11. The molar mass of SiO_2 is 60.1 g.

12. One mole of SO_3 consists of 6.02×10^{23} S atoms and 18.06×10^{23} O atoms.

13. When 36 g of water is decomposed, 4.0 g of hydrogen and 32 g of oxygen are produced.

38 / Atoms, Molecules, and Ions

14. Ammonium chloride crystals consist of an array of NH_4^+ and Cl^- ions.

15. The formula for sodium phosphate is Na_3P.

16. The best name for N_2O_5 is dinitrogen pentoxide.

17. The empirical formula of NO_2 is N_2O_4.

18. The percent composition of H_2O is 66% hydrogen and 33% oxygen.

SELF-TEST A

1. Complete the following isotope table:

Name	Symbol	Number of Protons	Number of Electrons	Number of Neutrons	Mass Number
Sodium	^{23}Na	11	___	12	23
___	^{40}Ar	___	___	22	___
Arsenic	^{75}As	___	___	___	___
Lead	___	___	___	___	___
___	___	___	___	20	___

2. a. What is the total number of fundamental particles (protons, neutrons, and electrons) in an atom of $^{56}_{26}Fe$.
   b. Approximately what fraction of the mass of the iron atom is due to electrons?

3. Take the mass of an atom to be the sum of the masses of its protons, neutrons, and electrons. Determine the percentages by mass of the protons, the neutrons, and the electrons in a $^{12}_{6}C$ atom.

4. If the arbitrary standard of the atomic mass unit scale had been defined such that a fluorine atom was 10 amu, what would be the atomic masses of oxygen and argon?

5. How many times heavier on the average is an atom of bromine than an atom of neon?

6. Calculate the molecular mass of the following:
   a. Carbon tetrachloride, CCl_4 b. Formaldehyde, H_2CO c. Xenon difluoride, XeF_2

7. Calculate the molar mass of the following:
   a. Na_2SO_4, sodium sulfate   d. NO_2, nitrogen dioxide
   b. $FeCl_3$, iron(III) chloride   e. N_2O_4, dinitrogen tetroxide
   c. $Ba(OH)_2$, barium hydroxide

8. What is the mass of 1 mol of each of the following?
   a. Cu   b. CuO   c. $CuSO_4$ d. $CuSO_4 \cdot 5H_2O$

9. How many O atoms are in each of the following?
   a. 1 mol ozone, O_3 b. 1 mol CuO   c. 1 formula unit of $CuSO_4$ d. 20 g $CuSO_4 \cdot 5H_2O$

10. What mass of $Ca_3(PO_4)_2$ would contain 0.50 mol Ca?

11. a. How many moles of $CaSO_4$ are there in 6.00×10^2 g $CaSO_4$?
    b. How many moles of oxygen are there in 6.00×10^2 g $CaSO_4$?
    c. How many oxygen atoms are there in 6.00×10^2 g $CaSO_4$?

12. How many antimony atoms are there in 5.00 mol Sb_2O_5?

13. a. What is the mass in grams of 1.5×10^{22} molecules of PCl_5?
    b. What is the mass in grams of 10^{12} carbon atoms?

14. Use the mole concept to answer the following questions:
    a. What is the mass of 17.5 mol of formaldehyde (H_2CO)?
    b. What is the mass of 7.5×10^{20} sulfur trioxide molecules (SO_3)?
    c. How many fluorine atoms are in 298 g of xenon tetrafluoride (XeF_4)?

15. Two compounds are formed when oxygen and sulfur combine. In one of them 0.50 g of oxygen combines with 0.50 g of sulfur. In the other compound 0.60 g of oxygen combines with 0.40 g of sulfur. Show that these data are consistent with the law of multiple proportions.

16. What is the empirical formula of each of the following compounds?
    a. $C_6H_8O_6$
    b. C_6H_6
    c. $MgCl_2$
    d. C_2H_2
    e. H_2O_2
    f. Hg_2Cl_2
    g. $HgCl$

17. Calculate the percent composition by mass of the following compounds:
    a. CO_2
    b. CO
    c. H_3AsO_3
    d. H_3AsO_4
    e. $NaNO_2$
    f. KNO_3
    g. $KHSO_4$
    h. $C_4H_4KO_7Sb$
    (tartar emetic)

18. Calculate the percentage of sodium in the following compounds:
    a. NaCl   b. $NaHCO_3$ c. Na_2CO_3 d. $NaC_2H_5CO_2$ e. $NaNO_3$

19. Determine the empirical formula of each compound from the given percent compositions:
    a. 46.7% N, 53.3% O
    b. 63.6% N, 36.4% O
    c. 55.3% K, 14.6% P, 30.1% O
    d. 26.6% K, 35.4% Cr, 38.1% O

20. Nitrogen tetroxide is a gas with a molar mass of 92.0 g. Its percent composition is 30.4% N and 69.6% O. What is the molecular formula?

21. NO_2 and N_2O_4 have the same empirical formula. Are they the same compound?

22. The compound glycerol contains three elements: C, H, and O. When a 0.740-g sample is burned in oxygen, 1.06 g CO_2 and 0.580 g H_2O are formed. What is the empirical formula of glycerol?

23. Write the formulas for the following compounds:
    a. potassium hydrogen carbonate
    b. magnesium nitrite
    c. sodium nitrate
    d. lithium nitride
    e. ammonium perchlorate
    f. strontium chloride
    g. calcium hypochlorite
    h. iron(II) fluoride
    i. mercury(II) sulfate
    j. barium sulfite
    k. zinc oxide
    l. dinitrogen oxide
    m. sodium carbonate
    n. copper sulfide

24. Name the following compounds:
    a. Na_2HPO_4
    b. $LiNO_3$
    c. $NaHCO_3$
    d. PCl_5
    e. CO
    f. HI(gas)
    g. HI(solution)
    h. NH_4I
    i. $Mg(OH)_2$
    j. Ag_2CO_3
    k. P_4O_{10}
    l. $Sr(NO_2)_2$
    m. $CaSO_4 \cdot 2H_2O$
    n. NaCN
    o. K_3N

25. Name the following acids:
    a. HNO_3
    b. HNO_2
    c. H_2SO_3
    d. H_2SO_4
    e. $HClO$
    f. $HClO_4$
    g. HCl
    h. H_3PO_4
    i. HCN

GENERAL PROBLEMS

26. The density of silver is 10.5 g/cm³. How many Ag atoms are present in a silver bar that measures 0.10 m × 0.05 m × 0.01 m?

27. The pesticide malathion has the chemical formula $C_{10}H_{19}O_6PS_2$.
    a. What is the molecular mass?
    b. The dose that is lethal to 50% of a human population is about 1.25 g per kilogram of body mass. How many molecules are in a dose lethal to a standard American adult male weighing 70 kg? To a standard female of 58 kg?

SELF-TEST B1

1. The element silver consists of two isotopes: ^{107}Ag, with an atomic mass of 106.905 amu and a natural abundance of 51.83%, and ^{109}Ag, with an atomic mass of 108.905 amu and natural abundance of 48.17%. Calculate the average atomic mass of silver.

2. What is the average mass in grams of a single atom of aluminum?

3. How many silver atoms are in 5.0 g Ag?

4. What is the mass of 2.0 mol of H_2?

5. How many moles of $NaNO_2$ are in 8.72 g $NaNO_2$?

6. How many formula units of $MgCl_2$ are present in 200 g $MgCl_2$?

7. How many molecules are in 25.0 g of methane (CH_4)?

8. Cyclohexane has the empirical formula CH_2. Its molecular mass is 84.16 amu. What is its molecular formula?

9. Mercury(I) nitrate contains the mercury(I) ion Hg_2^{2+}. Its molecular formula is $Hg_2(NO_3)_2$. Calculate its percent composition by mass.

10. Styrene has the molecular formula C_8H_8. How many times greater is its molecular mass than its empirical formula mass?

11. When 2.65 mg of the substance responsible for the green color on the yolk of a boiled egg is analyzed, it is found to contain 1.42 mg Fe and 1.23 mg S. What is the empirical formula of the compound?

SELF-TEST B2

1. The element silver exists in nature as two isotopes: ^{107}Ag has a mass of 106.905 amu and ^{109}Ag has a mass of 108.905 amu. From the average atomic mass given in the periodic table, calculate the natural abundances of the two silver isotopes.

2. If a single Al atom has an average mass of 4.48×10^{-23} g, what is the atomic mass of Al in atomic mass units?

3. What is the mass of 2.79×10^{22} atoms of Ag?

4. How many H_2 molecules are there in 4.0 g H_2?

5. What is the mass of 0.126 mol $NaNO_2$?

6. What is the mass of 1.26×10^{24} formula units of $MgCl_2$?

7. What is the mass of 9.40×10^{23} molecules of CH_4?

8. What is the empirical formula of cyclohexane, C_6H_{12}?

9. Determine the empirical formula of mercury(I) nitrate, which contains 76.4% Hg, 5.3% N, and 18.3% O.

10. The empirical formula of styrene is CH, and its molar mass is 104 g/mol. What is its molecular formula?

11. Calculate the masses of Fe and S in 2.65 mg Fe_2S_3.

ANSWERS

TRUE-FALSE QUESTIONS

1. True.

2. False. The total number of protons and neutrons gives the mass number.

3. False. The electron is the lightest of the three particles.

4. True.

5. True.

6. True.

7. False. The lightest stable isotope of carbon is assigned a mass of exactly 12.000... amu.

8. False. Chlorine consists of a mixture of isotopes with masses 35 and 37.

9. False. Ionic compounds are mixtures of cations and anions.

10. False. The molecular mass would be 44 amu; 44 g is the molar mass.

11. True.

12. True.

13. True.

14. True.

15. False. The formula is Na_3PO_4.

16. True.

17. False. The empirical formula is NO_2.

18. False. 11% H and 89% O by mass.

SELF-TEST A

1.

Name	Symbol	Number of Protons	Number of Electrons	Number of Neutrons	Mass Number
Sodium	^{23}Na	11	11	12	23
Argon	^{40}Ar	18	18	22	40
Arsenic	^{75}As	33	33	42	75
Lead	^{206}Pb	82	82	124	206*
Calcium	^{40}Ca	20	20	20	40*

*Arbitrarily chosen isotope; others are acceptable

2. a. 82   b. 0.00026

3. 50% proton mass, 50% neutron mass, 2.8×10^{-2}% electron mass

4. Oxygen would be 8.4 and argon would be 21.0.

5. 3.96 times

6. a. 153.81 amu   b. 30.17 amu   c. 169.3 amu

7. a. 142.04 amu   b. 162.21 amu   c. 171.3 amu   d. 46.0 amu   e. 92.0 amu

8. a. 63.546 g   b. 79.545 g   c. 159.6 g   d. 249.7 g

9. a. 18.1×10^{24} b. 6.02×10^{23} c. 4   d. 4.34×10^{23}

10. 51.7 g

11. a. 4.41 mol CaSO$_4$ b. 17.6 mol O   c. 1.06×10^{25} O atoms

12. 6.02×10^{24} Sb atoms

13. a. 5.2 g   b. 2.0×10^{-11} g

14. a. 525 g H$_2$CO   b. 0.10 g SO$_3$ c. 3.46×10^{24} F atoms

15. The ratio of the masses of oxygen combining with 1.0 g of sulfur is 1.5 g/1.0 g = 3/2, a ratio of small whole numbers.

16. a. C$_3$H$_4$O$_3$ b. CH   c. MgCl$_2$ d. CH   e. HO   f. HgCl   g. HgCl

17. a. 27.29% C, 72.71% O
    b. 42.88% C, 57.12% O
    c. 2.40% H, 59.5% As, 38.1% O
    d. 52.78% As, 2.13% H, 45.09% O
    e. 33.32% Na, 20.30% N, 46.38% O
    f. 38.67% K, 13.80% N, 47.47% O
    g. 28.71% K, 0.74% H, 23.55% S, 47.00% O
    h. 14.79% C, 1.24% H, 12.03% K, 34.47% O, 37.47% Sb

18. a. 39.34%   b. 27.37%   c. 43.39%   d. 23.94%   e. 27.07%

19. a. NO   b. N_2O c. K_3PO_4 d. $K_2Cr_2O_7$

20. N_2O_4

21. No, their molecules contain different numbers of N and O atoms.

22. $C_3H_8O_3$

23. a. $KHCO_3$ b. $Mg(NO_2)_2$ c. $NaNO_3$ d. Li_3N e. NH_4ClO_4
    f. $SrCl_2$ g. $Ca(OCl)_2$ h. FeF_2 i. $HgSO_4$ j. $BaSO_3$
    k. ZnO   l. N_2O m. Na_2CO_3 n. CuS

24. a. sodium hydrogen phosphate
    b. lithium nitrate
    c. sodium hydrogen carbonate
    d. phosphorus pentachloride
    e. carbon monoxide
    f. hydrogen iodide
    g. hydroiodic acid
    h. ammonium iodide
    i. magnesium hydroxide
    j. silver carbonate
    k. tetraphosphorus decoxide
    l. strontium nitrite
    m. calcium sulfite dihydrate
    n. sodium cyanide
    o. potassium nitride

25. a. nitric acid
    b. nitrous acid
    c. sulfurous acid
    d. sulfuric acid
    e. hypochlorous acid
    f. perchloric acid
    g. hydrochloric acid
    h. phosphoric acid
    i. hydrocyanic acid

26. 2.9×10^{24} Ag atoms

27. a. 330 amu   b. male, 1.6×10^{23} molecules; female, 1.3×10^{23} molecules

SELF-TEST B1

1. 107.87

2. 4.48×10^{-23} g

3. 2.79×10^{22}

4. 4.03 g

5. 0.126 mol

6. 1.26×10^{24}

7. 9.40×10^{23}

8. C_6H_{12}

9. 76.4% Hg, 5.3% N, 18.3% O.

10. 8

11. Fe_2S_3

SELF-TEST B2

1. 51.83% ^{107}Ag and 48.17% ^{109}Ag

2. 26.96 amu

3. 5.0 g

4. 1.2×10^{24}

5. 8.72 g

6. 200 g

7. 25 g

8. CH_2

9. $HgNO_3$

10. C_8H_8

11. 1.42 mg Fe, 1.23 g S

Chapter Three
CHEMICAL REACTIONS I: CHEMICAL EQUATIONS AND REACTIONS IN AQUEOUS SOLUTION

- Chemical Equations
- Properties of Aqueous Solutions
- Solubility Rules and Precipitation Reactions
- Acid-Base Reactions
- Redox Reactions and Oxidation Numbers
- Balancing Redox Reactions

CHEMICAL EQUATIONS

STUDY OBJECTIVES

You should be able to:
1. Write a chemical equation from a written description of the reaction that includes the names of reactants and products.
2. Balance a chemical equation when given a "skeletal" equation.

Balancing Equations. A chemical equation is a symbolic way to represent a chemical reaction. The formulas of the reactants are written on the left side of the equation, and the formulas of the products on the right side. Equations are written according to the following format.

1. Reactants and products are separated by an arrow that should be interpreted to mean "yields."
2. Plus signs are placed between reactants and between products. In general, a plus implies a mixing of the reactants or products. The plus sign can be interpreted to mean "and" or just "plus."
3. Abbreviations are sometimes used to indicate states of matter. These symbols, given in Table 3.1, are written in parentheses and follow the elements or compounds to which they refer. Symbols indicating the conditions of a reaction are placed above the arrow.

Table 3.1 Symbols Used in Writing Chemical Equations

Symbol	Meaning
$\rightarrow$	yields, produces
$\rightleftharpoons$	reversible reaction, equilibrium
(s)	solid phase
(l)	liquid phase
(g)	gas phase
(aq)	aqueous solution
$+$	added to, and
Δ	supply heat

4. Equations are balanced. That is, the number of atoms of a certain element that appear in the reactants must be equal to the number of atoms of that element appearing in the products. A balanced equation shows the same number of each type of atom on both sides of the equation. This means that mass is conserved in chemical reactions.

This format can be illustrated by the following decomposition reaction. When solid potassium chlorate ($KClO_3$) is heated, it decomposes to yield solid potassium chloride (KCl) and oxygen gas (O_2). The chemical equation is

$$KClO_3(s) \xrightarrow{\Delta} KCl(s) + O_2(g)$$

To balance the equation, we must adjust the coefficients that precede the formulas. Absence of a coefficient is understood to mean "one." The equation <u>as shown</u> means that one formula unit of $KClO_3$ decomposes to yield one formula unit of KCl and one molecule of O_2. Thus, the reactants contain 1 K atom (ion), 1 Cl atom, and 3 O atoms, and the products contain 1 K atom, 1 Cl atom, and 2 O atoms. But this equation is not balanced. It does not show the same number of O atoms on both sides.

When appropriate coefficients are inserted, the equation is balanced:

$$2KClO_3(s) \xrightarrow{\Delta} 2KCl(s) + 3O_2(g)$$

Reactants	Products
2 K atoms	2 K atoms
2 Cl atoms	2 Cl atoms
6 O atoms	6 O atoms

The balanced equation tells us that 2 formula units of $KClO_3$ decompose to yield 2 units of KCl and 3 molecules of O_2. Notice that if we multiply each coefficient by Avogadro's number, N_A, then $2 \times N_A$ units of $KClO_3$ becomes 2 mol of $KClO_3$. We also obtain 2 mol KCl and 3 mol O_2. Thus, the balanced equation also tells us that 2 mol of $KClO_3$ yields 2 mol KCl and 3 mol O_2.

Notice that a balanced equation results if we write

$$KClO_3 \rightarrow KCl + 3O$$

But this is incorrect! Changing the subscripts changes the formula, and so a different compound is being inserted into the equation. Here, we have changed the product from diatomic oxygen to oxygen atoms. These are two very different species. Never change the formulas of the reactants or products. Adjust only their coefficients. In the examples that follow we will present a few hints that will make balancing easier.

EXAMPLE 3.1 Balancing Chemical Equations

Balance the following displacement reaction. In this reaction zinc displaces H_2 from chemical combination with chlorine.

$$Zn(s) + HCl(aq) \rightarrow ZnCl_2(aq) + H_2(g)$$

METHOD OF SOLUTION

Step 1. As a general way to begin, *identify elements that occur in only two compounds in the equation*. In this case, each of the three elements Zn, H, and Cl occur in only two compounds.
Step 2. Of the elements that occur in only two compounds, *identify and balance first that element with the largest subscript*. Both Cl and H have the subscript 2, so balance either one of them first.

Let's balance H by placing the coefficient 2 before HCl:

$$Zn + 2HCl \rightarrow ZnCl_2 + H_2$$

Taking stock:

Reactants	Products
2 H atoms	2 H atoms
2 Cl atoms	2 Cl atoms

At this point both H and Cl are balanced. Further inspection reveals that Zn is also balanced. Thus the equation is balanced!

EXAMPLE 3.2 Balancing Chemical Equations

The reaction of a hydrocarbon with oxygen is an example of a combustion reaction. Balance the equation for the reaction of pentane with oxygen gas:

$$C_5H_{12} + O_2 \rightarrow CO_2 + H_2O$$

METHOD OF SOLUTION

Following step 1 given in the previous example, we note that both H and C appear in only two compounds. According to step 2, we note that of these, H has the largest subscript. Balance H first by placing the coefficient 6 before H_2O:

$$C_5H_{12} + O_2 \rightarrow CO_2 + 6H_2O$$

Reactants	Products
5 C atoms	1 C atom
12 H atoms	12 H atoms
2 O atoms	8 O atoms

Now balance the carbon atoms by placing a 5 in front of CO_2:

$$C_5H_{12} + O_2 \rightarrow 5CO_2 + 6H_2O$$

Reactants	Products
5 C atoms	5 C atom
12 H atoms	12 H atoms
2 O atoms	16 O atoms

Finally the oxygen atoms can be balanced. The products already have their coefficients determined, and so the reactants must be adjusted to supply 16 O atoms. Eight molecules of O_2 contain 16 O atoms, and so write the coefficient 8 before O_2:

$$C_5H_{12} + 8O_2 \rightarrow 5CO_2 + 6H_2O$$

Reactants	Products
5 C atoms	5 C atoms
12 H atoms	12 H atoms
16 O atoms	16 O atoms

EXAMPLE 3.3 Balancing Chemical Equations

Balance the following decomposition reaction:

$$KClO_3(s) \rightarrow KCl(s) + O_2(g)$$

METHOD OF SOLUTION

According to steps 1 and 2, we should balance O first. There are 3 O atoms on the reactant side and 2 on the product side. A third step is helpful in this situation. *Step 3:* For an element that occurs in only two substances in the equation, note whether *the number of atoms of that element is even on one side and odd on the other side of the equation*. If this is the case, *use two coefficients* which increase the number of atoms on each side of the equation to the *least common multiple*:

$$KClO_3 \rightarrow KCl + O_2$$

Reactants	Products
3 O atoms	2 O atoms

Least common multiple for balancing O is $3 \times 2 = 6$. The two coefficients are 2 and 3:

$$2KClO_3 \rightarrow KCl + 3O_2$$

Reactants	Products
6 O atoms	6 O atoms

Balancing the K and Cl atoms, we get

$$2KClO_3 \rightarrow 2KCl + 3O_2$$

EXAMPLE 3.4 Balancing Chemical Equations

Balance the following precipitation reaction in which sulfuric acid reacts with barium chloride in aqueous solution to yield hydrochloric acid and a precipitate of barium sulfate:

$$H_2SO_4(aq) + BaCl_2(aq) \rightarrow HCl(aq) + BaSO_4(s)$$

METHOD OF SOLUTION

Here it helps to recognize certain groups of atoms that maintain their identity during the reaction. In this equation SO_4^{2-}, the sulfate ion, appears in the reactants and products as a unit:

$$H_2SO_4 + BaCl_2 \rightarrow HCl + BaSO_4$$

Reactants		Products	
2 H atoms	1 Ba atom	1 H atom	1 Ba atom
1 sulfate ion	2 Cl atoms	1 Cl atom	1 sulfate ion

Either H or Cl could be balanced first. Balance the chlorine by placing the coefficient 2 in front of HCl:

$$H_2SO_4 + BaCl_2 \rightarrow BaSO_4 + 2HCl$$

A check of the H atoms shows they are now balanced along with the other elements.

PROPERTIES OF AQUEOUS SOLUTIONS

STUDY OBJECTIVES

You should be able to:
1. Define solute, solvent, and solution.
2. Distinguish between nonelectrolytes, weak electrolytes, and strong electrolytes.

Solutions. Many chemical reactions occur in aqueous solutions. Solutions are homogeneous mixtures of two or more substances. Solutions have two components, the *solute*, which is the substance present in the smaller amount, and the *solvent*, which is the substance present in the larger amount. A solute can be a solid, liquid, or gaseous substance. In aqueous solutions, water is the solvent.

Electrolytes and Nonelectrolytes. Many solutions of ionic compounds conduct electricity, whereas water itself essentially does not conduct. These solutions are called electrolyte solutions. For instance, both KCl(g) and HCl(s) dissociate into ions when dissolved in water:

$$KCl(aq) \rightarrow K^+(aq) + Cl^-(aq)$$

$$HCl(aq) \rightarrow H^+(aq) + Cl^-(aq)$$

It is these ions that carry an electrical current through the solution. Substances that ionize or dissociate in solution are called *electrolytes*. Thus a solution of HCl, for instance, contains H^+ ions and Cl^- ions, but *no* HCl molecules! *Strong* electrolytes are 100% dissociated into ions in solution.

A *nonelectrolyte* is a substance that does not ionize in aqueous solution. Such a solution does not conduct an electrical current. Sugar is an example of a nonelectrolyte. In solution, sugar exists as $C_6H_{12}O_6$ molecules. These are neutral (no charge) and cannot conduct an electrical current. All nonelectrolytes exist as molecules in aqueous solution.

Between electrolytes and nonelectrolytes are the weak electrolytes. These substances ionize only *partially* to yield a few ions capable of conducting a weak electrical current. Compounds that are incompletely dissociated into ions are called *weak electrolytes*. Acetic acid (CH_3COOH) is an example of a weak electrolyte. You will see later in Chapter 16 that in a solution of CH_3COOH about 99% of the CH_3COOH molecules are not ionized. Only 1% are ionized into H^+ and CH_3COO^- ions as shown below:

$$CH_3COOH(aq) \rightleftharpoons CH_3COO^-(aq) + H^+(aq)$$

Recall that a double arrow means the reaction is reversible. This means that the reaction can occur in the reverse direction as well as the forward direction. The reverse reaction removes ions from the solution and leads to incomplete ionization. Table 3.2 of the text lists examples of substances that are strong electrolytes, weak electrolytes, and nonelectrolytes.

EXAMPLE 3.5 Identifying Electrolytes and Nonelectrolytes

Identify the following substances as strong, weak, or nonelectrolytes:
a. CH_3OH (methyl alcohol)
b. NH_3
c. Na_2CO_3

METHOD OF SOLUTION

Refer to Table 3.2 in the text.
a. Various sugars and alcohols dissolve as neutral molecules. *Answer:* nonelectrolyte.
b. Ammonia is a weak base. See Section 3.4 of the text. *Answer:* weak electrolyte.
c. Sodium carbonate is an ionic compound. All ionic compounds dissociate completely in aqueous solution. *Answer:* strong electrolyte.

SOLUBILITY RULES AND PRECIPITATION REACTIONS

STUDY OBJECTIVES

You should be able to:
1. Write molecular, ionic, and net ionic equations for precipitation reactions.
2. Predict which product will precipitate when two electrolyte solutions are mixed.

Solubility Rules. The *solubility* of a solute is the amount that can be dissolved in a given quantity of solvent at a given temperature. For example, the solubility of $Pb(NO_3)_2$ is 56 g/100 mL H_2O at 20°C. The solubilities of ionic solids in water vary over a wide range of values. For convenience, we divide compounds into three categories called *soluble*, *slightly soluble*, and *insoluble*. Insoluble is a relative term and does not mean that no solute dissolves. Compounds are classified as *insoluble* if their solubility is less than 0.1 g/100 mL H_2O. On the other hand, *soluble* compounds are those whose solubilities are greater than 1 g/100 mL. The following "solubility rules" summarize the solubilities of various compounds in water at 25°C.

1. All alkali metal salts (Group 1A) are soluble.
2. All ammonium (NH_4^+) salts are soluble.
3. All salts containing nitrate (NO_3^-), chlorate (ClO_3^-), and perchlorate (ClO_4^-) are soluble.
4. Most hydroxides (OH^-) are insoluble, except those of the alkali metals and barium hydroxide, $Ba(OH_2)$. Calcium hydroxide, $Ca(OH_2)$, is slightly soluble.
5. Most compounds containing chlorides, bromides, and iodides are soluble, except those containing Ag^+, Hg_2^{2+}, and Pb^{2+} ions.
6. All carbonates (CO_3^{2-}), phosphates (PO_4^{3-}), and sulfides (S^{2-}) are insoluble, except those of ammonium (NH_4^+) and alkali metals.
7. Most sulfates (SO_4^{2-}) are soluble. Both Ag_2SO_4 and $CaSO_4$ are moderately soluble. Ba^{2+}, Hg^{2+}, and Pb^{2+} sulfates are insoluble.

Precipitation Reactions. When certain solutions of electrolytes are mixed, the cation of one solute may combine with the anion of the other solute to form an insoluble compound. This leads to the formation of a precipitate. A precipitate is an insoluble solid that settles from a solution. For instance, when a solution of $Pb(NO_3)_2$ is mixed with a solution of HCl, a precipitate of insoluble $PbCl_2(s)$ forms. This occurs because combining the solutions results in a new solution in which the concentration of dissolved $PbCl_2(aq)$ is greater than its solubility—a condition referred to as supersaturation. In this case, $PbCl_2(s)$ must precipitate. The balanced chemical equation is

$$Pb^{2+}(aq) + 2Cl^-(aq) \rightarrow PbCl_2(s)$$

This equation, representing the reaction of ions in the solution, is an example of a *net ionic equation*. If all ionic species in the mixed solution are written, we have an *ionic equation*:

$$Pb^{2+}(aq) + 2NO_3^-(aq) + 2H^+(aq) + 2Cl^-(aq) \rightarrow$$
$$PbCl_2(s) + 2H^+(aq) + 2NO_3^-(aq)$$

where H^+ and NO_3^- are called spectator ions because they do not react.

A molecular equation results when all the reactants and products are written as their usual chemical formulas:

$$Pb(NO_3)_2(aq) + 2HCl(aq) \rightarrow PbCl_2(s) + 2HNO_3(aq)$$

In order to predict whether a precipitate will form when two solutions are mixed, you need to know the solubility rules for ionic compounds.

EXAMPLE 3.6 Solubility Rules

According to the solubility rules, which of the following compounds is soluble in water?
a. $MgCO_3$
b. $AgNO_3$
c. $MgCl_2$
d. Na_3PO_4
e. KOH

METHOD OF SOLUTION

a. According to rule 6, $MgCO_3$ is insoluble.
b. According to rule 3, $AgNO_3$ is soluble.
c. According to rule 5, $MgCl_2$ is soluble.
d. According to rules 1 and 6, Na_3PO_4 is soluble.
e. According to rules 1 and 4, KOH is soluble.

EXAMPLE 3.7 Writing Ionic and Net Ionic Equations

Write balanced molecular, ionic, and net ionic equations for the reaction that occurs when a $BaCl_2$ solution is mixed with a Na_2SO_4 solution.

METHOD

The combined solution will contain Na^+, SO_4^{2-}, Ba^{2+}, and Cl^- ions. The two new possible ionic compounds are NaCl and $BaSO_4$. If one (or both) is insoluble, it (they) will precipitate. According to solubility rules 1 and 5, NaCl is soluble, and from rule 7, $BaSO_4$ is insoluble. The molecular equation should show $BaSO_4(s)$ as a product along with NaCl(aq). *Answer:* The molecular equation is

$$BaCl_2(aq) + Na_2SO_4(aq) \rightarrow BaSO_4(s) + 2NaCl(aq)$$

The ionic equation is

$$Ba^{2+}(aq) + 2Cl^-(aq) + 2Na^+(aq) + SO_4^{2-}(aq) \rightarrow BaSO_4(s) + 2Na^+(aq) + 2Cl^-(aq)$$

The net ionic equation omits spectator ions:

$$Ba^{2+}(aq) + SO_4^{2-}(aq) \rightarrow BaSO_4(s)$$

ACID-BASE REACTIONS

STUDY OBJECTIVES

You should be able to:
1. Write equations that show the reactions that occur when Arrhenius acids and bases dissolve in water.
2. Define acids and bases according to Brønsted's definition.
3. Write net ionic equations for various types of acid-base reactions.
4. Describe monoprotic, diprotic, and triprotic acids.

Arrhenius's Theory.
Acids and bases are general classifications of compounds that are based on their properties. In general, solutions of acids have a sour taste, change blue litmus to red, dissolve certain metals, and react with carbonates to produce carbon dioxide gas. On the other hand, bases are compounds that form solutions that have a bitter taste, feel slippery, and change red litmus to blue. The properties of an acid are neutralized by the addition of a base, and vice versa.

An acid-base theory or definition is an attempt to explain what happens to molecules of acidic and basic substances in solution that gives rise to their characteristic properties. Svante Arrhenius proposed that an acid is a substance that produces H^+ ions (protons) in aqueous solution, and a base is a substance that produces OH^- ions (hydroxide) in aqueous solution. Nitric acid is one example:

$$HNO_3(aq) \rightarrow H^+(aq) + NO_3^-(aq)$$

Calcium hydroxide is an example of a base:

$$Ca(OH)_2(aq) \rightarrow Ca^{2+}(aq) + 2OH^-(aq)$$

We can conclude that all acids and bases are electrolytes undergoing dissociation in aqueous solution. Arrhenius explained neutralization as the combination of H^+ and OH^- to form water:

$$H^+(aq) + OH^-(aq) \rightarrow H_2O(l)$$

Notice that the $H^+(aq)$ is called the hydrogen ion or a proton. To understand the name *proton*, remember that a hydrogen atom consists of a proton as the nucleus and an electron which travels around the nucleus. The positive ion results from the loss of the electron. Thus the hydrogen ion is just a proton.

Brønsted's Theory.
J. N. Brønsted proposed a more general theory of acid-base reactions than that of Arrhenius. According to Brønsted's theory, an *acid* is defined as a substance that is able to donate a proton, and a *base* is a substance that accepts a proton, in a chemical reaction. For example, when HCl gas dissolves in water, it can be viewed as donating a proton to molecules of the solvent. Thus HCl is a Brønsted acid:

$$\underset{\text{acid}}{HCl(aq)} + \underset{\text{base}}{H_2O(l)} \rightarrow H_3O^+(aq) + Cl^-(aq)$$

The water molecule is the proton acceptor; thus water is a Brønsted base in this reaction. All acid-base reactions, according to Brønsted, involve the transfer of a proton. In the above reaction H_3O^+ is called the *hydronium ion*. It is a close representation of the proton in solution. All protons are hydrated, which means that they are combined with water molecules.

When ammonia dissolves in water, it accepts a proton and so is classified as a Brønsted base. In this reaction water is the acid and ammonia is the base:

$$NH_3(aq) + H_2O(l) \rightleftharpoons NH_4^+(aq) + OH^-(aq)$$

Ammonia is a weak electrolyte because only a small fraction of dissolved NH_3 molecules are actually dissociated at any one time.

Monoprotic, Diprotic, and Triprotic Acids.
Some acids are capable of donating more than one proton to a base. Acids are referred to as monoprotic, diprotic, and triprotic depending on whether they can donate one, two, or three protons. Sulfuric acid (H_2SO_4) and carbonic acid (H_2CO_3) are diprotic acids. Phosphoric acid (H_3PO_4) is a triprotic acid.

Writing Equations for Acid-Base Reactions.
When writing equations for acid-base reactions occurring in aqueous solutions, the formula that appears in the equation will be that of the predominant

chemical form in the solution. Weak acids (HA) and bases (HB) are only a few percent ionized in solution, and so in an ionic equation their formulas will appear in the molecular form (HA and HB). On the other hand, strong acids and bases such as HCl and NaOH are 100% ionized, and so they will appear in the ionized form (H^+ and Cl^- and Na^+ and OH^-) in acid-base reactions.

Acid-base reactions are divided into four types. Examples of net ionic equations are given below.

- Reaction of a weak acid (HF) with a strong base (NaOH):

$$HF(aq) + OH^-(aq) \rightarrow F^-(aq) + H_2O(l)$$

Here we write the molecular formula of the weak acid, and for the strong base we write only OH^-. The Na^+ ion does not undergo any reaction.

- Reaction of a weak acid (HF) with a weak base (NH_3):

$$HF(aq) + NH_3(aq) \rightarrow F^-(aq) + NH_4^+(aq)$$

Both the weak acid and the weak base are written in the molecular form. The reaction leads to the formation of an ionic compound (NH_4F) which is a strong electrolyte.

- Reaction of a strong acid (HCl) with a weak base (NH_3):

$$H^+(aq) + NH_3(aq) \rightarrow NH_4^+(aq)$$

Cl^- is a spectator ion.

Reaction of a strong acid (HCl) with a strong base (NaOH):

$$H^+(aq) + OH^-(aq) \rightarrow H_2O(l)$$

Na^+ and Cl^- are spectator ions.

EXAMPLE 3.8 Writing Ionization Equations for Acids and Bases

Write the chemical equation describing the ionization reaction of each of the following Arrhenius acids and bases:
a. HI
b. KOH
c. $Mg(OH)_2$

METHOD OF SOLUTION

a. Hydroiodic acid is a strong acid and dissociates to yield hydrogen ions (protons) and iodide ions:

$$HI(aq) \rightarrow H^+(aq) + I^-(aq)$$

b. Potassium hydroxide is a strong base and dissociates into potassium ions and hydroxide ions:

$$KOH(aq) \rightarrow K^+(aq) + OH^-(aq)$$

c. Magnesium hydroxide is also a strong base and dissociates into two OH^- ions and one Mg^{2+} ion:

$$Mg(OH)_2(aq) \rightarrow Mg^{2+}(aq) + 2OH^-(aq)$$

REDOX REACTIONS AND OXIDATION NUMBERS

STUDY OBJECTIVES

You should be able to:
1. Define oxidation, reduction, half-reaction, oxidizing agent, and reducing agent.
2. Assign oxidation numbers to the atoms in molecules, simple ions, and polyatomic ions.

Oxidation and Reduction. In this section we review an extremely important type of chemical process called the oxidation-reduction reaction, or just redox reaction for short. In redox reactions electrons are transferred from one atom to another. *Oxidation* is a chemical change in which an atom, ion, or molecule loses electrons. *Reduction* is a chemical change in which an atom, ion, or molecule gains electrons. In the reaction of metallic sodium with chlorine gas, sodium is oxidized and chlorine is reduced:

$$2Na + Cl_2 \rightarrow 2Na^+ + 2Cl^-$$

An electron is transferred from the sodium atom to the chlorine atom. Oxidation and reduction always occur simultaneously.

In a redox reaction the substance oxidized is called the *reducing agent* because it supplies electrons which cause reduction in another substance. The substance being reduced is called the *oxidizing agent* because it acquires electrons from another substance. In the sodium-chlorine reaction, sodium is the reducing agent and chlorine is the oxidizing agent.

Redox reactions are conveniently separated into two *half-reactions*: the oxidation half-reaction and the reduction half-reaction. For the sodium-chlorine reaction they are

oxidation $\quad 2Na \rightarrow 2Na^+ + 2e^-$
reduction $\quad 2e^- + Cl_2 \rightarrow 2Cl^-$

Summation of the two half-reactions yields the overall redox reaction. Half-reactions are used for convenience, but you must not forget that oxidation and reduction occur simultaneously.

The concepts of oxidation and reduction have been extended from cases of complete electron gain or loss to reactions involving varying degrees of electron transfer. When tin ore is reduced to metallic tin by heating tin oxide with charcoal, tin ions are reduced and carbon is oxidized:

$$SnO_2(s) + C(s) \rightarrow Sn(l) + CO_2(g)$$

The reduction of tin is quite obvious, but in what sense is carbon oxidized? In the context of the original definition of oxidation, carbon is oxidized because it combines with oxygen. Carbon loses its electrons in the following sense. In charcoal, each carbon atom shares its four valence electrons equally with other carbon atoms. But in CO_2 the bonding electrons are shared unequally, lying closer to the oxygen atoms. Even though a C^{4+} is not formed in CO_2, the C atoms do relinquish their electrons to some extent to the two O atoms. Thus carbon is oxidized in the reaction. In this reaction oxygen is reduced.

Oxidation Numbers. The oxidation number or oxidation state is used to help keep account of electrons in chemical reactions. The *oxidation number* is the charge an atom would have in a molecule or in an ionic compound if electrons were completely transferred.

The oxidation number of an atom designates its positive or negative character. A positive oxidation number indicates that an atom has lost electrons. A negative oxidation number indicates that an atom has gained electrons. Oxidation numbers are useful in *writing formulas*, in *recognizing redox reactions*, and in *balancing redox reactions*. An element is oxidized in a reaction whenever its oxidation number increases as a result of the reaction. When the oxidation number of an element decreases in a reaction, it is reduced.

The following is a set of rules for assigning oxidation numbers:

1. *The oxidation number of an uncombined element is zero.* The atoms in free elements are neutral. Thus the atoms in Fe, Ne, O_2, H_2, and P_4, for example, have oxidation numbers of zero.
2. *The oxidation number of an element in a simple monatomic ion is the same as the charge of the ion.* The atoms in the monatomic ions Na^+, Mg^{2+}, and Al^{3+} have oxidation numbers of $+1$, $+2$, and $+3$, respectively. The oxidation number of chlorine in Cl^- is -1 and of sulfur in S^{2-} is -2.

3. *Certain elements have the same oxidation number in all (or almost all) of their compounds.*
   a. The oxidation number of oxygen is -2 in almost all of its compounds, except when it is in a peroxide, where it is -1.
   b. The oxidation number of hydrogen is $+1$, except when it is bonded to a metal in binary compounds. In metal hydrides, hydrogen has an oxidation number of -1. An example is AlH_3.
   c. Fluorine has an oxidation number of -1 in all of its compounds.
   d. The oxidation numbers of Group 1A and Group 2A metals in compounds are always $+1$ and $+2$, respectively. Aluminum is always $+3$.
4. *In a neutral compound, the algebraic sum of the oxidation numbers of all the atoms is zero.* For example, in sodium oxide (Na_2O)

$$2(\text{ox. no. Na}) + (\text{ox. no. O}) = 0$$

$$2(+1) + (-2) = 0$$

5. *The sum of the oxidation numbers of all atoms in a polyatomic ion is equal to the net charge of the ion.* For example, in nitrate ion (NO_3^-), which has a net charge of -1,

$$(\text{ox. no. N}) + 3(\text{ox. no. O}) = -1$$

$$(+5) + 3(-2) = -1$$

EXAMPLE 3.9 Determining Oxidation Numbers

Assign oxidation numbers to all the atoms in the following compounds and ions:
a. Na_2SO_4
b. $CuCl$
c. SO_3^{2-}

METHOD OF SOLUTION

a. In Na_2SO_4 we can assign oxidation numbers to Na and O because their oxidation numbers are almost always the same in all compounds (rule 3). Na is an alkali metal and is $+1$. Oxygen is -2. The oxidation number of sulfur varies from compound to compound. Apply rule 4, that the sum of the oxidation numbers of all atoms in the compound is zero:

$$2(\text{ox. no. Na}) + (\text{ox. no. S}) + 4(\text{ox. no. O}) = 0$$

$$2(+1) + (\text{ox. no. S}) + 4(-2) = 0$$

$$+(\text{ox. no. S}) = 8 - 2 = +6$$

b. In CuCl, the chlorine ion has an oxidation number of -1. Thus Cu must be $+1$ by rule 4.
c. In SO_3^{2-} we assume the oxidation number of oxygen is -2, and solve for the oxidation number of sulfur using rule 5:

$$(\text{ox. no. S}) + 3(\text{ox. no. O}) = -2$$

$$(\text{ox. no. S}) + 3(-2) = -2$$

$$(\text{ox. no. S}) = -2 + 6 = +4$$

EXAMPLE 3.10 Identifying Oxidation and Reduction

Determine whether iron in the reactant species is oxidized or reduced during the following reactions.
a. $2Fe + \frac{3}{2}O_2 + 2H_2O \rightarrow Fe_2O_3 \cdot 2H_2O$ (rust)
b. $FeO + CO \rightarrow Fe + CO_2$
c. $2Hg + 2Fe^{3+} + 2Cl^- \rightarrow Hg_2Cl_2 + 2Fe^{2+}$

METHOD OF SOLUTION

a. There are two ways to tell. First assign oxidation numbers to iron in Fe and in Fe_2O_3 (ignore the waters of hydration). The oxidation number changes from zero in Fe to $+3$ in Fe_2O_3. Since Fe becomes more positive, it must have lost electrons. Thus iron is oxidized. In the second way, when a metal combines with oxygen, its electrons are always transferred to the oxygen atom. All elements except fluorine are oxidized when they combine with O_2, and oxygen is reduced in these reactions.

b. Using the oxidation number change as a guide, the oxidation number of iron changes from $+2$ in FeO to zero in Fe. Therefore, iron is gaining electrons in this reaction, and so it is reduced.

c. The oxidation number of iron is more negative in the product, and so iron must gain an electron in this reaction. Fe is reduced.

BALANCING REDOX REACTIONS

STUDY OBJECTIVES

You should be able to:
1. Recognize redox reactions and identify the element oxidized and the element reduced.
2. Separate a redox equation into its half-reactions and balance the equation using the ion-electron method.

Redox Reactions. Oxidation-reduction reactions are often much more complex and difficult to balance than the types of reactions balanced previously in the chapter. In this section we will review how to recognize and balance redox reactions. In the previous section we noted that oxidation is the loss of electrons and reduction is the gain of electrons. Oxidation can also be defined as an *increase in oxidation number*. Likewise, reduction may be defined as a *decrease in oxidation number*:

$$2\overset{0}{Na} + \overset{0}{Cl_2} \rightarrow 2\overset{+1\ -1}{NaCl}$$

The numbers over the symbols of each element are oxidation numbers. The oxidation number of Na increases in the reaction by 1, indicating an electron has been lost. Oxidation of sodium has occurred. The oxidation number of Cl decreases by 1, indicating that Cl gained an electron. When the oxidation numbers of elements in a chemical equation change, the equation represents a redox reaction. In this reaction sodium is the reducing agent and chlorine (Cl_2) is the oxidizing agent.

Types of Redox Reactions. Redox reactions can be divided into four major types: combination reactions, decomposition reactions, displacement reactions, and disproportionation reactions.

Combination reactions In a combination reaction two reactants join together to give one product. The general equation is

$$A + B \rightarrow AB$$

element + element → compound

The above reaction of sodium with chlorine is an example. Whenever two elements combine, oxidation and reduction will occur as electrons are lost by the more metallic element and gained by the more nonmetallic element. When iron rusts, iron is oxidized and oxygen is reduced. You can always tell oxidation and reduction in an equation by assigning oxidation numbers and noting which elements change their oxidation numbers in the reaction.

$$4\overset{0}{Fe}(s) + 3\overset{0}{O_2}(g) \rightarrow 2\overset{+3\ -2}{Fe_2O_3}(s)$$

Decomposition reactions A decomposition reaction is just the reverse of a combination reaction. The general equation is

$$AB \rightarrow A + B$$

$$\text{compound} \rightarrow \text{element} + \text{element}$$

By reversing the rusting reaction above, we see that the oxidation number of iron decreases, and thus iron is reduced. The oxidation number of oxygen increases, and so oxygen is oxidized:

$$2\overset{+3\ -2}{\text{FeO}_3}(s) \rightarrow 4\overset{0}{\text{Fe}}(s) + 3\overset{0}{\text{O}}_2(g)$$

Displacement reactions In a displacement reaction, one element replaces another element in a compound. Many single displacement reactions take place in aqueous solution. The general equation is:

$$A + BC \rightarrow AC + B$$

$$\text{element} + \text{compound} \rightarrow \text{new compound} + \text{new element}$$

In this type of reaction an element (A) reacts with a compound (BC) and takes the place of one of the elements in the compound. The element previously a part of the compound (B) becomes a free element. We say that A has displaced B from the compound.
There are three types of displacement reactions:

1. *Hydrogen displacement.* Many reactive metals will displace hydrogen from water. Magnesium reacts with steam. Magnesium is oxidized and hydrogen is reduced:

$$\overset{0}{\text{Mg}}(s) + 2\overset{+1}{\text{H}}_2\text{O}(g) \rightarrow \overset{+2}{\text{Mg}}(\text{OH})_2(aq) + \overset{0}{\text{H}}_2(g)$$

2. *Metal displacement.* A metal in a compound can be displaced by another uncombined metal. These reactions occur in aqueous solution. An example is

$$\overset{0}{\text{Mg}}(s) + \overset{+2}{\text{Pb}}\text{SO}_4(aq) \rightarrow \overset{+2}{\text{Mg}}\text{SO}_4(aq) + \overset{0}{\text{Pb}}(s)$$

Magnesium is oxidized and hydrogen is reduced. Magnesium is the reducing agent and hydrogen is the oxidizing agent.

3. *Halogen displacement.* An uncombined halogen element can displace another halogen element from a compound. The power of these elements as oxidizing agents decreases as we read down in Group 7A. Thus bromine can oxidize iodine, but it cannot oxidize chlorine:

$$\overset{0}{\text{Br}}_2(l) + 2\overset{+1\ -1}{\text{NaI}}(aq) \rightarrow \overset{0}{\text{I}}_2(s) + 2\overset{+1\ -1}{\text{NaBr}}(aq)$$

Notice that bromine has displaced iodine from the compound NaI. Since the Na^+ ion is unchanged in this reaction, we could write a net ionic equation:

$$Br_2(l) + 2I^-(aq) \rightarrow I_2(s) + 2Br^-(aq)$$

Disproportionation In a disproportionation reaction, one reactant serves as both the oxidizing agent and the reducing agent. An element in the reactant is both oxidized and reduced in the reaction. What occurs is that some atoms of the element act as reducing agents and transfer their electrons to other atoms of the same element. An example is the reaction of chlorine with sodium hydroxide:

$$\overset{0}{\text{Cl}}_2(g) + 2\overset{-2\ +1}{\text{OH}^-}(aq) \rightarrow \overset{-2+1}{\text{OCl}^-}(aq) + \overset{-1}{\text{Cl}^-}(aq) + \overset{+1\ -2}{\text{H}_2\text{O}}(l)$$

Only the element chlorine has a change in oxidation number. Starting with an oxidation number of zero, some of the atoms change to a more positive state (+1), and some change to a more negative state

(−1). Thus chlorine has undergone disproportionation, a change in which it is both oxidized and reduced. The requirement for disproportionation is that an element have at least three possible oxidation states. Chlorine has many possible oxidation states, as shown in Figure 3.9 of the text. We will now review the steps involved in balancing redox equations.

Ion-Electron Method. In this method the overall reaction is divided into two half-reactions, one involving oxidation and the other reduction. Each half-reaction is balanced according to mass and charge. Before illustrating the steps involved, we will show how to separate the equation into balanced half-reactions. To separate the following equations, first identify which element is oxidized and which is reduced. Do this by writing oxidation numbers above each element:

$$\overset{+1\ -2}{H_2S} + \overset{+5\ -2}{NO_3^-} \rightarrow \overset{0}{S} + \overset{+2\ -2}{NO}$$

Note that S and N undergo oxidation number changes: S ($-2 \rightarrow 0$) and N ($+5 \rightarrow +2$). Sulfur is oxidized and nitrogen is reduced. Now we write the half-reactions:

oxidation $H_2S \rightarrow S$
reduction $NO_3^- \rightarrow NO$

Each half-reaction must be *balanced by mass*. In acidic solution, always add H^+ to balance hydrogen atoms, and add H_2O to balance oxygen atoms. Here we add H^+ to the oxidation half-reaction and H_2O to the reduction half-reaction:

oxidation $H_2S \rightarrow S + 2H^+$
reduction $NO_3^- \rightarrow NO + 2H_2O$

The oxidation half-reaction is now balanced by mass. The reduction half-reaction needs $4H^+$ on the left-hand side:

reduction $4H^+ + NO_3^- \rightarrow NO + 2H_2O$

At this point this half-reaction is also balanced according to mass.

Next the *ionic charges must be balanced* by adding electrons. In the oxidation half-reaction there is a charge of $+2$ on the right-hand side and zero on the left-hand side. To balance the charge, two electrons are added to the right:

oxidation $H_2S \rightarrow S + 2H^+ + 2e^-$

Notice that both sides of the equation have the same charge. In this case 0.

The reduction half-reaction can be charge balanced by adding three electrons to the left-hand side:

reduction $3e^- + 4H^+ + NO_3^- \rightarrow NO + 2H_2O$

Note that electrons are added to the reactant side of a reduction half-reaction and to the product side of an oxidation half-reaction.

The balanced redox equation is obtained by adding the two half-reactions. But first the number of electrons shown in the two half-reactions must be the same, since all the electrons lost during oxidation must be gained during reduction. Multiplying the oxidation half-reaction by 3 and the reduction half-reaction by 2 will make the number of electrons in the two half-reactions equal. That is, six electrons are given up during oxidation and six are gained during reduction:

$$3 \times (H_2S \rightarrow S + 2H^+ + 2e^-) = 3H_2S \rightarrow 3S + 6H^+ + 6e^-$$

$$2 \times (3e^- + 4H^+ + NO_3^- \rightarrow NO + 2H_2O) = 6e^- + 8H^+ + 2NO_3^- \rightarrow 2NO + 4H_2O$$

The sum of these half-reactions is the overall balanced redox equation:

$$3H_2S + 8H^+ + 2NO_3^- \rightarrow 3S + 6H^+ + 2NO + 4H_2O$$

The $6H^+$ on the right-hand side will cancel six of the eight H^+ on the left-hand side, yielding

$$3H_2S + 2H^+ + 2NO_3^- \rightarrow 3S + 2NO + 4H_2O$$

The following steps are useful in balancing redox reactions by the ion-electron method:

1. Write the skeletal equation containing the oxidizing and reducing agents and the products in ionic form. Separate the equation into two half-reactions.
2. Balance the atoms in each half-reaction separately. This is called a *mass balance*. For reactions in *acidic* medium, add H_2O to balance oxygen atoms and H^+ to balance hydrogen atoms. For reactions in *basic* medium, first balance the atoms as you would for an acidic solution. Then, for each H^+ ion, add an OH^- ion to both sides of the half-reaction. Whenever H^+ and OH^- appear on the same side, combine them to make H_2O.
3. Add electrons to one side of each half-reaction to equalize the charges. Electrons are added to the reactant side of a reduction half-reaction and to the product side of an oxidation half-reaction. The number of electrons added to one side of a half-reaction should make the total charge of that side equal to the charge on the other side. This procedure is called a *charge balance*.
4. Add the two half-reactions together. Before this can be done, the number of electrons shown in both half-reactions must be the same. The number of electrons in the two half-reactions can be equalized by multiplying one or both half-reactions by appropriate coefficients. Now add the two half-reactions, and check the final equation by inspection. Recall that a properly balanced equation consists of a set of the smallest possible whole numbers.

EXAMPLE 3.11   Balancing A Redox Reaction

Balance the following redox reaction by the ion-electron (half-reaction) method:

$$Cu + HNO_3 \rightarrow Cu^{2+} + NO_2 \quad \text{(in acidic solution)}$$

METHOD OF SOLUTION

Step 1. Separate the skeletal equation into half-reactions. Note that Cu is oxidized and nitrogen is reduced. The half reactions are

oxidation $Cu \rightarrow Cu^{2+}$

reduction $HNO_3 \rightarrow NO_2$

Step 2. The oxidation half-reaction is already balanced by mass. Balance the reduction half-reaction by adding one H_2O to the right-hand side and one H^+ to the left-hand side:

reduction $H^+ + HNO_3 \rightarrow NO_2 + H_2O$

Step 3. To balance according to charge, add electrons to one side of each half-reaction:

oxidation $Cu \rightarrow Cu^{2+} + 2e^-$

reduction $e^- + H^+ + HNO_3 \rightarrow NO_2 + H_2O$

Step 4. First equalize the number of electrons in the two half-reactions by multiplying the reduction half-reaction by 2:

$$2e^- + 2H^+ + 2HNO_3 \rightarrow 2NO_2 + 2H_2O$$

Thus two electrons are supplied by oxidation and two are consumed by reduction. Now add the two half-reactions to obtain a balanced overall equation. The electrons on both sides will cancel:

oxidation $\quad\quad\quad\quad\quad\quad\quad\quad Cu \rightarrow Cu^{2+} + 2e^-$

reduction $\quad\quad 2e^- + 2H^+ + 2HNO_3 \rightarrow 2NO_2 + 2H_2O$
$\quad\quad\quad\quad\quad\quad \overline{Cu + 2H^+ + 2HNO_3 \rightarrow Cu^{2+} + 2NO_2 + 2H_2O}$

TRUE-FALSE QUESTIONS

1. To balance this equation,

    $$NH_3(g) + O_2(g) \rightarrow NO(g) + H_2O(l)$$

    we can write

    $$NH_3(g) + O_2(g) \rightarrow NO(g) + H_3O(l)$$

2. The following equation is balanced:

    $$B_{10}H_{14} + 3CH_3OH \rightarrow 10B(OCH_3)_3 + 22H_2$$

3. A solution of ethyl alcohol in water will conduct electricity.

4. Na_3PO_4 and K_3PO_4 should be soluble in water, but $Mg_3(PO_4)_2$ and $Ca_3(PO_4)_2$ are insoluble.

5. When solutions of K_2CO_3 and $NaNO_3$ are mixed, no precipitate forms.

6. The salt formed when HBr is neutralized by $Ca(OH)_2$ is CaBr.

7. Hydrogen ions do not exist free in aqueous solution.

8. Acetic acid (CH_3COOH) is a tetraprotic acid.

9. The oxidation number of Cl in ClO_4^- is -1.

10. The oxidation number of the oxygen atoms in O_3 is zero.

11. The oxidation number of hydrogen in MgH_2 is $+2$.

12. In the following reaction, aluminum is reduced:

    $$2Al + Fe_2O_3 \rightarrow Al_2O_3 + 2Fe$$

13. In the above reaction, Fe_2O_3 is the oxidizing agent.

SELF-TEST A

1. Write a balanced equation for each of the following word equations:
    a. chlorine + sodium iodide $\rightarrow$ sodium chloride + iodine
    b. copper(II) bromide + sodium sulfide $\rightarrow$ copper(II) sulfide + sodium bromide

c. calcium chloride + sodium phosphate → calcium phosphate + sodium chloride
d. methane (CH_4) + diatomic oxygen → carbon dioxide + water
e. magnesium nitride + water → magnesium hydroxide + ammonia
f. copper(II) oxide + ammonia → copper + nitrogen gas + water

2. Balance the following equations by inspection:
   a. $S(s) + O_2(g) \rightarrow SO_3(g)$
   b. $Fe_2O_3(s) + HCl(aq) \rightarrow FeCl_3(aq) + H_2O(l)$
   c. $NO_2(g) + H_2O(l) \rightarrow HNO_3(aq) + NO(g)$
   d. $KOH(aq) + H_2SO_4(aq) \rightarrow K_2SO_4(aq) + H_2O(l)$
   e. $C_4H_{10}(g) + O_2(g) \rightarrow CO_2(g) + H_2O(l)$

3. Identify each of the following substances as a strong electrolyte, a weak electrolyte, or a nonelectrolyte:
   a. NaBr   b. HCl   c. CH_3COOH d. NH_3 e. glucose

4. Write balanced *net ionic equations* for the reaction between
   a. $HCl(aq)$ and $Mg(OH)_2(aq)$
   b. $CH_3COOH(aq)$ and $NaOH(aq)$
   c. $NH_3(aq)$ and $HBr(aq)$
   d. $NH_3(aq)$ and $CH_3COOH(aq)$

5. Write balanced net ionic equations for the reactions that occur when equal volumes of 0.10 M solutions of the following solutes are mixed:
   a. $Ba(NO_3)_2$ and Na_2CO_3
   b. RbCl and $AgNO_3$
   c. $Pb(NO_3)_2$ and K_2S

6. Characterize the following compounds as soluble or insoluble in water:
   a. $Mg(OH)_2$ d. $CaCO_3$
   b. AgCl   e. $Pb(NO_3)_2$
   c. $BaSO_4$ f. Na_2CO_3

7. According to the solubility rules, which of the following compounds are soluble in water?
   a. Na_2SO_4 b. $BaSO_4$ c. $CuSO_4$ d. $PbSO_4$ e. Ag_2SO_4

8. Assign oxidation numbers to the underlined elements in the following molecules and ions:
   a. $\underline{N}O_2^-$ g. $K\underline{Mn}O_4$
   b. $\underline{N}H_3$ h. $\underline{C}H_4$
   c. $H\underline{Cl}O$ i. $Na_2\underline{S}_2O_3$
   d. $H\underline{Cl}O_3$ j. $\underline{Hg}_2Cl_2$
   e. $\underline{Cl}O_4^-$ k. $\underline{S}_4O_6^{2-}$
   f. $H_2\underline{S}O_3$

9. Identify which of the following reactions are oxidation-reduction reactions:
   a. $Mg(s) + HCl(aq) \rightarrow MgCl_2(aq) + H_2(g)$
   b. $HCl(aq) + NH_3(aq) \rightarrow NH_4Cl(aq)$
   c. $Mg(s) + CO_2(g) \rightarrow MgO(s) + C(s)$
   d. $Pb^{2+}(aq) + S^{2-} \rightarrow PbS(s)$

10. In the following reactions identify the elements that are oxidized and those that are reduced. Identify the oxidizing agents and reducing agents.
    a. $S + O_2 \rightarrow SO_2$
    b. $BrO_3^- + 6I^- + 6H^+ \rightarrow 3I_2 + 3H_2O + Br^-$
    c. $IO_3^- + 3I^- + 6H^+ \rightarrow 3I_2 + 3H_2O$
    d. $As + H^+ + NO_3^- + H_2O \rightarrow H_3AsO_3 + NO$
    e. $3H_2 + N_2 \rightarrow 2NH_3$
    f. $Ag^+ + Cl^- \rightarrow AgCl(s)$
    g. $H^+ + OH^- \rightarrow H_2O$

11. Balance the following redox reactions by the ion-electron method:
   a. $H_2C_2O_4 + MnO_4^- \rightarrow Mn^{2+} + CO_2$ (acidic solution)
   b. $H_2O_2 + I^- \rightarrow I_2$ (acidic solution)
   c. $Cr_2O_7^{2-} + H_3AsO_3 \rightarrow Cr^{3+} + H_3AsO_4$ (acidic solution)
   d. $Cl_2 \rightarrow ClO_4^- + Cl^-$ (basic solution)
   e. $P_4 \rightarrow PH_3 + HPO_3^{2-}$ (basic solution)

SELF-TEST B1

1. According to the solubility rules, which of the following compounds are insoluble in water?
   a. NH_4CO_3 b. $AgBr$ c. $CaCO_3$ d. $FeCl_3$ e. $Ba(OH)_2$ f. ZnS

2. Which of the following compounds are strong electrolytes?
   a. KOH b. Sucrose   c. MgI_2 d. Methyl alcohol   e. HBr f. $AgCl$

3. Identify which of the following reactions are oxidation-reduction reactions:
   a. $Ca(s) + HCl(aq) \rightarrow CaCl_2(aq) + H_2(g)$
   b. $HNO_3(aq) + NaOH(aq) \rightarrow NaNO_3(aq) + H_2O(l)$
   c. $H_2O_2 + 2HI \rightarrow I_2 + 2H_2O$
   d. $Pb^{2+}(aq) + S^{2-} \rightarrow PbS(s)$

4. Assign oxidation numbers to the underlined elements in the following molecules and ions:
   a. $\underline{C}O_3^{2-}$
   b. $\underline{Cr}O_4^{2-}$
   c. $H_2\underline{C}_2O_4$
   d. $Na_2\underline{O}_2$
   e. $\underline{S}_8$
   f. $\underline{Sn}H_4$

5. In the following reactions identify the oxidizing agents and the element that is reduced:
   a. $3Ag + 4HNO_3 \rightarrow 3AgNO_3 + NO + 2H_2O$
   b. $6Fe^{2+} + Cr_2O_7^{2-} + 14H^+ \rightarrow 6Fe^{3+} + 2Cr^{3+} + 7H_2O$
   c. $Cu^{2+} + Zn \rightarrow Cu + Zn^{2+}$

SELF-TEST B2

1. According to the solubility rules, which of the following compounds are soluble in water?
   a. NH_4CO_3 b. $AgBr$ c. $CaCO_3$ d. $FeCl_3$ e. $Ba(OH)_2$ f. ZnS

2. Which of the following compounds are nonelectrolytes?
   a. KOH b. Sucrose   c. MgI_2 d. Methyl alcohol   e. HBr f. $AgCl$

3. Identify which of the following reactions are acid-base reactions:
   a. $Ca(s) + HCl(aq) \rightarrow CaCl_2(aq) + H_2(g)$
   b. $HNO_3(aq) + NaOH(aq) \rightarrow NaNO_3(aq) + H_2O(l)$
   c. $H_2O_2 + 2HI \rightarrow I_2 + 2H_2O$
   d. $Pb^{2+}(aq) + S^{2-} \rightarrow PbS(s)$

4. Assign oxidation numbers to the underlined elements in the following molecules and ions:
   a. $\underline{C}O_2$
   b. $K_2\underline{Cr}O_4$
   c. $\underline{C}H_4$
   d. $Na_2\underline{O}$
   e. $H_2\underline{S}$
   f. $Ca\underline{H}_2$

5. In the following reactions identify the reducing agents and the element that is oxidized:
   a. $3Ag + 4HNO_3 \rightarrow 3AgNO_3 + NO + 2H_2O$
   b. $6Fe^{2+} + Cr_2O_7^{2-} + 14H^+ \rightarrow 6Fe^{3+} + 2Cr^{3+} + 7H_2O$
   c. $Cu^{2+} + Zn \rightarrow Cu + Zn^{2+}$

ANSWERS

TRUE-FALSE QUESTIONS

1. False. Never change the subscripts; change the coefficients only.
2. False. $B_{10}H_{14} + 30CH_3OH \rightarrow 10B(OCH_3)_3 + 22H_2$.
3. False. Ethyl alcohol does not ionize in aqueous solution.
4. True.
5. True.
6. False. Its formula is $CaBr_2$.
7. True.
8. False. Monoprotic. The H atoms bonded to carbon are not ionized.
9. False. +7.
10. True.
11. False. In metal hybrids the oxidation number of H is -1.
12. False. Oxidized.
13. True.

SELF-TEST A

1. a. $Cl_2 + 2NaI \rightarrow 2NaCl + I_2$
   b. $CuBr_2 + Na_2S \rightarrow CuS + 2NaBr$
   c. $3CaCl_2 + 2Na_3PO_4 \rightarrow Ca_3(PO_4)_2 + 6NaCl$
   d. $CH_4 + 2O_2 \rightarrow CO_2 + 2H_2O$
   e. $Mg_3N_2 + 6H_2O \rightarrow 3Mg(OH)_2 + 3NH_3$
   f. $3CuO + 2NH_3 \rightarrow 3Cu + N_2 + 3H_2O$
2. a. $2S(s) + 3O_2(g) \rightarrow 2SO_3(g)$
   b. $Fe_2O_3(s) + 6HCl(aq) \rightarrow 2FeCl_3(aq) + 3H_2O(l)$
   c. $3NO_2(g) + H_2O(l) \rightarrow 2HNO_3(aq) + NO(g)$
   d. $2KOH(aq) + H_2SO_4(aq) \rightarrow K_2SO_4(aq) + 2H_2O(l)$
   e. $2C_4H_{10}(g) + 13O_2(g) \rightarrow 8CO_2(g) + 10H_2O(l)$
3. a. strong   b. strong   c. weak   d. weak   e. nonelectrolyte
4. a. $CH_3COOH + OH^- \rightarrow H_2O + CH_3COO^-$
   b. $H^+ + OH^- \rightarrow H_2O$
   c. $NH_3 + H^+ \rightarrow NH_4^+$
   d. $NH_3 + CH_3COOH \rightarrow NH_4^+ + CH_3COO^-$
5. a. $Ba^{2+}(aq) + CO_3^{2-}(aq) \rightarrow BaCO_3(s)$
   b. $Ag^+(aq) + Cl^-(aq) \rightarrow AgCl(s)$
   c. $Pb^{2+}(aq) + S^{2-}(aq) \rightarrow PbS(s)$
6. a. insoluble   b. insoluble   c. insoluble   d. insoluble   e. soluble   f. soluble
7. a, c, and e
8. a. +3   b. -3   c. +1   d. +5   e. +7   f. +4   g. +7   h. -4
   i. +2   j. +1   k. +2.5
9. a and c
10. a. S is oxidized and O is reduced. S is the reducing agent and O_2 is the oxidizing agent.
    b. I is oxidized and Br is reduced. I^- is the reducing agent and BrO_3^- is the oxidizing agent.
    c. I is oxidized and I is reduced. I^- is the reducing agent and IO_3^- is the oxidizing agent.
    d. As is oxidized and N is reduced. As is the reducing agent and NO_3^- is the oxidizing agent.
    e. H is oxidized and N is reduced. H_2 is the reducing agent and N_2 is the oxidizing agent.
    f. This is not a redox reaction.
    g. This is not a redox reaction.
11. a. $5H_2C_2O_4 + 2MnO_4^- + 6H^+ \rightarrow 2Mn^{2+} + 10CO_2 + 8H_2O$
    b. $H_2O_2 + 2I^- + 2H^+ \rightarrow I_2 + 2H_2O$
    c. $Cr_2O_7^{2-} + 3H_3AsO_3 + 8H^+ \rightarrow 2Cr^{3+} + 3H_3AsO_4 + 4H_2O$
    d. $4Cl_2 + 8OH^- \rightarrow ClO_4^- + 7Cl^- + 4H_2O$
    e. $P_4 + 4OH^- + 2H_2O \rightarrow 2PH_3 + 2HPO_3^{2-}$

SELF-TEST B1

1. b, c, and f
2. a, c, e, and f
3. a and c
4. a. +4   b. +6   c. +3   d. −1   e. 0   f. −1
5. a. HNO_3, N   b. $Cr_2O_7^{2-}$, Cr   c. Cu^{2+}, Cu

SELF-TEST B2

1. a, d, and e
2. b and d
3. b
4. a. +4   b. +6   c. −4   d. −2   e. −2   f. −1
5. a. Ag, Ag   b. Fe^{2+}, Fe   c. Zn, Zn

Chapter Four
CHEMICAL REACTIONS II: MASS RELATIONSHIPS

- Stoichiometry: Amounts of Reactants and Products
- Solution Concentration
- Precipitation Reactions and Gravimetric Analysis
- Acid-Base Titrations
- Redox Titrations

STOICHIOMETRY: AMOUNTS OF REACTANTS AND PRODUCTS

STUDY OBJECTIVES

You should be able to:
1. Calculate the mass of a substance that is produced or consumed in a chemical reaction, given the mass of any other species involved in the reaction.
2. Determine which reactant is the limiting reagent in a chemical reaction.
3. Calculate the percent yield of a product.

Mass Relationships in Reactions. An extremely important feature of balanced chemical equations is that they contain information about the relative masses of reactants in the reaction. The balanced equation can be used to decide how many grams of one substance will be needed to react with a given mass of another or how many grams of product can be produced by the reaction of a specific mass of a reactant. For instance, the balanced equation

$$2Al + Fe_2O_3 \rightarrow Al_2O_3 + 2Fe$$

$$\text{2 mol} \quad \text{1 mol} \quad\quad \text{1 mol} \quad \text{2 mol}$$

$$\text{54.0 g} \quad \text{159.7 g} \quad\quad \text{102.0 g} \quad \text{111.7 g}$$

states that 2 mol of Al will react completely with 1 mol of Fe_2O_3 and will yield 1 mol of Al_2O_3 and 2 mol of Fe. Converting these molar amounts to grams, we find that 54.0 g Al reacts with 159.7 g Fe_2O_3, producing 102.0 g Al_2O_3 and 111.7 g Fe.

It is important to realize that the mass relationships can be determined only from the balanced equation, which is written in terms of moles. The quantitative relationships between elements and compounds in chemical reactions is a part of chemistry called stoichiometry. In all stoichiometry problems the balanced chemical equation provides the "bridge" that allows us to relate the amount of reactant with a given amount of another, or that is required to yield a given amount of a product.

Conversion Factors. To calculate the amount of a specific product that is formed when a given amount of one reactant is consumed, we need conversion factors. In the preceding equation, note that

2 mol of Al is consumed whenever 1 mol of Fe_2O_3 reacts and that whenever 1 mol of Al_2O_3 is formed, 2 mol of Al must react. We can write the following equivalences:

$$2 \text{ mol Al} \Leftrightarrow 1 \text{ mol Fe}_2\text{O}_3 \qquad 1 \text{ mol Al}_2\text{O}_3 \Leftrightarrow 2 \text{ mol Al}$$

The symbol $\Leftrightarrow$ means "is stoichiometrically equivalent to." This symbol is used because the equivalency is not universally true. These amounts are equivalent only in the context of this specific reaction, and this allows us to write conversion factors, such as

$$\frac{2 \text{ mol Al}}{1 \text{ mol Fe}_2\text{O}_3}, \quad \frac{1 \text{ mol Al}_2\text{O}_3}{2 \text{ mol Al}}, \quad \text{and} \quad \frac{2 \text{ mol Fe}}{2 \text{ mol Al}}$$

which apply to this reaction only. One of these conversion factors will be used in the following example.
How many grams of Fe_2O_3 will react with 15.0 g Al?

$$2Al + Fe_2O_3 \rightarrow Al_2O_3 + 2Fe$$

Three steps are required. Since the balanced equation relates the number of moles of Fe_2O_3 to moles of Al, first (1) convert the grams of Al to moles of Al using its atomic mass. Then (2) use a conversion factor (1 mol Fe_2O_3 per 2 mol Al) to find the number of moles of Fe_2O_3 that will react. Finally (3) the moles of Fe_2O_3 are expressed in grams using its formula mass:

$$2Al + Fe_2O_3 \rightarrow Al_2O_3 + 2Fe$$

step 2
mol A $\rightarrow$ mol Fe_2O_3

step 1 ↑   step 3 ↓

g Al given   g Fe_2O_3 required

1. $15.0 \text{ g Al} \times \dfrac{1 \text{ mol Al}}{27.0 \text{ g Al}} = 0.555 \text{ mol Al}$
2. $0.555 \text{ mol Al} \times \dfrac{1 \text{ mol Fe}_2\text{O}_3}{2 \text{ mol Al}} = 0.278 \text{ mol Fe}_2\text{O}_3$
3. $0.278 \text{ mol Fe}_2\text{O}_3 \times \dfrac{159.7 \text{ g Fe}_2\text{O}_3}{1 \text{ mol Fe}_2\text{O}_3} = 44.4 \text{ g Fe}_2\text{O}_3$

This calculation can be set up and carried out in one equation:

$$15.0 \text{ g Al} \times \frac{1 \text{ mol Al}}{27.0 \text{ g Al}} \times \frac{1 \text{ mol Fe}_2\text{O}_3}{2 \text{ mol Al}} \times \frac{159.7 \text{ g Fe}_2\text{O}_3}{1 \text{ mol Fe}_2\text{O}_3} = 44.4 \text{ g Fe}_2\text{O}_3$$

Summary Once you have the balanced chemical equation, you can work all stoichiometry problems using the three steps of the mole method:

Step 1. Convert the amounts of given substances into moles if necessary.
Step 2. Use the coefficients in the balanced equation to calculate the number of moles of the sought or unknown substances in the problem.
Step 3. Convert the number of moles of the needed substance into mass units if required. Always check your answer for reasonableness.

Limiting Reagents. Usually, when a reaction is carried out, the reactants are not present in the exact ratio prescribed by the balanced chemical equation. In this situation, one reagent will be completely consumed before the other runs out. When this occurs, no more product can be made, and the reaction will stop. The reagent which is consumed first is called the *limiting reagent*. The other reagent is said to be present in excess. See Figure 4.1.

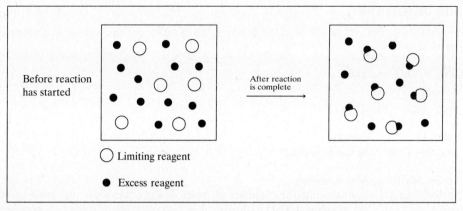

Figure 4.1. The limiting reagent reacts completely while some of the excess reagent remains unreacted.

Take, for example, the reaction

$$2Al + Fe_2O_3 \rightarrow Al_2O_3 + 2Fe$$

If 1.2 mol of Al and 1 mol of Fe_2O_3 are mixed together, the reagent that "runs out" first is Al, because the 1.2 mol of Al will consume 0.6 mol of Fe_2O_3. When the reaction ceases, 0.4 mol of Fe_2O_3 will be left unreacted. Fe_2O_3 is said to be present in *excess*. In this situation because the amount of Al limits the amount of product that can be produced, Al is called the *limiting reagent*. In any stoichiometry problem, it is important to determine which reactant is the limiting one, in order to correctly predict the yields of products.

Note, in this case, that the limiting reagent was not the reagent in lesser molar amount. It is helpful to compare the mole ratio of the reactants given in the problem to the ratio required by the balanced equation (the stoichiometric ratio):

$$\text{given ratio} = 1.2 \text{ mol Al to 1 mol } Fe_2O_3$$

$$\text{stoichiometric ratio} = 2 \text{ mol Al to 1 mol } Fe_2O_3$$

In this case the given ratio of Al to Fe_2O_3 is *less than* the stoichiometric ratio. Therefore, there is not enough Al to react with the given amount of Fe_2O_3, Aluminum is the limiting reagent.

Percent Yield. In practice, it is not unusual for the actual yield to be less than the calculated, or "theoretical," yield. The theoretical yield is calculated by assuming that all of the limiting reagent reacts according to the balanced equation. The percent yield is defined as

$$\% \text{ yield} = \frac{\text{actual yield}}{\text{theoretical yield}} \times 100\%$$

Limiting reagents and theoretical yields are illustrated in Examples 4.2 and 4.3.

EXAMPLE 4.1 The Mole Method

Sulfur dioxide can be removed from stack gases by reaction with quicklime (CaO):

$$SO_2(g) + CaO(s) \rightarrow CaSO_3(s)$$

If 975 kg of SO_2 is to be removed from stack gases by the above reaction, how many kilograms of CaO are required?

METHOD OF SOLUTION

First be certain that the equation is correctly balanced. Then follow the three steps as outlined in the above section:

$$SO_2 + CaO \rightarrow CaSO_3$$

$$\text{mol } SO_2 \xrightarrow{\text{step 2}} \text{mol } CaO$$

step 1↑ step 3↓
kg SO_2 given   kg CaO required

1. Convert the 975 kg SO_2 to moles.
2. Use the chemical equation to find the number of moles of CaO that reacts per mole of SO_2.
3. Convert moles of CaO to grams to CaO.

CALCULATION

First let's convert kg of SO_2 to g of SO_2:

$$975 \text{ kg } SO_2 \times \frac{10^3 \text{ g}}{1 \text{ kg}} = 9.75 \times 10^5 \text{ g } SO_2$$

Write out an equation which restates the problem:

$$? \text{ g CaO} = 9.75 \times 10^5 \text{ g } SO_2$$

String the conversion factors from the three steps one after the other.

$$? \text{ g CaO} = 9.75 \times 10^5 \text{ g } SO_2 \times \frac{1 \text{ mol } SO_2}{64.1 \text{ g } SO_2} \times \frac{1 \text{ mol CaO}}{1 \text{ mol } SO_2} \times \frac{56.1 \text{ g CaO}}{1 \text{ mol CaO}}$$

$$= 8.53 \times 10^5 \text{ g CaO } (853 \text{ kg CaO})$$

EXAMPLE 4.2 Limiting Reagent

Phosphine (PH_3) burns in oxygen (O_2) to produce phosphorus pentoxide (P_2O_5) and water:

$$2PH_3(g) + 4O_2(g) \rightarrow P_2O_5(s) + 3H_2O(l)$$

How many grams of P_2O_5 will be produced when 17.0 g of phosphine is mixed with 16.0 g of O_2 and reaction occurs?

METHOD OF SOLUTION

When the amounts of both reagents are given, it is possible that one reagent will be completely used up before the other. The limiting reagent will be completely consumed and will determine the amount of products. To find the limiting reagent, first find the number of moles of each reagent available.

CALCULATION

$$17.0 \text{ g } PH_3 \times \frac{1 \text{ mol } PH_3}{34.0 \text{ g } PH_3} = 0.500 \text{ mol } PH_3$$

$$16.0 \text{ g } O_2 \times \frac{1 \text{ mol } O_2}{32.0 \text{ g } O_2} = 0.500 \text{ mol } O_2$$

The balanced equation gives the stoichiometric ratio of moles of O_2 required to react per mole of PH_3:

$$\text{stoichiometric ratio} = \frac{2 \text{ mol } O_2}{1 \text{ mol } PH_3}$$

The given ratio of O_2 to PH_3 is found by comparing the given amounts:

$$\text{given ratio} = \frac{0.50 \text{ mol } O_2}{0.50 \text{ mol } PH_3} = \frac{1 \text{ mol } O_2}{1 \text{ mol } PH_3}$$

Comparison of the two ratios indicates that the given amount of O_2 is not enough to react with all of the PH_3. Thus O_2 is the limiting reagent. When all of the O_2 is consumed, some PH_3 will be left unreacted. The theoretical yield of P_2O_5 is calculated by assuming complete reaction of the limiting reagent:

$$\text{g } O_2 \rightarrow \text{mol } O_2 \rightarrow \text{mol } P_2O_5 \rightarrow \text{g } P_2O_5$$

$$? \text{ g } P_2O_5 = 0.500 \text{ mol } O_2 \times \frac{1 \text{ mol } P_2O_5}{4 \text{ mol } O_2} \times \frac{141.9 \text{ g } P_2O_5}{1 \text{ mol } P_2O_5}$$

$$= 17.7 \text{ g } P_2O_5$$

EXAMPLE 4.3 Percent Yield

In the reaction of 4.0 mol of N_2 with 6.0 mol of H_2, a chemist obtained 1.6 mol of NH_3. What is the percent yield of NH_3?

$$3H_2 + N_2 \rightarrow 2NH_3$$

METHOD OF SOLUTION

To calculate the percent yield, you must first calculate the theoretical yield. The theoretical yield of NH_3 is controlled by the limiting reagent. From the balanced equation we see that the stoichiometric ratio is 3 mol of H_2 to 1 mol of N_2. The given amounts correspond to a given mole ratio of 1.5 mol of H_2 to 1.0 mol of N_2. Therefore, there is not enough H_2 to react with all the N_2. This makes H_2 the limiting reagent. In this case, 6.0 mol of H_2 will react with 2.0 mol of N_2, yielding 4.0 mol of NH_3:

$$\text{theoretical yield} = 6.0 \text{ mol } H_2 \times \frac{2 \text{ mol } NH_3}{3 \text{ mol } H_2} = 4.0 \text{ mol } NH_3$$

The percent yield is found by dividing the actual yield by the theoretical yield:

$$\% \text{ yield } NH_3 = \frac{\text{actual yield } NH_3}{\text{theoretical yield } NH_3} \times 100\%$$

$$= \frac{1.6 \text{ mol } NH_3}{4.0 \text{ mol } NH_3} \times 100\%$$

$$= 40\%$$

SOLUTION CONCENTRATION

STUDY OBJECTIVES

You should be able to:
1. Define molarity and calculate the molarity of a solute in solution.
2. Determine the amount of a solute needed to prepare a solution of specific concentration.
3. Describe the steps involved in the dilution of a solution of known concentration and calculate the concentration of the diluted solution.

Solution Concentration. Solutions are characterized by their concentration, that is, the amount of solute dissolved in a given amount of solvent. The most common unit of concentration used for aqueous solutions is *molarity*. The molarity is the number of *moles of solute per liter of the solution*:

$$\text{molarity} = \frac{\text{moles of solute}}{\text{liters of solution}}$$

To prepare a 0.50 molar (0.50 M) solution of KCl, for instance, we first measure out 0.50 mol of solid KCl, that is, 37.3 g. The KCl is then added to a 1.0-L volumetric flask. This is a flask with a calibrated ring etched around the neck to mark the point at which the flask contains 1.0 L liquid. When water is added to this mark, the flask will contain 1.0 L of the KCl solution and will contain 0.50 mol of KCl.

The number of moles of KCl in given amounts of the above solution is easy to find. For instance, 0.10 L of 0.50 M KCl will contain

$$0.10 \text{ L soln} \times \frac{0.50 \text{ mol KCl}}{1 \text{ L soln}} = 0.050 \text{ mol KCl}$$

Here molarity is being used as a conversion factor (moles per liter). In general,
solution volume (V) × *molarity* (M) = *moles of solute*.

Dilution. Dilution is a procedure used to prepare a less concentrated solution from a more concentrated solution. Stock solutions of laboratory acids and ammonia, in particular, are purchased in gallon jugs as highly concentrated aqueous solutions. The key to understanding dilution is to remember that adding more water to a given amount of solution does not change the number of moles of solute present in solution. See Figure 4.2.

moles of solute before dilution = moles of solute after dilution.

Since the moles of solute equals the solution volume (V) times the molarity (M), we have

$$\text{moles}_i = \text{moles}_f$$

$$M_i V_i = M_f V_f$$

where the subscripts i and f stand for initial solution and final solution, respectively. The concentration after dilution is

$$M_f = M_i \left(\frac{V_i}{V_f} \right)$$

Note that V_f is always larger than V_i.

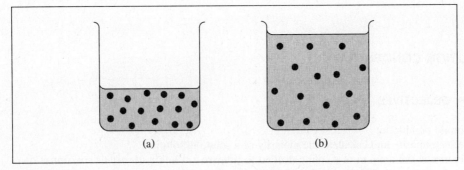

Figure 4.2. After the original solution in (*a*) is diluted, (*b*) the total number of moles of solute remains unchanged.

Electrolytes and Nonelectrolytes. It is important to keep in mind that molarity refers only to the amount of solute originally dissolved in water and does not reflect any subsequent processes such as the dissociation of a salt or the ionization of an acid. The symbols 1.0 M HCl and 1.0 M NaCl mean that 1.0 mol of the solutes HCl and NaCl were dissolved to make 1 L of solution. It, however, does not say anything about the actual chemical species in the solution such as the ions: H^+, Na^+, and Cl^-. Since HCl and NaCl are strong electrolytes, there are no molecules of HCl or NaCl in these solutions.

EXAMPLE 4.4 Calculation of Molarity

What is the molarity of a solution made from 32.1 g KNO_3 dissolved in enough water to make 500 mL of solution?

METHOD OF SOLUTION

First write the equation that defines molarity:

$$\text{molarity} = \frac{\text{moles } KNO_3}{\text{liters of solution}}$$

Then substitute into the equation the number of moles of KNO_3 in 32.1 g and the volume of solution.

CALCULATION

$$\text{mol } KNO_3 = 32.1 \text{ g } KNO_3 \times \frac{1 \text{ mol } KNO_3}{111.1 \text{ g } KNO_3}$$

$$= 0.289 \text{ mol } KNO_3$$

Convert 500 mL to 0.500 L:

$$\text{molarity} = \frac{0.289 \text{ mol } KNO_3}{0.500 \text{ L}} = 0.578 \ M$$

This is normally written 0.578 M KNO_3.

EXAMPLE 4.5 Use of Molarity

How many moles of solute are in 250 mL of 0.100 M KCl?

METHOD OF SOLUTION

$$? \text{ mol KCl} = 250 \text{ mL}$$

To obtain the number of moles, we will use the concentration as a conversion factor:

$$0.100 \ M \text{ KCl} \quad \text{means} \quad \frac{0.100 \text{ mol KCl}}{1 \text{ L soln}}$$

When the volume (in liters) is multiplied by the concentration (mol/L), we obtain moles of solute. The volume of 250 mL must be expressed in liters or else the units will not cancel.

$$V \times M = \text{moles of solute}$$

CALCULATION

$$? \text{ mol KCl} = 0.250 \text{ L} \times \frac{0.100 \text{ mol KCl}}{1 \text{ L}} = 0.025 \text{ mol KCl}$$

EXAMPLE 4.6 Use of Molarity

What volume of a concentrated (12 M) hydrochloric acid stock solution is required to prepare 800 mL of 0.120 M HCl?

METHOD OF SOLUTION

The number of moles of HCl in the 800 mL solution must come from the stock solution. Therefore,

$$\text{mol HCL}_{initial} = \text{mol HCl}_{final}$$

And since moles = MV, we can write

$$M_i V_i = M_f V_f$$

Rearranging to solve for the initial volume, we get

$$V_i = V_f \frac{M_f}{M_i}$$

CALCULATION

$V_i = ?$

$V_f = 800$ mL (or 0.800 L)

$M_f = 0.120\ M$

$M_i = 12\ M$

$$V_i = 0.800 \text{ L} \times \frac{0.120\ M}{12\ M} = 0.008 \text{ L (or 8 mL)}$$

COMMENT

Note that 0.800 L of 0.120 M HCl contains 0.0960 mol HCl, the exact quantity supplied by 0.008 L of 12 M HCl.

PRECIPITATION REACTIONS AND GRAVIMETRIC ANALYSIS

STUDY OBJECTIVES

You should be able to:
1. Use the balanced chemical equation to predict the yield of a precipitate given information about the amounts of the reactants.
2. Use data about the mass of a precipitate to calculate the concentration of an unknown.

Gravimetric Analysis. Quantitative analysis is the determination of the amount or concentration of a substance in a sample. Precipitation reactions form the basis of a type of chemical analysis called gravimetric analysis. A gravimetric analysis experiment involves the formation of a precipitate and measurement of its mass. By knowing the mass and chemical formula of the precipitate, we can calculate the mass (or concentration) of the cation or anion component in the original sample.

Stoichiometry of a Precipitation Reaction.
First, however, we will look at a precipitation reaction. When solutions of $AgNO_3$ and NaCl are mixed, a precipitate of AgCl forms because silver chloride is insoluble in water. The equation is

$$AgNO_3(aq) + NaCl(aq) \rightarrow AgCl(s) + NaNO_3(aq)$$

If 0.275 L of 0.100 M $AgNO_3$ are mixed with 0.135 L of 1.00 M NaCl, how many grams of AgCl would be formed? This example is just like the limiting reagent problems you worked earlier except that now the reaction occurs in solution. Therefore, let's find the number of moles of each reactant from $M \times V$:

$$\text{mol } AgNO_3 = 0.275 \text{ L} \times \frac{0.100 \text{ mol}}{\text{L}} = 0.0275 \text{ mol } AgNO_3$$

$$\text{mol NaCl} = 0.135 \text{ L} \times \frac{1.00 \text{ mol}}{\text{L}} = 0.135 \text{ mol NaCl}$$

The given ratio is

$$\frac{0.135 \text{ mol NaCl}}{0.0275 \text{ mol } AgNO_3} = 4.91 : 1$$

Since $AgNO_3$ and NaCl react in a 1:1 ratio, NaCl must be present in excess. The reaction will stop when all the $AgNO_3$ (specifically the Ag^+) has reacted. Since 1 mol of AgCl forms for each mol of $AgNO_3$, then the yield of AgCl will be 0.0275 mol. The amount of AgCl(s) formed is

$$\text{? g AgCl} = 0.0275 \text{ mol AgCl} \times \frac{143 \text{ g AgCl}}{1 \text{ mol AgCl}} = 3.93 \text{ g AgCl}$$

In gravimetric analysis the goal is to turn this procedure around, that is, to isolate and weigh the precipitate from a precipitation reaction. This makes the mass of precipitate a known quantity. Then we calculate the number of moles and concentration of the limiting reagent:

Given the amount of limiting reagent, calculate the yield of a precipitate,

or

Given the yield of a precipitate, calculate the amount of limiting reagent.

EXAMPLE 4.7 Solution Stoichiometry

Calculate the number of moles of NaCl that must be added to 0.752 L of 0.150 M $AgNO_3$ solution in order to precipitate all the Ag^+ ions as AgCl.

METHOD OF SOLUTION

First we need to determine the chemical equation for the precipitation reaction. Recall that NaCl is a strong electrolyte and so dissociates entirely into Na^+ and Cl^- ions in solution. $AgNO_3$ is also a strong electrolyte, yielding Ag^+ and NO_3^- ions. At first the mixed solution contains Ag^+, NO_3^-, Na^+, and Cl^- ions. Since AgCl is insoluble, the net ionic reaction is

$$Ag^+ + Cl^- \rightarrow AgCl$$

74 / Chemical Reactions II: Mass Relationships

The ionic equation is

$$Ag^+(aq) + NO_3^-(aq) + Na^+(aq) + Cl^-(aq) \rightarrow AgCl(s) + Na^+(aq) + NO_3^-(aq)$$

And the molecular equation is

$$AgNO_3(aq) + NaCl(aq) \rightarrow AgCl(s) + NaNO_3(aq)$$

CALCULATION

From the molecular equation we see that 1 mol of NaCl ≎ 1 mol of $AgNO_3$. Therefore, if we find the number of moles of $AgNO_3$, then we will also know the number of moles of NaCl required to react with it. From the volume and concentration of $AgNO_3$, we know that

$$\text{mol } AgNO_3 = V \times M$$

$$= 0.752 \text{ L} \times \frac{0.150 \text{ mol}}{1 \text{ L}} = 0.113 \text{ mol } AgNO_3$$

$$\text{mol NaCl} = \text{mol } AgNO_3 = 0.113 \text{ mol NaCl}$$

COMMENT

In this case 0.113 mol of NaCl supplies enough Cl^- ion to react with the given amount of Ag^+ ion.

EXAMPLE 4.8 Gravimetric Analysis

Calculate the molar concentration of Ba^{2+} ions in a 500.0-mL sample of an unknown aqueous solution if 2.47 g $BaSO_4$ is formed upon the addition of excess Na_2SO_4.

METHOD OF SOLUTION

(1) First write the chemical equation for the precipitation reaction. (2) Then use the mass of the precipitate (the product) to calculate the number of moles of Ba^{2+} originally present. (3) Finally divide by the volume of the original Ba^{2+} solution to determine the concentration of Ba^{2+}. Barium sulfate, being insoluble, is formed in the following reaction:

$$Ba^{2+}(aq) + SO_4^{2-}(aq) \rightarrow BaSO_4(s)$$

Therefore, the number of moles of Ba^{2+} originally equals the number of moles of $BaSO_4$ precipitate.

CALCULATION

$$? \text{ mol } BaSO_4 = 2.47 \text{ g } BaSO_4 \times \frac{1 \text{ mol } BaSO_4}{233.4 \text{ g } BaSO_4} = 0.0106 \text{ mol } BaSO_4$$

$$\text{mol } Ba^{2+} = \text{mol } BaSO_4 = 0.0106 \text{ mol } Ba^{2+}$$

To find the original concentration of Ba^{2+} ion, substitute into the equation for molarity:

$$\text{molarity of } Ba^{2+} = \frac{\text{mol } Ba^{2+}}{\text{L soln}} = \frac{0.0106 \text{ mol}}{0.500 \text{ L soln}}$$

$$= 0.0212 \text{ } M \text{ } Ba^{2+}$$

ACID-BASE TITRATIONS

STUDY OBJECTIVES

You should be able to:
1. Calculate the amount of a base needed to neutralize a given amount of acid.
2. Use titration data to calculate the concentration of an unknown acid or base solution.

Acid-Base Titrations. An important type of reaction that occurs in solution is the neutralization reaction between acids and bases. Recall, according to the Arrhenius definition, that an acid is a substance that dissociates in aqueous solution to yield hydrogen ions, H^+:

    hydrochloric acid $HCl(aq) \rightarrow H^+(aq) + Cl^-(aq)$
    nitric acid $HNO_3(aq) \rightarrow H^+(aq) + NO_3^-(aq)$

A base is a substance that dissociates in aqueous solution to form hydroxide ions, OH^-:

    potassium hydroxide $KOH(aq) \rightarrow K^+(aq) + OH^-(aq)$
    magnesium hydroxide $Mg(OH)_2(aq) \rightarrow M^{2+}(aq) + 2OH^-(aq)$

In an acid-base neutralization reaction the hydrogen ion from the acid and the hydroxide ion from the base combine to form water. The metal ion from the base and the nonmetal ion from the acid constitute a solution of a salt. The acid-base neutralization reaction of hydrochloric acid and magnesium hydroxide is represented as

$$2HCl(aq) + Mg(OH)_2(aq) \rightarrow MgCl_2(aq) + 2H_2O(l)$$
      acid   base   a salt

   A titration is a procedure for determining the concentration of an unknown acid (or base) solution using a known (standardized) concentration of a base (or acid) solution. In the titration of an acid solution of unknown concentration, a known volume of a standardized base solution is added to the acid until the acid is just neutralized. The point at which exactly enough base has been added to neutralize the acid is called the *equivalence point*. Because the concentration (M) and the required volume (V) of the base are known, the number of moles of base needed to neutralize the acid can be calculated. The number of moles of acid that were neutralized is readily calculated using the mole ratios from the balanced neutralization reaction. If the volume of acid solution was measured, then the concentration of acid is

$$\text{acid concentration} = \frac{\text{moles of acid neutralized}}{\text{liters of acid solution}}$$

EXAMPLE 4.9 Stoichiometry of an Acid-Base Reaction

What volume of 0.900 M HCl is required to completely neutralize 25.0 g of $Ca(OH)_2$?

METHOD OF SOLUTION

Write and balance the neutralization equation:

$$2HCl + Ca(OH)_2 \rightarrow CaCl_2 + 2H_2O$$

76 / *Chemical Reactions II: Mass Relationships*

Since the balanced equation relates moles of one reactant to moles of another reactant, first determine the number of moles of Ca(OH)$_2$ in 25.0 g.

CALCULATION

$$\text{g Ca(OH)}_2 \rightarrow \text{mol Ca(OH)}_2 \rightarrow \text{mol HCl} \rightarrow \text{volume HCl}$$

1. Determine the number of moles of calcium hydroxide. The formula mass of Ca(OH)$_2$ is 74.1 g/mol:

$$25.0 \text{ g Ca(OH)}_2 \times \frac{1 \text{ mol Ca(OH)}_2}{74.1 \text{ g Ca(OH)}_2} = 0.337 \text{ mol Ca(OH)}_2$$

2. Now determine the number of moles of HCl required to neutralize the base. One mole of Ca(OH)$_2$ is neutralized by 2 mol HCl:

$$0.337 \text{ mol Ca(OH)}_2 \times \frac{2 \text{ mol HCl}}{1 \text{ mol Ca(OH)}_2} = 0.674 \text{ mol HCl}$$

3. Finally determine the volume of acid that contains the required number of moles of HCl:

$$0.674 \text{ mol HCl} \times \frac{1 \text{ L soln}}{0.900 \text{ mol HCl}} = 0.749 \text{ L soln}$$

EXAMPLE 4.10 Acid-Base Titration

A volume of 128 mL of 0.650 M Ba(OH)$_2$ was required to completely neutralize 50.0 mL of nitric acid solution (HNO$_3$). What was the concentration of the acid solution?

METHOD OF SOLUTION

This is similar to the other problems. Our goal is to fill in the equation for molarity:

$$\text{molarity of nitric acid} = \frac{\text{mol HNO}_3}{\text{L acid soln}}$$

First we need the balanced chemical equation:

$$\text{Ba(OH)}_2(aq) + 2\text{HNO}_3(aq) \rightarrow \text{Ba(NO}_3)_2(aq) + 2\text{H}_2\text{O}(l)$$

$$(\text{vol} \times \text{conc}) \rightarrow \text{mol Ba(OH)}_2 \rightarrow \text{mol HNO}_3 \rightarrow \text{conc HNO}_3$$

CALCULATION

1. The number of moles of Ba(OH)$_2$ required to neutralize the HNO$_3$ was

$$0.128 \text{ L soln} \times \frac{0.650 \text{ mol Ba(OH)}_2}{1 \text{ L soln}} = 0.0832 \text{ mol Ba(OH)}_2$$

2. The number of moles of HNO$_3$ that reacted was

$$0.0832 \text{ mol Ba(OH)}_2 \times \frac{2 \text{ mol HNO}_3}{1 \text{ mol Ba(OH)}_2} = 0.166 \text{ mol HNO}_3$$

3. The concentration of the acid was

$$\text{molarity} = \frac{\text{mol HNO}_3}{1 \text{ L soln}} = \frac{0.166 \text{ mol}}{0.050 \text{ L}} = 3.33 \ M \text{ HNO}_3$$

REDOX TITRATIONS

STUDY OBJECTIVE

You should be able to:
1. Use the results of a redox titration to calculate the concentration of an unknown solution.

Redox Titrations. In redox titrations the oxidizing agent and reducing agent react by a known reaction with exact stoichiometry. If the amount of one substance reacting is known, the amount of the other substance that reacts can be calculated. When the concentration of the oxidizing agent solution is known, this solution is placed in a buret. A known volume of the solution of unknown concentration of reducing agent is placed in the titration flask. For example, consider the titration of arsenic(III) of unknown concentration with standard potassium permanganate ($KMnO_4$):

$$5H_3AsO_3 + 2MnO_4^- + 6H^+ \rightarrow 2Mn^{2+} + 5H_3AsO_4 + 3H_2O$$

The $KMnO_4$ solution containing the purple permanganate ion is placed in a buret. The solution is added drop by drop to a known volume of H_3AsO_3 solution in a flask. The resulting mixture remains colorless as long as MnO_4^- is consumed by the reaction. When enough permanganate has been added to oxidize all the H_3AsO_3, the next slight excess of $KMnO_4$ gives a pale pink color to the solution and signals the endpoint. At that point the number of moles of $KMnO_4$ reacted is equal to its concentration times the volume added from the buret:

$$\text{mol MnO}_4^- = \text{volume} \times \text{molarity}$$

In this case the number of moles of H_3AsO_3 originally present can be calculated because we know the mole ratio from the balanced equation:

$$(\text{mol MnO}_4^-) \times \frac{5 \text{ mol H}_3\text{AsO}_3}{2 \text{ mol MnO}_4^-} = \text{mol H}_3\text{AsO}_3$$

The concentration of H_3AsO_3 is

$$\text{molarity} = \frac{\text{mol H}_3\text{AsO}_3}{\text{L soln}}$$

EXAMPLE 4.11 A Redox Titration

In acid solution 31.0 mL of 0.150 M $KMnO_4$ was required to completely oxidize 50.0 mL of H_3AsO_3. What is the concentration of the H_3AsO_3 solution?

METHOD

The balanced equation was given previously:

$$5H_3AsO_3 + 2MnO_4^- + 6H^+ \rightarrow 2Mn^{2+} + 5H_3AsO_4 + 3H_2O$$

78 / *Chemical Reactions II: Mass Relationships*

We need to find the number of moles of H_3AsO_3 that reacted and then divide by the original volume. Three steps are required.

1. The number of moles of $KMnO_4$ required to oxidize all the H_3AsO_3 to H_3AsO_4:

$$31.0 \text{ mL} \times \frac{10^{-3} \text{L}}{1 \text{ mL}} \times \frac{0.15 \text{ mol}}{1 \text{ L}} = 4.65 \times 10^{-3} \text{ mol } KMnO_4$$

$$= 4.65 \times 10^{-3} \text{ mol } MnO_4^-$$

2. The number of moles of H_3AsO_3 oxidized:

$$4.65 \times 10^{-3} \text{ mol } MnO_4^- \times \frac{5 \text{ mol } H_3AsO_3}{2 \text{ mol } MnO_4^-} = 1.16 \times 10^{-2} \text{ mol } H_3AsO_3$$

3. The concentration of H_3AsO_3:

$$\text{molarity} = \frac{\text{mol } H_3AsO_3}{\text{liters soln}}$$

$$= \frac{1.16 \times 10^{-2} \text{ mol}}{50.0 \text{ mL}} \times \frac{1 \text{ mL}}{10^{-3} \text{ L}} = 0.232 \, M$$

EXAMPLE 4.12 A Redox Titration

To oxidize the Fe(II) present in 1.682 g of a compound called "Mohr's salt," 36.8 mL of unknown potassium permanganate solution was needed. Calculate the molarity of the original $KMnO_4$ solution. Mohr's salt is $Fe(NH_4)_2(SO_4)_2 \cdot 6H_2O$ (molal mass = 392.14 g/mol).

METHOD

Write a balanced redox equation for the electron transfer reaction that occurs between Fe^{2+} and MnO_4^-. The Fe^{2+} is oxidized to Fe^{3+} and MnO_4^- is reduced to Mn^{2+}:

$$5Fe^{2+} + MnO_4^- + 8H^+ \rightarrow Mn^{2+} + 5Fe^{3+} + 4H_2O$$

Remember, in a titration if you know the number of moles of one reactant consumed, you can calculate the moles of the other reagent consumed. First calculate the number of moles of Mohr's salt and then the number of moles of Fe^{2+}. From the balanced equation we can determine the number of moles of MnO_4^-. Divide this by the original volume of $KMnO_4$ solution:

g Mohr's salt → mol Mohr's salt → mol Fe → mol MnO_4^- → concentration MnO_4^-

CALCULATION

The number of moles of Mohr's salt can be calculated:

$$1.682 \text{ g Mohr's salt} \times \frac{1 \text{ mol}}{392.14 \text{ g Mohr's salt}} = 4.289 \times 10^{-3} \text{ mol Mohr's salt}$$

There is 1 mol of Fe^{2+} per mole of Mohr's salt:

$$\text{mol } Fe^{2+} = \text{mol Mohr's salt} = 4.289 \times 10^{-3} \text{ mol}$$

The number of moles of MnO_4^- reduced is

$$4.289 \times 10^{-3} \text{ mol } Fe^{2+} \times \frac{1 \text{ mol } MnO_4^-}{5 \text{ mol } Fe^{2+}} = 8.579 \times 10^{-4} \text{ mol } MnO_4^-$$

Since mol MnO_4^- = mol $KMnO_4$, then

$$8.579 \times 10^{-4} \text{ mol } MnO_4^- = 8.579 \times 10^{-4} \text{ mol } KMnO_4$$

The concentration of the $KMnO_4$ solution is found by dividing the moles of $KMnO_4$ by the volume of $KMnO_4$ solution:

$$\text{molarity of } KMnO_4 = \frac{8.579 \times 10^{-4} \text{ mol } KMnO_4}{36.8 \text{ mL}} \times \frac{1 \text{ mL}}{10^{-3} \text{ L}}$$

$$= 2.33 \times 10^{-2} \, M \; KMnO_4$$

TRUE-FALSE QUESTIONS

1. The equation

    $$Li_3N(s) + 3H_2O(l) \rightarrow NH_3(g) + 3LiOH(aq)$$

    tells us that when 3.0 g of H_2O react, 3.0 g of LiOH are formed.

2. The limiting reagent is always the reactant present in the smallest molar amount.

3. For the equation

    $$2C + O_2 \rightarrow 2CO$$

    when 1 mol of C is mixed with 2 mol of O_2, carbon is the limiting reagent and oxygen is present in excess.

4. Molarity is defined as moles of solute divided by liters of solvent.

5. The term *gravimetric analysis* refers to experiments in which a measurement of the volume of a precipitate is used to determine the concentration of an unknown.

6. For all acid-base titrations; mol acid = mol base.

7. The equivalence point of an acid-base titration is the point at which stoichiometrically equivalent quantities of acid and base have been mixed together.

8. To determine the concentration of an unknown in a redox titration, it must be an oxidizing or a reducing agent.

SELF-TEST A

1. Complete the following table for the reaction:

   $2SO_2 + O_2 \rightarrow 2SO_3$

	mol SO_2	g O_2	mol SO_3	g SO_3
a.	1.50			
b.		20.0		
c.			5.21	

2. The discovery of oxygen occurred from the decomposition of mercury(II) oxide, HgO. How many grams of O_2 would be produced by the reaction of 24.2 g of the oxide?

3. The "hypo" used in photographic developing can be made by the reaction

   $Na_2CO_3 + 2Na_2S + 4SO_2 \rightarrow 3Na_2S_2O_3 + CO_2$

   a. How many grams of Na_2CO_3 (sodium carbonate) are required to produce 321 g of sodium redox ("hypo") $Na_2S_2O_3$?
   b. How many grams of Na_2S are required to react with 25.0 g Na_2CO_3 if SO_2 is present in excess?

4. Carbon dioxide in the air of a spacecraft can be removed by its reaction with lithium hydroxide:

   $CO_2(g) + 2LiOH(s) \rightarrow Li_2CO_3(s) + H_2O(l)$

   On the average, a person will exhale about 1 kg of CO_2/day. How many kilograms of LiOH are required to react with 1.0 kg of CO_2?

5. Hydrofluoric acid (HF) can be prepared according to the following equation:

   $CaF_2 + H_2SO_4 \rightarrow 2HF + CaSO_4$

   a. How many grams of HF can be prepared from 75.0 g H_2SO_4 and 63.0 g CaF_2?
   b. How many grams of the excess reagent will remain after the reaction ceases?
   c. If the actual yield of HF is 26.2 g, what is the percent yield?

6. The reaction of iron ore follows the equation

   $2Fe_2O_3 + 3C \rightarrow 4Fe + 3CO_2$

   a. How many grams of Fe can be produced from a mixture of 200 g Fe_2O_3 and 300 g C?
   b. If the actual yield of Fe is 110 g, what is the percent yield of iron?

7. What is the molarity of a solution consisting of 11.8 g of NaOH dissolved in enough water to make exactly 300 mL of solution?

8. How many grams of NaCl are present in 45.0 mL of 1.25 M NaCl?

9. How many liters of 0.50 M glucose ($C_6H_{12}O_6$) solution will contain exactly 100 g glucose?

10. If 30 mL of 0.80 M KCl is mixed with water to make a total volume of 0.400 L, what is the final concentration of KCl?

11. When aqueous solutions of $Pb(NO_3)_2$ and Na_2SO_4 are mixed, a precipitate of $PbSO_4$ is formed. Calculate the mass of $PbSO_4$ formed when 655 mL of 0.150 M $Pb(NO_3)_2$ and 525 mL of 0.0751 M Na_2SO_4 are mixed.

12. Calculate the molar concentration of Pb^{2+} ions in 500.0 mL unknown aqueous solution if 1.07 g $PbSO_4$ is formed upon the addition of excess Na_2SO_4.

13. How many milliliters of 0.10 M H_2SO_4 would be required to neutralize 2.5 mL of 1.0 M NaOH?

14. In a titration experiment, a student finds that 23.6 mL of 0.755 M H_2SO_4 solution is required to completely neutralize 30.0 mL NaOH solution. Determine the concentration of the NaOH solution.

15. If 10 mL of 1.0 M HCl are required to neutralize 50 mL of a NaOH solution, how many mL of 1.0 M H_2SO_4 will neutralize another 50 mL of the same NaOH solution?

16. Calculate the volume of 0.0300 M Ce^{4+} required to reach the equivalence point when titrating 41.0 mL of 0.0200 M $C_2O_4^{2-}$ (oxalate ion):

    $Ce^{4+} + C_2O_4^{2-} \rightarrow Ce^{3+} + CO_2$ (acidic solution)

17. A sample of iron ore weighing 1.824 g is analyzed by converting the Fe to Fe^{2+} and then titrating with standard potassium dichromate:

    $6Fe^{2+} + Cr_2O_7^{2-} + 14H^+ \rightarrow 6Fe^{3+} + 2Cr^{3+} + 7H_2O$

    If 37.21 mL of 0.0213 M $K_2Cr_2O_7$ was required to reach the equivalence point, what was the percentage of iron in the ore?

18. $KMnO_4$ solution can be standardized against As_2O_3. A 0.2661-g sample of As_2O_3 is dissolved in acidic solution:

    $As_2O_3 + 2H_2O \rightarrow 2H_3AsO_3$

    If 38.22 mL $KMnO_4$ solution is required to react with this amount of As_2O_3, what is the molarity of the $KMnO_4$ solution?

    $2MnO_4^- + 6H^+ + 5H_3AsO_3 \rightarrow 2Mn^{2+} + 5H_3AsO_4 + 3H_2O$

SELF-TEST B1

1. Silicon tetrachloride ($SiCl_4$) can be prepared by heating silicon in chlorine gas:

    $Si(s) + 2Cl_2(g) \rightarrow SiCl_4(l)$

    How many moles of $SiCl_4$ are produced when 4.24 mol of Cl_2 gas reacts?

2. The following reaction can be used to prepare hydrogen gas:

    $CaH_2 + 2H_2O \rightarrow Ca(OH)_2 + 2H_2$

    How many grams of H_2 would result from the reaction of 100 g CaH_2 with an excess of H_2O?

3. Given 5.00 mol KOH and 2.00 mol H_3PO_4, how many moles of K_3PO_4 can be prepared?

    $H_3PO_4 + 3KOH \rightarrow K_3PO_4 + 3H_2O$

4. How many grams of HCl are contained in 250 mL of 0.500 M HCl?

5. What is the molarity of a solution consisting of 11.8 g of NaOH dissolved in enough water to make 300 mL of solution?

6. When 147 mL of 4.25 M NH_3 solution is mixed with 353 mL of H_2O, what is the concentration of the final solution?

7. Describe how you would prepare 2.50×10^2 mL of 0.666 M KOH solution starting with a 5.0 M KOH solution.

8. Hydrofluoric acid (HF) can be prepared according to the reaction

    $CaF_2 + H_2SO_4 \rightarrow 2HF + CaSO_4$

    In one experiment 42.0 g CaF_2 was treated with excess H_2SO_4 and a yield of 14.2 g of HF was obtained. Calculate the percent yield of HF.

9. If 18.7 mL of H_2SO_4 solution was required to completely neutralize 2.10 g KOH, calculate the molarity of the sulfuric acid solution.

10. In order to precipitate AgCl, excess $AgNO_3$ was added to 10.0 mL of a solution containing Cl^-. If 0.339 g of AgCl was formed, what was the concentration of Cl^- in the original solution?

11. What volume of 0.210 M H_2SO_4 solution is needed to exactly neutralize 50.0 mL of 0.082 M NaOH?

12. What volume of 0.200 M $K_2Cr_2O_7$ will be required to oxidize 4.0 g of H_3AsO_3?

    $14H^+ + Cr_2O_7^{2-} + 3H_3AsO_3 \rightarrow 2Cr^{3+} + 3H_3AsO_4$

SELF-TEST B2

1. Consider the reaction of silicon with chlorine gas:

    $Si(s) + 2Cl_2(g) \rightarrow SiCl_4(l)$

    How many grams of Cl_2 must react in order to produce 2.12 mol $SiCl_4$?

2. To obtain a yield of 9.50 g H_2, how many grams of CaH_2 must react according to the equation $CaH_2 + 2H_2O \rightarrow Ca(OH)_2 + 2H_2$?

3. Given 5.00 mol of KOH and 2.00 mol H_3PO_4 that react to yield 1.67 mol K_3PO_4 according to the equation

    $H_3PO_4 + 3KOH \rightarrow K_3PO_4 + 3H_2O$

    Which was the limiting reagent and how many moles of the excess reagent remain unreacted?

4. If 4.56 g HCl(g) are dissolved in enough water to make 250 mL of solution, what is the molarity of HCl?

5. How many milliliters of 0.983 M NaOH solution will contain exactly 11.8 g NaOH?

6. How many milliliters of 4.25 M NH_3 solution and how many milliliters of H_2O are needed to prepare 500 mL of 1.25 M NH_3?

7. How much water would you add to dilute 33.3 mL of 5.0 M KOH to form a 0.666 M KOH solution?

8. Hydrofluoric acid (HF) can be prepared according to the reaction

    $CaF_2 + H_2SO_4 \rightarrow 2HF + CaSO_4$

    In one experiment 42.0 g CaF_2 was treated with excess H_2SO_4 and a 66% yield of HF was obtained. What was the yield of HF in grams?

9. How many milliliters of 1.00 M H_2SO_4 solution is required to neutralize 2.10 g KOH?

10. Ten milliliters of a 0.236 M NaCl solution is to be treated with excess $AgNO_3$ solution to precipitate the Cl^-. Write a net ionic equation for the reaction, and calculate the yield of AgCl in grams.

11. What volume of 0.0824 M NaOH solution is needed to neutralize 9.8 mL of 0.210 M H_2SO_4?

12. In an experiment, 52.0 mL of $K_2Cr_2O_7$ solution was required to oxidize 4.0 g of H_3AsO_3. What was the molarity of the dichromate solution?

ANSWERS

TRUE-FALSE QUESTIONS

1. False. When 3.0 mol H_2O react, 3.0 mol LiOH are formed.
2. False. The limiting reagent is the one that runs out first in a reaction.
3. True.
4. False. Molarity is moles of solute per liter of solution.
5. False. You measure the mass of the precipitate.
6. False. It depends on the mole ratio. Sometimes 2 mol acid = 1 mol base and vice versa.
7. True.
8. True.

SELF-TEST A

1.

	mol SO_2	g O_2	mol SO_3	g SO_3
a.	1.50	24.0	1.50	120
b.	1.25	20.0	1.25	100
c.	5.21	83.4	5.21	417

2. 1.79 g O_2
3. a. 71.8 g Na_2CO_3 b. 36.8 g Na_2S
4. 1.09 kg
5. a. H_2SO_4 is the limiting reagent; 30.6 g HF
   b. 3.27 g CaF_2
   c. 85.6% yield
6. a. Fe_2O_3 is the limiting reagent; 1.40 g Fe
   b. 78.6% yield
7. 0.983 M NaOH
8. 3.29 g NaCl
9. 1.1 L
10. 0.060 M KCl
11. 12.0 g $PbSO_4$
12. 7.06×10^{-3} M Pb^{2+}
13. 12 mL of 0.10 M H_2SO_4

14. 1.19 M NaOH
15. 5.0 mL
16. $2Ce^{4+} + C_2O_4^{2-} \rightarrow 2Ce^{3+} + 2CO_2$ = 54.6 mL Ce^{4+} soln
17. 14.56% Fe
18. 0.0282 M $KMnO_4$

SELF-TEST B1

1. 2.12 mol
2. 9.50 g
3. 1.67 mol
4. 4.56 g
5. 0.983 M
6. 1.25 M
7. Take 33.3 mL of 5.0 M KOH and dilute to a final volume of 250 mL.
8. 66%
9. 1.00 M
10. 0.236 M
11. 9.8 mL
12. 52.9 mL

SELF-TEST B2

1. 301 g
2. 100 g
3. The limiting reagent was KOH; 0.33 mol H_3PO_4 was unreacted.
4. 0.50 M HCl
5. 300 mL
6. 147 mL NH_3 and 353 mL H_2O
7. 250 L − 33.3 mL = 217 mL
8. 14.2 g
9. 18.7 mL
10. 0.338 g, $Ag^+(aq) + Cl^-(aq) \rightarrow AgCl(s)$
11. 50 mL
12. 0.200 M

Chapter Five
THE GASEOUS STATE

- Properties of Gases
- The Gas Laws
- Stoichiometry Involving Gases
- Kinetic Molecular Theory of Gases
- Nonideal Gases

PROPERTIES OF GASES

STUDY OBJECTIVES

You should be able to:
1. List the properties that are characteristic of gases.
2. Interconvert pressure readings among the common units used to express pressure.

Characteristics of Gases. All matter exists as either a solid, a liquid, or a gas, depending on the temperature and the pressure. Of the three states, the simplest to describe is the gaseous state. Gases have certain general characteristics.

1. *Expansion*. Gases expand indefinitely to fill the space available to them.
2. *Indefinite shape*. Gases fill all parts of a container evenly and so have no definite shape of their own.
3. *Compressibility*. Gases are the most compressible of the states of matter.
4. *Mixing*. Two or more gases will mix evenly and completely when confined to the same container.
5. *Low density*. Gases have much lower densities than liquids and solids. Typically, densities of gases are about $\frac{1}{1000}$ those of liquids and solids.

Gas Pressure. One of the most obvious properties of a gas, to someone who has inflated bicycle or automobile tires, is its pressure. *Pressure* is force per unit area. Any exertion of force applied to an area can be expressed as pressure:

$$\text{pressure} = \frac{\text{force}}{\text{area}}$$

Some bicycle tires must be inflated to 60 psi. That's 60 pounds per square inch of tire surface. Recall that *weight* is the force with which an object is pulled vertically downward by gravity. Thus, a pound is a measure of force, and pounds per square inch is a unit of pressure.

A large number of units are used to measure pressure. The *standard atmosphere* is the average pressure exerted by the atmosphere at sea level. In common units this is 14.7 psi. This pressure is enough to support a column of mercury 760 mm high at 0°C. The pressure that supports 1 mm of Hg is called a *torr*. In the SI, pressure is measured in a unit called the *pascal* (Pa), which is defined as the pressure from a force of one newton exerted on one square meter of area:

$$1 \text{ Pa} = 1 \text{ N/m}^2$$

It takes 1.013×10^5 Pa to equal 1 atm.

Since pascals are so small, two somewhat more convenient units of pressure are the kilopascal and the bar:

$$1 \text{ kPa} = 10^3 \text{ Pa}$$

$$1 \text{ bar} = 10^5 \text{ Pa}$$

Notice that

$$1.013 \text{ bars} = 1 \text{ atm}$$

The following is a list of pressure units in terms of the standard atmosphere:

$$1 \text{ atmosphere} = 14.7 \text{ psi}$$
$$= 760.0 \text{ mm Hg} = 760.0 \text{ torr}$$
$$= 33.9 \text{ ft H}_2\text{O}$$
$$= 1.013 \times 10^5 \text{ Pa} = 101.3 \text{ kPa}$$
$$= 1.013 \text{ bars}$$

Most chemists seem to prefer to express pressure in units of millimeters of mercury or atmospheres.

Densities of Gases. The *density* of a gas is, of course, its mass per unit volume. Because the density of a gas will vary with its temperature and pressure, a set of standard conditions must be used to compare gas densities. Together, 0°C and 1 atm are called *standard temperature* and *pressure*, or STP. Table 5.1 shows the densities of several gases at STP. By comparison with the density for $H_2O(l)$, you can see that the density of a gas at STP is roughly $\frac{1}{1000}$ that of a liquid. Also, notice that the density of a gas is proportional to its molecular mass.

Table 5.1 Densities of Several Gases at STP

Gas	Density (g/L)
He	0.177
N_2	1.251
Air	1.293
O_2	1.429
Cl_2	3.214
SF_6	6.602
$H_2O(l)$	1000

EXAMPLE 5.1 Converting Pressure Units

The atmospheric pressure on Mars at the *Viking I* landing site was 7.3 mbars. Express this pressure in atmospheres, millimeters of mercury, and pascals.

METHOD OF SOLUTION

Treat this like any other factor-label problem. First state the problem:

$$? \text{ atm} = 7.3 \text{ mbars}$$

Then our road map is

millibars → bars → atmospheres

Substituting in the conversion factors from the earlier list, we get

$$? \text{ atm} = 7.3 \text{ mbars} \times \frac{10^{-3} \text{ bar}}{1 \text{ mbar}} \times \frac{1 \text{ atm}}{1.013 \text{ bars}}$$

$$= 7.2 \times 10^{-3} \text{ atm}$$

$$? \text{ mm Hg} = 7.2 \times 10^{-3} \text{ atm} \times \frac{760 \text{ mm Hg}}{1 \text{ atm}}$$

$$= 5.5 \text{ mm Hg}$$

$$? \text{ Pa} = 7.2 \times 10^{-3} \text{ atm} \times \frac{1.013 \times 10^5 \text{ Pa}}{1 \text{ atm}}$$

$$= 730 \text{ Pa}$$

THE GAS LAWS

STUDY OBJECTIVES

You should be able to:
1. State Boyle's, Charles's, and Avogadro's laws and know what variables are held constant for each law. Solve problems involving these laws.
2. Solve problems involving changes in the conditions of a gas, using the ideal gas equation.
3. Determine the density of an ideal gas of known molecular mass, and vice versa.
4. Solve problems using Dalton's law of partial pressures.

The Ideal Gas Equation. The properties of a gas that we are most often concerned with are pressure (P), volume (V), absolute temperature (T), and amount (number of moles, n). The experimentally observed relationship between these properties is called the *ideal gas law*:

$PV = nRT$

where R is the ideal gas constant equal to 0.0821 L · atm/K · mol. The dots appearing in the units are to remind us that liter and atmosphere are in the numerator and Kelvin and mole are in the denominator. The units of R may also be written as $R = 0.0821$ L atm K^{-1} mol^{-1}.

The ideal gas law has great importance in the study of gases. It does not contain information that is characteristic of any particular gas. Rather, it is a generalization applicable to most gases at pressures up to about 10 atm and at temperatures above 0°C. An *ideal gas* is one whose behavior agrees with that predicted by the ideal gas law.

In the textbook, the ideal gas law is shown to result from combining Boyle's, Charles's, and Avogadro's laws. Here we will take the reverse approach and show how the ideal gas law reduces to each of these laws under certain conditions. It is helpful when discussing the volume changes of a gas to visualize that we are considering a gas trapped in a container with a movable barrier so that its volume is not fixed. See Figure 5.1.

Boyle's law states that when the temperature is held constant, the volume of a given amount of gas is inversely proportional to the applied pressure.

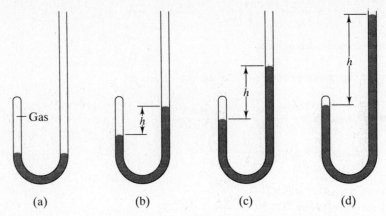

Figure 5.1. Apparatus for studying the pressure-volume relationship of a gas. In (a) the pressure of the gas is equal to the atmospheric pressure. By adding Hg(1), the pressure on the gas increases from (a) to (d) as more Hg is added. The total pressure of the gas is equal to the pressure due to the height, h, of Hg *plus* the atmospheric pressure.

The ideal gas law states

$$PV = nRT$$

As long as we do not add or remove gas, the number of moles of gas (n) is a constant. If T is also held constant, then the term nRT must be a constant. Let's call the constant k_1:

$$nRT = k_1$$

Since $PV = nRT$, PV must equal a constant:

$$PV = k_1 \quad \text{(Boyle's law)}$$

This is the equation of Boyle's law. On rearranging, the equation becomes

$$V = \frac{k_1}{P}$$

and we see, as Boyle's law states, that the volume of a gas is inversely proportional to the pressure.

For changes in pressure from P_1 to P_2, we can apply Boyle's law. Since the product PV is a constant, the value of PV at the initial pressure P_1 will equal PV at the final pressure P_2:

$$P_1V_1 = P_2V_2 \quad (n \text{ and } T \text{ are constant})$$

Charles's law states that at constant pressure the volume of a gas is directly proportional to its absolute temperature. Again, as long as n remains unchanged and if P is also constant, the ideal gas law can be rearranged to place all the constants on one side of the equation. Here we see that V/T equals a constant:

$$\frac{V}{T} = \frac{nR}{P} = k_2$$

Rearranging, we get Charles's law:

$$V = k_2 T$$

This equation shows that V is proportional to the *absolute* temperature T. Since V/T is equal to a constant, then V_1/T_1 at an initial temperature T_1 and volume V_1 will equal the quotient V_2/T_2 at a final temperature T_2 and volume V_2:

$$\frac{V_1}{T_1} = \frac{V_2}{T_2} \quad (n \text{ and } P \text{ are constants})$$

Avogadro's law says that at the same temperature and pressure, equal volumes of different gases contain the same number of molecules (or moles). This is another way of saying that the volume of a gas is proportional to the number of moles present. Taking the ideal gas equation and setting the constant terms T and P on the same side of the equation, we get

$$V = \frac{nRT}{P}$$

Thus

$$V = nk_3 \quad (\text{Avogadro's law})$$

where $k_3 = RT/P =$ constant. Again, the ideal gas law reduces to "one of the gas laws" when, as in this case, pressure and temperature are held constant.

A particularly useful quantity is the volume occupied by 1 mol of ideal gas at STP. Substituting into the preceding equation gives

$$V = \frac{nRT}{P} = \frac{(1 \text{ mol})(0.0821 \text{ L} \cdot \text{atm}/\text{K} \cdot \text{mol})(273 \text{ K})}{1 \text{ atm}}$$

$$= 22.4 \text{ L}$$

The molar volume of an ideal gas is 22.4 L at STP.

In addition to these laws, a useful relationship for changes in pressure, volume, and temperature of *a fixed number of moles* of gas is

$$\frac{P_1 V_1}{T_1} = \frac{P_2 V_2}{T_2}$$

Gas Density. The density of a gaseous compound is $d = m/V$, where m is the mass of the gas in grams. Recall the number of moles n of gas is $n = m/\mathcal{M}$, where $\mathcal{M}$ is the molar mass. Substituting for n in the ideal gas equation gives

$$PV = \frac{m}{\mathcal{M}} RT$$

Rearranging gives

$$\frac{m}{V} = \frac{P\mathcal{M}}{RT}$$

and so the density of a gas that follows the ideal gas law is

$$d = \frac{P\mathcal{M}}{RT}$$

Note that the density is directly proportional to its molar mass, as we noted earlier. Thus, a measurement of the density of a gas allows the calculation of its molar mass. Note also that the density increases as gas pressure increases, and density decreases as temperature increases. The principle of the hot air balloon is that hot air has a lesser density than cold air. A hot air balloon floats on the more dense colder air surrounding it, just as a piece of wood floats on water.

Dalton's Law of Partial Pressures. Each component gas of a mixture of gases uniformly fills the containing vessel. Each component exerts the same pressure as it would if it occupied that volume alone. Thus, the total pressure of a mixture is the sum of the individual pressures, called the *partial pressures*, of each component:

$$P_t = P_A + P_B + P_C + \cdots$$

Also, the total number of moles n_t is the sum of the number of moles of each component:

$$n_t = n_A + n_B + n_C + \cdots$$

The ratio n_A/n_t, the ratio of the number of moles of component A to the total number of moles of all components, is called the mole fraction of component A. The symbol X_A is used for the mole fraction of component A. According to Dalton's law, the partial pressure of a gas is proportional to its mole fraction in the mixture:

$$P_A = \frac{n_A}{n_t} P_t = X_A P_t$$

Whenever gases are collected by displacement of water, the total gas pressure is the sum of the partial pressure of the collected gas and the partial pressure of water vapor:

$$P_t = P_{gas} + P_{H_2O}$$

Consequently, the partial pressure of the collected gas is $P_{gas} = P_t - P_{H_2O}$, where the partial pressure of water depends on the temperature and corresponds to the vapor pressure of water.

EXAMPLE 5.2 Gas Law Calculation

When 2.0 L of chlorine gas (Cl_2) at STP is warmed at constant pressure to 100°C, what is the new volume?

METHOD OF SOLUTION

The volume of a fixed amount of a gas at constant pressure is proportional to the absolute temperature according to Charles's law, where $T_1 = 273$ K and $T_2 = (100°C + 273°C)$ K $= 373$ K:

$$\frac{V_1}{T_1} = \frac{V_2}{T_2}$$

CALCULATION

Rearranging gives

$$V_2 = V_1 \times \frac{T_2}{T_1}$$

where V_1, the initial volume, is 2.0 L. Substituting, we get

$$2.0 \text{ L} \times \frac{373 \text{ K}}{273 \text{ K}} = 2.7 \text{ L}$$

COMMENT

Note that we expect $V_2 > V_1$ because of the temperature increase. This provides us with a simple check of our answer. If we calculated that $V_2 < V_1$, we would know our answer was wrong.

EXAMPLE 5.3 Application of Gas Laws

Given 10.0 L of neon gas at 5°C and 630 mm Hg, calculate the new volume at 400°C and 2.5 atm.

METHOD OF SOLUTION

First note that both the pressure and the temperature of the gas are changed but the number of moles is constant. Write the ideal gas equation with all the constant terms on one side:

$$\frac{PV}{T} = nR = \text{a constant}$$

We see that PV/T is a constant. Therefore

$$\frac{P_1 V_1}{T_1} = \frac{P_2 V_2}{T_2}$$

where the subscript 2 refers to the final state and the subscript 1 refers to the initial state.

CALCULATION

Rearranging to solve for V_2 gives

$$V_2 = V_1 \times \frac{P_1}{P_2} \times \frac{T_2}{T_1}$$

where

$$P_1 = 630 \text{ mm Hg} \times \frac{1 \text{ atm}}{760 \text{ mm Hg}} = 0.829 \text{ atm}$$

$$P_2 = 2.5 \text{ atm}$$

$$T_1 = (273°C + 5°C) \text{ K} = 278 \text{ K}$$

$$T_2 = (273°C + 400°C) \text{ K} = 673 \text{ K}$$

Substituting yields

$$V_2 = 10.0 \text{ L} \times \frac{0.829 \text{ atm}}{2.5 \text{ atm}} \times \frac{673 \text{ K}}{278 \text{ K}}$$

$$= 8.0 \text{ L}$$

COMMENT

Note that both P_1 and P_2 must be expressed in the same units and both temperatures in kelvins.

EXAMPLE 5.4 Ideal Gas Law

What volume will be occupied by 0.833 mol of fluorine (F_2) at 645 mm Hg and 15.0°C?

METHOD OF SOLUTION

The ideal gas law relates the volume of a gas to its temperature, pressure, and number of moles.

CALCULATION

$$V = \frac{nRT}{P}$$

First convert the pressure into atmospheres:

$$P = 645 \text{ mm Hg} \times \frac{1 \text{ atm}}{760 \text{ mm Hg}} = 0.849 \text{ atm}$$

$$= \frac{(0.833 \text{ mol})(0.821 \text{ L} \cdot \text{atm/K} \cdot \text{mol})(288 \text{ K})}{0.849 \text{ atm}} = 23.2 \text{ L}$$

COMMENT

Even though it takes longer to write out the units for each term in the equation, doing so is very helpful because when the units cancel and leave you with the desired units (in this case liters), you are more likely to get the right answer.

EXAMPLE 5.5 Molar Mass of a Gas

A gaseous compound has a density of 1.69 g/L at 25°C and 714 torr. What is its molar mass?

METHOD OF SOLUTION

The density of an ideal gas is directly proportional to its molecular mass. We use the equation

$$d = \frac{P\mathcal{M}}{RT}$$

CALCULATION

Convert pressure into units of atmospheres:

$$P = 714 \text{ torr} \times \frac{1 \text{ atm}}{760 \text{ torr}} = 0.939 \text{ atm}$$

Rearranging and substituting, we get

$$\mathcal{M} = \frac{dRT}{P} = \frac{(1.69 \text{ g/L})(0.0821 \text{ L} \cdot \text{atm/K} \cdot \text{mol})(298 \text{ K})}{0.939 \text{ atm}} = 44.0 \text{ g/mol}$$

EXAMPLE 5.6 Partial Pressures

Oxygen gas is collected over water at 30°C and the total pressure is 645 mm Hg:
a. What is the partial pressure of oxygen? The vapor pressure of water at 30°C is 31.8 mm Hg.
b. What are the mole fractions of oxygen and water vapor?

METHOD OF SOLUTION FOR a

Mixtures of gases obey Dalton's law of partial pressures, which says that the total pressure is the sum of the partial pressures of oxygen and water vapor:

$$P_t = P_{O_2} + P_{H_2O}$$

CALCULATION FOR a

$$P_{O_2} = P_t - P_{H_2O} = 645 \text{ mm Hg} - 31.8 \text{ mm Hg}$$

$$= 613 \text{ mm Hg}$$

METHOD OF SOLUTION FOR b

Recall that the partial pressures of O_2 and H_2O are related to their mole fractions,

$$P_{O_2} = X_{O_2} P_t \qquad P_{H_2O} = X_{H_2O} P_t$$

CALCULATION FOR b

Therefore, on rearranging, we get

$$X_{O_2} = \frac{P_{O_2}}{P_t} \qquad X_{H_2O} = \frac{P_{H_2O}}{P_t}$$

$$X_{O_2} = 613/645 = 0.950 \quad \text{and} \quad X_{H_2O} = 31.8/645 = 0.049$$

COMMENT

Note also that the sum of the mole fractions is 1.0 within the number of significant figures given:

$$X_{O_2} + X_{H_2O} = 0.950 + 0.049 = 0.999$$

STOICHIOMETRY INVOLVING GASES

STUDY OBJECTIVE

You should be able to:
1. Calculate the volume of any gaseous reactant consumed or of any gaseous product generated in a chemical reaction, given the mass of any species involved in the reaction.

Earlier we saw that stoichiometry involves relationships between masses and moles of the reactants and of the products of a chemical reaction. The ideal gas equation furnishes a way to convert between moles and the volume of a gas. The reaction of a known volume of a given gas will yield a specific volume of a product gas. When the balanced equation is known, the volumes of the two gases can be related by the following path:

volume of a given gas at specified P and T → moles of gaseous species → moles of desired species → volume of the desired gas at P and T

EXAMPLE 5.7 Volume of a Gaseous Product

Calculate the volume of ammonia gas, measured at 645 torr and 21°C, that is produced by the complete reaction of 25.0 g of quicklime (CaO) with excess ammonium chloride (NH_4Cl) solution.

METHOD OF SOLUTION

First write the balanced chemical equation:

$$CaO(s) + 2NH_4Cl(aq) \rightarrow 2NH_3(g) + CaCl_2(aq) + H_2O(l)$$

Here we must determine the number of moles of NH_3 formed in the reaction and convert this into the volume of an ideal gas. Three steps are required: (1) convert grams of CaO to moles of CaO, (2) convert moles of CaO to moles of NH_3 produced, and (3) use the ideal gas equation to calculate the volume of NH_3:

$$g\ CaO \xrightarrow{1} mol\ CaO \xrightarrow{2} mol\ NH_3 \xrightarrow{3} volume\ NH_3$$

We already know how to find conversion factors 1 and 2. Therefore let's first find the number of moles of NH_3 formed.

CALCULATION

State the problem:

$$?\ mol\ NH_3 = 25.0\ g\ CaO$$

$$?\ mol\ NH_3 = 25.0\ g\ CaO \times \frac{1\ mol\ CaO}{56.1\ g\ CaO} \times \frac{2\ mol\ NH_3}{1\ mol\ CaO} = 0.891\ mol\ NH_3$$

The volume of 0.891 mol NH_3 can be calculated from the ideal gas equation:

$$V = \frac{nRT}{P}$$

Convert pressure into units of atmospheres before substituting into the ideal gas equation:

$$P = 645\ torr \times \frac{1\ atm}{760\ torr} = 0.849\ atm$$

$$V = \frac{0.891\ mol\ (0.0821\ L \cdot atm/K \cdot mol)(294\ K)}{0.849\ atm} = 25.3\ L$$

EXAMPLE 5.8 Gas Stoichiometry

Calculate the volume of methane (CH_4) at STP required to completely consume 3.50 L of oxygen at STP.

METHOD OF SOLUTION

As always, write the balanced chemical equation:

$$CH_4(g) + 2O_2(g) \rightarrow CO_2(g) + H_2O(l)$$

According to Avogadro's law, the volume of a gas is proportional to the number of moles of gas, at constant temperature and pressure. Thus, the stoichiometric ratio

$$\frac{1\ mol\ CH_4}{2\ mol\ O_2}$$

can also be written in terms of volume:

$$\frac{1\ L\ CH_4}{2\ L\ O_2}$$

CALCULATION

State the problem:

$$?\ L\ CH_4 = 3.50\ L\ O_2$$

Map out the conversion steps:

volume $O_2 \rightarrow$ volume CH_4

$$? \text{ L CH}_4 = 3.50 \text{ L O}_2 \times \frac{1 \text{ L CH}_4}{2 \text{ L O}_2}$$

$$= 1.75 \text{ L CH}_4$$

KINETIC MOLECULAR THEORY OF GASES

STUDY OBJECTIVES

You should be able to:
1. Describe the assumptions on which the kinetic molecular theory is based.
2. Calculate the average molecular speeds of given gases.
3. Describe diffusion and effusion and calculate the relative rates of these processes for two given gases.

Postulates of the Kinetic Molecular Theory. In Chapter 1 you learned that scientific laws are concise statements of mathematical equations that relate the results of a large number of experiments. Attempts to account for, or explain, a law are theoretical. The theory that has been very successful in accounting for the properties or gases is called the *kinetic molecular theory of gases*.

The main assumptions that explain the behavior of an ideal gas are as follows:

1. Gases consist of molecular particles. These molecules are much smaller than the distances between them. The volume occupied by a molecule is considered negligible.
2. Gas molecules are in constant motion, undergoing frequent collisions with each other and with the container walls.
3. Although energy may be transferred from one molecule to another in a collision, the total energy of all molecules remains constant. In other words, the collisions are elastic.
4. The average kinetic energy of a molecule in a gas is

   $$\overline{KE} = \tfrac{1}{2}\overline{mu^2}$$

   where m is the mass of the molecule and u is its speed. The quantity $\overline{u^2}$ is the mean square speed. The average kinetic energy is proportional to the absolute temperature:

   $$\overline{KE} \propto T \quad \text{or} \quad \overline{KE} = kT$$

   where k is a proportionality constant.
5. Gas molecules behave as independent particles; all attractive and repulsive intermolecular forces are negligible.

The manner in which the kinetic theory explains many of the properties of gases and the gas laws is discussed in Section 5.7 of the textbook. According to the kinetic theory, gas molecules are in constant, random motion. The pressure exerted by a gas on the container walls is assumed to arise from impacts on the walls by gas molecules. During impacts, force is exerted on the wall area. The gas pressure depends on the volume, the temperature, and the amount of gas. The kinetic molecular theory explains these observations in the following ways:
1. *Boyle's law*, $V = k_1/P$. As the volume of the container is increased, there are fewer impacts of gas molecules with the wall per second; thus, the gas pressure is lowered.

2. *Charles's law*, $V = k_2 T$. As the temperature of a gas increases, the average kinetic energy and speed of the gas molecules increase. Thus, the collision frequency and the force of impact with the wall increase. If one of the walls is a movable piston, then the volume of the container will expand until balanced by the external force.
3. *Gay-Lussac's law*, $P = k'T$. When a gas is heated in a container with a fixed volume, the gas molecules will impact more forcefully with the wall, thus increasing the pressure.
4. *Avogadro's law*, $V = k_3 n$. As more gas molecules are added to the container, the number of impacts per second with the wall increases, and so the pressure increases correspondingly. If the volume of the container is not fixed, then the volume will increase.

Speeds of Gas Molecules. Experiments show that the molecules of a gas do not all move at the same speed. Instead, the speeds are distributed over a range. The distribution, called the *Maxwell speed distribution* (Figure 5.17 in the textbook), always has the same general shape. The highest point on the curve marks the most probable speed. Some molecules move with speeds much less than the most probable speed, while others move with speeds much greater than this speed. According to Maxwell, the root-mean-square speed, u_{rms}, of gas molecules with molar mass $\mathcal{M}$ at absolute temperature T is given by the equation

$$u = \sqrt{\frac{3RT}{\mathcal{M}}}$$

where $R = 8.314$ J/K·mol and the molar mass $\mathcal{M}$ must be expressed in kilograms per mole in order for the units to come out in meters per second. This equation reveals that gas molecules, on the average, travel at very high speeds. For instance, at 25°C the root-mean-square speed of He atoms is 1.36×10^3 m/s, and for N_2 molecules it is 5.15×10^2 m/s. These values correspond to 3040 mi/h and 1150 mi/h, respectively. The speed of sound in a gas is about the same as the root-mean-square speed of the gas molecules. The speed of sound through He gas is much greater than through air.

The kinetic molecular theory predicts that the average kinetic energy of the molecules of a gas is determined *only* by the temperature. The average kinetic energy is

$$\overline{KE} = \tfrac{1}{2}m\overline{u^2} = kT$$

and has the same value for all gases at the same temperature. This means that gas molecules with higher molar masses will have slower average speeds, while lighter molecules will have higher average speeds.

Diffusion and Effusion. Diffusion and effusion are illustrated in Figures 5.19 and 5.20 in the textbook. *Diffusion* is the movement of components of a system that arises from concentration differences. If the concentration is not uniform in a mixture of two gases, the gases diffuse into each other until the composition is uniform. It is the random motion of molecules that eventually eliminates differences in concentration within a system.

Effusion is the passage of a gas under pressure through a very small hole into a volume maintained at a lower pressure. Thomas Graham observed experimentally that the rate of effusion is inversely proportional to the square root of the molar mass $\mathcal{M}$ of the gas. Comparing two gases with molar masses $\mathcal{M}_1$ and $\mathcal{M}_2$, he observed effusion rates r_1 and r_2. According to Graham's law, the ratio of effusion rates is given by

$$\frac{r_1}{r_2} = \sqrt{\frac{\mathcal{M}_2}{\mathcal{M}_1}}$$

The kinetic molecular theory provides a ready explanation of Graham's law. At constant temperature the average kinetic energy of molecules of gas 1 is the same as the average kinetic energy of molecules of gas 2, but at any given temperature the lighter gas molecules have greater root-mean-square speeds than molecules of the heavier gas. The ratio of the root-mean-square speeds of two gases is derived in

Section 5.7 of the text:

$$\frac{(u_{rms})_1}{(u_{rms})_2} = \sqrt{\frac{\mathcal{M}_2}{\mathcal{M}_1}}$$

This relationship is equivalent to Graham's law and tells us that the rate of effusion is proportional to the speed of gas molecules

---

EXAMPLE 5.9 Root-Mean-Square Speed

Calculate the root-mean-square speed of gaseous argon atoms at STP.

METHOD OF SOLUTION

The root-mean-square speed is given by the equation

$$u_{rms} = \sqrt{\frac{3RT}{\mathcal{M}}}$$

where $R = 8.314$ J/K $\cdot$ mol; T is the absolute temperature, 273 K; and $\mathcal{M}$ is the molar mass of Ar in kilograms, 0.0399 kg/mol. Note that standard pressure is irrelevant here because the average kinetic energy depends only on the temperature. Before substituting into the equation, we recall that 1 J = 1 kg m^2/s^2.

CALCULATION

$$u_{rms} = \sqrt{\frac{3(8.314 \text{ J/K} \cdot \text{mol})(273 \text{ K})}{(39.9 \times 10^{-3} \text{ kg/mol})} \times \frac{1 \text{ kg m}^2/\text{s}^2}{1 \text{ J}}}$$

$$= (17.1 \times 10^4 \text{ m}^2/\text{s}^2)^{1/2} = 413 \text{ m/s}$$

COMMENT

The speed 413 m/s is equivalent to 923 mi/h.

EXAMPLE 5.10 Relative Rates of Effusion

What is the ratio of the effusion rates for hydrogen (H_2) and hydrogen deuteride (HD)?

METHOD OF SOLUTION

According to Graham's law of effusion, the rate is inversely proportional to the square root of the molar mass:

$$r = \sqrt{\frac{1}{\mathcal{M}}}$$

and for two gases,

$$\frac{r_1}{r_2} = \sqrt{\frac{\mathcal{M}_2}{\mathcal{M}_1}}$$

Substituting in the molar masses of H_2 and HD gives

$$\frac{r_{H_2}}{r_{HD}} = \sqrt{\frac{3.0 \text{ g/mol}}{2.0 \text{ g/mol}}} = 1.2$$

NONIDEAL GASES

STUDY OBJECTIVES

You should be able to:
1. List two molecular properties responsible for deviations from ideal gas behavior and identify the conditions under which gases will be most likely to behave in a nonideal manner.
2. Calculate the pressure of a real gas as predicted by the van der Waals equation.

Nonideal Gas Behavior. According to the kinetic theory, molecules of a gas are quite independent of one another and do not exert attractive or repulsive forces on one another. Such molecules would obey the ideal gas law. For an ideal gas, a plot of PV/nRT versus P at constant temperature should be a horizontal line, with $PV/nRT = 1$ at all values of P. As Figure 5.24 of the text shows, this is true for real gases only at low pressures (less than ~ 10 atm).

Another way to observe nonideality of gases is to lower the temperature. Cooling a gas decreases the average kinetic energy of its molecules. When cooled enough, all gases will condense to the liquid. This suggests that intermolecular forces of attraction exist between molecules of real gases. Condensation occurs when the average kinetic energy is not great enough for molecules to break away from intermolecular attractive forces. The same forces account for nonideal behavior in gases.

Van der Waals Equation. Van der Waals modified the ideal gas law to account for two molecular properties that affect gas behavior: intermolecular forces and finite molecular volume. According to van der Waals, the pressure of a real gas will be lower than that of an ideal gas because attraction to neighboring molecules tends to lessen the impact a real molecule makes with the wall. We can express this in mathematical terms:

$$P_{real} = P_{ideal} - \frac{an^2}{V^2}$$

where an^2/V is the appropriate correction. Both n and V are defined as before. The proportionality constant a is different for each gas. The magnitude of a is a measure of the strength of intermolecular attractive forces and increases with increasing molecular mass and increasing complexity of molecular structure (Table 5.4 in the textbook).

The other correction concerns the volume of the gas. According to the assumption of the kinetic theory, molecules are volumeless points. Since every real molecule does occupy a small, though finite, volume, a certain volume within a container of volume V is excluded, or not available, to gas molecules. The effective volume is $V - nb$, where b is a constant that represents the volume of a molecule. The term nb is a correction for the finite volume of a mole of molecules.

The van der Waals equation is written in the form of the ideal gas equation:

$$P_{ideal} V_{effective} = nRT$$

$$\underbrace{\left(P_{real} + \frac{an^2}{V^2}\right)}_{\text{corrected pressure}} \underbrace{(V - nb)}_{\text{corrected volume}} = nRT$$

EXAMPLE 5.11 Van der Waals Equation

The molar volume of isopentane (C_5H_{12}) is 1.0 L at 503 K and 30.0 atm:
a. Does isopentane behave as an ideal gas?
b. Given that $a = 17.0$ L^2 atm/mol^2 and $b = 0.136$ L/mol, calculate the pressure of isopentane as predicted by the van der Waals equation.

METHOD OF SOLUTION

a. According to the ideal gas equation, the pressure of 1 mol of a gas at 503 K that occupies 1.0 L is

$$P = \frac{nRT}{V} = \frac{(1.0 \text{ mol})(0.082 \text{ L} \cdot \text{atm/K} \cdot \text{mol})(503 \text{ K})}{1.0 \text{ L}}$$

$$= 41.3 \text{ atm}$$

This calculated result differs considerably from the observed pressure of 30 atm. In fact, the percent error, which is the difference between the two values divided by the actual pressure, is

$$\% \text{ error} = \frac{41.3 \text{ atm} - 30.0 \text{ atm}}{30.0 \text{ atm}} \times 100\%$$

$$= 37.7\%$$

We conclude that under these conditions C_5H_{12} behaves in a nonideal manner.

b. In this case, write the van der Waals equation:

$$\left(P_{\text{real}} + \frac{an^2}{V^2}\right)(V - nb) = nRT$$

and substitute into it, but first calculate the correction terms:

$$\frac{an^2}{V^2} = \frac{(17.0 \text{ L}^2 \cdot \text{atm/mol}^2)(1.0 \text{ mol})^2}{1.0 \text{ L}^2} = 17.0 \text{ atm}$$

$$nb = 1.0 \text{ mol} (0.136 \text{ L/mol}) = 0.136 \text{ L}$$

$$nRT = (1.0 \text{ mol})(0.0821 \text{ L} \cdot \text{atm/K} \cdot \text{mol})(503 \text{ K}) = 41.3 \text{ L atm}$$

Now substitute using the van der Waals equation:

$$(P + 17.0 \text{ atm})(1.0 \text{ L} - 0.136 \text{ L}) = 41.3 \text{ L atm}$$

$$(P + 17.0 \text{ atm})(0.864) = 41.3 \text{ atm}$$

$$P + 17.0 \text{ atm} = 47.8 \text{ atm}$$

$$P = 30.8 \text{ atm}$$

Thus, the pressure calculated by the van der Waals equation is much closer to the actual value of 30.0 atm. The percent error is only 2.6%.

TRUE-FALSE QUESTIONS

1. The gaseous state is the least compressible of the states of matter.

2. The pressure exerted by a column of mercury 60 cm high and 1 cm in diameter is the same as the pressure exerted by another Hg column of the same height but with a diameter of 2 cm.

3. The SI unit of pressure is the newton.

4. A pressure of one standard atmosphere can support a column of water with a height of 33.9 ft.

5. As the temperature is raised from 100°C to 200°C, the volume of an ideal gas will double at constant pressure.

6. One mole of an ideal gas occupies 22.4 L at 1 atm and 25°C.

7. The density of a gas is inversely proportional to its molecular mass.

8. The volume of a gas is inversely proportional to its pressure.

9. The partial pressure of a component in a gas mixture is proportional to its mole fraction.

10. In the reaction $2A(g) + B(g) \rightarrow 2 C(g)$, if 2 L of B react, then 4 L of C will be formed.

11. Consider two gases, A and B. Gas A has a molar mass of 50 g/mol, and gas B has a molar mass of 100 g/mol. At STP, 100 g of A will occupy the same volume as 50 g of B.

12. According to the kinetic molecular theory, all molecules of a certain gas at temperature T have the same kinetic energy.

13. According to Graham's law, CH_4 will diffuse faster than N_2.

14. The strength of intermolecular forces between CO_2 molecules is greater than between CH_4 molecules.

15. Deviations from ideal behavior are more pronounced at low pressure than at high pressure.

SELF-TEST A

1. A barometer reads 695 mm Hg. Calculate the pressure in units of

    a. atm   b. torr   c. bars   d. Pa   e. psi

2. What is the pressure of the gas trapped in the apparatus shown below when the atmospheric pressure is 0.950 atm?

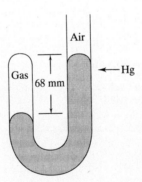

3. The pressure of H_2 gas in a 0.50-L cylinder is 1775 psi at 70°F. What volume would the gas occupy at 1 atm and the same temperature?

4. If 30.0 L of oxygen is cooled at constant pressure from 200°C to 0°C, what would be the new volume of oxygen?

5. A balloon has a volume of 1500 L of He at 0.925 atm and 23°C. At an altitude of 20 km the temperature is −50°C and the pressure is 0.25 bar. What is the volume of this balloon at 20 km?

6. The Martian atmosphere is mostly CO_2 at a pressure of 5.5 mm Hg at a temperature of −31.4°C. What is the density of the atmosphere?

7. How many grams of chlorine (Cl_2) occupy a 0.716-L cylinder when the pressure is 10.9 atm at 30°C?

8. Fill in the blank spaces in the table.

	P	V	n	T
a.	7.25 atm	40.0 L	10.5 mol	———
b.	451 torr	150 mL	2.50×10^{-3} mol	———
c.	14.2 atm	12.0 L	———	325 K
d.	152 kPa	120 mL	———	25°C
e.	2.50 atm	———	12.0 mol	501 K
f.	2280 torr	———	2.00×10^2 mol	450°C
g.	———	22.4 L	1.25 mol	301 K
h.	———	10.0 mL	0.0625 mol	25°C

9. Calculate the volume occupied by 15.2 g of CO_2 at 0.74 atm and 24°C.

10. What is the density of uranium hexafluoride gas (UF_6) at STP?

11. When 1.48 g of mercuric chloride is vaporized in a 1.00-L bulb at 680 K, the pressure is 225 mm Hg. What are the molar mass and molecular formula of mercuric chloride vapor?

12. Determine the molecular mass of chloroform gas if a sample weighing 0.495 g is collected as a vapor (gas) in a flask of volume 127 cm^3 at 98°C. The pressure of the chloroform vapor at this temperature in the flask was determined to be 754 mm Hg.

13. A 150-mL sample of O_2 gas is collected over water at 20°C and 758 torr. What volume will the same sample of oxygen occupy at STP when it is dry?

14. A sample of nitrogen gas is bubbled through liquid water at 25°C and then collected in a volume of 750 cm^3. The total pressure of the gas, which is saturated with water vapor, is found to be 740 mm at 25°C. The vapor pressure of water at this temperature is 24 mm. How many moles of nitrogen are in the sample?

15. The volume of carbon monoxide gas (CO) collected over water at 25°C was 680 cm^3 with a total pressure of 752 mm Hg. The vapor pressure of water at 25°C is 23.8 mm Hg. Determine the partial pressure and mole fraction of CO in the container.

16. The partial pressures of N_2, O_2, and Ar in dry air are 570, 153, and 6 torr, respectively. What are the mole fractions of these three gases?

17. A mixture of 40.0 g of O_2 and 40.0 g of He has a total pressure of 0.900 atm. What are the partial pressures of O_2 and He in the mixture?

18. a. What volume of CO_2 at 1 atm and 225°C would be produced by the reaction of 12.0 g $NaHCO_3$?

   $2NaHCO_3 + H_2SO_4 \rightarrow Na_2SO_4 + 2CO_2 + 2H_2O$

   b. On cooling to 20°C, what volume would the CO_2 occupy?

19. How many liters of ammonia at 10 atm and 500°C can be produced by the reaction of 6.0 g of hydrogen with excess N_2?

$$3H_2(g) + N_2(g) \rightarrow 4NO(g) + 6H_2O(l)$$

20. In the oxidation of ammonia,

$$4NH_3(g) + 5O_2(g) \rightarrow 4NO(g) + 6H_2O(l)$$

how many liters of O_2, measured at 18°C and 1.10 atm, must be used to produce 50 L of NO at the same condition?

21. How many liters of oxygen, measured at STP, are required for the complete combustion of 72.0 g of hexane (C_6H_{14}), a component of gasoline?

22. In an effusion experiment it required 45 s for a certain number of moles of an unknown gas to pass through a small orifice into a vacuum. Under the same conditions it required 18 s for the same number of moles of O_2 to effuse. Find the molar mass of the unknown gas.

23. What is the relative rate of effusion of neon atoms compared to the rate of effusion of O_2 molecules?

24. If the average molecular speed of a N_2 molecule is 475 m/s at 25°C, what is the average speed of a He molecule at 25°C?

25. In each of the following pairs, which gas would you expect to deviate more than the value $PV/nRT = 1$ expected for an ideal gas?
    a. N_2 or SF_6 b. He or O_2 c. CO_2 or SO_2

26. Calculate the pressure of 200 mol NH_3 in a 10.0-L container at 500°C using
    (a) the ideal gas law
    (b) the van der Waals equation

GENERAL PROBLEMS

27. A 0.356-g sample of $XH_2(s)$ reacts with water according to the following equation:

$$XH_2(s) + 2H_2O(l) \rightarrow X(OH)_2(s) + 2H_2(s)$$

The hydrogen evolved is collected over water at 23°C and occupies a volume of 431 mL at 746 mm Hg total pressure. Find the number of moles of H_2 produced and the atomic weight of X ($VP_{H_2O} = 21$ mm Hg).

28. A H_2 gas thermometer has a volume of 100.0 cm^3 when immersed in an ice-water bath at 0°C. When immersed in boiling liquid Cl_2, the volume of the H_2 at the same pressure was 87.2 cm^3. Determine the temperature of the boiling point of Cl_2 in kelvins and degrees Celsius.

29. a. The buoyant force on a balloon is proportional to the mass of air it displaces. What mass of air is displaced by a weather balloon at an altitude where the pressure is 212 mm Hg and the temperature is -35°C? Initially the balloon contained 10 mol He.
    b. The lift of the balloon is the difference between its mass and that of the displaced air. What is the lift of the above balloon?

30. A certain noble gas compound contains 68.8% Kr and 31.2% F. Its density at STP is 5.44 g/L. What is the molecular formula of the compound?

SELF-TEST B1

1. A sample of gas has a volume of 200 cm^3 at 25°C and 700 torr. What pressure must be applied to permit expansion of the gas to 500 cm^3 at the same temperature?

2. If 2.00 L of oxygen at −15°C is allowed to warm to 25°C at constant pressure, what is the new volume of oxygen gas?

3. The gas pressure in an aerosol can is 1.5 atm at 25°C. If the gas is an ideal gas, what pressure would develop in the can if it were heated to 450°C?

4. What pressure in atmospheres will be exerted by 12 kmol of methane (CH$_4$) when stored at 22°C in a 3000-L tank?

5. What is the density of H$_2(g)$ at 35°C and 650 torr?

6. How many O$_2$ molecules occupy a 1.0-L flask at 75°C and 777 mm Hg?

7. When 2.96 g of mercuric chloride is vaporized in a 1.00-L bulb at 680 K, the pressure is 450 mm. What is the molecular weight and molecular formula of mercuric chloride vapor?

8. Calculate the average speed of ozone molecules (O$_3$) high in the stratosphere, where the temperature is −83°C.

9. If 21 s is required for a certain number of moles of N$_2$ gas to effuse through a small orifice into a vacuum, how many seconds would be required for the same number of moles of XeF$_4$ gas to effuse under the same conditions?

10. The discovery of oxygen occurred from the decomposition of mercury(II) oxide:

    $$2HgO \rightarrow 2Hg + O_2(g)$$

    What volume of oxygen would be produced by the reaction of 35.2 g of the oxide if the gas being measured is at STP?

SELF-TEST B2

1. A sample of gas has a volume of 200 cm^3 at 25°C and 700 mm Hg. If the pressure is reduced to 280 mm Hg, what volume would the gas occupy at the same temperature?

2. Two liters of oxygen gas at −15°C is heated and the volume expands. At what temperature will the volume reach 2.31 L?

3. A gas in a confined volume has a pressure of 3.64 atm at 450°C. What would be the pressure of the gas if it were cooled to 25°C?

4. What is the temperature of a 3000-L gas cylinder if it contains 12,000 mol of methane gas at 96.4 atm (or 1420 lb/cm^2)?

5. At what temperature will H$_2(g)$ have a density of 0.0677 g/L if the pressure is 650 mm Hg?

6. What is the pressure exerted by 2.16 × 10^{22} molecules of O$_2$ in a 1.0-L container at 75°C?

7. How many grams of HgCl$_2$ must be vaporized in a 1.00-L bulb at 680 K to give a pressure of 450 mm Hg?

8. High in the stratosphere molecules of ozone (O_3) have an average speed of 290 m/s. What is the temperature corresponding to this average speed?

9. In an effusion experiment, 57 s was required for a certain number of moles of a gaseous compound of xenon and fluorine, of unknown formula, to pass through a small orifice into a vacuum. Under the same conditions, 21 s was required for the same number of moles of N_2 to effuse. Determine the molar mass of the unknown gas and suggest its formula.

10. The discovery of oxygen occurred from the decomposition of mercury(II) oxide:

$$2HgO \rightarrow 2Hg + O_2(g)$$

How many grams of mercury(II) oxide must be decomposed to yield 1.82 L of O_2 gas at 1 atm and 298 K.

ANSWERS

TRUE-FALSE QUESTIONS

1. False. Gases are the most compressible state.
2. True.
3. False. The SI unit of pressure is the pascal.
4. True.
5. False. The volume is directly proportional to the kelvin temperature, not the Celsius temperature.
6. False. The molar volume is 22.4 L at 0°C, not at 25°C.
7. False. Directly proportional to molecular mass.
8. True.
9. True.
10. True.
11. False. At the same temperature, 50 g of A will occupy the same volume as 100 g of B.
12. False. Gas molecules have the same *average* kinetic energy.
13. True.
14. True. Compare van der Waals *a* values.
15. False. Deviations are more pronounced at high pressure.

SELF-TEST A

1. a. 0.914 atm   b. 695 torr   c. 0.926 bar   d. 9.26×10^4 Pa   e. 13.4 psi
2. 790 mm Hg
3. 60 L
4. 17.3 L
5. 5240 L
6. 0.016 g/L
7. 22.3 g
8. a. 336 K   b. 434 K   c. 6.39 mol   d. 7.36×10^{-3} mol   e. 197 L   f. 3960 L
   g. 1.38 atm   h. 153 atm
9. 11.4 L
10. 15.7 g/L
11. 279 g/mol, $HgCl_2$
12. 120 g/mol
13. 0.136 L
14. 0.0289 mol N_2
15. CO: 728 mm Hg; $X = 0.968$
16. 0.782 N_2, 0.210 O_2, 0.008 Ar
17. $P_{O_2} = 0.1$ atm $P_{He} = 0.8$ atm
18. a. 5.84 L   b. 3.44 L

19. 12.6 L
20. 62.5 L
21. $C_6H_{14} + 9\tfrac{1}{2}O_2 \rightarrow 6CO_2 + 7H_2O$; 178 L O_2
22. 200 g/mol
23. $r_{Ne}/r_{O_2} = 1.25$
25. 1360 m/s
25. a. SF_6 b. O_2 c. SO_2
26. a. 1270 atm   b. 3250 atm
27. 0.0169 mol H_2, 40.1 g/mol
28. 238 K and $-35°C$
29. a. $V = 700$ L, thus 290 g air   b. 290 g air -40 g He = 250 g lift
30. KrF_2

SELF-TEST B1

1. 280 torr
2. 2.31 L
3. 3.64 atm
4. 96.4 atm
5. 0.0677 g/L
6. 2.16×10^{22}
7. 279 g/mol, $HgCl_2$
8. 290 m/s
9. 57 s
10. 1.82 L

SELF-TEST B2

1. 500 cm^2
2. 25°C
3. 1.5 atm
4. 295 K
5. 308 K
6. 777 mm Hg
7. 2.96 g
8. 190 K
9. 207 g/mol, XeF_4
10. 35.2 g

Chapter Six
THERMOCHEMISTRY

- Energy Changes in Chemical Reactions
- Enthalpy Changes and Hess's Law
- Calorimetry
- Standard Enthalpies of Formation
- Heat of Solution
- The First Law of Thermodynamics

ENERGY CHANGES IN CHEMICAL REACTIONS

STUDY OBJECTIVES

You should be able to:
1. Define heat, work, and energy.
2. Distinguish between exothermic and endothermic reactions.
3. Describe a state function.

Thermochemistry. Almost every chemical reaction occurs with either the absorption or evolution of energy. Recall that energy (E) is defined as the ability to do work. When a system does work (w) or evolves heat (q), its capacity to do additional work decreases, which means that its energy has decreased. In this chapter we are concerned with the heat absorbed or released in chemical reactions. This topic is called *thermochemistry*.

In the study of thermochemistry, certain terms are defined very carefully and are used in a precise way. A thermodynamic *system* is the part of the universe that is selected for study. It may be a beaker containing a mixture of reactants, a bacteria colony, or the entire earth. The boundary of the system must be carefully defined. The *surroundings* are the part of the universe that interacts with the system. Systems can interact with the surroundings by the exchange of heat, work, and matter.

Heat is energy that is transferred because of a temperature difference between the system and the surroundings. Heat always flows from a hotter to a colder body. *Work* is energy lost or gained by a system by mechanical means rather than by heat conduction. Systems do not "contain" work or heat. What they do contain is energy. The SI unit for amount of energy is the joule (J). Since both work and heat are energy in transition, they also have units of joules. A non-SI unit for heat is the calorie. One calorie (cal) is the amount of heat required to raise the temperature of 1 gram of water from 14.5°C to 15.5°C. Heat or energy in calories can be converted to joules:

$$1 \text{ cal} = 4.184 \text{ J}$$

$$1 \text{ kcal} = 4.184 \text{ kJ}$$

Chemical reactions that evolve heat to the surroundings are called *exothermic reactions*. The combustion of wood, coal, or hydrogen is an exothermic reaction. During an exothermic reaction the surroundings warm up. When heat is absorbed from the surroundings by a reaction system, the reaction is referred to as *endothermic*. The melting of ice and the boiling of water are endothermic processes. During an endothermic reaction the surroundings cool down.

According to the *first law of thermodynamics*, the energy of the universe is constant. In any process, chemical or otherwise, energy is neither created nor lost. This means that the energy evolved as heat during chemical reactions must come from somewhere. Because atoms and molecules are made up of charged particles, they contain potential (stored) energy as a result of the relative positions of atomic nuclei and electrons in molecules. During a chemical reaction, the positions of electrons and atomic nuclei change, and the product molecules may contain more or less chemical potential energy than the reactants, as illustrated in Figure 6.3(a) in the textbook. The heat evolved results from the loss of potential energy as reactants are converted to products. Chemical energy is a type of potential energy. It is only released when substances undergo chemical change.

In an endothermic reaction the products have greater chemical potential energy than the reactants [Figure 6.3(b) textbook]. The heat absorbed brings about a corresponding increase in the potential energy of the system.

State Functions. Internal energy (E) is an example of a *state function*. This means that the internal energy of the system depends only on the conditions (the state) of the system and not on how it got that way. Changes in internal energy (ΔE) during a chemical change depend only on the state and identity of the products and the state and identity of the reactants. The value of ΔE is independent of the path of the reaction.

Consider the energy profile for the exothermic reaction of hydrogen and chlorine shown in Figure 6.1:

$$H_2 + Cl_2 \rightarrow 2HCl$$

When 1 mol of H_2 and 1 mol of Cl_2 react to form 2 mol of HCl, the loss in chemical potential energy (ΔE) is always the same regardless of how the reaction is carried out. The potential energy lost (step 1) by the system is transferred to the surroundings as heat. If we complete a cycle by returning to the initial state (step 2),

$$2HCl \rightarrow H_2 + Cl_2$$

this second process must absorb an amount of energy equal to that lost in step 1. The total energy change in the system for the cycle (step 1 + step 2) must be zero. Other state functions are temperature, pressure, and as we will see, enthalpy.

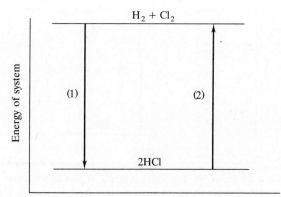

Figure 6.1. A potential energy diagram showing the changes in energy of the system. Energy lost by the system is transferred to the surroundings as heat.

ENTHALPY CHANGES AND HESS'S LAW

STUDY OBJECTIVES

You should be able to:
1. Use the thermochemical equation to calculate the heat evolved or absorbed when a given amount of reactant is converted to products.
2. Use Hess's law to calculate $\Delta H°$ for a reaction that is the sum of several given chemical reactions.

Enthalpy Changes. Most reactions that you carry out in the general chemistry laboratory occur at constant pressure because the atmospheric pressure does not change appreciably in three or four hours. The heat evolved or absorbed (q) by a reaction carried out at a constant pressure is equal to the enthalpy change of the system ΔH, or the enthalpy of reaction:

$$q = \Delta H$$

Heats of reaction are reported as ΔH rather than as q. Enthalpy H is a property of the system. Like energy, enthalpy is a state function. Recall that Δ denotes a change:

$$\Delta H = H_2 - H_1$$

Here H_1 is the enthalpy of the reactants and H_2 is that of the products. Changes in enthalpy are often illustrated by a diagram. Figure 6.2 shows graphically the enthalpy change for the following reaction:

$$SO_2(g) + \tfrac{1}{2}O_2(g) \rightarrow SO_3(g)$$

A thermochemical equation is a balanced chemical equation with a term added to show the enthalpy changed ΔH. Two examples are

$$2HgO(s) \rightarrow 2Hg(l) + O_2(g) \qquad \Delta H° = 181.4 \text{ kJ}$$

$$2SO_2(g) + O_2(g) \rightarrow 2SO_3(g) \qquad \Delta H° = -198 \text{ kJ}$$

There are several points to keep in mind when interpreting thermochemical equations.

1. For an exothermic reaction heat is evolved, and the enthalpy of the system decreases. Enthalpy change ΔH has a negative sign. It is positive for an endothermic reaction because the system has more enthalpy after it has absorbed heat.

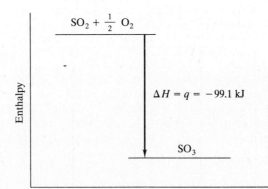

Figure 6.2. Enthalpy is a state function. The heat evolved in this exothermic reaction at constant pressure is equal to the enthalpy change ΔH. This enthalpy diagram shows the change in enthalpy of the system as the reactants are converted to products.

2. The enthalpy change in kilojoules applies to the reaction of the amount of substance shown in the balanced equation. Thus, when 2 mols of HgO decomposes as shown, 181.4 kJ of heat must be absorbed. The value of ΔH must always pertain to the equation and to the way the equation is balanced. The equation provides the following conversion factors:
181.4 kJ per reaction of 2 mol HgO
181.4 kJ per reaction of 2 mol Hg
181.4 kJ per reaction of 1 mol O_2
From this we see that the actual heat evolved or absorbed by a chemical reaction is an *extensive* property; that is, it depends on the amount of reactants:
3. The superscript ° refers to the standard states of the substances at the temperature concerned. In the first reaction above, the value $\Delta H° = 181.4$ kJ applies when HgO in its most stable crystalline form at 25°C decomposes into Hg in its standard state at 25°C, which is a liquid, and oxygen in its standard state at 25°C, which is a gas.
4. Both energy and enthalpy are *state properties*. That is, ΔE and ΔH are independent of the path followed in a reaction. The magnitudes of ΔE and ΔH depend only on the identity of the reactants (initial state) and the products (final state). If we write the reverse of a chemical reaction, the magnitude of ΔH is the same (as we saw in Fig. 6.1), but its sign is reversed. The reaction

$$4Al(s) + 3O_2(g) \rightarrow 2Al_2O_3(s) \qquad \Delta H° = -3340 \text{ kJ}$$

is exothermic, but the reverse reaction is endothermic:

$$2Al_2O_3(s) \rightarrow 4Al(s) + 3O_2(g) \qquad \Delta H° = 3340 \text{ kJ}$$

Hess's Law of Heat Summation. In 1840 G. H. Hess pointed out that the overall enthalpy change for a chemical reaction at constant pressure is the same regardless of the intermediate steps involved. For example, consider the combustion reaction

$$S(s) + \tfrac{3}{2}O_2(g) \rightarrow SO_3(g) \qquad \Delta H°_{rxn} = -395.2 \text{ kJ}$$

The enthalpy changes for two other reactions involving SO_2 and resulting in SO_3 are known:

$$S(s) + O_2(g) \rightarrow SO_2(g) \qquad \Delta H_1 = \Delta H°_{rxn} = -296.1 \text{ kJ}$$

$$SO_2(g) + \tfrac{1}{2}O_2(g) \rightarrow SO_3(g) \qquad \Delta H_2 = \Delta H°_{rxn} = -99.1 \text{ kJ}$$

If we treat these two equations algebraically and sum them and their enthalpy changes, we have an example of *Hess's law of heat summation*:

$$S(s) + O_2(g) \rightarrow SO_2(g) \qquad \Delta H_1 = -296.1 \text{ kJ}$$

$$\underline{SO_2(g) + \tfrac{1}{2}O_2(g) \rightarrow SO_3(g) \qquad \Delta H_2 = -99.1 \text{ kJ}}$$

$$S(s) + \tfrac{3}{2}O_2(g) \rightarrow SO_3(g) \qquad \Delta H_3 = \Delta H_1 + \Delta H_2 = -395.2 \text{ kJ}$$

Here we see that the sum $\Delta H_1 + \Delta H_2$ gives the same enthalpy change, -395.2 kJ, as previously observed for the overall reaction. Whether the combustion of sulfur occurs directly to SO_3 or by a path through the intermediate SO_2, the overall enthalpy change is the same.

Hess's law makes it possible to calculate the enthalpy changes for reactions that cannot be studied directly. This is illustrated in Example 6.1 and again in a following section.

Summary of Thermochemistry Principles

1. The change in enthalpy ΔH is directly proportional to the amount of reactant consumed.
2. The ΔH for the forward reaction is equal in magnitude, but opposite in sign, to ΔH for the reverse reaction.
3. Hess's law: If a reaction can be regarded as the sum of two or more steps, ΔH for the overall reaction must be the sum of the enthalpy changes for all steps.

EXAMPLE 6.1 Thermochemical Equations

The thermochemical equation for the combustion of propane is

$$C_3H_8(g) + 5O_2(g) \rightarrow 3CO_2(g) + 4H_2O \quad \Delta H^\circ_{rxn} = -2220 \text{ kJ}$$

a. How many kilojoules of heat are released when 0.50 mol of propane reacts?
b. How much heat is released when 88.2 g of propane reacts?

METHOD OF SOLUTION

The heat released by an exothermic reaction is an extensive property. Therefore q depends on the amount of propane consumed. The equation indicates that 2220 kJ of heat is evolved per mole of propane burned. The conversion factor is

$$\frac{-2220 \text{ kJ}}{1 \text{ mol } C_3H_8}$$

CALCULATION FOR a

$$q = \text{mol } C_3H_8 \times \frac{\text{kJ}}{\text{mol } C_3H_8}$$

$$= 0.50 \text{ mol } C_3H_8 \times \frac{-2220 \text{ kJ}}{1 \text{ mol } C_3H_8}$$

$$= -1110 \text{ kJ}$$

CALCULATION FOR b

Since we know the heat of reaction per mole, we should first convert the number of grams of C_3H_8 to moles:

$$q = 88.2 \text{ g } C_3H_8 \times \frac{1 \text{ mol } C_3H_8}{44.1 \text{ g } C_3H_8} \times \frac{-2220 \text{ kJ}}{1 \text{ mol } C_3H_8}$$

$$= -4440 \text{ kJ}$$

EXAMPLE 6.2 Hess's Law

The standard enthalpy change for the combustion of 1 mol of ethanol is

$$C_2H_5OH(l) + 3O_2(g) \rightarrow 2CO_2(g) + 3H_2O(l) \quad \Delta H^\circ_{rxn} = -1367 \text{ kJ}$$

What is ΔH°_{rxn} for the following reaction:

$$C_2H_5OH(l) + 3O_2(g) \rightarrow 2CO_2(g) + 3H_2O(g)$$

in which the H_2O is formed as a gas rather than as a liquid? The heat of vaporization of water is

$$H_2O(l) \rightarrow H_2O(g) \quad \Delta H_{vap} = 44 \text{ kJ}$$

METHOD OF SOLUTION

We can imagine a two-step path for this reaction. In the first step, ethanol undergoes combustion to form liquid H_2O, followed by the second step in which $H_2O(l)$ is vaporized. The sum of the two steps gives the desired overall reaction, and according to Hess's law, the sum of the two ΔH's gives the overall ΔH.

CALCULATION

$$C_2H_5OH(l) + 3O_2(g) \rightarrow 2CO_2(g) + 3H_2O(l) \qquad \Delta H°_{rxn} = -1367 \text{ kJ}$$

$$\underline{3H_2O(l) \rightarrow 3H_2O(g) \qquad \Delta H°_{rxn} = 132 \text{ kJ}}$$

$$C_2H_5OH(l) + 3O_2(g) \rightarrow 2CO_2(g) + 3H_2O(g) \qquad \Delta H°_{rxn} = -1235 \text{ kJ}$$

CALORIMETRY

STUDY OBJECTIVES

You should be able to:
1. Calculate the heat evolved or absorbed when an object or substance undergoes a known temperature change.
2. Determine the enthalpy of reaction using data from both constant-pressure and constant-volume calorimeter experiments.

Specific Heat. The device used to measure the amount of heat evolved or absorbed in a reaction is called a *calorimeter*. All calorimetry is based on the temperature rise observed in some medium, usually water. The *specific heat* (s) of a substance is the amount of heat required to raise the temperature of *one gram* of the substance by 1.0°C. A related term is the *heat capacity* (C), which is the amount of heat required to raise the temperature of a given mass of substance by 1°C. The amount of heat flowing in or out of a substance is

$$q = C \Delta t$$

Here Δt is the temperature change given by $t_f - t_i$, where t_f is the final temperature and t_i is the initial temperature:

$$q = C(t_f - t_i)$$

In terms of the specific heat,

$$q = ms \Delta t$$

where m is the mass in grams of the object absorbing the heat and $C = ms$.

Constant-Volume Calorimetry. Bomb calorimeters, as shown in Figure 6.9 of the textbook, are used to measure the heat evolved in combustion reactions of foods and fuels. The stainless steel "bomb" is loaded with a small amount of substance and O_2 at 30 atm of pressure and is immersed in a known amount of water. The heat evolved during combustion is absorbed by the water and the calorimeter assembly (bomb and contents, can, thermometer, etc.). Measurements of Δt of the water and calorimeter are related to the heat evolved by the reaction q and to ΔH.

We saw previously that the enthalpy change of a system is equal to the heat evolved at constant pressure q:

$$\Delta H = q$$

Reactions in a bomb occur under constant-volume conditions rather than constant-pressure conditions, and so the heat evolved is not quite equal to ΔH:

$$\Delta H \cong q$$

However, for most reactions the difference is small and, for our purpose, can be neglected. For instance, for the combustion of 1 mol of pentane, the difference is only 7 kJ out of 3500 kJ. Therefore, we often can use

$$q \cong \Delta H$$

without significant error.

To measure q, the heat evolved by the reaction, we must measure the heat absorbed by both the bomb calorimeter and the surrounding water. Let q_{rxn} be the amount of heat from the reaction.

$$q \text{ of reaction} = -(q \text{ of bomb calorimeter} + q \text{ of water})$$

$$q_{rxn} = -(q_{bomb} + q_{water})$$

This equation says that the heat evolved by the reaction is equal to the heat gained by the bomb calorimeter and the water but has the opposite sign. The negative sign takes into account that the heat gained by the bomb and the water was evolved by an exothermic reaction. The bomb and surrounding water make up the calorimeter. The heat capacity of both the bomb calorimeter and the surrounding water must be known. Substituting the equations for heat flow given above yields

$$q_{rxn} = -[C_{bomb}(\Delta t) + m_{water}s(\Delta t)]$$

Here C_{bomb} is the heat capacity of the bomb and s is the specific heat of water. Then ΔH, the enthalpy change of the system, is essentially equal to q_{rxn}; $q_{rxn} \cong \Delta H$.

Constant-Pressure Calorimeter. A constant-pressure calorimeter is open to the atmosphere, and the measurement of heat given off by a reaction is a direct measure of the enthalpy change:

$$q = \Delta H$$

Such a device is shown in Figure 6.11 (textbook). A reaction occurring in the solution liberates heat (q) that is absorbed by the solution (which is mostly water), the stirrer, the thermometer, and the container walls. These latter three parts are considered the calorimeter, and so

$$q = -(q_{calorimeter} + q_{water})$$

EXAMPLE 6.3 Determining the Calorimeter Constant

The combustion of benzoic acid is often used as a standard source of heat for calibrating combustion bomb calorimeters. The heat of combustion of benzoic acid has been accurately determined to be 26.42 kJ/g. When 0.8000 g of benzoic acid was burned in a calorimeter containing 950 g of water, a temperature rise of 4.08°C was observed. What is the heat capacity of the bomb calorimeter (the calorimeter constant)?

METHOD OF SOLUTION

The combustion of 0.8000 g of benzoic acid produces a known amount of heat:

$$q_{rxn} = 0.8000 \text{ g} \times \frac{-26.42 \text{ kJ}}{1 \text{ g}} = -21.14 \text{ kJ} = -21.14 \times 10^3 \text{ J}$$

And since Δt and the amount of water are known, C_{bomb}, the heat capacity of the calorimeter, can be calculated. The total heat of combustion is absorbed by the bomb calorimeter and water:

$$q_{rxn} = -(q_{bomb} + q_{water})$$

$$q_{rxn} = -(C_{bomb} \Delta t + ms \Delta t)$$

The heat absorbed by the water was $q_{water} = ms \Delta t$, where

m = 950 g water

s = 4.184 J/g ·°C

Δt = 4.08°C

Substitution yields

$$q_{water} = 950 \text{ g } (4.184 \text{ J/g} \cdot °C)(4.08°C) = 16.20 \times 10^3 \text{ J}$$

The bomb calorimeter must have absorbed the difference between the 21.14×10^3 J released and the 16.20×10^3 J absorbed by the water:

$$q_{rxn} = -(q_{bomb} + q_{water})$$

$$q_{bomb} = -q_{rxn} - q_{water} = -(-21.14 \times 10^3 \text{ J}) - 16.20 \times 10^3 \text{ J}$$

$$q_{bomb} = 4.94 \times 10^3 \text{ J}$$

The heat capacity of the calorimeter is

$$C_{bomb} = q_{bomb}/\Delta t = 4.94 \times 10^3 \text{ J}/4.08°C$$

$$C_{bomb} = 1210 \text{ J}/°C$$

EXAMPLE 6.4 Determining the Heat of Combustion

The thermochemical equation for the combustion of pentane (C_5H_{12}) is

$$C_5H_{12}(l) + 8O_2(g) \rightarrow 5CO_2(g) + 6H_2O(l) \qquad \Delta H° = -3509 \text{ kJ}$$

From the following information calculate the heat of combustion, $\Delta H°$, and compare your result to the value given in the above equation. The 0.5521 g of C_5H_{12} was burned in the presence of excess O_2 in a bomb calorimeter. The heat capacity of the calorimeter was 1.800×10^3 J/°C, and the temperature of the calorimeter and 1.000×10^3 g of water rose from 21.22°C to 25.70°C.

METHOD OF SOLUTION

The heat evolved by the combustion reaction is absorbed by the water and the calorimeter assembly:

$$q_{rxn} = -[q_{bomb} + q_{water}]$$

where

$$q_{rxn} = -(C_{bomb} \Delta t + ms \Delta t)$$

CALCULATION

$$\Delta t = 25.70°C - 21.22°C = 4.48°C$$

Substituting yields

$$q_{bomb} = -\frac{1.800 \times 10^3 \text{ J}}{°C} \times (4.48°C) = 8.06 \times 10^3 \text{ J}$$

$$q_{water} = 1.000 \times 10^3 \text{ g} \times \frac{4.184 \text{ J}}{\text{g} \cdot °C} \times (4.48°C) = 1.87 \times 10^3 \text{ J}$$

$$q_{rxn} = -[8060 \text{ J} + 18{,}700 \text{ J}]$$

$$q_{rxn} = -26{,}800 \text{ J} = -26.8 \text{ kJ} \quad \text{(three significant figures)}$$

Here $\Delta H°$ refers to the reaction of 1 mol of pentane. The 26,800 J was evolved by 0.5521 g C_5H_{12}. The molar mass of pentane is 72.15 g:

$$\Delta H° = q$$

$$= \frac{-26.8 \text{ kJ}}{0.5521 \text{ g}} \times \frac{72.15 \text{ g } C_5H_{12}}{1 \text{ mol}}$$

$$= -3500 \text{ kJ/mol} = -3.50 \times 10^3 \text{ kJ/mol}$$

COMMENT

To three significant figures, the result compares well with the value given in the thermochemical equation. Small differences will arise because the heat evolved at constant volume is not quite the same as the heat evolved at constant pressure (which is equal to ΔE). Only the heat evolved at constant pressure exactly equals ΔH, the enthalpy change.

STANDARD ENTHALPIES OF FORMATION

STUDY OBJECTIVES

You should be able to:
1. Define standard enthalpy of formation.
2. Calculate $\Delta H°_{rxn}$ for any reaction with the aid of a table of enthalpies of formation.

Formation Reactions. In a formation reaction 1 mol of a compound is formed from the elements in their stable forms. For methane such a reaction is $C(s) + 2H_2(g) \rightarrow CH_4(g)$. The *standard enthalpy of formation* $\Delta H°_f$ is the heat liberated or absorbed when 1 mol of a compound is formed from the elements in their standard states at constant pressure and at the temperature of concern. Enthalpies of formation at 25°C and 1 atm pressure are tabulated in Appendix 2 (textbook). We will discuss their use shortly, but first let us look at their source.

In most cases, enthalpies of formation cannot be determined by a direct calorimetry experiment. For instance, the formation reaction of hydrogen sulfide (H_2S) will not go cleanly in a calorimeter, and many other compounds besides H_2S are formed when sulfur and hydrogen react. In a case like this, Hess's law, rather than direct calorimetry, is used to determine $\Delta H°_f$:

$$S(s) + H_2(g) \rightarrow H_2S(g) \quad \Delta H°_f = ?$$

The enthalpies of reaction for the following reactions involving sulfur and hydrogen are known from combustion bomb calorimetry:

$$S(s) + O_2(g) \rightarrow SO_2(g) \qquad \Delta H°_{rxn} = -296.1 \text{ kJ}$$

$$H_2(g) + \tfrac{1}{2}O_2(g) \rightarrow H_2O(l) \qquad \Delta H°_{rxn} = -285.8 \text{ kJ}$$

$$H_2S(g) + \tfrac{3}{2}O_2(g) \rightarrow SO_2(g) + H_2O(l) \qquad \Delta H°_{rxn} = -561.7 \text{ kJ}$$

These equations can be arranged so that their sum is the formation reaction of hydrogen sulfide:

$$S + O_2 \rightarrow SO_2$$

$$H_2 + \tfrac{1}{2}O_2 \rightarrow H_2O$$

$$SO_2 + H_2O \rightarrow H_2S(g) + \tfrac{3}{2}O_2$$

Notice that the combustion of the hydrogen sulfide equation has been reversed. The other equations are unchanged in this example. Therefore, we reverse the sign of the third reaction:

$$S(s) + O_2(g) \rightarrow SO_2(g) \qquad \Delta H^\circ_{rxn} = -296.1 \text{ kJ}$$

$$H_2(g) + \tfrac{1}{2}O_2(g) \rightarrow H_2O(l) \qquad \Delta H^\circ_{rxn} = -285.8 \text{ kJ}$$

$$\underline{SO_2(g) + H_2O(l) \rightarrow H_2S(g) + \tfrac{3}{2}O_2(g) \qquad \Delta H^\circ_{rxn} = 561.7 \text{ kJ}}$$

$$S(s) + H_2(g) \rightarrow H_2S(g) \qquad \Delta H^\circ_f = -20.2 \text{ kJ}$$

Summation of the three reaction steps yields the net reaction, and summation of the ΔH° values yields the enthalpy change of the net reaction. Example 6.5 in the text provides another example of the indirect method. The value of ΔH°_f for a compound depends on the temperature, pressure, and physical state of the compound. Appendix 2 in the textbook lists standard enthalpies of formation at 25°C for 58 different compounds.

Rather than measuring a standard enthalpy change ΔH°_{rxn} for every chemical reaction, chemists have devised a way to use the enthalpy of formation values of compounds involved in a reaction. Enthalpies of formation can be used to calculate the standard enthalpy change for any reaction for which ΔH°_f is known for all reactants and products. The relationship is

$$\Delta H^\circ_{rxn} = \Sigma n \, \Delta H^\circ_f \text{ (products)} - \Sigma m \, \Delta H^\circ_f \text{ (reactants)}$$

where Σ (sigma) means "the sum of" and n and m are the stoichiometric coefficients. So here you simply add together all the ΔH°_f of the products and add together all the ΔH°_f of the reactants and subtract the reactant sum from the product sum. Example 6.5 illustrates the use of this equation and shows how the stoichiometric coefficients are entered into the equation.

EXAMPLE 6.5   Using Enthalpies of Formation

Using ΔH°_f values in Appendix 2 of the textbook, calculate the standard enthalpy change for the incomplete combustion of ethane (C_2H_6):

$$C_2H_6(g) + \tfrac{5}{2}O_2(g) \rightarrow 2CO(g) + 3H_2O(l)$$

METHOD OF SOLUTION

The enthalpy change for this chemical reaction in terms of enthalpies of formation is

$$\Delta H^\circ_{rxn} = \Sigma n \, \Delta H^\circ_f \text{ (products)} - \Sigma m \, \Delta H^\circ_f \text{ (reactants)}$$

$$\Delta H^\circ_{rxn} = [2 \text{ mol} \times \Delta H^\circ_f(CO) + 3 \text{ mol} \times \Delta H^\circ_f(H_2O)]$$

$$- [1 \text{ mol} \times \Delta H^\circ_f(C_2H_6) - \tfrac{5}{2} \text{ mol} \times \Delta H^\circ_f(O_2)]$$

To avoid cumbersome notation, the physical states of the reactants and products were omitted from this equation. Be careful to obtain from Appendix 2 the appropriate value of ΔH°_f. Note that the coefficients from the chemical equation are equal to the number of moles of each substance and that all the terms involving ΔH°_f of a *reactant* are preceded by a negative sign. Next, look up and insert values.

CALCULATION

$$\Delta H^\circ_{rxn} = [2\,\text{mol}(-110.5\,\text{kJ/mol}) + 3\,\text{mol}(-285.8\,\text{kJ/mol})] - [1\,\text{mol}(-84.68\,\text{kJ}) + \tfrac{5}{2}(0)]$$

$$= (-221.0\,\text{kJ}) + (-857.4\,\text{kJ}) - (-84.68\,\text{kJ}) = -221.0\,\text{kJ} - 857.4\,\text{kJ} + 84.68\,\text{kJ}$$

$$= -993.7\,\text{kJ}$$

HEAT OF SOLUTION

STUDY OBJECTIVES

You should be able to:
1. Define the heat of solution and the heat of dilution.
2. Describe the energy changes that occur during the solution process and apply Hess's law to calculate the enthalpy of solution.

Heat of Solution. The *heat of solution* of a solute is the enthalpy change associated with the dissolution of 1 mol of solute in a certain amount of solvent. The solution process can be represented by a thermochemical equation. In the case of hydrogen chloride gas dissolving in water to form hydrochloric acid, the thermochemical equation is

$$HCl(g) + H_2O(l) \rightarrow HCl(aq) \qquad \Delta H^\circ_{soln} = -75.1\,\text{kJ}$$

Values for the heat of solution ΔH_{soln} of several solutes in H_2O are given in Table 6.1.

Table 6.1 Heats of Solution

Solute	ΔH°_{soln} (kJ/mol)
NaOH	−44.5
Ca(OH)$_2$	−11.7
NaBr	−0.60
NaCl	+4.0
NH$_4$NO$_3$	+26.2

The energy changes that occur when ionic solids dissolve in water can be understood by imagining the solution process to take place in two steps. We realize that when a compound such as NaBr dissolves in water, the Na$^+$ and Br$^-$ ions must break away from the crystal lattice and enter the aqueous phase where they are surrounded by water molecules. According to Hess's law, a two-step process that starts with solute plus solvent and ends up with a solution will have the same overall ΔH as ΔH_{soln}.

First, we imagine a step in which the lattice is broken up and the ions enter the gas phase:

$$NaBr(s) \rightarrow Na^+(g) + Br^-(g)$$

The energy change for this process is called the *lattice energy U*. All ionic crystals have a characteristic lattice energy. This is the energy required to completely separate all the ions in 1 mol of solid into ions in the gas phase. For NaBr(s), $U = 735$ kJ. Second, the gaseous Na$^+$ and Br$^-$ ions enter the aqueous phase, where they are hydrated. During hydration, Na$^+$ and Br$^-$ ions are surrounded by molecules of the aqueous solvent. An ion in water is attracted to the polar water molecules, and a quantity of energy is released that is called the hydration energy ΔH_{hydr}:

$$Na^+(g) + Br^-(g) \xrightarrow{H_2O} Na^+(aq) + Br^-(aq)$$

According to Hess's law, the overall energy change for the two steps will be the heat of solution:

$$\begin{array}{ll} NaBr(s) \rightarrow Na^+(g) + Br^-(g) & U \\ Na^+(g) + Br^-(g) \xrightarrow{H_2O} Na^+(aq) + Br^-(aq) & \Delta H_{hydr} \\ \hline NaBr(s) \rightarrow Na^+(aq) + Br^-(aq) & \Delta H_{soln} \end{array}$$

where $\Delta H_{soln} = U + \Delta H_{hydr}$.

Heat of Dilution. Dilution of concentrated laboratory reagents such as hydrochloric and sulfuric acids is accompanied by sufficient evolution of heat to make the procedure hazardous. The *heat of dilution* is the heat released or absorbed when additional solvent is added to a solution. If heat is released during the preparation of a certain solution, heat will also be released if that solution is diluted. This idea holds true for cases where heat is absorbed during a solution process as well, for then heat is also absorbed during dilution.

EXAMPLE 6.6 Heat of Hydration

The heat of solution of NaBr has been determined to be -0.60 kJ/mol. The lattice energy of NaBr is 735 kJ/mol. Determine the heat of hydration of NaBr and write the equation for the hydration reaction.

METHOD OF SOLUTION

Recall that the heat of solution is the sum of the lattice energy (U) and the heat of hydration (ΔH_{hydr}):

$$\Delta H_{soln} = U + \Delta H_{hydr}$$

CALCULATION

Rearranging so as to solve for the heat of hydration yields

$$\Delta H_{hydr} = \Delta H_{soln} - U$$

$$= -0.60 \text{ kJ/mol} - 735 \text{ kJ/mol}$$

$$= -736 \text{ kJ/mol}$$

where ΔH_{hydr} refers to the heat liberated or absorbed when the gas phase ions are dissolved in water. The equation is

$$Na^+(g) + Br^-(g) \xrightarrow{H_2O} Na^+(aq) + Br^-(aq)$$

COMMENT

For NaBr, U and ΔH_{hydr} have essentially the same values but are opposite in sign. Therefore, when NaBr dissolves in water, the energy U required to break up the crystal lattice is "paid back" by the hydration of the Na^+ and Br^- ions. Also, NaBr does not dissolve in nonpolar solvents. Molecules of nonpolar solvents can only interact weakly with ions in the crystal lattice, and so "solvation" is not sufficient to compensate for the lattice energy.

THE FIRST LAW OF THERMODYNAMICS

STUDY OBJECTIVES

You should be able to:
1. Calculate the work done when a gas expands.
2. Write the equation for the first law of thermodynamics and give the sign conventions for q and w.
3. Relate ΔE to ΔH.

Work of Expansion. Energy can be transferred from one system to another by means of either heat flow or work. Calculations involving the heat liberated by chemical reactions were performed previously in this chapter. There are many kinds of work, but in chemistry we are mainly concerned with work done by electricity and work done by expanding gases. This chapter is concerned with the work of expansion. Electrical energy is discussed in Chapter 19.

In the reaction

$$2HgO(s) \rightarrow 2Hg(l) + O_2(g)$$

the products occupy more volume than the reactants by virtue of the fact that gases occupy approximately 1000 times more volume per mole than do solids and liquids. The oxygen "expanding" from an open reaction vessel must do work in pushing back the atmosphere.

Mechanical work is expressed mathematically as

$$\text{work} = Fd$$

where F is the force exerted and d is the distance a mass is moved by the force. To investigate the relationship of pressure-volume changes to work, consider a cylinder with a piston of cross-sectional area A (Figure 6.16 text). The work done by expansion through a distance d is

$$w = Fd$$

Dividing and multiplying by A gives

$$w = \frac{F}{A}(A \times d)$$

The pressure of a gas (P) is equal to the force per unit area (F/A). And $A \times d$ is the volume change (ΔV) that occurs. Therefore, the work of expansion is

$$w = P\,\Delta V$$

In order for expansion to take place, the pressure of the gas within the cylinder (internal pressure) must be greater than the external pressure. In this equation, P is the opposing pressure (or external pressure), not the internal pressure.

Heat is energy that is transferred because of a temperature difference between the system and the surroundings. Heat always flows from a hotter to a colder body. Systems do not "contain" work or heat. What they do contain is energy. The SI unit for the amount of energy is the joule (J). Since both work and heat are energy in transition, they also have units of joules.

The First Law. A system contains a certain quantity of energy (E) called its internal energy. The first law of thermodynamics is merely the statement that energy is conserved during any process. In other words, the total energy of the universe is constant. A useful form of the first law is

$$\Delta E = q + w$$

where ΔE is a change in internal energy of the system, w is the work done on the system, and q is the heat absorbed by the system. A change in a system's internal energy may also be written as

$$\Delta E = E_f - E_i$$

Here ΔE is the change in a state function and depends only on the initial and final states, E_i and E_f. The change in energy of a system depends only on how much heat flows into ($+q$) or out of ($-q$) the system and on how much energy is gained when work is done on the system ($+w$) or is lost when the system does work on the surroundings ($-w$). These sign conventions are summarized below.

The Sign Conventions. Thermodynamic quantities always have two parts. The number gives the amount of change, and the sign indicates the direction of flow. Here we use the convention that the sign reflects the system's point of view rather than the surroundings' point of view. The symbol q denotes the heat transferred to the system. *When q is positive*, it signifies that heat has been transferred to the system (endothermic) and that it contributes to an increase in the internal energy of the system. *When q is negative*, it signifies that heat has flowed out of the system (exothermic) with a corresponding decrease in internal energy. This heat is absorbed by the surroundings.

The symbol w denotes the work *done* on the system by the surroundings. *When w is positive*, it signifies that work has been done on the system and has contributed to an increase in the internal energy. When the volume of a system is compressed by the surroundings, work is done on the system. *When w is negative*, it signifies that work has been done *by* the system on the surroundings. If the system does work on the surroundings, its internal energy must decrease. When the volume of a system expands, the system pushes back the surroundings and work is done *by* the system. These conventions are the result of taking the system's point of view. Energy entering the system is positive. Energy leaving the system is negative.

Thermal Energy. A system at a higher temperature has more thermal energy than when it is at a lower temperature. Thermal energy is a contribution to the internal energy (E) of a system. How does a system contain thermal energy? This form of energy is associated with the thermal motion of molecules. You learned in Chapter 5 that according to the kinetic molecular theory the average kinetic energy of molecules increases as temperature increases. Molecules have kinetic energy because they translate (move from one place to another), rotate (a tumbling motion), and vibrate. When vibrating, atoms within a molecule act as if they are attached by springs. As heat is added to a system, the molecular motions increase, becoming more energetic.

State Functions. Internal energy (E) is an example of a *state function*. This means that the internal energy of the system depends only on the conditions (the state) of the system and not on how it got that way. Changes in internal energy ΔE (chemical potential energy and thermal energy) during a chemical change depend only on the state and identity of the products and the state and identity of the reactants. Energy change (ΔE) is independent of the path of the reaction. *Heat is not a state function* because the heat associated with a given change depends on the path. You cannot write $\Delta q = q_f - q_i$. Example 6.6 in the text shows that work, like heat, is not a state function.

Enthalpy. Two important conditions under which chemical reactions can be carried out are constant volume and constant pressure. The energy change is

$$\Delta E = q + w = q - P \Delta V$$

where $w = -P \Delta V$ and is the work done on the system. Under conditions of *constant volume*, $\Delta V = 0$. Therefore, $P \Delta V = 0$ and $w = 0$, and so

$$\Delta E = q$$

Under conditions of *constant pressure*, when the system expands against the atmosphere,

$$\Delta E = q_p + w = q_p - P \Delta V$$

$$q_p = \Delta E + P \Delta V$$

Earlier in the chapter we said that the heat evolved in a reaction at constant pressure is called the enthalpy change ΔH:

$$\Delta H = q_p$$

From the equation for q_p above

$$\Delta H = \Delta E + P\Delta V$$

Enthalpy is a state function

$$H = E + PV$$

and so ΔH is equal to the difference between the initial and final enthalpies of the system:

$$q_p = \Delta H = H_2 - H_1$$

EXAMPLE 6.7 Work of Expansion

A gas initially at a pressure of 10.0 atm and having a volume of 5.0 L is allowed to expand at constant temperature against a constant external pressure of 4.0 atm until the new volume is 12.5 L. Calculate the work done by the gas on the surroundings.

METHOD OF SOLUTION

In this problem the system does work on the surroundings as it expands, and by convention (Table 6.6 in the text), the sign is negative:

$$w = -P\Delta V = -P(V_2 - V_1)$$

where P is the pressure opposing the expansion and ΔV is the change in volume of the system.

CALCULATION

$$w = -4.0 \text{ atm}(12.5 \text{ L} - 5.0 \text{ L})$$
$$= -30 \text{ L} \cdot \text{atm}$$

This quantity can be expressed in units of joules:

$$w = -30 \text{ L} \cdot \text{atm} \times \frac{101.3 \text{ J}}{1 \text{ L} \cdot \text{atm}} = -3.0 \times 10^3 \text{ J}$$

EXAMPLE 6.8 Work Done on the System

Calculate the work done on the system when 6.0 L of a gas is compressed to 1.0 L by a constant external pressure of 2.0 atm.

METHOD OF SOLUTION

The work done is

$$w = -P\Delta V = -P(V_2 - V_1)$$
$$= -2.0 \text{ atm}(1.0 \text{ L} - 6.0 \text{ L}) = +10 \text{ L} \cdot \text{atm}$$
$$= 1.0 \times 10^3 \text{ J}$$

COMMENT

Obtaining a positive value for work means that in a compression, work is done *on* the system by the surroundings. The positive value for work means that the system gains energy.

EXAMPLE 6.9 The First Law of Thermodynamics

A gas is allowed to expand at constant temperature from a volume of 10.0 L to 20.0 L against an external pressure of 1.0 atm. If the gas also absorbs 250 J of heat from the surroundings, what are the values of q, w, and ΔE?

METHOD OF SOLUTION

The work done by the system is

$$w = -P\,\Delta V = -P(V_2 - V_1)$$

$$= -1.0 \text{ atm}(20.0 \text{ L} - 10.0 \text{ L}) = -10 \text{ L} \cdot \text{atm}$$

$$= -1.0 \times 10^3 \text{ J}$$

The amount of heat absorbed was 250 J:

$$q = 250 \text{ J}$$

The first law of thermodynamics is

$$\Delta E = q + w$$

$$= 250 \text{ J} - 1000 \text{ J}$$

$$= -750 \text{ J}$$

COMMENT

In this example, the system did more work than the energy absorbed as heat; therefore the internal energy E decreased.

TRUE-FALSE QUESTIONS

1. The following thermochemical equation describes an exothermic reaction:

    $$\tfrac{1}{2}N_2(g) + \tfrac{1}{2}O_2(g) \rightarrow NO(g) \qquad \Delta H^\circ_{rxn} = 90.4 \text{ kJ/mol}$$

2. The enthalpy change ΔH is equal to the heat absorbed or released by a process carried out at constant volume.

3. If

    $$HgO(s) \rightarrow Hg(l) + \tfrac{1}{2}O_2(g) \qquad \Delta H^\circ_{rxn} = 90.7 \text{ kJ}$$

    then

    $$2Hg(l) + O_2(g) \rightarrow 2HgO(s) \qquad \Delta H^\circ_{rxn} = -181 \text{ kJ}$$

4. Hess's law is a consequence of the fact that enthalpy is a state function.

5. All substances that absorb 100 J of heat will show the same temperature increase Δt.

6. To calculate the enthalpy change for a combustion reaction carried out in a calorimeter, we must measure the mass of the sample, the mass of water, and the heat capacity of the calorimeter.

7. The enthalpy of formation of $H_2(g)$ at 25°C and 1 atm pressure is zero.

8. The enthalpy of formation of $I_2(g)$ at 25°C and 1 atm pressure is zero.

9. The enthalpy of formation of methane (CH_4) is equal in magnitude but opposite in sign to its enthalpy of combustion.

10. Energy, work, and heat all have the same units and are all state functions.

11. According to the sign conventions, $+q$ signifies that heat flows into the system.

12. The heat absorbed by a system during a process carried out at constant pressure is equal to $\Delta E + P \Delta V$.

SELF-TEST A

1. Define energy, work, and heat. Consider a 9-V "alkaline" battery. Which of these three does it contain? Name a device that uses a battery to do work. Name a device that uses a battery to generate heat.

2. What is a state function? Name three state functions used in thermochemistry.

3. Given the thermochemical equation

$$SO_2(g) + \tfrac{1}{2}O_2(g) \rightarrow SO_3(g) \qquad \Delta H°_{rxn} = -99 \text{ kJ}$$

how much heat is liberated when
a. $\tfrac{1}{2}$ mol of SO_2 reacts
b. 2 mol of SO_2 reacts

4. Given the thermochemical equation

$$H_2 + I_2 \rightarrow 2HI \qquad \Delta H°_{rxn} = 52 \text{ kJ}$$

what is ΔH for the reaction

$$HI \rightarrow \tfrac{1}{2}H_2 + \tfrac{1}{2}I_2 \qquad \Delta H°_{rxn} = ?$$

5. Use the following reactions and their associated enthalpy changes:

$$2H(g) \rightarrow H_2(g) \qquad \Delta H°_{rxn} = -436 \text{ kJ}$$

$$Br_2(g) \rightarrow 2Br(g) \qquad \Delta H°_{rxn} = +224 \text{ kJ}$$

$$H_2(g) + Br_2(g) \rightarrow 2HBr(g) \qquad \Delta H°_{rxn} = -72 \text{ kJ}$$

Calculate $\Delta H°_{rxn}$ in kilojoules for the reaction

$$HBr(g) \rightarrow H(g) + Br(g)$$

6. Nitrogen and oxygen react according to the following thermochemical equation:

   $$N_2(g) + O_2(g) \rightarrow 2NO(g) \quad \Delta H^\circ_{rxn} = 180 \text{ kJ}$$

   How many kilojoules of heat are absorbed when 50 g of N_2 reacts with excess O_2 to produce 107 g of NO?

7. How much energy is required to raise the temperature of 180 g of graphite from 25°C to 500°C? The specific heat of graphite is 0.720 J/g·°C.

8. The quantity 0.500 g of ethanol [$C_2H_5OH(l)$] was burned in a bomb calorimeter containing 2000 g of water. The heat capacity of the bomb calorimeter was 950 J/°C, and the temperature rise was found to be 1.6°C.
   a. Write a balanced equation for the combustion of ethanol.
   b. Calculate the amount of heat transferred to the calorimeter.
   c. Calculate ΔH for the reaction as written in part a.

9. The enthalpy of combustion of sulfur is

   $$S(\text{rhombic}) + O_2(g) \rightarrow SO_2(g) \quad \Delta H^\circ = -296 \text{ kJ}$$

   What is the enthalpy of formation of SO_2?

10. The combustion of methane occurs according to the equation

    $$CH_4(g) + 2O_2(g) \rightarrow CO_2(g) + 2H_2O(l) \quad \Delta H^\circ_{rxn} = -890 \text{ kJ}$$

    Use enthalpies of formation for CO_2 and H_2O to determine the enthalpy of formation of methane.

11. During expansion of its volume from 1.0 L to 10.0 L against a constant external pressure of 2.0 atm, a gas absorbs 200 J of energy as heat. Calculate the change in internal energy of the gas.

12. How much work must be done on or by the system in a process in which the internal energy remains constant and 322 J of heat is transferred from the system?

SELF-TEST B1

1. Given the thermochemical equation

   $$SO_2(g) + \tfrac{1}{2}O_2(g) \rightarrow SO_3(g) \quad \Delta H^\circ_{rxn} = -99.0 \text{ kJ}$$

   how much heat is evolved when 75 g SO_2 undergoes combustion?

2. From the following enthalpies of reaction, calculate the enthalpy of combustion of methane (CH_4) with F_2:

   $$CH_4(g) + 4F_2(g) \rightarrow CF_4(g) + 4HF(g)$$

   $$C(\text{graphite}) + 2H_2(g) \rightarrow CH_4(g) \quad \Delta H^\circ_{rxn} = -75 \text{ kJ}$$

   $$C(\text{graphite}) + 2F_2(g) \rightarrow CF_4(g) \quad \Delta H^\circ_{rxn} = -933 \text{ kJ}$$

   $$H_2(g) + F_2(g) \rightarrow 2HF(g) \quad \Delta H^\circ_{rxn} = -542 \text{ kJ}$$

3. How much heat is absorbed by 80 g of iron when its temperature is raised from 25°C to 500°C? The specific heat of iron is 0.444 J/g·°C.

4. A reaction used for rocket engines is

$$N_2H_4(l) + 2H_2O_2(l) \rightarrow N_2(g) + 4H_2O(l)$$

What is the enthalpy of reaction in kilojoules? The enthalpies of formation are

$\Delta H_f^\circ(N_2H_4) = 95.1$ kJ

$\Delta H_f^\circ(H_2O_2) = -187.8$ kJ

$\Delta H_f^\circ(H_2O) = -285.8$ kJ

5. Benzoic acid ($C_6H_5CO_2H$) is commonly used as a standard for calibrating bomb calorimeters. Its enthalpy of combustion has been determined accurately to be -3226.7 kJ/mol. When 1.0236 g of benzoic acid was burned in a bomb calorimeter, the temperature of the calorimeter and the 1000 g of water surrounding it rose from 20.66°C to 24.47°C. What is the heat capacity of the calorimeter?

6. The reaction that occurs when a typical fat, glyceryl trioleate, is metabolized by the body is

$$C_{57}H_{104}O_6(s) + 80O_2(g) \rightarrow 57CO_2(g) + 52H_2O(l) \quad \Delta H_{rxn}^\circ = -3.35 \times 10^4 \text{ kJ}$$

How much heat is evolved when 1 g of this fat is completely oxidized?

7. Using data from Appendix 2 in the text, calculate the standard enthalpy of formation of octane (C_8H_{18}) given the following combustion reaction:

$$2C_8H_{18}(l) + 25O_2(g) \rightarrow 16CO_2(g) + 18H_2O(l) \quad \Delta H_{rxn}^\circ = -11{,}020 \text{ kJ}$$

8. To determine the heat capacity of a bomb calorimeter, a student added 150 g of water at 50.0°C to the bomb. The bomb initially was at 20.0°C. The final temperature of the water and the bomb was 32.0°C. What is the heat capacity of the bomb in J/°C?

9. Given that the lattice energy of LiCl is 828 kJ/mol and the heat of hydration of Li^+ and Cl^- ions is -865 kJ/mol, calculate the heat of solution of LiCl(s).

10. Calculate the work done on a gas when 22.4 L of the gas is compressed to 2.24 L under a constant external pressure of 10.0 atm.

11. A system does 975 J of work on its surroundings while at the same time it absorbs 625 J of heat. What is the change in energy ΔE for the system?

SELF-TEST B2

1. How many grams of SO_2 must be burned to yield 116 kJ of heat? The combustion reaction is

$$SO_2(g) + \tfrac{1}{2}O_2(g) \rightarrow SO_3(g) \quad \Delta H^\circ = -99.0 \text{ kJ}$$

2. From the following heats of combustion with fluorine, calculate the enthalpy of formation of CH_4:

$CH_4(g) + 4F_2(g) \rightarrow CF_4(g) + 4HF(g) \quad \Delta H_{rxn}^\circ = -1942$ kJ

$C(\text{graphite}) + F_2(g) \rightarrow CF_4(g) \quad \Delta H_{rxn}^\circ = -933$ kJ

$H_2(g) + F_2(g) \rightarrow 2HF(g) \quad \Delta H_{rxn}^\circ = -542$ kJ

3. A piece of iron initially at a temperature of 25°C absorbs 16.9 kJ of heat. If its mass is 80 g, calculate the final temperature. The specific heat of iron is 0.444 J/g·°C.

4. Given the enthalpy change for the following reaction:

   $N_2H_4(l) + 2H_2O_2(l) \rightarrow N_2(g) + 4H_2O(l)$ $\Delta H^°_{rxn} = -862.7$ kJ

   and the following $\Delta H^°_f$'s:

   $\Delta H^°_f(N_2H_4) = 95.1$ kJ

   $\Delta H^°_f(H_2O) = -285.8$ kJ

   Calculate the enthalpy of formation of $H_2O_2(l)$.

5. Benzoic acid ($C_6H_5CO_2H$) is commonly used as a standard for calibrating bomb calorimeters. Its enthalpy of combustion has been determined accurately to be -3226.7 kJ/mol. How many degrees of temperature rise would be expected in a bomb calorimeter assembly and 1000 g of water if 1.0236 g of benzoic acid was burned in the calorimeter (the heat capacity of the calorimeter is 2.91 kJ/°C)?

6. For the reaction

   $C_{57}H_{104}O_6(s) + 80O_2(g) \rightarrow 57CO_2(g) + 52H_2O(l)$ $\Delta H^°_{rxn} = -33.5 \times 10^3$ kJ

   how much energy would have to be expended in order of to get rid of 1 pound of this fat?

7. Using the data in Appendix 2, calculate the heat of combustion of octane according to the balanced equation

   $C_8H_{18}(l) + \frac{25}{2}O_2(g) \rightarrow 8CO_2(g) + 9H_2O(l)$

8. A stainless steel bomb calorimeter assembly initially at 20.0°C with a heat capacity of 941 J/°C was immersed in 150 g water at 50°C. Calculate the final temperature of the calorimeter and the water.

9. Given that the heat of solution of LiCl is -37.1 kJ/mol and that the lattice energy (U) for LiCl is 828 kJ/mol, calculate the heat of hydration (ΔH_{hydr}) of Li^+ and Cl^- ions.

10. Calculate the work done on a gas when 2.24 L of the gas expands to 22.4 L against a constant external pressure of 1.0 atm.

11. A system does 975 J of work on the surroundings. How much heat does the system absorb at the same time if its energy charge is -350 J?

ANSWERS

TRUE-FALSE QUESTIONS

1. False. Because $\Delta H^°_{rxn}$ is positive, the reaction is endothermic.
2. False. The heat absorbed or evolved at constant pressure is equal to ΔH.
3. True.
4. True.
5. False. All substances do not have the same specific heat.
6. False. Δt must be known.
7. True.
8. False. The gas phase of I_2 is not the standard state at 25°C.
9. False. The formation reaction involves the reaction of the elements C and H to form CH_4, whereas combustion is the reaction of CH_4 with O_2.

10. False. The amounts of work and heat depend on how the process is carried out. Of these, only energy is a state function.
11. True.
12. True.

SELF-TEST A

1. The battery contains energy, which is the capacity to do work. Work is energy lost or gained by mechanical means, and heat is energy transferred because of a temperature difference. A toy truck or car can be moved (displaced) when energy in the battery is used to do work. A flashlight uses energy in the battery to produce a hot glowing filament in the bulb, which transfers heat to the surroundings.
2. A state function is a property of a system that depends only on its present condition (state) and not on how it got that way. Energy, enthalpy, and temperature are state functions.
3. a. 50 kJ   b. 198 kJ
4. $\Delta H^{\circ}_{rxn} = -26$ kJ
5. $\Delta H^{\circ}_{rxn} = 366$ kJ
6. 321 kJ ($q = 321$ kJ)
7. 61.6 kJ
8. a. $C_2H_5OH + 3O_2 \rightarrow 2CO_2 + 2H_2O$
   b. $q = 2000(4.18)(1.6) + 950(1.6) = 14,900$ J
   c. $\Delta H^{\circ}_{rxn} = -1370$ kJ
9. $\Delta H^{\circ}_f(SO_2) = -296$ kJ
10. $\Delta H^{\circ}_f(CH_4) = -75$ kJ
11. $\Delta E = -1618$ J
12. $w = +322$ J

SELF-TEST B1

1. 116 kJ
2. $\Delta H^{\circ}_{rxn} = -1942$ kJ
3. 16.9 kJ
4. -862.7 kJ
5. 2.91 kJ/°C
6. 37.8 kJ/g
7. -210 kJ
8. 941 J/°C
9. 37.1 kJ/mol
10. $w = +202$ L atm
11. $\Delta E = -350$ J

SELF-TEST B2

1. 75 g
2. -75 kJ/mol
3. 500°C
4. -187.8 kJ
5. 3.81°C
6. 1.72×10^4 kJ
7. $\Delta H^{\circ}_{rxn} = -5.51 \times 10^3$ kJ/mol
8. 32°C
9. -865 kJ
10. $w = -20.21$ atm
11. $q = 625$ J

Chapter Seven
QUANTUM THEORY AND THE ELECTRONIC STRUCTURE OF ATOMS

- Radiant Energy
- The Bohr Atom
- Wave-Particle Duality
- Quantum Mechanics and the Hydrogen Atom
- Electron Configurations

RADIANT ENERGY

STUDY OBJECTIVES

You should be able to:
1. Use the equations relating wavelength, frequency, and the speed of light.
2. Describe a quantum and be able to calculate the energy of a quantum emitted by an atom.
3. Explain how Einstein accounted for the frequency dependence of the photoelectric effect.

To understand some of the experiments that illuminated the behavior of electrons in atoms requires some appreciation of the nature of electromagnetic radiation. Radiation is energy that is transmitted through space in the form of waves. Electromagnetic radiation consists of an electric field component and a magnetic field component that are perpendicular to each other and to their direction of propagation (see Figure 7.3 of the text). The distance between identical points on successive waves, such as the wave crests, is called the wavelength, λ (lambda). All electromagnetic waves travel at the same speed through a vacuum. This speed, known as the *speed of light*, is 3.00×10^8 m/s and has the symbol c. The frequency of a wave is the number of wave crests that pass a given point per second. Its symbol is ν (nu), and the unit of frequency is cycles per second, which is written as /s or s^{-1}. The SI unit for one cycle per second is called a hertz (Hz). These three quantities, c, λ, and ν, are related by the equation

$$c = \lambda \nu$$

Figure 7.4 in the textbook shows that the wavelengths of electromagnetic waves have been observed to vary from as short as 5×10^{-4} nm for some γ rays to more than 100 m for radio waves. The visible spectrum ranges from 400 to 700 nm and is only a tiny fraction of the complete span of electromagnetic waves.

Note that rearranging the preceding equation gives

$$\nu = \frac{c}{\lambda}$$

According to this equation, the frequency increases as the wavelength decreases. Thus, radio waves with their long wavelengths have low frequencies, and γ rays with their short wavelengths have extremely high frequencies. The radio waves of an AM station broadcasting at a frequency of 1000 kilohertz (1000 kHz), or 10^6 s^{-1}, emits electromagnetic waves with a wavelength of 300 m.

Quantum Theory. In 1900 Max Planck solved a long-standing riddle when he provided an explanation for the emission spectrum of heated objects. Heated bodies emit radiation usually in the infrared. However, a body will emit visible radiation if it is hot enough. According to classical physics, an object can acquire or lose any amount of energy no matter how small. But all attempts by scientists to explain the wavelengths of electromagnetic radiation emitted by heated bodies using classical theory have failed. In Planck's quantum theory, however, there is a lower limit to the smallest increment of energy (a quantum) that an atom can gain or lose. In addition, the total amount of energy gained or lost by the atom is some multiple of this increment of energy. These discrete amounts of energy are called quanta. The energy of a quantum (E) emitted by an atom in a heated object is given by Planck's equation:

$$E = h\nu$$

where h is Planck's constant, 6.63×10^{-34} joule second, or J s; and ν is the frequency of the radiation. With the use of this equation, Planck was able to explain the entire spectrum of radiation emitted by heated bodies.

In 1905 Albert Einstein used Planck's quantum hypothesis to explain the photoelectric effect. When ultraviolet light is directed onto a clean metal electrode, it causes electrons to be ejected from the metal surface. If the frequency of the light is varied, the significant finding is that below a certain frequency no photoelectrons are observed. However, above this frequency, called the *threshold frequency*, the number of photoelectrons and their kinetic energy increase with increasing frequency of the radiation. Therefore, the energy absorbed from a light beam is proportional to its frequency, $E \propto \nu$. To Einstein, the significance of the light frequency was that it was a measure of the energy of particlelike entities of light called *quanta*. He assumed that the energy content E of such a bundle, or quantum of energy, is related to its frequency by Planck's equation. Quanta were later renamed *photons* and have an energy

$$E_{photons} = h\nu$$

Einstein's interpretation of the existence of the threshold frequency was that a quantum of light would interact with one electron on the surface of the metal. This electron is held to the metal by a force called the binding energy of the metal (BE). Below the threshold frequency, a photon does not have enough kinetic energy to overcome the binding energy and eject an electron from the metal surface, but photons associated with light above the threshold frequency do. As the frequency of light is increased, photons become more energetic ($h\nu$) and transfer more energy to the electrons. Since the binding energy is constant, then as the photon energy increases, more of its energy is available to become kinetic energy (KE) of the photoelectrons:

$$KE = h\nu - BE$$

Planck's and Einstein's use of the quantum concept demonstrated that light, in addition to having properties of a wave, has properties of particles as well. This behavior is known as the *dual nature of light*.

EXAMPLE 7.1 Wavelength and Frequency

Domestic microwave ovens generate microwaves with a frequency of 2.450 GHz. What is the wavelength of this microwave radiation?

METHOD OF SOLUTION

The equation relating the wavelength to the frequency is

$$c = \lambda \nu$$

where c is the speed of light.

CALCULATION

Rearranging and substituting, we get

$$\lambda = \frac{c}{\nu} = \frac{3.0 \times 10^8 \text{ m/s}}{2.450 \text{ GHz}} \times \frac{1 \text{ GHz}}{10^9 / \text{s}}$$

$$= 0.122 \text{ m}$$

EXAMPLE 7.2 Planck's Equation

The orange light given off by a sodium vapor lamp has a wavelength of 589 nm. What is the energy of a single photon of this radiation?

METHOD OF SOLUTION

The energy of a quantum is related to its frequency. Since wavelength is given here, substitute c/λ into Planck's equation for ν:

$$E_{\text{photon}} = h\nu = \frac{hc}{\lambda}$$

CALCULATION

Introduce values for Planck's constant and the speed of light into the equation. The units will not cancel correctly unless the wavelength is converted from nanometers into meters:

$$E = \frac{(6.63 \times 10^{-34} \text{ J} \cdot \text{s})(3.0 \times 10^8 \text{ m/s})}{(589 \times 10^{-9} \text{ m})}$$

$$= 3.38 \times 10^{-19} \text{ J}$$

COMMENT

We can compare this energy to that of a mole of substance. The energy of a mole of photons is the product of $E_{\text{photon}} \times N_A$, where N_A is Avogadro's number.

$$E = (3.38 \times 10^{-19} \text{ J})(6.02 \times 10^{23}/\text{mol})$$

$$= 203{,}000 \text{ J/mol} = 203 \text{ kJ/mol}$$

THE BOHR ATOM

STUDY OBJECTIVES

You should be able to:
1. List the two assumptions made in Bohr's model of the hydrogen atom.
2. Calculate the wavelength of radiation.

Emission Spectra. Heated objects and gases excited by an electrical current give off electromagnetic radiation in a process called emission. By passing emitted light through a prism, an emission spectrum consisting of the wavelength of emitted electromagnetic radiation is produced. A *continuous* spectrum contains all the wavelengths between the longest and shortest emitted by that substance. "White" light results from the emission of all wavelengths in the visible part of the spectrum.

The light emitted from excited atoms in the gas phase exhibits a *line spectrum* rather than continuous spectrum. A line spectrum consists of several discrete wavelengths. The observed wavelengths are characteristic of the element and can serve to identify the presence of the element in a sample.

Emission Spectrum of Hydrogen. A portion of the line spectrum of hydrogen falls in the visible region and is called the Balmer series. Figure 7.8(b) of the text shows that the prominent visible lines are at 656 nm, 486 nm, and 434 nm. In all, five spectral series are known for hydrogen. The Lyman series is in the ultraviolet region, and the Paschen, Brackett, and Pfund series are in the infrared region.

Bohr Model. In 1913 Niels Bohr produced his famous model of the hydrogen atom. The great achievement of this model was that it explained the source and the observed wavelengths of lines in the spectrum of the H atom. Bohr postulated a "solar system" model for the H atom in which the electron traveled in circular orbits about the proton. The basic assumptions of the Bohr's model were as follows:

1. The electron moves in a circular orbit about the nucleus. Of the infinite number of possible orbits only certain orbits with distinct radii are allowed. As long as the electron remains in an orbit, its energy remains constant.
2. The energy of the electron increases the farther its orbit lies from the nucleus. When energy is absorbed by the atom, the electron must jump to a higher energy orbit. When an electron transition occurs from a higher to a lower energy orbit, radiation is emitted by the atom.

Bohr was able to calculate the radii of the allowed orbits and their energies. The energies of the H atom are given by

$$E_n = -R_H \left(\frac{1}{n^2}\right)$$

where R_H is the Rydberg constant, which in units of joules has the value 2.18×10^{-18} J. The minus sign in the equation simply means that the energy of the hydrogen atom is lower than that of a free proton and a free electron, for which the energy is zero. It does not signify negative energy. The quantity n is an integer $n = 1, 2, 3, \ldots$ and is called the principal quantum number. Inserting values for n into the energy equation gives the energies of all allowed orbits shown in Figure 7.11 of the text.

The allowed radii of the hydrogen atom are given by $r_n = 52.9$ pm (n^2), where pm refers to the unit picometer (1 pm = 10^{-12} m). As n increases, so do the orbit radius and the energy of the atom.

Bohr explained that the emission of light, when an electrical current passes through a gas, is due to the formation excited atoms. He proposed that in the emission process an electron drops from a higher to a lower orbit. During this transition the atom emits radiation. For energy to be conserved, the energy lost as radiation must equal the difference in energy between the initial energy level E_i and the final energy level E_f. Let the quantum number n_i represent the higher energy level and n_f represent the final level. Then the change in energy of the atom ΔE is

$$\Delta E = E_f - E_i$$

where, according to Bohr's equation, the energy change for the atom is

$$\Delta E = R_H \left(\frac{1}{n_i^2} - \frac{1}{n_f^2}\right)$$

Bohr's stroke of genius was to equate ΔE to the energy of a photon of emitted light:

$$E_{\text{photon}} = h\nu = \frac{hc}{\lambda}$$

$$= \Delta E_{\text{atom}}$$

$$h\nu = R_H \left(\frac{1}{n_i^2} - \frac{1}{n_f^2}\right)$$

and the frequency of light emitted should be given by:

$$\nu = \frac{R_H}{h}\left(\frac{1}{n_i^2} - \frac{1}{n_f^2}\right)$$

The wavelengths of emitted light can be calculated from $\lambda = c/\nu$ if the frequency is known.

With this model of the hydrogen atom, Bohr was able to predict the wavelengths of all the literally hundreds of observed lines in the spectrum of hydrogen!

EXAMPLE 7.3 The Hydrogen Atom

a. What is the energy difference in joules between the $n = 2$ and $n = 3$ principal energy levels of the hydrogen atom?

b. What is the wavelength of the light produced when the electron transition $n = 3 \rightarrow n = 2$ occurs?

METHOD OF SOLUTION FOR a

The energy of the atom is quantized and depends on the orbit of the electron. Each orbit is assigned a principal quantum number n:

$$E_n = -R_H\left(\frac{1}{n^2}\right)$$

The difference in energy between levels $n = 2$ and $n = 3$ is

$$\Delta E = E_3 - E_2 = \left[-R_H\left(\frac{1}{3^2}\right)\right] - \left[-R_H\left(\frac{1}{2^2}\right)\right]$$

CALCULATION FOR a

Substituting for R_H gives

$$\Delta E = E_3 - E_2 = \left[-2.18 \times 10^{-18}\,\text{J}(\tfrac{1}{9})\right] - \left[-2.18 \times 10^{-18}\,\text{J}(\tfrac{1}{4})\right]$$

$$= -0.242 \times 10^{-18}\,\text{J} + 0.545 \times 10^{-18}\,\text{J}$$

$$= 0.303 \times 10^{-18}\,\text{J} = 3.03 \times 10^{-19}\,\text{J}$$

METHOD OF SOLUTION FOR b

The energy lost by the atom appears as a photon of radiation with a characteristic frequency and wavelength:

$$\Delta E_{\text{atom}} = E_{\text{photon}} = h\nu = h\frac{c}{\lambda}$$

CALCULATION FOR b

$$h\nu = E_{\text{photon}} = 3.03 \times 10^{-19}\,\text{J}$$

$$\nu = \frac{3.03 \times 10^{-19}\,\text{J}}{6.63 \times 10^{-34}\,\text{J}\cdot\text{s}} = 4.57 \times 10^{14}\,\text{s}^{-1}$$

and

$$\lambda = \frac{c}{\nu} = \frac{3.00 \times 10^8}{4.57 \times 10^{14}\,\text{s}^{-1}}\,\text{m}\cdot\text{s}^{-1} = 656 \times 10^{-9}\,\text{m}$$

$$= 656\,\text{nm}$$

WAVE-PARTICLE DUALITY

STUDY OBJECTIVE

You should be able to:
1. Calculate the wavelength of a particle, given its momentum.

Matter Waves. In 1925 Louis de Broglie had the novel idea that since light had been found to have a dual nature with properties of both waves and particles, then particles of *matter* should have a dual nature and exhibit wave properties as well. He deduced that the wavelength λ of a particle depends on its mass m and velocity u:

$$\lambda = \frac{h}{mu}$$

Experimental evidence of the wave properties of small particles was provided in 1927 by diffraction experiments with electron beams. Diffraction is a phenomenon that can be explained only by wave motion. Since electrons were observed to produce diffraction patterns, then electrons must behave as waves. According to de Broglie's equation, the wavelength of a particle increases as its mass decreases, so wave properties (diffraction) are most prominent for the electron and are the basic principle behind the electron microscope. Diffraction patterns of neutron beams have also been observed and serve as a means to learn about the structures of molecules. As the mass of a particle reaches macroscopic size (say, $> 10^{-12}$ g), its wavelength becomes extremely short and wave properties cannot be observed.

EXAMPLE 7.4   De Broglie Wavelength

When an atom of Th-232 undergoes radioactive decay, an alpha particle which has a mass of 4.0 amu is ejected from the Th nucleus with a velocity of 1.4×10^7 m/s. What is the de Broglie wavelength of the alpha particle?

METHOD OF SOLUTION

The de Broglie wavelength depends on the mass and velocity of the particle. The equation is

$$\lambda = \frac{h}{mu}$$

Because Planck's constant has units of joule seconds, and $1\ \text{J} = 1\ \text{kg m}^2\ \text{s}^{-2}$, the mass of the alpha particle must be expressed in kilograms. Since 1 mol of alpha particles has a mass of 4.0 g, then the mass of one particle is

$$\frac{?\ \text{kg}}{\text{particle}} = \frac{4.0 \times 10^{-3}\ \text{kg/mol}}{6.02 \times 10^{23}\ \text{particles/mol}} = 6.6 \times 10^{-27}\ \text{kg/particle}$$

CALCULATION

The wavelength of this alpha particle is

$$\lambda = \frac{h}{mu} = \frac{(6.63 \times 10^{-34}\ \text{J s}) \times \left(\frac{1\ \text{kg m}^2/\text{s}^2}{1\ \text{J}}\right)}{(6.6 \times 10^{-27}\ \text{kg})(1.4 \times 10^7\ \text{m/s})}$$

Notice that all the units cancel except meters. The wavelength is

$$\lambda = 7.2 \times 10^{-15}\ \text{m}$$

COMMENT

The wavelength is smaller than the diameter of the thorium nucleus, which is about 2×10^{-14} m.

QUANTUM MECHANICS AND THE HYDROGEN ATOM

STUDY OBJECTIVES

You should be able to:
1. Describe the use of the terms *probability*, *electron density*, and *orbital* in the wave mechanical model of the hydrogen atom.
2. Describe the significance of the three quantum numbers that correspond to an orbital in an atom.
3. Describe the shapes of *s*, *p*, and *d* orbitals.
4. State the Pauli exclusion principle.

Quantum Mechanics. In 1926 Erwin Schrödinger incorporated all the wave and particle properties into a form of mechanics called *wave mechanics* or *quantum mechanics*. He formulated the now famous Schrödinger wave equation and obtained a set of mathematical functions called *wave functions*, ψ. For any set of coordinates within the atom, ψ has a numerical value. The square of the wave function ψ^2 is related to the probability that an electron will be found in a small region (volume) in space. Where ψ^2 is large, the probability of finding the electron is high. Where ψ^2 is small, the probability of finding the electron is low.

It is helpful to use terms like *electron clouds* or *electron density* (Figure 7.22 in the text) to represent the probability concept. The term *electron cloud* must not be interpreted to mean that one electron occupies the whole of the space occupied by this cloud at one time. Rather, the electron is considered a point charge, and the density of the cloud at a specific point gives only the probability of finding the electron at that point.

As a result of the probability concept, we no longer speak of distinct Bohr orbits. In the wave mechanical model of an atom, the Bohr orbit is replaced with an *orbital* or *atomic orbital*. Orbitals are regions in an atom within which electrons have a high probability, usually 90%, of being found. Each atomic orbital has a characteristic shape and energy.

Atomic Orbitals. An orbital is a graphical representation of the electron probability around the nucleus. The density of the probability cloud at each region in space (see Figure 7.22 in the text) represents the probability of finding the electron there. Orbitals are often drawn using equal-probability contours. If we go far enough from the nucleus in a particular direction, we come to a point at which 90% of the time the electron, when it is in that direction, will be inside that point. Connecting all such points produces an equal-probability boundary surface such that 90% of the time the electron will be inside that surface no matter what direction from the nucleus we consider. This graphical representation gives the shape of the orbital.

In general chemistry we are interested in the shapes of only three types of orbitals. The characteristics of *s*, *p*, and *d* orbitals are given in Table 7.1, and the shapes of *s* and *p* orbitals are shown in Figure 7.1.

Table 7.1 Symbols and Shapes of Orbitals

Orbital	Designation	Number of Orbitals per Shell	Shape of 90% Boundary Surface
s	ns	1	spherical
p	np	3	dumbbell (2 lobes)
d	nd	5	4 lobes

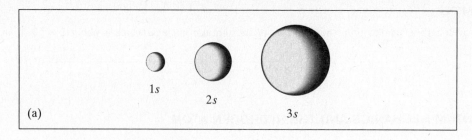

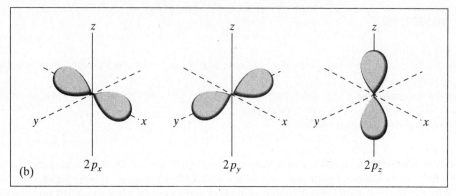

Figure 7.1. The shapes of s and p orbitals. The surfaces shown enclose 90% of the electron density. (a) All s orbitals are spherical and increase in size as the principal quantum number is increases. (b) The p subshell consists of three p orbitals. Except for their orientations in space, these orbitals have identical shape and energy.

Quantum Numbers. Both the energy and the probability distribution are described by a set of numbers called quantum numbers. The *principal quantum number n* determines the energy of a hydrogen atom according to the formula $E_n = -R_H(1/n^2)$. This number can have any integral value, $n = 1, 2, 3, \ldots$. The larger the value of n, the greater the energy and the farther the electron is from the nucleus on the average. When $n = 1$, the electron in the hydrogen atom is said to be in the *ground state*. All other values of n correspond to excited states. When $n = \infty$, the highest possible energy, the atom has lost the electron and exists as a H^+ ion.

The angular momentum quantum number ℓ is related to the shape of the orbital. Within each principal energy level one or more orbital types will exist. Thus any one principal quantum level may have more than one value for ℓ. For a given n, ℓ has integral values from 0 to $n - 1$; $\ell = 1, 2, 3, \ldots, n - 1$. Each value of ℓ defines a group of orbitals composing a specific sublevel or subshell

The shape of atomic orbitals will be discussed in the next section. For now each orbital is specified by a letter designation. Orbitals for which $\ell = 0$ are also referred to as s orbitals. Electrons in s orbitals have zero units of angular momentum. The letters used to designate orbitals with different ℓ values are

ℓ	0	1	2	3	4	$\cdots$
Orbital	s	p	d	f	g	$\cdots$

The magnetic quantum number m_ℓ describes the orientation of an orbital in space. The value of m_ℓ may have integral values from $-\ell$ to $+\ell$, including zero:

$$m_\ell = -\ell, -\ell + 1, \ldots, 0, \ldots, \ell - 2, \ell - 1, +\ell$$

The number of m_ℓ values is very important because it designates the number of orbitals within a subshell. For example, when $\ell = 2$, then $m_\ell = -2, -1, 0, 1, 2$. We see there are five m_ℓ values. The d orbitals always occur in groups of five orbitals. These five d orbitals all have the same energy but have five different orientations in space. They make up the d subshell. A *subshell* is a group of orbitals designated by a particular ℓ value within a shell. Table 7.2 shows the number of orbitals in the most common subshells.

Table 7.2 The Number of Orbitals in Various Subshells

Type of Orbital	ℓ Value	m_ℓ Values	Number of Orbitals
s	0	0	1
p	1	-1, 0, 1	3
d	2	-2, -1, 0, 1, 2	5
f	3	-3, -2, -1, 0, 1, 2, 3	7

The *spin quantum number* m_s designates one of two possible spin directions for electrons. The two possible values of m_s are $+\frac{1}{2}$ and $-\frac{1}{2}$. The *Pauli exclusion principle* states that no two electrons in an atom can have all four quantum numbers the same. Accordingly, this principle limits the numbers of electrons each atomic orbital can hold to no more than two electrons; one electron will have $m_s = +\frac{1}{2}$ and the other $m_s = -\frac{1}{2}$.

Let's take an electron in a particular orbital. Say a $3p_x$ orbital. The three quantum numbers associated with this orbital are $n = 3$, $\ell = 1$, and $m_\ell = -1$. Any electrons in this orbital must have these identical n, ℓ, and m_ℓ quantum numbers. The two sets of quantum numbers for electrons in this orbital differ in their m_s values:

Electron 1: $n = 3$, $\ell = 1$, $m_\ell = -1$, $m_s = +\frac{1}{2}$.

Electron 2: $n = 3$, $\ell = 1$, $m_\ell = -1$, $m_s = -\frac{1}{2}$.

The result of this qualitative treatment of wave mechanics gives three essential pieces of information about electrons in atoms:

1. Quantum numbers (n, ℓ, m) and their relationships to each other
2. Electron probability clouds or orbitals
3. The maximum number of electrons per orbital

EXAMPLE 7.5 Quantum Numbers

If the principal quantum number of an electron is $n = 2$, what values could the following have?
a. Its ℓ quantum number
b. Its m_ℓ quantum number
c. Its m_s quantum number

METHOD OF SOLUTION

a. Recall that the angular momentum quantum number ℓ has values that depend on the value of the principal quantum number n; $\ell = 0, 1, 2, \ldots, n - 1$. *Answer:* When $n = 2$, $\ell = 0, 1$.
b. The magnetic quantum number m_ℓ depends on the value of ℓ where $m_\ell = -\ell, -\ell + 1, \ldots, 0, \ldots, \ell + 1, \ell$. *Answer:* When $\ell = 1$, $m_\ell = -1, 0, 1$; and when $\ell = 0$, $m_\ell = 0$.
c. *Answer:* The m_s quantum number for a single electron could be either $+\frac{1}{2}$ or $-\frac{1}{2}$.

EXAMPLE 7.6 Quantum Numbers and Orbitals

List all the possible orbitals associated with the principal energy level $n = 5$.

METHOD OF SOLUTION

The type of orbital is given by the angular momentum quantum number. When $n = 5$; $\ell = 0, 1, 2, 3, 4$. *Answer:* These correspond to subshells consisting of s, p, d, and f orbitals.

EXAMPLE 7.7 Quantum Numbers

For the following sets of quantum numbers for electrons, indicate which quantum numbers—n, ℓ, or m_ℓ—could not occur and state why.
a. 3, 2, 2
b. 2, 2, 2
c. 2, 0, −1

METHOD OF SOLUTION

a. When $n = 3$; $\ell = 0, 1, 2$. When $\ell = 2$; $m_\ell = -2, -1, 0, 1, 2$. *Answer:* Thus the set in a can occur.
b. When $n = 2$; $\ell = 0, 1$. *Answer:* The set in b cannot occur because $\ell \neq 2$ when $n = 2$. Comment: This also means that the $n = 2$ energy level cannot have d orbitals.
c. When $n = 2$; $\ell = 0, 1$. When $\ell = 0$, $m_\ell = 0$. *Answer:* The set in c cannot occur because $m_\ell \neq -1$ when $\ell = 0$.

EXAMPLE 7.8 Quantum Numbers and Orbitals

What values can m_ℓ take for
a. A $3d$ orbital
b. A $2s$ orbital

METHOD OF SOLUTION

The value of m_ℓ depends only on ℓ, and so first convert each orbital designation to the corresponding value of ℓ:
a. For a d orbital $\ell = 2$; therefore, possible m_ℓ values are $-2, -1, 0, 1$, and 2.
b. For an s orbital $\ell = 0$; therefore $m_\ell = 0$.

ELECTRON CONFIGURATIONS

STUDY OBJECTIVES

You should be able to:
1. Compare the order of orbital energies in the hydrogen atom to the order in a many-electron atom.
2. Use the Aufbau principle to write the electron configuration of an element.
3. Use Hund's rule to construct orbital diagrams for the electron configurations of elements.

The Energies of Orbitals. The way electrons are distributed among the orbitals of an atom is its *electron configuration* and *electronic structure*. Each of the principal energy levels (called shells) of an atom can be divided into subshells, which are sets of orbitals. The s subshell has one s orbital, the p subshell has three orbitals, the d subshell has five orbitals, and so on. In addition each subshell has an electron capacity based on a maximum of two electrons per orbital. The orbitals available to a shell and the subshell capacities are summarized in Table 7.3.

In a hydrogen atom all of the orbitals in a particular energy level (same n value) have the same energy as shown in Figure 7.2(a). In a hydrogen atom the electron resides in the $1s$ orbital because it has the lowest possible energy in that orbital. The ground state electron configuration of H is $1s^1$, as shown

Table 7.3 Allowed Numbers of Orbitals per Energy Level

Principal Quantum Number	Orbital Shape ℓ	Orbital Name	Number of Orbitals per Subshell	Electron Capacity per Subshell
1	0	$1s$	1	2
2	0	$2s$	1	2
2	1	$2p$	3	6
3	0	$3s$	1	2
3	1	$3p$	3	6
3	2	$3d$	5	10
4	0	$4s$	1	2
4	1	$4p$	3	6
4	2	$4d$	5	10
4	3	$4f$	7	14

below:

$$\text{principal energy level} \rightarrow 1s^1 \begin{array}{l} \leftarrow \text{number of electrons per subshell} \\ \leftarrow \text{orbital occupied} \end{array}$$

Atoms of elements other than hydrogen have more than one electron. In these atoms each principal energy level is split into several energies. These energies correspond to the orbitals of one type having slightly different energies than orbitals of another type. This splitting is shown in Figure 7.2(b). The energy of an orbital in a many-electron atom depends on both n and ℓ. For a given value of n, the energy of orbitals increases as l increases. Therefore, for atoms other than hydrogen atoms, the $2p$ orbital has a higher energy than the $2s$ orbital.

The order of increasing energy for orbitals within a given principal energy level is

$$s < p < d < f < \cdots$$

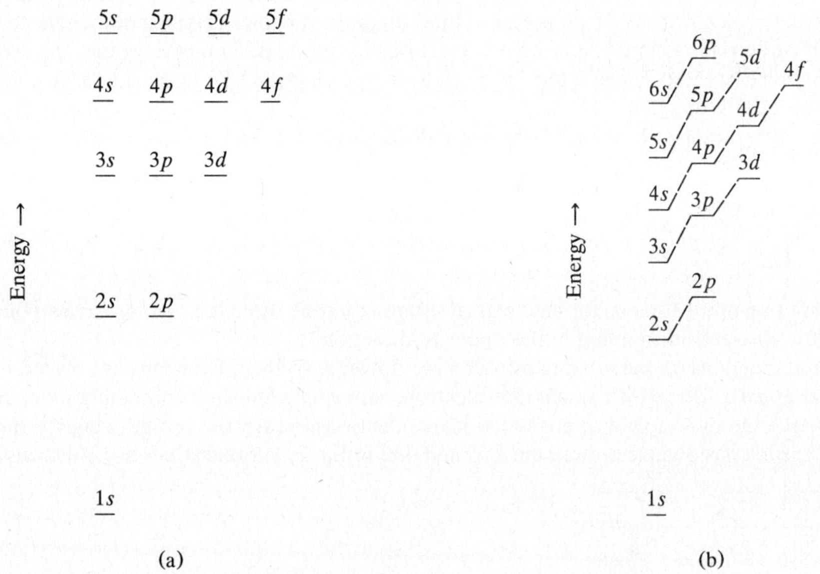

Figure 7.2. (a) In the hydrogen atom all the orbitals in the same principal energy level have the same energy. (b) In a many-electron atom the orbitals of a principal energy level have slightly different energies.

This trend results from the fact that s orbitals have a greater electron density near the nucleus than do p orbitals, and p orbitals have a greater electron density near the nucleus than do d orbitals. In a many-electron atom, the energies of the orbitals increase as shown as follows (Figure 7.2):

$$1s < 2s < 2p < 3s < 3p < 4s < 3d < 4p < 5s \cdots$$

Note also that the higher principal energy levels (n values) have more subshells, and as these split in energy, some of these subshells can actually be lower in energy than subshells from a lower principal energy level. Thus the energy of a $4s$ orbital is very close to the energy of a $3d$ orbital in a many-electron atom. In potassium and calcium atoms the energy of the $4s$ orbital is less than the energy of the $3d$ orbital.

Electron Configurations. To write the electron configuration for the ground state of a many-electron atom, recall that the order in which subshells are occupied is from lowest energy to highest energy. The order shown in Figure 7.2(b) can easily be remembered by drawing Figure 7.27 of the textbook. The general approach of building up the atoms of the elements by adding a proton to the nucleus and an electron to an orbital is called the *Aufbau* or *building-up principle*.

Helium has two electrons both of which can occupy the $1s$ orbital. The electron configuration is He $1s^2$. A lithium atom has three electrons. According to the Pauli principle, we can place only two electrons in the $1s$ orbital; the third electron must enter the next higher orbital, the $2s$ orbital. Its electron configuration is Li $1s^2 2s^1$. The electron configurations of the second-period elements are

Be $1s^2 2s^2$

With Be the $1s$ and $2s$ orbitals are filled, and so additional electrons must enter the $2p$ orbitals. The p subshell is being filled over the next six elements:

B $1s^2 2s^2 2p^1$
C $1s^2 2s^2 2p^2$
N $1s^2 2s^2 2p^3$
O $1s^2 2s^2 2p^4$
F $1s^2 2s^2 2p^5$
Ne $1s^2 2s^2 2p^6$

Orbital Diagrams. The above electron configurations tend to hide information about the number of electrons in any one outer orbital. The $2p$ subshell consists of three $2p$ orbitals: $2p_x$, $2p_y$, and $2p_z$. Thus its capacity is 6 electrons. If we use an orbital diagram, a choice arises about where to place the electrons. In carbon, for example, are the two electrons in the p subshell both in the $2p_x$ orbital or are they distributed so that the $2p_x$ and the $2p_y$ each have one electron? To make the choice, first we need to describe an orbital diagram.

An orbital diagram uses boxes to designate individual orbitals and groups of boxes to designate subshells:

☐ ☐ ☐ ☐ ☐
$1s$ $2s$ $2p$

An arrow pointing up (↑) stands for an electron spinning in one direction, and an arrow pointing down (↓) stands for an electron spinning in the opposite direction.

The orbital diagrams for carbon and nitrogen are drawn as follows. Electrons are placed into orbitals according to *Hund's rule*, which states that electrons entering a subshell containing more than one orbital will have the most stable arrangement if the electrons occupy the orbitals singly rather than in pairs. Thus carbon has one electron in the $2p_x$ and one in the $2p_y$, rather than two electrons in the $2p_x$:

C [↑↓] [↑↓] [↑|↑|]
    $1s$ $2s$ $2p$

N [↑↓] [↑↓] [↑|↑|↑]
    $1s$ $2s$ $2p$

According to Hund's rule, nitrogen atoms have the electrons in the $2p$ subshell distributed with one each in the $2p_x$, $2p_y$, and $2p_z$.

The orbital diagram indicates the number of unpaired electrons in an atom. The presence, or absence, of unpaired electrons is detected experimentally by the behavior of a substance when placed in a magnetic field. *Paramagnetism* is the property of attraction to a magnetic field. *Diamagnetism* is the property of repulsion by a magnetic field. The atoms of paramagnetic substances contain unpaired electrons. Those of diamagnetic substance contain only paired electrons. Using the three principles that have been given, the electron configurations of the ground states of most atoms can be estimated.

Noble Gas Core Abbreviations. The electron configurations of the noble gas elements can be used as abbreviations when writing electron configurations. Neon has the configuration Ne $1s^2 2s^2 2p^6$. All elements beyond neon have this configuration for the first 10 electrons. We use the symbol [Ne] to represent the configuration of the 10 electrons in neon and call it a neon core. Similarly 18 electrons arranged in the ground state of an argon atom ($1s^2 2s^2 2p^6 3s^2 3p^6$) are called an argon core, written [Ar]. [Kr] and [Xe] cores can also be used. In practice we select the noble gas that most nearly precedes the element being considered. Therefore, calcium and strontium could be written in full or abbreviated as follows:

Ca $1s^2 2s^2 2p^6 3s^2 3p^6 4s^2$ or [Ar]$4s^2$
Sr $1s^2 2s^2 2p^6 3s^2 3p^6 4s^2 3d^{10} 4p^6 5s^2$ or [Kr]$5s^2$

EXAMPLE 7.9 Electron Configurations

Which of the following electron configurations would correspond to ground states and which to excited states?
a. $1s^2 2s^2 2p^1$
b. $1s^2 2p^1$
c. $1s^2 2s^2 2p^1 3s^1$

METHOD OF SOLUTION

a. In the above configuration electrons occupy the lowest possible energy states. *Answer:* It corresponds to a ground state.
b. In this configuration electrons do not occupy the lowest possible energy states. The $2s$ orbital, which lies lower than the $2p$, is vacant, and the last electron is in the $2p$ orbital. *Answer:* This is an excited state.
c. In this case the last electron is in the $3s$ orbital while the lower energy $2p$ subshell is not completely filled. *Answer:* This is an excited state.

EXAMPLE 7.10 Electron Configurations

Write the electron configuration for a potassium atom.

METHOD OF SOLUTION

Potassium has 19 electrons. The first 10 of these will occupy the same orbitals as neon, which is given above: $1s^2 2s^2 2p^6$. This leaves 9 electrons to account for. Keeping in mind the order of increasing orbital energies, the next 2 electrons fill the $3s$ orbital. Then 6 electrons can be placed into the $3p$ subshell. This leaves 1 electron. According to the order of orbital energies, the $3d$ does not fill next. Rather, the $4s$ orbital is lower in energy than the $3d$. The last electron in potassium enters the $4s$ orbital. The electron configuration of potassium is

K $1s^2 2s^2 2p^6 3s^2 3p^6 4s^1$

or in terms of the argon core abbreviation,

K [Ar]$4s^1$

EXAMPLE 7.11   Orbital Diagrams

a. Write the electron configuration for arsenic.
b. Draw its orbital diagram.
c. Are As atoms diamagnetic or paramagnetic?

METHOD OF SOLUTION

a. From the atomic number of As we see there are 33 electrons per arsenic atom. From Figure 7.29 in the text, the order of filling orbitals is $1s$, $2s$, $2p$, $3s$, $3p$, $4s$, and so on. Placing electrons in the lowest energy orbitals until they are filled, we find that the first 18 electrons are arranged $1s^2 2s^2 2p^6 3s^2 3p^6$, which corresponds to an Ar core. The next 2 enter the $4s$, and the next 10 enter the $3d$. This leaves 3 electrons for the $4p$ subshell. *Answer:* The electron configuration of As is

As [Ar]$4s^2 3d^{10} 4p^3$

b. All of the orbitals are filled except for the $4p$ orbitals. The electrons must be placed into the $4p$ orbitals in accordance with Hund's rule. The orbital diagram for the outer orbitals is

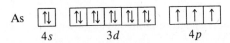

c. *Answer:* Arsenic atoms are paramagnetic because they contain 3 unpaired electrons.

TRUE-FALSE QUESTIONS

1. The greater the frequency of light, the faster the speed of light.

2. The energy of a photon is inversely proportional to the wavelength of the radiation.

3. Bohr postulated that radiation is emitted when an electron transition occurs from a lower to a higher energy orbit.

4. The quantum mechanical model of the H atom confirms that Bohr's orbits are reasonable but shows that the energies of the principal energy levels are wrong.

5. The value of the magnetic quantum number depends on ℓ.

6. An orbital for which $\ell = 2$ is a d orbital.

7. The m_ℓ quantum number can have only one of two possible values, $-\frac{1}{2}$ or $+\frac{1}{2}$.

8. No two electrons in an atom can have the same set of four quantum numbers.

9. The d subshell consists of a set of six orbitals.

10. In a many-electron atom the energies of the p_x, p_y, and p_z orbitals are not the same.

11. The electron configuration for nitrogen can be written as $1s^2 2s^2 2p^3$ or $1s^2 2s^2 2p_x^1 2p_y^1 2p_z^1$.

12. The configuration $1s^2 2s^2 2p^5 3s^2$ would be an excited state of sodium.

SELF-TEST A

1. A certain AM radio station broadcasts at a frequency of 600 kHz. What is the wavelength of these radio waves in meters?

2. How long would it take a radio wave of frequency 5.5×10^5 Hz to travel from the planet Venus to Earth (28 million miles)?

3. Calculate the energy required to remove an electron from a hydrogen atom in the ground state.

4. Which of the following has the greatest ionization energy?
   a. H   b. He$^+$ c. Li^{2+}

5. The average kinetic energy of a neutron at 25°C is 6.2×10^{-21} J. What is its de Broglie wavelength? The mass of a neutron is 1.008 amu. *Hint:* kinetic energy $= \frac{1}{2}mu^2 = (mu)^2/2m$.

6. Deduce the possible sets of four quantum numbers for an electron that arise when $n = 2$.

7. What is the maximum number of electrons that can occupy an $n = 3$ energy level? The $n = 4$ energy level?

8. Which of the following subshells has a capacity of 10 electrons?
   a. $5s$ b. $2p$ c. $4p$ d. $3d$ e. $6s$

9. How many orbitals are occupied by one or more electrons in a germanium atom?

10. Complete the sentence: An electron with $\ell = 2$ must
    a. Have $m_\ell = -2$
    b. Be in an $n = 3$ energy level
    c. Be in a p orbital
    d. Be in a d orbital

11. Write quantum numbers for
    a. An electron in a $2s$ orbital
    b. The outermost electrons in Ge

12. Write the electron configuration for the following atoms: Ar, Se, and Ag.

13. How many electrons are represented by the abbreviation [Kr]?

14. Which of the following is the correct orbital diagram for chromium?

    a. 3d [↑↓|↑↓| | |] 4s [↑↓]
    b. 3d [↑|↑|↑|↑|]   4s [↑↓]
    c. 3d [↑↓|↑↓|↑↓|↑↓|] 4s [ ]
    d. 3d [↑|↑|↑|↑|↑]   4s [↑]

15. Which of the following atoms has the greatest number of unpaired electrons?
    a. Ti   b. Ag   c. O   d. P   e. K

GENERAL PROBLEMS

16. If the energy required to ionize 1 mol of H atoms were used to raise the temperature of water, what mass of water could have its temperature increased by 50°C?

17. If 5% of the energy supplied to an incandescent light bulb is radiated as visible light, how many "visible" photons per second are emitted by a 100-watt bulb? Assume the wavelength of all visible light to be 560 nm. Given: 1 watt = 1 J/s.

18. The light-sensitive compound in most photographic films is silver bromide (AgBr). Assume, when film is exposed, that the light energy absorbed dissociates the molecule into atoms. (The actual process is more complex.) If the energy of dissociation of AgBr is 100 kJ/mol, find the wavelength of light that is just able to dissociate AgBr.

SELF-TEST B1

1. What is the wavelength of electromagnetic radiation with frequency 5.0×10^{14} s^{-1}?

2. What is the energy of a photon of radiation having a frequency of 6.2×10^{14} s^{-1}?

3. The red line in the spectrum of lithium occurs at 670.8 nm. What is the energy of a photon of this light? What is the energy of 1 mol of these photons?

4. What wavelength of radiation will be emitted when an electron in a hydrogen atom jumps from the $n = 5$ to the $n = 1$ principal energy level? Name the region of the electromagnetic spectrum corresponding to this wavelength.

5. Which of the following sets of quantum numbers are not allowed for describing an electron in an orbital in a hydrogen atom?

	n	ℓ	m_ℓ	m_s
a.	3	2	-3	$\frac{1}{2}$
b.	2	3	0	$-\frac{1}{2}$
c.	2	1	0	$-\frac{1}{2}$

6. Which choice is a possible set of quantum numbers for the last electron added to make up an atom of gallium (Ga) in its ground state?

	n	ℓ	m_ℓ	m_s
a.	4	2	0	$-\frac{1}{2}$
b.	4	1	0	$\frac{1}{2}$
c.	4	2	-2	$-\frac{1}{2}$
d.	3	1	$+1$	$\frac{1}{2}$
e.	3	0	0	$-\frac{1}{2}$

7. Write the electron configuration for the following atoms: Sb, V, and Pb.

8. How many unpaired electrons do nitrogen atoms have?

SELF-TEST B2

1. What is the frequency of light that has a wavelength of 600 nm?

2. A photon has an energy of 4.11×10^{-19} J. What is the frequency of this radiation?

3. The energy of a mole of photons is 179 kJ. Calculate the wavelength of the radiation.

4. A hydrogen emission line in the ultraviolet region of the spectrum at 95.2 nm corresponds to a transition from a higher energy level n to the $n = 1$ level. What is the value of n for the higher energy level?

5. Which of the following are incorrect designations for an atomic orbital?
   a. $3f$ b. $4s$ c. $2d$ d. $4f$

6. What orbital does the last electron added to complete an atom of gallium (Ga) in its ground state enter?
   a. $3p$ b. $3d$ c. $4s$ d. $4p$ e. $3f$

7. What element has atoms with the electron configuration $[Xe]6s^2 4f^{14} 5d^{10} 6p^2$?

8. What second-period element has atoms in the ground state with three unpaired electrons?

ANSWERS

TRUE-FALSE QUESTIONS

1. False. The speed of light is constant.
2. True.
3. False. For radiation to be emitted an electron must go from a higher to a lower energy orbit.
4. False. The energy levels in both models have the same energy.
5. True.
6. True.
7. False. m_ℓ depends on ℓ and has values $\ell, \ell - 1, \ldots, 0, \ldots, -\ell + 1, -\ell$.
8. True.
9. False. A set of five d orbitals.
10. False. Their energies are the same.
11. True.
12. True.

SELF-TEST A

1. 500 m
2. 150 s
3. 2.18×10^{-18} J
4. Li^{2+}
5. $\lambda = 95.2$ nm; ultraviolet
6.

n	ℓ	m_ℓ	m_s
2	0	0	$-\frac{1}{2}$
2	0	0	$+\frac{1}{2}$
2	1	-1	$-\frac{1}{2}$
2	1	-1	$+\frac{1}{2}$
2	1	0	$-\frac{1}{2}$
2	1	0	$+\frac{1}{2}$
2	1	$+1$	$-\frac{1}{2}$
2	1	$+1$	$+\frac{1}{2}$

7. 18
8. d
9. 16
10. d
11. a. $n = 2$, $\ell = 0$, $m_\ell = 0$, $m_s = \frac{1}{2}$
    b. $n = 4$, $\ell = 0$, $m_\ell = 0$, $m_s = \frac{1}{2}$
    $n = 4$, $\ell = 0$, $m_\ell = 0$, $m_s = -\frac{1}{2}$
    $n = 4$, $\ell = 1$, $m_\ell = -1$, $m_s = \frac{1}{2}$
    $n = 4$, $\ell = 1$, $m_\ell = 0$, $m_s = \frac{1}{2}$
12. Ar $1s^2 2s^2 2p^6 3s^2 3p^6$
    Se [Ar]$4s^2 3d^{10} 4p^4$
    Ag [Kr]$5s^1 4d^{10}$
13. 36
14. d
15. d
16. 6280
17. 1.4×10^{19} photons/s
18. 1200 nm

SELF-TEST B1

1. 600 nm
2. 4.11×10^{-19} J
3. 2.97×10^{-19} J/photon, 179 kJ/mol
4. 95.2 nm, ultraviolet
5. a and b
6. c
7. Sb [Kr]$5s^2 4d^{10} 5p^3$; V [Ar]$4s^2 3d^3$; Pb [Xe]$6s^2 4f^{14} 5d^{10} 6p^2$
8. 3

SELF-TEST B2

1. 5.0×10^{14} s^{-1}
2. 6.2×10^{14} s^{-1}
3. 671 nm
4. $n = 5$
5. a and c
6. d
7. Pb
8. N

Chapter Eight
PERIODIC RELATIONSHIPS AMONG THE ELEMENTS

- Classification of Elements by Properties
- Electron Configuration and the Periodic Table
- Periodic Trends Explained
- Types of Elements

CLASSIFICATION OF ELEMENTS BY PROPERTIES

STUDY OBJECTIVES

You should be able to:
1. Discuss the basis for the arrangement of elements in the periodic tables by Meyer and Mendeleev.
2. State the periodic law.

History of the Periodic Table. After a number of the elements had been discovered, certain similarities and differences in their properties became apparent. For example, sodium and potassium are metallic solids, soft enough to be sliced with a knife. They both have low melting points and low densities. They are also similar in their reactions. They both react with halogens to form salts that have the same general formula, MX.

The elements chlorine and bromine also resemble each other. Both are highly colored. Chlorine is a yellow gas, and bromine is a brown liquid at room temperature. Both exist as diatomic molecules, and both react readily with many metals, causing oxidation. During the mid-1800s a number of chemists tried to classify and arrange the elements according to their properties.

In 1869 Dimitri Mendeleev and Lothar Meyer, working independently, reported very similar ways to classify the elements. Both arranged elements in rows by increasing atomic mass and grouped elements with similar chemical and physical properties in vertical columns. Mendeleev stated the *periodic law*: *When the elements are arranged according to increasing atomic mass, their chemical and physical properties show periodic variations*. In other words, similarities in properties recur at periodic intervals of atomic mass. For example, periodicity in the atomic radius and in the first ionization energy is illustrated in Figures 8.8 and 8.14 of the textbook.

In several instances Mendeleev did not know of an element with the correct atomic mass to fit into a column of elements of similar properties. In these cases he left gaps in the table. He insisted that these gaps corresponded to as yet undiscovered elements. He was able to predict the properties of these missing elements by following trends within the group. The close agreement between his predictions and the properties of the elements when discovered later was remarkable.

Because of the usefulness of his table in guiding the discovery of new elements, Mendeleev is remembered as the designer of the periodic table of the elements. In the twentieth century, the periodic law was updated by Henry Moseley, who pointed out that the properties of the elements are determined by their atomic number, not their atomic mass. The mass reversals, such as the one at Ar/K (Section 8.1 text), were puzzling exceptions to Mendeleev's law. Moseley's work, with its emphasis on atomic number, meant that these reversals were no longer considered a problem.

Table 8.1 Properties of the Halogens

Group 7A Element	Melting Point	Boiling Point	Formula of Sodium Salt	Reactivity toward H_2
F_2	−223	−188	NaF	Explosive
Cl_2	−101	−34	NaCl	Initiated by light at 25°C
Br_2	−7	+59	NaBr	Slow at 25°C
I_2	+114	+187	NaI	Requires high temperatures

The horizontal rows of elements with atomic number increasing by 1 from left to right are called *periods*. The vertical columns with atomic number and atomic mass increasing from top to bottom are called *groups or families*. Elements within a group have similar, but not identical, properties. Table 8.1 shows how properties of the halogens vary gradually within a group.

EXAMPLE 8.1 Predicting Properties of Elements

Predict the missing value for the following:

a.
Element	Density (g/cm³)
Ca	1.55
Sr	?
Ba	3.5

b.
Element	Density (g/cm³)
Cl	0.099
Br	0.114
I	?

METHOD OF SOLUTION

a. From the positions of these elements in Group 2A of the periodic table, the density of Sr could be estimated to be halfway between the values for Ca and Ba. The average of 1.55 and 3.5 g/cm³ is 2.5 g/cm³. The observed value is 2.6 g/cm³.

b. In this case we must extrapolate rather than interpolate as above because I is farther down in Group 7A than Cl and Br. We could assume the change in the radius from Br to I will be the same as the change from Cl to Br. The difference between Cl and Br is 0.015 nm. Adding 0.015 to 0.114 gives the estimated radius for I as 0.129 nm. The observed radius of I is 0.133 nm.

EXAMPLE 8.2 Element Groups

Given the following formulas of compounds:

Sodium sulfate Na_2SO_4

Magnesium oxide MgO

Aluminum chloride $AlCl_3$

write the formulas of the following compounds using the periodic table:
a. Calcium oxide
b. Potassium sulfate
c. Gallium bromide

METHOD OF SOLUTION

a. Notice that Ca is in the same chemical group as Mg. If one Mg ion combines with one oxide ion, then one Ca ion will combine with one oxide ion. *Answer:* The formula is CaO
b. If sodium sulfate is Na_2SO_4, then potassium, being in the same chemical group as sodium, should form a sulfate with the same ratio of potassium atoms to sulfate ions. *Answer:* The formula is K_2SO_4
c. Gallium is in Group 3A along with Al, and bromine is a halogen, as is Cl. Thus if aluminum chloride has the formula $AlCl_3$, gallium will combine with any three halogen atoms. *Answer:* The formula of gallium bromide is $GaBr_3$.

ELECTRON CONFIGURATIONS AND THE PERIODIC TABLE

STUDY OBJECTIVES

You should be able to:
1. Give the locations in the periodic table of the representative, transition, and inner transition elements.
2. Write the electron configuration of any element from the position of the element in the periodic table.

Chemical Groups. Why should elements in a group have similar chemical properties? The modern theory of atomic structure provides an answer to this question. First let's write the electron configurations for elements in any vertical column in the periodic table. Table 8.2 shows the electron configurations of atoms of the Group 7A elements.

Table 8.2 Group 7A Electron Configurations

F	$1s^2 2s^2 2p^5$
Cl	$[Ne]3s^2 3p^5$
Br	$[Ar]4s^2 3d^{10} 4p^5$
I	$[Kr]5s^2 4d^{10} 5p^5$

The table shows that the electron configuration of the outermost principal energy levels of Group 7A elements contains seven electrons arranged $ns^2 np^5$, where n is the principal quantum number of the outermost energy level. Examination shows that the halogen group is not unique. The electron configuration of the outermost electrons of elements in any one group is the same.

The chemical similarities of the Group 7A elements can be explained quite simply by their electron configurations: *All the elements in a group have the same electron configuration in their outermost principal energy level.* The electrons in the outermost energy level are called *valence* electrons. The similarity of the valence electron configuration holds true for elements in the other groups as well.

Representative Elements. The modern periodic table is arranged according to the type of subshells being filled with electrons (Figure 8.2 in the textbook). Groups 1A through 7A of the periodic table include elements that have an incompletely filled *set* of *s* or *p* orbitals. These are called *representative elements*, or main group elements. The groups are given numerals from 1 to 7, followed in most books by the letter A, which stands for a group of representative elements. The noble gases have completely filled outer *ns* and *np* subshells, $ns^2 np^6$, except for helium.

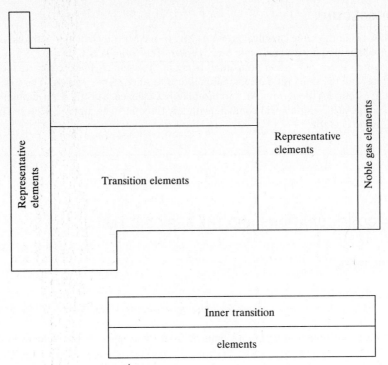

Figure 8.1. General groups of elements.

Transition Elements. The elements in the center of the periodic table, whose groups are labeled B, are the *transition metals*. Transition elements have an inner principal energy level which is only partially filled. In these elements it is the d subshell that is incompletely filled. In the fourth period, either the atoms or the ions of the elements Sc through Cu have incompletely filled $3d$ subshells. In the fifth period, the atoms or the ions of the elements Y to Cd have incompletely filled $4d$ subshells. Because their d subshells are completed, the elements Zn, Cd, and Hg (Group 2B) are often not considered true transition elements.

Below the periodic table, there are two sets of 14 elements each depicted as horizontal rows. These groups are generally known as *inner transition* elements but are also referred to as lanthanides (atomic numbers 58–71) and actinides (atomic numbers 90–103). For these elements an inner f subshell is incompletely filled. Figure 8.1 shows the positions in the periodic table of the groups of elements discussed above.

The Symbols A and B. Students beginning the study of chemistry in the mid-1990s may find a contradiction between periodic tables in textbooks and those on the walls of their classrooms. The letters A and B are used differently on the European and U.S. versions of the table. Traditionally in the United States, groups are labeled with an A to signify that they are representative elements, and those marked B are the transition element groups. Elements in Group 6A, for example, have 6 valence electrons. In the new numbering system for groups the groups are just numbered from 1 to 18 across the table. Group 6A is the same as 16. The number 16 has no physical significance. We will continue to use the letters A and B as they have been used in the United States.

Electron Configurations. The chemical properties of an atom are related to the configuration of the atom's valence electrons. The periodic table can be used to determine the electron configuration of any element. For representative elements the group number gives the number of electrons in the outermost principal energy level (valence electrons), as shown in Table 8.3.

Table 8.3 Configuration of Outermost Principal Energy Level

Group	1A	2A	3A	4A	5A	6A	7A	8A
Configuration	ns^1	ns^2	ns^2np^1	ns^2np^2	ns^2np^3	ns^2np^4	ns^2np^5	ns^2np^6

To determine the number of electrons in the d subshell of a transition element, we must count over from the first transition metal group, 3B, to the element of interest. Once the outermost electron configurations are known, work backward through the main groups until reaching a noble gas core (see Example 8.4).

Cations and Anions. The electron configurations of ions can also be readily described. For instance, a K^+ ion has 18 electrons, and its configuration is compared below to that of a K atom, which has 19 electrons:

$$K \quad 1s^2 2s^2 2p^6 3s^2 3p^6 4s^1 \quad \text{or} \quad [Ar]\,4s^1$$

$$K^+ \quad 1s^2 2s^2 2p^6 3s^2 3p^6 \quad \text{or} \quad [Ar]$$

Recall that the abbreviation [Ar] represents the configuration of the first 18 electrons, which corresponds to the configuration for an Ar atom. Potassium, used here as an example, is characteristic of all representative elements in that the stable ions of these elements have electron configurations that are isoelectronic with a noble gas configuration, ns^2np^6. The K^+ ion is isoelectronic with Ar. In fact, K^+, Ca^{2+}, Cl^-, S^{2-}, and Ar are all isoelectronic. The term *isoelectronic* refers to ions and atoms that have the *same number* of electrons and the same electron configurations. See Example 8.5, parts c and d, for the electronic configurations of transition metal ions.

When forming a cation from an atom of a transition metal, recall that *most* of the transition metal atoms do not acquire a noble gas electron configuration for their cations. Atoms of transition metal elements generally have two electrons in the s orbital of the highest principal energy level n and a partially filled d subshell of the $n-1$ principal energy level (see Table 7.3 text). The electrons most easily lost are those in the outermost principal energy level, the ns. Indeed many of the transition metals form $+2$ ions. Loss of additional electrons from the $(n-1)d$ subshell yields ions with charges greater than a $+2$ charge. See Example 8.5, parts c and d.

Some of the heavy, representative metals exhibit an *inert pair effect*. For example, Sn ions can be Sn^{2+} and Sn^{4+}. When a tin atom ($[Kr]4d^{10}5s^25p^2$) loses all four valence electrons, it becomes a Sn^{4+} ion ($[Kr]4d^{10}$). The tendency for tin to form a Sn^{2+} ($[Kr]4d^{10}5s^2$) ion is an example of the inert pair effect. The pair of electrons in the $5s$ orbital are somewhat unreactive. Tin and other heavy metals in Groups 3A, 4A, and 5A display this tendency to lose only electrons from p orbitals.

EXAMPLE 8.3 Electron Configurations and the Periodic Table

Refer only to the periodic table:
a. Write the electron configuration for the outermost principal energy level for any Group 5A element.
b. Identify the group whose outermost electrons have the configuration ns^2np^1.

METHOD OF SOLUTION

a. The elements in Group 5A are representative elements and so have all inner subshells filled, and the outermost electrons occupy s and p orbitals. The numeral 5 corresponds to five outermost electrons. The first two occupy the ns subshell, and the next three the np subshell. *Answer:* The electron configuration of any Group 5A element is ns^2np^3.
b. Among the representative elements, those in the first group to the right of the transition metals, Group 3A, have the electron configuration ns^2np^1.

EXAMPLE 8.4 Electron Configurations Using the Periodic Table

Write the electron configuration for
a. S
b. Ni

METHOD OF SOLUTION

a. First locate the element sulfur in Group 6A. It is a representative element with 16 electrons. Being in Group 6A it has six valence electrons with a configuration ns^2np^4. Sulfur is in the third period; therefore $n = 3$. The outermost electron configuration is $3s^23p^4$. The inner 10 electrons occupy the same orbitals as Ne. *Answer:* The abbreviated electron configuration of sulfur is

$$S\ [Ne]3s^23p^4$$

The unabbreviated configuration is S $1s^22s^22p^63s^23p^4$.

b. Nickel is a transition element and appears in period 4. Thus the last electrons are entering an inner $(n - 1)d$ subshell, the $3d$. Nickel is in the 8th column of the d block. This means there are 8 electrons in the $3d$ subshell, $3d^8$. Since Ni is of higher atomic number than K and Ca, the $4s$ orbital is filled, $4s^2$. For its inner electrons, Ni has an argon core of 18 electrons.
Answer: Ni $[Ar]4s^23d^8$.

EXAMPLE 8.5 Electron Configurations of Ions

Write the electron configurations of the following ions:
a. Mg^{2+}
b. S^{2-}
c. Fe^{2+}
d. Fe^{3+}

METHOD OF SOLUTION

a. A Mg atom has $Z = 12$, so the atom has 12 electrons, but the Mg^{2+} ion contains only 10 electrons. Write the configuration for Mg:

$$Mg\ 1s^22s^22p^63s^2 \quad \text{or} \quad [Ne]\ 3s^2$$

Then remove the two electrons in the outermost energy level to form the 2 + ion:

$$Mg^{2+}\ 1s^22s^22p^6 \quad \text{or} \quad [Ne]$$

The Mg^{2+} ions are *isoelectronic* with a Ne atom. That is, they have the same number of electrons.

b. A S atom has 16 electrons and has the electron configuration $1s^22s^22p^63s^23p^4$. The sulfur atom becomes the negative S^{2-} ion by gaining two electrons. Adding two electrons to the outermost subshell gives

$$S^{2-}\ 1s^22s^22p^63s^23p^6 \quad \text{or} \quad [Ar]$$

c. As before, write the configuration of the atom:

$$Fe\ [Ar]4s^23d^6$$

When forming a cation from an atom of a transition metal, electrons are always removed, first from the ns orbital and then from the $(n - 1)d$ orbital. The first two electrons lost are removed from the outermost principal energy level (highest n value), which is the $4s$ orbital. This gives

Fe^{2+} $[Ar]3d^6$

d. Next, remove one more electron from Fe^{2+}. This gives

Fe^{3+} $[Ar]3d^5$

COMMENT

The stable ions of all but a few representative elements are isoelectronic to a noble gas. Keep in mind that most transition metals can form more than one cation and that, for the most part, these ions are not isoelectronic with the preceding noble gases.

PERIODIC TRENDS EXPLAINED

STUDY OBJECTIVES

You should be able to:
1. Describe the terms *ionization energy* and *electron affinity*.
2. Account for the trend in the magnitudes of the atomic radius, ionic radius, ionization energy, and electron affinity for the elements across a period or within a group of the periodic table.
3. Predict the trend in atomic and ionic radii within an isoelectronic series.

Atomic Size. Figure 8.6 of the textbook shows that the atomic radius decreases, in general, as the atomic number increases across any period of the periodic table. Thus, the alkali metals have the largest atoms and the noble gases the smallest. Furthermore, within a group the atomic radius increases as the atomic number increases, as shown in Figure 8.2.

Another trend in chemical behavior is the *diagonal relationship*. This refers to similarities in some properties between a second-row element and the third-row element that is one space to the right in the table. Two such elements are on a diagonal (Figure 8.16 in the text). For example, Li and Mg are on a diagonal, and they have similar atomic radii. If you start at Li and go first to Be, the atomic radius decreases. Next, on moving down to Mg, the radius will increase again. If the decrease and increase are about the same, then Li and Mg will have approximately the same radius (155 pm for Li versus 160 pm for Mg). This relationship does not hold for all properties of two diagonally related elements.

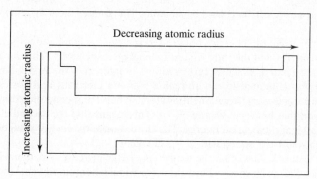

Figure 8.2. Periodic trends in the atomic radius.

Ionic Radii. Because atoms and ions of the same element have different numbers of electrons, we should expect atomic radii and ionic radii to have different values. *The radii of cations are smaller than those of the corresponding neutral atoms*. This should be expected because positive ions are formed by removing one or more electrons from the outermost shell. Since these electrons are farthest from the nucleus, their absence will make the cation significantly smaller. In addition, loss of an electron causes a decrease in the amount of electron-electron repulsion, which also causes the cation to be smaller than the neutral atom.

In contrast, *the radii of anions are larger than those of the corresponding neutral atoms*. When an electron is added to an atom to form an anion, there is an increase in the electron-electron repulsion. This causes electrons to spread out as much as possible, and so anions have a larger radius than the corresponding atoms. The radii of several metal and nonmetal ions are compared with their atomic radii in Table 8.4.

Table 8.4 Atomic and Ionic Radii of Elements in Groups 1A and 7A

Element	Group 1A Radius (pm)		Element	Group 7A Radius (pm)	
	Atomic	Ionic		Atomic	Ionic
Li	123	60	F	64	136
Na	154	95	Cl	99	181
K	203	133	Br	111	195
Rb	216	148	I	128	216
Ca	174	99	O	73	140

We can also compare radii within an isoelectronic series. In the series in Table 8.5 *the radii decrease steadily as atomic number increases*. Each species has 18 electrons arranged in an argon configuration, and so the electron-electron repulsion is about the same in each ion. The reason for the decrease in radius is that the nuclear charge increases steadily within this series. This causes the electrons to be attracted more strongly toward the nucleus and the radius to contract as the nuclear charge increases.

Table 8.5 Atomic and Ionic Radii in an Isoelectronic Series

Species	S^{2-}	Cl^-	Ar	K^+	Ca^{2+}
Radius (pm)	219	181	154	133	99

Ionization Energy. The minimum energy required to remove an electron from the ground state of an atom in the gas phase is its *ionization energy*. The magnitude of the ionization energy is a measure of how strongly the outermost electron is held by an atom. The greater the ionization energy, the stronger the electron is held.

For a many-electron atom, the energies required to remove a second and a third electron are called the second ionization energy and the third ionization energy, respectively. The third ionization energy is always greater than the second, which in turn is always greater than the first ionization energy. The explanation of this trend is related to the attraction between electrons and positive ions. The first electron removed must come away from a neutral atom. When the second electron is removed, it must move away from an ion that has a net charge of +1. This means that the second electron is held more strongly, leading to a higher ionization energy. The third ionization energy is higher yet because the electron must be removed from an ion with a +3 charge.

Table 8.11 in the textbook shows that the ionization energy increases when moving from left to right across a period. Thus, the alkali metals have the lowest ionization energy and the noble gases the highest. Within a group the ionization energy decreases as atomic number increases (Figure 8.3).

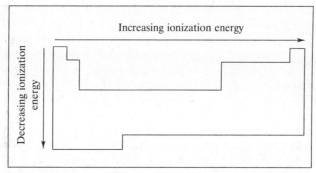

Figure 8.3. Periodic trends in ionization energy.

Electron Affinity. Another property of atoms is one that relates to their ease in forming negative ions. The electron affinity is the energy change that occurs when an electron is added to a neutral gaseous atom in its ground state:

$$F(g) + e^- \rightarrow F^-(g) \quad \text{electron affinity} = -333 \text{ kJ/mol}$$

The greater the attraction of an atom for an electron, the more negative the electron affinity value. This is an exothermic reaction, so energy is released when a stable negative ion is formed. The more energy is released, the more stable is the ion compared to the atom.

Table 8.5 of the text shows the electron affinity values of elements arranged according to their positions in the periodic table. Figure 8.14 (textbook) is a plot of electron affinity values for elements in period 2. A clear trend in electron affinity values of elements within a period is not evident. However, the electron affinities of the Group 6A and 7A elements are much more negative than those of other elements. Within groups of the periodic table the values usually do not follow a clear trend. In Groups 2A and 5A, electron affinity becomes more negative as atomic number increases. Looking down Groups 6A and 7A, the electron affinity values go through a minimum at S and Cl, respectively, and then increase again. One general observation we can make is that the electon affinities of nonmetal elements, except for the noble gases, are more negative than those of the metallic elements. Thus, nonmetal atoms tend to attract an additional electron much more strongly than metal atoms do. Figure 8.4 shows the electron affinity values of the first 20 elements.

Effective Nuclear Charge. The explanation of the trends in atomic size and ionization energy is based on the concept of the *effective nuclear charge* (Z_{eff}). The outermost electrons in an atom do not feel the full positive charge of the nucleus. Electrons in inner levels, lying between the nucleus and the outermost electrons, tend to shield the outermost electrons from the nuclear charge. In a sodium atom, for example, the inner 10 electrons ($1s^2 2s^2 2p^6$) shield the outer $3s$ electron from the positive charge of 11 protons. The Z_{eff} experienced by the $3s$ electron is only about $+1$. The Z_{eff} is equal to the nuclear charge Z minus the shielding constant σ. The shielding constant is essentially equal to the number of inner shell electrons. Thus, $Z_{eff} = Z - \sigma$.

Atomic Radii. The Z_{eff} affects the atomic radius in the following way. In proceeding across a period, one proton at a time is added to the nucleus, and one electron is added to the outermost orbital. *Electrons within the same energy level do not effectively shield each other from the nucleus.* Thus, for magnesium the inner 10 electrons shield the electrons in the $3s$ orbitals from the $12+$ charge of the nucleus, resulting in an effective nuclear charge of $+2$. One of the $3s$ electrons does not effectively shield the other $3s$ electron because they are in the same orbital. Continuing from left to right across the third period, all the outermost electrons are in the $n = 3$ energy level, and the Z_{eff} increases by about one unit per element:

$$\xrightarrow{\text{increasing } Z_{eff}}$$

Na(+1) Mg(+2) Al(+3) Si(+4) P(+5) S(+6) Cl(+7) Ar(+8)

154 / *Periodic Relationships among the Elements*

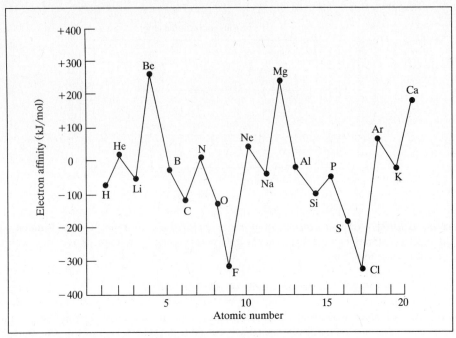

Figure 8.4. Electron affinities of the first 20 elements. The more negative the electron affinity, the greater the attraction an atom has for an extra electron.

This means a greater force of attraction is experienced by the outermost electrons of atoms of elements on the right side of the periodic table than by those on the left side. Consequently, the atoms gradually decrease in size as we proceed from left to right across a period.

In proceeding from one element to another down a group, each successive element has its outer electrons in a principal energy level with a larger n value. The effective nuclear charge experienced by the outermost electrons is essentially the same for all elements within a group. The result of this is that the size of the outer orbital is affected most by the value of n. As you will recall, the size of an orbital increases as n increases. Therefore, atomic radius increases from top to bottom in a group.

Ionization Energy Trends. The importance of the ionization energy is that it correlates closely with electron configuration and chemical properties. When considering ionization energy, it helps to recall that the first electron removed is the one farthest from the nucleus. Within a group of elements the ionization energy decreases with increasing atomic number. This trend is explained in the following way. When looking down a group of elements the effective nuclear charge is essentially constant. However, what does change is the energy level of the outermost electron. The outermost electron resides in increasingly higher energy levels. Thus, as atomic number increases, the outermost electron is held more weakly and the ionization energy decreases.

The trend within a period should also be related to the effective nuclear charge. In moving across the periodic table from left to right, the size of the atoms decreases due to the increase in effective nuclear charge. The outer electrons become more tightly held as we move from left to right, and the ionization energy must increase.

One characteristic of metals is the relative ease with which electrons can be removed from their atoms. Thus, the lower the ionization energy, the more metallic the element. The metals are located on the left-hand side of the periodic table. Also, elements in all groups exhibit increased metallic behavior with increased atomic number.

Electron Affinity Trend. The trend in electron affinities is also related to Z. The attraction of a nucleus for an additional electron depends on how far the available orbital is from the nucleus. The small atoms then have the greater electron affinities (larger negative values). In general, electron affinities become more negative as we read from left to right across the periodic table. The halogens

Table 8.6 Atomic Radius (pm)

Sc	Ti	V	Cr	Mn	Fe	Co	Ni	Cu	Zn
144	132	122	119	118	117	116	115	118	121

have the most negative electron affinities. The noble gases, with filled s and p subshells, have no tendency to add an electron because the next available orbital is an s orbital in the next highest energy level.

A characteristic of nonmetals is their tendency to acquire electrons. Thus, nonmetals typically have negative electron affinities. The nonmetals are located in the upper right-hand portion of the periodic table.

Transition Elements. Within a row of transition elements, in contrast to the representative elements, there is only a slight decrease in atomic radius when reading from left to right, as shown in Table 8.6.

This is so because electrons are being added to an inner d subshell. In period 4, for instance, the outermost electrons occupy a $4s$ orbital, but each successive electron is added to an inner $3d$ subshell. As we move across the period, the increasing nuclear charge is effectively shielded by the increase in the number of $3d$ electrons. Thus the outer $4s$ electrons within a series of the transition elements experience almost a constant Z_{eff}.

EXAMPLE 8.6   Trends in Atomic Radius

Which one of the following has the smallest atomic radius?
a. Li
b. Na
c. Be
d. Mg

METHOD OF SOLUTION

Atomic radii increase from top to bottom in a group; therefore Li atoms are smaller than Na atoms, and Be atoms are smaller than atoms of Mg. Next compare Be and Li. The atomic radius decreases from left to right within a period of the periodic table; thus Be atoms are smaller than Li atoms as well as the other choices given.

EXAMPLE 8.7

In each of the following pairs, choose the ion with the *largest* ionic radius:
a. K^+ or Na^+
b. K^+ or Ca^{2+}
c. K^+ or Cl^-

METHOD OF SOLUTION

a. K^+ and Na^+ are in the same group in the periodic table. The outer electrons in K^+ occupy the third principal energy level, and those in Na^+ occupy the second. *Answer:* K^+ has the greater ionic radius.
b. K^+ and Ca^{2+} belong to an isoelectronic series; both have 18 electrons. Calcium ions with their greater nuclear charge attract their electrons more strongly than K^+ ions and so are smaller. *Answer:* K^+ is larger.
c. Again K^+ and Cl^- are isoelectronic species. The ion with the greater nuclear charge (K^+) will be smaller. *Answer:* Cl^- has the greater ionic radius.

EXAMPLE 8.8 Trends in Ionization Energy

Which one of the following has the highest ionization energy?
a. K
b. Br
c. Cl
d. S

METHOD OF SOLUTION

Ionization energy increases from left to right within a period. Thus, the value for Cl is greater than for S, and the value for Br is greater than for K. In comparing Cl and Br, Cl has the higher ionization energy value because ionization energy decreases from top to bottom within a group.

EXAMPLE 8.9 Ionization Energy

Why does sodium have a lower first ionization energy than lithium?

METHOD OF SOLUTION

When a lithium atom is ionized, the electron most easily removed comes from a $2s$ orbital, whereas in sodium the electron most easily removed comes from a $3s$ orbital. Because of screening effects, the effective nuclear charge Z_{eff} experienced by these electrons is $+1$ in both atoms. The main reason for a difference in ionization energy in these atoms is the larger distance of separation between the electron and the nucleus in the case of Na atoms. As the principal quantum number n increases, so does the average distance of the electron from the nucleus. Consequently, the electron becomes easier to remove as you read down a group. *Answer:* Na will have a lower ionization energy than lithium.

EXAMPLE 8.10 Ionization Energies

The first, second, and third ionization energies for calcium are

$$I_1 = 590 \text{ kJ/mol} \quad I_2 = 1145 \text{ kJ/mol} \quad I_3 = 4900 \text{ kJ/mol}$$

Explain why so much more energy is required to remove the third electron from Ca, as compared to removal of the first and second electrons. *Answer:* The trend $I_3 > I_2 > I_1$ is observed because I_3 is the energy needed to remove an electron from a $+2$ ion, I_2 is the energy necessary to remove an electron from a $+1$ ion, and I_1 is the energy needed to remove an electron from an atom. However, another factor must be involved to explain the very large difference between I_3 and I_2. The first two electrons (valence electrons) are removed from the $3s$ orbital. But the third electron must be removed from the inner $2p$ subshell. The $n = 2$ principal energy level lies much closer to the nucleus than the $n = 3$ energy level; therefore, its electrons are held much more strongly.

TYPES OF ELEMENTS

STUDY OBJECTIVES

You should be able to:
1. Describe typical properties of metals, nonmetals, and metalloids.
2. Identify the following chemical groups in the periodic table: alkali metals, alkaline earths, halogens, noble gases, and coinage metals.

Metals. All elements are classified in three broad groups according to their chemical and physical properties. These are the metals, nonmetals, and metalloids. Within this classification, certain groups of elements are known by common names, such as the alkali metals, alkaline earth metals, coinage metals, halogens, and noble gases. In the section below, the three broad groups and their characteristics are described first, followed by the groups with common names and their characteristics.

Most of the elements are metals. In general, they appear in the left and center of the periodic table. All metallic elements except mercury are solids at 25°C. They have a lustrous appearance, are good conductors of heat and electricity, can be hammered or rolled into sheets (a property referred to as malleability), and can be drawn into wires (a property referred to as ductility). Metals generally have high densities and high melting points.

Metals have low ionization energies, low electron affinities, and relatively large atomic radii. All metallic elements combine with nonmetals such as oxygen and chlorine to form salts. The most reactive metals are at the left of the periodic table. The transition metals are less reactive than the Group 1A and 2A metals. Elements within a chemical group are more metallic as atomic mass increases.

Nonmetals. Eighteen of the elements are nonmetals. These elements are on the right side of the periodic table. At 25°C, eleven are gases, one (bromine) is a liquid, and the rest are brittle solids. Typically, their densities and melting points are low.

The atoms of nonmetal elements have high ionization energies and high electron affinities. Nonmetals combine with metals to form ionic compounds and with other nonmetal elements to form molecular compounds. The nonmetals, except hydrogen, are located on the upper right-hand side of the periodic table. Recall that hydrogen is a gas and exists as diatomic molecules, as do a number of nonmetals.

Metalloids. Several elements have some properties that are characteristic of metals and some that are like those of nonmetals. These elements are called *metalloids*. Many periodic tables show a zig-zag line separating the metals from nonmetals. The elements that border this line on both sides are metalloids (except for Al, which is a metal). They include boron, silicon, germanium, arsenic, antimony, and tellurium. The metalloids have ionization energies and electron affinity values intermediate between metals and nonmetals.

Alkali Metals. Certain groups of elements are known by common names (Figure 8.5). Some of the characteristics of the alkali metals, alkaline earth metals, coinage metals, halogens, and noble gases are described below.

The elements in Group 1A, with the exception of hydrogen, are called alkali metals. The ns electron in the highest principal energy level is well shielded from the nucleus and is easily lost. Their low ionization energies make the alkali metals the most active family of metals. These elements are found in nature as +1 ions in chemical combination with nonmetal ions and polyatomic ions. The densities of these elements are low, in part because of their large radii. Li, Na, and K are even less dense than water.

Alkaline Earths. Group 2A elements are called *alkaline earth* metals. The outermost electron configuration is ns^2. Although these two are not as easily lost as the ns^1 electron in alkali metal atoms, they are readily given up. Alkaline earth metals exist in nature as +2 ions in chemical combination with negative ions.

Halogens. Group 7A elements are called the *halogens*. None of these elements is ever found free in nature. All have high electron affinities and so tend to acquire one electron to form −1 ions. Ionic compounds containing these anions are salts. Indeed, the name halogen means "salt former." These nonmetals exist in the elemental form as diatomic molecules. At 25°C, F_2 and Cl_2 are gases, Br_2 is a liquid, and I_2 is a solid.

Noble Gases. Group 8A elements make up the *noble gas* elements. The term *noble* here means nonreactive. He, Ne, and Ar do not form any chemical compounds. Since 1960, a number of Kr and Xe compounds have been synthesized. The chemical inactivity of the noble gases is the result of the ns^2np^6

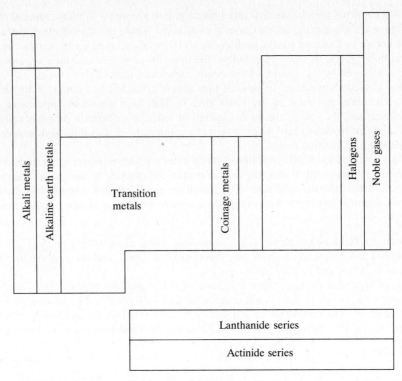

Figure 8.5. Common names of several groups of elements.

electron configuration of the outermost electrons in their atoms. The noble gas elements have zero electron affinity and high ionization energies.

Coinage Metals. The elements of Group 1B—copper, silver, and gold—are generally nonreactive and are called the *noble metals* or *coinage metals*. These metals are excellent conductors of heat and electricity.

Atoms of the coinage metals have 1 electron in the outer s subshell and 10 electrons in the underlying d subshell. The electron configurations of their outermost electrons are $ns^1(n-1)d^{10}$. Note that this is an exception to the configuration expected from the Aufbau principle, which predicts $ns^2(n-1)d^9$. It is as if the $(n-1)d$ orbitals "borrowed" an electron from the higher energy ns orbital. For the coinage metal atoms the "borrowed" electron is used to complete an inner subshell. Apparently the completed d subshell has an enhanced stability that corresponds to a lowering of the energy of the atom. Thus $ns^1(n-1)d^{10}$ is a lower energy configuration than $ns^2(n-1)d^9$.

In copper, silver, and gold this one outer electron is held much more tightly than the outermost electron in the alkali metal atoms. This can be most easily seen by comparing ionization energies, as shown in Table 8.7.

Table 8.7 Comparison of Ionization Energies of Coinage Metals and Alkali Metals

Period	Element	I (kJ/mol)
4	K	419
	Cu	745
5	Rb	403
	Ag	731
6	Cs	375
	Au	890

EXAMPLE 8.11 Properties of Group 4A Elements

Write the electron configurations for the atoms of all elements in Group 4A. Specify whether the element is a metal, a nonmetal, or a metalloid.

METHOD OF SOLUTION

The outermost principal energy level of atoms in Group 4A contains four electrons. The general electron configuration for elements in this group is ns^2np^2. Their electron configurations are

C $1s^22s^22p^2$	nonmetal
Si [Ne]$3s^23p^2$	metalloid
Ge [Ar]$3d^{10}4s^24p^2$	metalloid
Sn [Kr]$4d^{10}5s^25p^2$	metal
Pb [Xe]$4f^{14}5d^{10}6s^26p^2$	metal

EXAMPLE 8.12 Properties of Potassium and Chlorine

Compare the magnitudes of electron affinity and ionization energy for the alkali metal potassium with the halogen chlorine. Comment on their relative abilities to form ions.

METHOD OF SOLUTION

From Figure 8.13 (text) the electron affinities of K and Cl are -48 and -348 kJ/mol, respectively. Figure 8.11 (text) gives the ionization energies of K and Cl as 419 and 1251 kJ/mol, respectively. These values suggest that the metal atom K can lose an electron and form a K^+ ion much more easily than can the nonmetal Cl. And the nonmetal atom Cl can attract an electron to form a Cl^- much more readily than can the metallic K atom.

TRUE-FALSE QUESTIONS

1. There was no periodic table before electron configurations of the elements were known.

2. Mendeleev had no explanation for the gaps in his periodic table.

3. The chemical and physical properties of elements vary in a periodic manner when the elements are arranged according to increasing atomic mass.

4. The representative elements include all the metals and nonmetals.

5. The electron configuration of the outermost electrons of the noble gas elements is ns^2np^6.

6. Carbon is a representative element.

7. Gallium is a nonmetal.

8. Electron affinity is the energy change when an element such as $Br_2(l)$ acquires one electron to become an ion:
   $$\tfrac{1}{2}Br_2(l) + e^- \rightarrow Br^-(l)$$

9. As a rule, the metallic elements have higher ionization energies than the nonmetallic elements.

10. Electrons in the outermost shell of a many-electron atom do not feel the full attraction of the positively charged nucleus.

160 / *Periodic Relationships among the Elements*

11. The atomic radius increases with increasing atomic mass within a group, but the ionization energy decreases.

12. The effective nuclear charge increases in a group as atomic mass increases.

13. The atomic radii of representative elements vary more than those of transition elements in a period of the periodic table.

SELF-TEST A

1. The element francium is extremely rare, and very little is known about its chemical and physical properties. Use the following data to estimate its density and ionization energy:

	K	Rb	Cs
Density (g/cm^3)	0.86	1.53	1.87
Ionization energy (kJ/mol)	419	403	375

2. The element technetium ($Z = 43$) does not occur on earth. The densities of Mo and Ru are 10.2 and 12.4 g/cm^3, respectively. Estimate the density of Tc.

3. The periodic table has been extended to include the transuranium elements ($Z > 92$). What elements would element number 104 be similar to in chemical properties?

4. Which is the general electron configuration for the outermost electrons of elements in Group 4A?
   a. ns^1 b. ns^2 c. ns^2np^4 d. ns^2np^2 e. $ns^2np^6nd^7$

5. In what group of the periodic table is each of the following elements found?
   a. $1s^22s^22p^6$
   b. $[Ar]4s^1$
   c. $[Xe]6s^24f^{14}5d^5$
   d. $[Ne]3s^23p^5$

6. Use the periodic table to write the electron configurations of Cr, Sb, and Pb.

7. How many valence electrons does an arsenic atom have?

8. Successive ionization energies—first, second, third, etc.—always show an increasing trend: $I_1 < I_2 < I_3 < I_n$. For aluminum atoms, which ionization energy value will show an exceptionally large increase over the preceding ionization energy value?
   a. second   b. third   c. fourth   d. fifth   e. sixth

9. Write the electron configurations for the following ions:
   a. Ca^{2+} b. Se^{2-} c. Cl^- d. Mn^{2+} e. Co^{3+} f. Sc^{3+}

10. An Ar atom is isoelectronic with which one of the following?
    a. Ne   b. K   c. Sc^{3+} d. Cl^{2-} e. Na^+

11. Which atom should have the largest radius?
    a. Br   b. Cl   c. Se   d. Ge   e. C

12. Which is the larger ion or atom in each pair?
    a. I^- or Cs^+ b. Ne or K^+ c. Mg or Mg^{2+}

13. Which atom should have the greatest ionization energy?
    a. Se   b. Te   c. Na   d. Si   e. S

14. Why does atomic radius decrease in going from left to right across a row in the periodic table?

15. Which two from the following would be most likely to have similar ionization energies? B, C, Si, Al, Ar.

SELF-TEST B1

1. Without referring to the periodic table, write the electron configuration of the element with atomic number 21.

2. How many valence electrons does an atom of phosphorus have?

3. Write the outer electron configuration for the halogen elements.

4. Based on periodic trends, which one of the following elements has the greatest ionization energy? Cl, K, S, Se, Br.

5. Which of the following has the largest radius? Na^+, Mg^{2+}, Al^{3+}, S^{2-}, Ar.

6. Identify an isoelectronic pair among the following: Na^+, Ar, K^+, Ne, Se^{2-}.

7. Which of the following atoms has both a large ionization energy and a large negative electron affinity? K, Ne, Br, Fe, N.

8. Of the following, which is the most metallic element? V, Ge, Se, As, Zn.

SELF-TEST B2

1. Write the electron configuration of Sc by first finding its position in the periodic table.

2. Write the symbol of an element with five valence electrons.

3. What group of elements has the outer electron configuration ns^2np^5?

4. Based on periodic trends, which one of the following elements has the smallest ionization energy? Cl, K, S, Se, Br.

5. Which of the following has the largest radius? S^{2-}, Ar, Se^{2-}, O^{2-}, Al^{3+}.

6. How many electrons does each of the following species have? Na^+, Ar, K^+, Ne, Se^{2-}.

7. Which one of the following elements has both a low ionization energy and a small negative electron affinity? K, Ne, Br, Fe, N.

8. Which of the following is the least metallic element? V, Ge, Al, As, Ca.

ANSWERS

TRUE-FALSE QUESTIONS

1. False. The elements were first grouped according to their properties.
2. False. His famous gaps were predicted to be missing elements.
3. False. According to increasing atomic number.
4. False. The transition and inner transition elements are metals but are not representative elements.
5. True.
6. True.
7. False. Metal.
8. False. It is an atom, Br(g), in the gas phase that acquires the electron, not a molecule, and not a liquid.
9. False. Metallic elements have low ionization energies.

162 / *Periodic Relationships among the Elements*

10. True.
11. True.
12. False. No, it stays about the same.
13. True.

SELF-TEST A

1. $I \cong 350$ kJ; density $\cong 2.0$ g/cm^3
2. Density $\cong 11.3$ g/cm^3
3. Ti, Zr, and Hf
4. d.
5. a. Group 8A, noble gases c. A transition metal
   b. Group 1A, alkali metals d. Group 7A, halogens
6. Cr [Ar]$4s^2 3d^4$; the observed configuration is Cr [Ar]$4s^1 3d^5$, which is an exception to the rules we have developed. Sb [Kr]$5s^2 4d^{10} 5p^3$ Pb [Xe]$6s^2 4f^{14} 5d^{10} 6p^2$
7. 5
8. c
9. a. Ca^{2+} [Ar] d. Mn^{2+} [Ar]$3d^5$
   b. Se^{2-} [Kr] e. Co^{3+} [Ar]$3d^6$
   c. Cl$^-$ [Ar] f. Sc^{3+} [Ar]
10. c.
11. d.
12. a. I$^-$ b. K$^+$ c. Mg
13. e.
14. The effective nuclear charge increases from left to right because the nuclear charge increases by one proton at a time, and as electrons are added to the atom, they go into the same principal energy level, where they do not shield each other effectively. As Z_{eff} increases, the electrons are pulled closer to the nucleus and the atomic radius decreases correspondingly.
15. B ($I = 801$ kJ) and Si ($I = 786$ kJ) diagonal relationship

SELF-TEST B1

1. $1s^2 2s^2 2p^6 3s^2 3p^6 4s^2 3d^1$
2. 5
3. $ns^2 np^5$
4. Cl
5. S^{2-}
6. Na$^+$ and Ne
7. Br
8. V

SELF-TEST B2

1. Sc [Ar]$4s^2 3d^1$
2. P
3. Halogens (Group 7A)
4. K
5. Se^{2-}
6. 10
7. K
8. As

Chapter Nine
CHEMICAL BONDING I: BASIC CONCEPTS

- Lewis Symbols and the Octet Rule
- Ionic Bonding and the Lattice Energy
- Covalent Bonding and Lewis Structures
- Electronegativity
- Formal Charge
- Exceptions to the Octet Rule
- Bond Energies

LEWIS SYMBOLS AND THE OCTET RULE

STUDY OBJECTIVES

You should be able to:
1. Draw Lewis symbols for atoms and ions.
2. State the octet rule and describe its origin.

Lewis Symbols. Ionic and covalent bonds are two important types of chemical bonds that will be discussed later in this chapter. First we will consider the electron configurations discussed in the previous chapters as a starting point to discuss chemical bonding in general. A Lewis dot symbol of an element consists of the chemical symbol with one or more dots placed around it. Each dot corresponds to a valence electron—valence electrons are those in the outermost principal energy level. The orbital diagram and the Lewis symbol for the fluorine atom are shown in Figure 9.1. Fluorine has seven electrons in its outermost principal energy level ($n = 2$). Five are in the $2p$ subshell, and two are in the $2s$.

The Lewis dot symbols and the electron configurations of the outermost electrons of elements in the third period are as follows:

Na·	·Mg·	·Ȧl·	·Ṡi·	·P̈·	:S̈·	:C̈l·	:Ä̇r:
$3s^1$	$3s^2$	$3s^23p^1$	$3s^23p^2$	$3s^23p^3$	$3s^23p^4$	$3s^23p^5$	$3s^23p^6$

Note that for representative elements the number of valence electrons is the same as the group number.

Octet Rule. The Lewis symbols of the noble gas group show eight electrons corresponding to filled s and p subshells. In our study of the periodic table we saw that the outer electron configuration is related to the chemical and physical properties of an element. G. N. Lewis reasoned that when atoms enter into

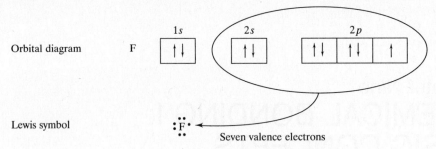

Figure 9.1. Orbital diagram and Lewis symbol for the fluorine atom.

chemical combination, they become more stable. He proposed that atoms gain or lose electrons until they have the same number of valence electrons as noble gas atoms, that is, eight. The *octet rule* states that an atom tends to gain, lose, or share electrons until it has eight electrons in the valence shell.

This rule proves to be very useful because it allows the prediction of the correct charge for all the monoatomic ions formed by the representative elements. The representative metals form positive ions by losing one or more electrons until a noble gas core is achieved:

$$Na[Ne]3s^1 \rightarrow Na^+[Ne] + e^-$$

$$Mg[Ne]3s^2 \rightarrow Mg^{2+}[Ne] + 2e^-$$

$$Al[Ne]3s^23p^1 \rightarrow Al^{3+}[Ne] + 3e^-$$

The electron configurations of the above ions (Na^+, Mg^{2+}, and Al^{3+}) are all the same, namely, $1s^22s^22p^6$. All three have eight electrons in the outer shell ($n = 2$), and all satisfy the octet rule. Species that have identical electron configurations are said to be *isoelectronic*. All three ions are isoelectronic with the noble gas neon.

The nonmetal atoms form negative ions by acquiring electrons until a noble gas core is achieved:

$$S[Ne]3s^23p^4 + 2e^- \rightarrow S^{2-}[Ar]$$

$$Cl[Ne]3s^23p^5 + e^- \rightarrow Cl^-[Ar]$$

Sulfide and chloride ions then have net charges of -2 and -1 because the atoms needed $2e^-$ and $1e^-$, respectively, to complete an octet. Both of these ions are isoelectronic with the noble gas argon. Their dot symbols are

$:\!\ddot{\underset{..}{S}}\!:^{2-}$ and $:\!\ddot{\underset{..}{Cl}}\!:^{-}$

EXAMPLE 9.1 Lewis Dot Symbols

Write Lewis symbols for the following elements:
a. Ca
b. O, S, and Se

METHOD OF SOLUTION

The Lewis symbol of an element consists of the element symbol surrounded by between one and eight dots, representing valence electrons.
a. Ca is in Group 2A. It has two electrons in the outermost energy level. These are its valence electrons.
   Answer: The Lewis symbol is

   $\cdot Ca \cdot$

b. O, S, and Se are all in the same group, 6A. *Answer:* Each has six valence electrons:

$$:\ddot{\underset{..}{O}}\cdot \qquad :\ddot{\underset{..}{S}}\cdot \qquad :\ddot{\underset{..}{Se}}\cdot$$

EXAMPLE 9.2 Lewis Dot Symbols for Ions

Write Lewis symbols for the following ions:
a. Ca^{2+}
b. Se^{2-}

METHOD OF SOLUTION

a. Removing two electrons (dots) from $\cdot Ca \cdot$ gives simply Ca^{2+} as the Lewis symbol of a calcium ion. No electrons are shown around the Ca^{2+} symbol because this ion has lost its valence-shell electrons.
b. Adding two electrons to $:\ddot{Se}\cdot$ gives $:\ddot{Se}:^{2-}$ as the Lewis symbol for a selenide ion.

IONIC BONDING AND THE LATTICE ENERGY

STUDY OBJECTIVES

You should be able to:
1. Describe how Coulomb's law is related to the ionic bond.
2. Predict the trend in atomic and ionic radii within an isoelectronic series.
3. List the energy terms that influence the tendency for two elements to form an ionic compound.
4. Use the Born-Haber cycle to calculate the magnitude of the lattice energies of ionic solids.

Coulomb's Law. The force that gives rise to the ionic bond is the coulombic attraction existing between a positive ion and a negative ion. This force of attraction is given by Coulomb's law:

$$F = k\frac{Q_1 Q_2}{r^2}$$

where Q_1 and Q_2 are the charges on the two ions, r is the distance between them, and k is the proportionality constant. Here we will not be using the value of k. This equation predicts that the force of attraction between two ions, and thus the strength of the ionic bond, increases as the net charges of the ions increase. The force of attraction also increases as the distance between ions decreases.

Chemical bonding results when the energy of two interacting atoms is lowered. The potential energy (E) of two charged particles is directly proportional to their charges and inversely proportional to their distance of separation (r):

$$E = k\frac{Q_1 Q_2}{r}$$

When one ion is positive and the other negative, E will be negative. Thus, bringing two oppositely charged particles closer together lowers their energy. The lower the value of the potential energy, the more stable the pair of ions.

The factors that govern the stability of ion pairs in a crystal are the *magnitude of their charges* and the *distance between ionic centers*. The distance between ionic centers depends on the radii of the individual ions, and is discussed in the following section.

Lattice Energy. In an ionic crystal the cations and anions are arranged in an orderly three-dimensional array, as illustrated in Figure 2.10 of the textbook. In a sodium chloride or rock-salt type of crystal, the anion has six nearest-neighbor cations. The ions are packed in such a way as to maximize attraction and minimize repulsion.

The lattice energy provides a measure of the attraction between ions and the strength of the ionic bond. *Lattice energy* is the energy required to separate the ions in 1 mol of a solid ionic compound into gaseous ions. Gaseous ions are far enough apart from one another that they do not interact. For example, the lattice energy of NaF(s) is equal to 908 kJ/mol. This means 908 kJ is required to vaporize 1 mol NaF(s) and form 1 mol Na^+ ions and F^- ions in the gas phase:

$$NaF(s) \rightarrow Na^+(g) + F^-(g) \quad \Delta H° = 908 \text{ kJ}$$

The lattice energy is related to the stability of the ionic solid. The larger the lattice energy, the more stable the solid.

The value of the lattice energy depends on the charges of the ions and the ionic radii, in accordance with Coulomb's law. This can be seen in Table 9.1 by comparing the sodium halides. As the sum of the two ionic radii increases in going from NaF to NaI, the separation of ionic centers ($r = r_1 + r_2$) increases and the lattice energy decreases. The effect of ionic charge on lattice energy can be seen by comparing the lattice energies of NaF and CaO in Table 9.1. The larger value for CaO is due to the stronger attraction of the Ca^{2+} ion for the O^{2-} ion.

Table 9.1 Lattice Energies

	$r_1 + r_2$ (pm)	Lattice Energy (kJ/mol)	Melting Point (°C)
NaF	231	908	1012
NaCl	276	788	801
NaBr	290	736	747
NaI	314	686	660
CaO	239	3540	2580

The lattice energies are reflected in the melting points of ionic crystals. During melting, the ions gain enough kinetic energy to overcome the potential energy of attraction, and they move away from each other. The higher the melting point, the more energy the ions need to separate from one another. Therefore, as the lattice energy increases, so does the melting point.

Factors Favoring Formation of Ionic Bonds. It is important to identify any properties of atoms that affect their ability to form ionic compounds. In the formation of ions from atoms, the metal atom loses an electron and the nonmetal atom gains an electron:

$$Li(g) \rightarrow Li^+(g) + e^-$$

$$e^- + F(g) \rightarrow F^-(g)$$

The overall change is

$$Li(g) + F(g) \rightarrow Li^+(g) + F^-(g)$$

One factor favoring the formation of an ionic compound is the ease with which the metal atom loses an electron. A second factor is the tendency of the nonmetal atom to gain an electron. Thus ionic compounds tend to form between elements of low ionization energy and those of high electron affinity. The alkali metals and alkaline earth metals have low ionization energies. They tend to form ionic compounds with the halogens and Group 6A elements, both of which have high electron affinities.

A third factor is the lattice energy. Electrostatic attraction of the ions results in large amounts of energy being released when two kinds of gaseous ions are brought together to form a crystal lattice.

$$\text{Li}^+(g) + \text{F}^-(g) \rightarrow \text{LiF}(s)$$

In general, the smaller the ionic radius and the greater the ionic charge, the greater the lattice energy.

Calculation of the Lattice Energy Using the Born-Haber Cycle.
Lattice energies cannot be measured directly and must be calculated using Hess's law. In Chapter 6 you learned that if a reaction can be broken down into a series of steps, the overall enthalpy of reaction is equal to the sum of the enthalpy changes for the individual steps. The series of steps used to calculate the lattice energies of ionic solids is called the Born-Haber cycle. Each step is one you've seen previously in relation to the properties of atoms. Here we will illustrate the calculation of the lattice energy U for potassium chloride:

$$\text{KCl}(s) \rightarrow \text{K}^+(g) + \text{Cl}^-(g) \qquad U = ?$$

The Born-Haber cycle starts by taking the overall equation to be the reaction in which the ionic compound is formed from the elements in their standard states. In this example the enthalpy change is the same as the enthalpy of formation of potassium chloride from potassium and chlorine:

$$\text{K}(s) + \tfrac{1}{2}\text{Cl}_2(g) \rightarrow \text{KCl}(s) \qquad \Delta H^\circ_{\text{overall}} = \Delta H^\circ_f (\text{KCl})$$

This reaction is then envisioned to occur by a number of steps.

1. First potassium sublimes:

$$\text{K}(s) \rightarrow \text{K}(g) \qquad \Delta H^\circ_1 = \Delta H_{\text{sub}}$$

2. Next, diatomic chlorine is dissociated into Cl atoms. The required energy is the bond dissociation energy for $\tfrac{1}{2}$ mol of Cl — Cl bonds:

$$\tfrac{1}{2}\text{Cl}_2(g) \rightarrow \text{Cl}(g) \qquad \Delta H^\circ_2 = \tfrac{1}{2}\text{BE}(\text{Cl} - \text{Cl})$$

Up to now the focus has been on making the atoms of the elements potassium and chlorine. Next these will be made into the appropriate ions.

3. Now, 1 mol of potassium atoms is ionized:

$$\text{K}(g) \rightarrow \text{K}^+(g) + e^- \qquad \Delta H^\circ_3 = I$$

4. To form a Cl$^-$ ion, an electron is added to the chlorine atom. This will release an energy equal to the electron affinity of chlorine, EA:

$$\text{Cl}(g) + e^- \rightarrow \text{Cl}^-(g) \qquad \Delta H^\circ_4 = \text{EA}$$

5. Finally, 1 mol of gaseous K$^+$ and 1 mol of gaseous Cl$^-$ ions are combined to make 1 mol of KCl(s). An amount of energy equal to the lattice energy will be released. The lattice energy must have the same magnitude as ΔH°_5 but an opposite sign:

$$\text{K}^+(g) + \text{Cl}^-(g) \rightarrow \text{KCl}(s) \qquad \Delta H^\circ_5 = -U$$

168 / Chemical Bonding I: Basic Concepts

The summation of these changes gives the overall reaction above:

$$K(s) + \tfrac{1}{2}Cl_2(g) \rightarrow KCl(s) \qquad \Delta H^\circ_{overall} = \Delta H^\circ_f(KCl)$$

Therefore, in general,

$$\Delta H^\circ_{overall} = \Delta H^\circ_1 + \Delta H^\circ_2 + \Delta H^\circ_3 + \Delta H^\circ_4 + \Delta H^\circ_5$$

or

$$\Delta H^\circ_f(KCl) + \Delta H_{sub} + \tfrac{1}{2}BE + I(\text{metal element}) + EA(\text{nonmetal}) + U$$

The lattice energy U of ΔH°_5 can be calculated from

$$-U = \Delta H^\circ_5 = \Delta H^\circ_{overall} - [\Delta H^\circ_1 + \Delta H^\circ_2 + \Delta H^\circ_3 + \Delta H^\circ_4]$$

EXAMPLE 9.3 Predicting Melting Points

Which member of the pair will have the higher melting point?
a. NaCl or CaO
b. NaCl or NaI

METHOD OF SOLUTION

a. According to Coulomb's law, the doubly charged ions in CaO will attract each other more strongly than do the singly charged ions in NaCl. *Answer:* CaO will have a higher melting temperature.
b. According to Coulomb's law, the closer the centers of two ions can approach each other, the stronger the attraction between them. The sum of the ionic radii in NaCl is smaller than in NaI. *Answer:* NaCl will have the higher melting point.

EXAMPLE 9.4 The Born-Haber Cycle

Given the following data, calculate the lattice energy of potassium chloride:

Enthalpy of sublimation of potassium	90.0 kJ
Dissociation energy (BE) of Cl — Cl	242.7 kJ
Ionization energy (I) of K	419 kJ
Electron affinity (EA) of Cl	-348 kJ
ΔH°_f KCl(s)	-435.9 kJ

METHOD OF SOLUTION

From the discussion in the above section, we can calculate the lattice energy U (or ΔH°_5) from

$$-U = \Delta H^\circ_5 = \Delta H^\circ_{overall} - [\Delta H^\circ_1 + \Delta H^\circ_2 + \Delta H^\circ_3 + \Delta H^\circ_4]$$

Next, we identify the appropriate energy changes:

$$K(s) \rightarrow K(g) \qquad \Delta H^\circ_1 = \Delta H_{sub} = 90.0 \text{ kJ}$$

$$\tfrac{1}{2}Cl_2(g) \rightarrow Cl(g) \qquad \Delta H^\circ_2 = \tfrac{1}{2}BE = 121.4 \text{ kJ}$$

$$K(g) \rightarrow K^+(g) + e^- \qquad \Delta H^\circ_3 = I = 419 \text{ kJ}$$

$$Cl(g) + e^- \rightarrow Cl^-(g) \qquad \Delta H^\circ_4 = EA = -348 \text{ kJ}$$

$$K(s) + \tfrac{1}{2}Cl_2(g) \rightarrow KCl(s) \qquad \Delta H^\circ_{overall} = -435.9 \text{ kJ}$$

Then we substitute into the equation:

$$-U = \Delta H_5^\circ = \Delta H_{overall}^\circ - [\Delta H_1^\circ + \Delta H_2^\circ + \Delta H_3^\circ + \Delta HJ_4^\circ]$$

$$-U = \Delta H_5^\circ = -435.9 \text{ kJ} - [90.0 \text{ kJ} + 121.4 \text{ kJ} + 419 \text{ kJ} + (-348 \text{ kJ})]$$

$$-U = \Delta H_5^\circ = -435.9 \text{ kJ} - 282.4 \text{ kJ} = -718 \text{ kJ}$$

$$U = 718 \text{ kJ}$$

COVALENT BONDING AND LEWIS STRUCTURES

STUDY OBJECTIVES

You should be able to:
1. Describe covalent bonding in molecules by drawing their Lewis structures.
2. Draw resonance structures for molecules or polyatomic ions.

Lewis Structures. Molecules are held together by bonds resulting from the sharing of electrons between two atoms in such a way as to follow the octet rule. A simple *covalent bond* is formed when two atoms in a molecule share a pair of electrons.

The formation of a covalent bond in hydrogen chloride can be represented with Lewis structures as

$$\text{H}\cdot + \cdot\ddot{\underset{..}{\text{Cl}}}: \rightarrow \text{H}:\ddot{\underset{..}{\text{Cl}}}: \quad \text{or} \quad \text{H} - \ddot{\underset{..}{\text{Cl}}}:$$

where the dash represents a covalent bond, or a pair of electrons shared by both H and Cl. By sharing the electron pair, both the hydrogen atom and chlorine atom satisfy the octet rule. The stability of this bond results from both atoms acquiring a noble gas configuration. Notice that hydrogen is an exception to the octet rule. Rather than achieving an octet, it needs only two electrons to achieve a filled outer energy level. The H atom becomes isoelectronic with helium. The electron pairs on the Cl atom that are not involved in bonding are called *lone pairs*, *unshared pairs*, or *nonbonding electrons*. The bonding in methane and carbon tetrachloride is shown in Figure 9.2. The circles represent the valence shells of the atoms. They help to point out that each atom achieves an octet of valence electrons by sharing one or more pairs of electrons.

In some cases two or three pairs of electrons are shared by two atoms in order to reach an octet. In these molecules, *multiple bonds* exist. A *double bond* is a covalent bond in which two pairs of electrons

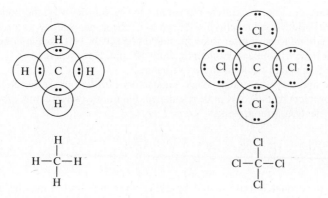

Figure 9.2. Sharing the electron pairs in CH_4 and CCl_4.

are shared between two atoms, as between C and O in formaldehyde:

H_2CO (formaldehyde)   H:C̈::O   or   H—C(H)=O

In general, atoms joined by a double bond lie closer together than atoms joined by a single bond. The C = O bond length is shorter than the C — H bond length.

Nitrogen molecules (N_2) contain a triple bond:

:N:::N: or   :N≡N:

Lewis structures represent the covalent bonding and location of unshared electron pairs within molecules and polyatomic ions. The steps for writing Lewis structures are as follows:

1. Arrange the atoms in a reasonable skeletal form, placing the unique atom in the center. Determine what atoms are bonded to each other.
2. Count the valence electrons. For polyatomic anions remember to add one electron for each unit of charge.
3. Connect the central atom to the others with single bonds. First add unshared pairs to the atoms bonded to the central atom to complete their octets (except for hydrogen, of course). Then, keeping in mind that the maximum number of electrons is the number counted in step 2, add the remaining unshared pairs to the central atom.
4. If the octet rule is satisfied for each atom and the total number of electrons is correct, stop here since the structure can be considered correct. If the octet rule is not met for the central atom, go on to step 5.
5. In some cases there is a shortage of valence electrons. To complete the octet of the central atom, write double or triple bonds between the central atom and the surrounding atoms. To make a double bond, move one of the unshared pairs from the surrounding atom to make the additional bond.
6. Repeat steps 4 and 5.

For application of this procedure see Examples 9.6 and 9.7.

The Concept of Resonance. It sometimes happens that a satisfactory electron dot formula for a molecule or polyatomic ion cannot be drawn. When such a situation arises, a special procedure is invoked to arrive at a Lewis structure. For the nitrite ion, for instance, the following structure shows the correct number of valence electrons and satisfies the octet rule. The brackets are used to indicate that the −1 charge belongs to the entire nitrite ion, not to just one atom in the structure:

$$[:\ddot{O} - \ddot{N} = \ddot{O}]^-$$

However, the structure does not accurately represent what is known about the bond lengths of the N — O bonds in NO_2^-. Both bond lengths are known to be the same, whereas according to the structure we expect the double bond to be shorter than the single bond. A N — O single-bond length should be about 136 pm, and a N = O double-bond length about 122 pm. However, the two bond lengths in NO_2^- are equal and are intermediate between these two values.

It turns out to be impossible to draw a single satisfactory Lewis structure for NO_2^-. Chemists have gotten around this problem by the concept of resonance. First draw two structures for NO_2^- that reflect different choices of electron arrangements:

$$[:\ddot{O} - \ddot{N} = \ddot{O}]^- \leftrightarrow [\ddot{O} = \ddot{N} - \ddot{O}:]^-$$

The nitrite ion is not adequately represented by either structure—but it may be described by a combination of these structures. This combination structure, which cannot be drawn, is called a *resonance hybrid* of the contributing structures. The symbol ↔ indicates that the structures shown are *contributing* or *resonance structures*.

In applying the concept of resonance, we assume that NO_2^- is a hybrid, or an average of the two structures. Thus, the N — O bonds are intermediate between single and double bonds. The term resonance was perhaps a poor choice as regards a hybrid, because the word implies to some that the real molecule flips from one structure to the other. *This is not what is implied here*. The point is that the properties of NO_2^- (and of other examples) cannot be accounted for by a single Lewis structure. Instead, the nitrite ion has properties as if it were a hybrid of two resonance structures.

EXAMPLE 9.5 Drawing a Lewis Structure

Draw the Lewis structure for hydrazine, N_2H_4. How many unshared electron pairs (lone pairs) are there on each N atom?

METHOD OF SOLUTION

1. Arrange the atoms in a reasonable skeletal form. H atoms form only one bond and so must be located on the outside of the atom.

    ```
    H   N   N   H
        H   H
    ```

2. Count the valence electrons. Each N atom has 5 valence electrons and each H atom has 1. There are $2(5) + 4(1) = 14$ valence electrons.
3. Connect the atoms with single bonds:

    ```
    H—N—N—H
      |   |
      H   H
    ```

    Normally we would add unshared pairs to complete all octets of surrounding atoms, but in this case the H atoms only need the two electrons shared in the bond to the N atom. Count the number of electrons used: 5 pairs = 10 valence electrons. Now add unshared pairs to complete the octets of the N atoms:

    ```
    H—N̈—N̈—H
      |   |
      H   H
    ```

4. Count the electrons: 7 pairs = 14 valence electrons. This is the same number as given in step 2.

COMMENT

Note that each N atom has one unshared electron pair.

EXAMPLE 9.6 Drawing a Lewis Structure with Resonance

Draw the Lewis structure for the nitrate ion, NO_3^-. All three N — O bonds are equivalent.

METHOD OF SOLUTION

1. Arrange the atoms:

    ```
    O   N   O
        O
    ```

2. Count the valence electrons:

   $3(6) + 5 + 1 = 24$ electrons

   (the ionic charge)

3. Connect the central atom to the others with single bonds:

   $$O-N-O$$
   $$\,\,\,\,\,\,\,\,\,|$$
   $$\,\,\,\,\,\,\,\,\,O$$

   Add unshared pairs to the atoms bonded to the central atom:

   $$:\ddot{O}-N-\ddot{O}:$$
   $$\,\,\,\,\,\,\,\,\,\,\,\,\,|$$
   $$\,\,\,\,\,\,\,\,\,\,\,:\ddot{O}:$$

4. Count the electron pairs used. Since 12 electron pairs = 24 electrons, there are not enough electrons to add any to the central atom.
5. Move one of the unshared pairs from the surrounding atom to make a double bond:

   $$\left[\begin{array}{c}\ddot{O}=N-\ddot{O}:\\|\\:\ddot{O}:\end{array}\right]^-$$

   This structure, while correct, does not show that all three N — O bonds are equivalent. Three contributing structures can be drawn:

   $$\left[\begin{array}{c}\ddot{O}=N-\ddot{O}:\\|\\:\ddot{O}:\end{array}\right]^- \leftrightarrow \left[\begin{array}{c}:\ddot{O}-N-\ddot{O}:\\||\\:O:\end{array}\right]^- \leftrightarrow \left[\begin{array}{c}:\ddot{O}-N=\ddot{O}\\|\\:\ddot{O}:\end{array}\right]^-$$

   The Lewis structure of the nitrate ion is a resonance hybrid of these contributing structures.

COMMENT

Another molecule with three resonance structures is SO_3. Note that SO_3 has 24 valence electrons and is isoelectronic with NO_3^-.

ELECTRONEGATIVITY

STUDY OBJECTIVES

You should be able to:
1. Describe the trends in electronegativity within the periodic table.
2. Explain what is meant by a polar bond and predict the relative polarity of bonds.
3. Classify bonds in given substances as ionic, polar covalent, or covalent.

Electronegativity. Chemical bonds are rarely purely covalent or completely ionic. Rather, most bonds exhibit some characteristics of both. In the previous chapter we saw that atoms of the elements exhibit varying tendencies in their ability to attract and hold free electrons in the gas phase. In other

Table 9.2 Electronegativities of Second- and Third-Row Representative Elements

Second row	Li	Be	B	C	N	O	F
	1.0	1.5	2.0	2.5	3.0	3.5	4.0
Third row	Na	Mg	Al	Si	P	S	Cl
	0.9	1.2	1.5	1.8	2.1	2.5	3.0

words, electron affinity values show periodic variations. The term *electronegativity* is used to describe the ability of an atom *within a molecule* to attract a shared electron pair toward itself. Linus Pauling developed a method for determining the relative electronegativities of the elements. These values are given in Figure 9.8 of the text. Pauling assigned the value 1.0 to Li and 4.0 to F. The values for second- and third-row elements are given in Table 9.2. Electronegativity values exhibit periodic behavior. In general, electronegativities increase from left to right across a period and decrease within a group from top to bottom, as shown in Figure 9.3.

As a consequence of the differing abilities of atoms in a bond to attract the shared-electron pair, most electron pairs are not shared equally. This imbalance causes an electron pair to shift slightly toward the more electronegative atom, giving rise to a *polar covalent bond*. In the HCl molecule, for instance, the electronegativity χ of Cl is 3.0, and for H it is 2.1. Delta (Δ) is the electronegativity difference.

$$\Delta = \chi_{Cl} - \chi_H$$

$$= 3.0 - 2.1 = 0.9$$

The chlorine with its higher electronegativity attracts the electron pair more strongly. This makes the Cl atom slightly negative and H slightly positive:

$$\delta + \quad \delta -$$

$$H - Cl$$

Here δ denotes a partial charge, that is, a charge less than 1.0, as it would be in an ion.

Pure covalent bonding, which is the equal sharing of electron pairs, occurs only in homonuclear diatomic molecules. Examples are H_2, N_2, and Cl_2. In a diatomic molecule with both atoms the same, Δ must be zero, and the bonding electron pair is shared equally. Bonds of this type are described as *nonpolar covalent bonds*, or *pure covalent bonds*.

In bonds involving different atoms, Δ will depend on the individual electronegativities. Bonds between atoms such that $\Delta < 2.0$ are classified as polar covalent, or simply polar bonds. When $\Delta \geq 2.0$, a bond is mostly ionic, for in this case one atom so outdoes the other at attracting electron pairs that electrons can be considered to be completely transferred to the more electronegative atom.

Here we see that nonpolar covalent bonds and ionic bonds represent extreme situations in bonding. To refer to a bond as being "ionic" or "covalent" is an oversimplification. Sometimes the term *percent*

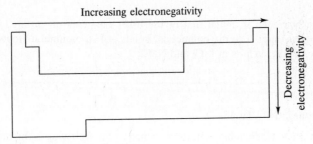

Figure 9.3. Periodic trends in electronegativity.

Table 9.3 Bond Character of Some Common Bonds

	Δ
Ionic compound	
NaCl	2.1
NaBr	1.9
NaI	1.6
KCl	2.2
KBr	2.0
$MgCl_2$	1.8
$MgBr_2$	1.6
Generally considered ionic	
AgCl	1.1
$AlCl_3$	1.5
Polar covalent bonds	
C — Cl	0.5
N — H	0.9
O — H	1.4
S — H	0.4
S — O	1.0
N — O	0.5

ionic character is used to describe the polar nature of the bond. A pure covalent bond has zero percent ionic character, while a purely ionic bond would be described as having 100% ionic character.

Bonds between Group 1A or 2A metals and the halogens are classified as ionic because of the large Δ values obtained. See Table 9.3 for examples.

EXAMPLE 9.7 Bond Polarity

What is the order of increasing ionic character for C — O, C — H, and O — H?

METHOD OF SOLUTION

As the electronegativity difference Δ increases, the bond becomes more polar and its ionic character increases. Using Figure 9.8 of the textbook, we can determine the differences:

For C — O $\Delta = 3.5 - 2.5 = 1.0$
For C — H $\Delta = 2.5 - 2.1 = 0.4$
For O — H $\Delta = 3.5 - 2.1 = 1.4$

Answer: Ionic character increases in the order

C — H < C — O < O — H

COMMENT

The electronegativity of hydrogen is unlike that of the other elements of Group 1A. In terms of electronegativity, hydrogen is similar to the nonmetals. Bonds of H to nonmetal atoms are polar covalent bonds rather than ionic bonds, such as in LiCl and NaCl.

EXAMPLE 9.8 Electronegativity Trends

Using the trends within the periodic table, determine which of the following is the most electronegative element: As, Se, or S.

METHOD OF SOLUTION

Se and As are in the same period, and so the one farther to the right has the higher electronegativity. That one is Se. Now compare S and Se. They are in the same group. The one nearer the top of the group has the greater electronegativity, which is sulfur.

EXAMPLE 9.9 Types of Bonds

For the following pairs of elements, label all bonds between them as ionic, polar covalent, or pure covalent:
a. Rb and Br
b. S and S
c. C and N

METHOD OF SOLUTION

a. The value of Δ for a Rb — Br bond is 2.0, and so RbBr is ionic.
b. For S — S, $\Delta = 0.0$, and so the bond is a pure covalent bond.
c. In the periodic table, carbon and nitrogen are adjacent to each other in period 2. The one on the right is N; it is more electronegative, and so we expect a polar covalent bond. We can check this with the Δ value; $\Delta = 0.5$.

FORMAL CHARGE

STUDY OBJECTIVE

You should be able to:
1. Draw Lewis structures of molecules showing the formal charges on the atoms.

The concept of formal charge provides a rational basis for choosing the more plausible Lewis structure from among several possibilities. Formal charge can also be used to pick plausible contributing resonance structures. The *formal charge* is the charge that an atom seems to have in a Lewis structure. When determining the formal charge, all nonbonding electrons count as belonging entirely to the atom in which they are found. All bonding electrons are divided equally between the bonded atoms. Thus the formal charge of an atom in a Lewis structure is the number of valence electrons in an isolated atom minus the number of electrons assigned to that atom in a molecule. The formula for the formal charge of an atom is

$$\text{formal charge} = \begin{pmatrix} \text{number of} \\ \text{valence} \\ \text{electrons} \end{pmatrix} - \begin{pmatrix} \text{number of} \\ \text{nonbonding} \\ \text{electrons} \end{pmatrix} - \frac{1}{2} \begin{pmatrix} \text{number of} \\ \text{bonding} \\ \text{electrons} \end{pmatrix}$$

It will be good to keep in mind that the formal charge is really more a property of a structural formula than that of the species the formula represents. Formal charges do not indicate actual charge separations in the molecule.

Two possible Lewis structures for BF_3 are

The formal charges in (1) are

> The boron atom: formal charge = $3 - 0 - \frac{1}{2}(6) = 0$
>
> The fluorine atom: formal charge = $7 - 6 - \frac{1}{2}(2) = 0$

The formal charges in (2) are

> The boron atom: formal charge = $3 - 0 - \frac{1}{2}(8) = -1$
>
> The fluorine atom (double bonded):   formal charge = $7 - 4 - \frac{1}{2}(4) = +1$
>
> The other fluorine atoms: formal charge = $7 - 6 - \frac{1}{2}(2) = 0$

The rule that is used to establish the more plausible structure is: *A Lewis structure in which there are no formal charges is preferred over one where formal charges are present*. Thus structure (1) is preferred over structure (2).

When formal charges are assigned in a Lewis structure, the sum of the formal charges must be zero in a neutral molecule. For a polyatomic ion the formal charges must add up to the charge of the ion. Thus for the chlorite ion (ClO_2^-),

$$[:\ddot{\underset{..}{O}}-\ddot{\underset{..}{Cl}}-\ddot{\underset{..}{O}}:]^-$$

the formal charges are

$$[:\overset{-}{\ddot{\underset{..}{O}}}-\overset{+}{\ddot{\underset{..}{Cl}}}-\overset{-}{\ddot{\underset{..}{O}}}:]^-$$

The example brings out another feature of the formal charge concept. *The most plausible Lewis structures will be those with negative formal charges on the more electronegative atoms*. Since oxygen is more electronegative than chlorine, the formal negative charges on oxygen in this structure are more reasonable than in some other structure that would palce a positive charge on the oxygen atoms.

EXAMPLE 9.10   Assigning Formal Charges

Assign formal charges to the atoms in the following Lewis structures:
a. $:C \equiv O:$
b. $\ddot{\underset{..}{O}} = \ddot{S} - \ddot{\underset{..}{O}}:$

METHOD OF SOLUTION

The formula used to calculate the formal charge of an atom is

$$\begin{pmatrix}\text{formal} \\ \text{charge}\end{pmatrix} = \begin{pmatrix}\text{number of} \\ \text{valence} \\ \text{electrons}\end{pmatrix} - \begin{pmatrix}\text{number of} \\ \text{nonbonding} \\ \text{electrons}\end{pmatrix} - \frac{1}{2}\begin{pmatrix}\text{number of} \\ \text{bonding} \\ \text{electrons}\end{pmatrix}$$

a. For the carbon atom, formal charge = $4 - 2 = \frac{1}{2}(6) = -1$
   For the oxygen atom, formal charge = $6 - 2 - \frac{1}{2}(6) = +1$
b. For the sulfur atom, formal charge = $6 - 2 - \frac{1}{2}(6) = +1$
   For the oxygen atom on the right, formal charge = $6 - 6 - \frac{1}{2}(2) = -1$
   For the oxygen atom on the left, formal charge = $6 - 4 - \frac{1}{2}(4) = 0$

COMMENT

Some chemists don't approve of the CO structure given in part a because it places a positive formal charge on the more electronegative oxygen atom.

EXCEPTIONS TO THE OCTET RULE

STUDY OBJECTIVE

You should be able to:
1. Describe three types of molecules that are exceptions to the octet rule.

Lewis structures can be drawn for many compounds with the aid of the octet rule; however, structures of some compounds do not follow the rule. The textbook points out three types of molecules that are exceptions to the octet rule. These are molecules which have an *incomplete octet*, molecules which have an *odd number of electrons*, and molecules in which the central atom has an *expanded octet*.

Two common oxides of nitrogen, NO and NO_2, have *odd numbers of electrons*. Since an even number of electrons is required for complete pairing, the octet rule cannot be satisfied. Two additional odd-electron molecules that are known to exist in our atmosphere for very short periods of time are OH (hydroxyl radical) and HO_2 (hydroperoxyl radical):

$$\cdot \ddot{\text{O}} - \text{H} \qquad \text{H} - \ddot{\text{O}} - \ddot{\text{O}} \cdot$$

Species such as these are paramagnetic due to the presence of unpaired electrons and are called *free radicals*.

The boron halides BX_3 are well-known examples of molecules with an *incomplete octet*. They are sometimes called *electron-deficient* molecules. They are all planar molecules in which the boron atom has only six valence electrons. Thus they have incomplete octets:

$$\text{X} \diagdown \text{B} \diagup \text{X}$$
$$|$$
$$\text{X}$$

Normally when there is a shortage of electrons, we can draw a double-bonded structure as shown below for BF_3:

$$:\ddot{\text{F}}:$$
$$\|$$
$$\text{B}$$
$$\diagup \quad \diagdown$$
$$:\ddot{\text{F}}: \quad :\ddot{\text{F}}:$$

However, experiments indicate that each B — F bond is a single bond, as shown in the first dot structure. Also, the assignment of formal charges indicates that a structure with a double bond would have adjacent formal charges. As was discussed above, this is unfavorable. In addition, the more electronegative F atom would have a positive charge, rather than the preferred negative charge for fluorine.

Molecules exhibiting an *expanded octet* (having more than eight valence electrons) require the presence of nonmetal atoms from the third period or beyond in the periodic table. Second-period elements never exceed the octet rule. Third-period elements are just as likely to exceed the octet rule as they are to follow it. Where, for example, PCl_3 obeys the octet rule, gaseous PCl_5 has a phosphorus atom that is joined by single bonds to five chlorine atoms. The phosphorus atom has 10 electrons in its valence shell. This is called an expanded octet:

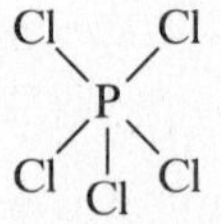

```
   Cl      Cl
     \   /
      P
    / | \
  Cl  |  Cl
      Cl
```

The central atoms in SF_4 and SF_6 and in the interhalogen compounds ClF_5, BrF_5, and IF_7 exhibit expanded octets.

In Chapter 10, you will need to draw a number of dot structures of molecules exhibiting expanded octets. When the central atom is from the third period or beyond, complete the octets of the surrounding atoms first and then complete the central atom. *If extra electron pairs remain*, place them on the central atom. See Example 9.12 below.

These exceptions seem to be telling us that an atom with a completed octet is not necessary for covalent bonding to occur. Within the Lewis framework, it is really the sharing of electron pairs that leads to the covalent bond. A shared pair of electrons acts to attract both atoms.

EXAMPLE 9.11   Exceptions to the Octet Rule

Draw Lewis structures for
a. GaI_3
b. NO_2 (all bonds are equivalent)
c. ClF_3

METHOD OF SOLUTION
a. Gallium is in Group 3A of the periodic table, a group well known for its electron-deficient elements. GaI_3 has $3 + 3(7) = 24$ valence electrons:

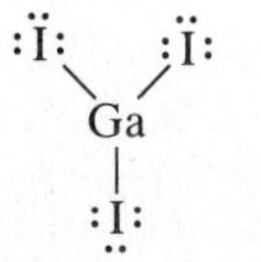

```
  :Ï:     :Ï:
     \   /
      Ga
      |
     :Ï:
```

This structure shows 24 electrons, the correct number. Ga, with three electron pairs, is electron deficient.

b. NO_2 has 17 valence electrons. With an odd number of electrons, it cannot obey the octet rule. The best we can do is start with 18 valence electrons as in NO_2^- and then remove one from the nitrogen atom (because it is the unique atom). Two contributing structures are necessary:

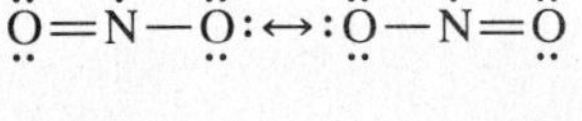

$$\ddot{O}=\dot{N}-\ddot{\underset{..}{O}}: \leftrightarrow :\ddot{\underset{..}{O}}-\dot{N}=\ddot{O}$$

c. ClF_3 has 28 valence electrons. Twenty-six electrons are required to complete the octets of the four atoms. The remaining two electrons are placed on the central Cl atom because chlorine is in the third period and has vacant $3d$ orbitals that can hold electrons in addition to an octet. Chlorine is said to have an expanded octet:

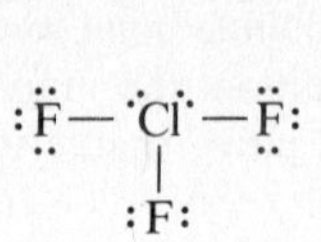

```
 :F̈—C̈l̇—F̈:
      |
     :F̈:
```

BOND ENERGIES

STUDY OBJECTIVES

You should be able to:
1. Determine the average bond energy of a bond using given enthalpies of reaction.
2. Use a table of bond energies to estimate ΔH for a reaction.

Average Bond Energies. Occasionally the enthalpy change ΔH is needed for a reaction for which enthalpy of formation data do not exist. One approach that allows an estimate of $\Delta H°$ uses the concept of bond energy. Consider the gas-phase atomization process

$$CH_4(g) \rightarrow C(g) + 4H(g) \quad \Delta H°_{rxn} = 1664 \text{ kJ}$$

In this reaction, four C — H bonds are broken; therefore we can define the average C — H bond energy as one-fourth of $\Delta H°_{rxn}$ for the reaction. Hence the average C — H bond energy in CH_4 is 416 kJ/mol.

The actual bond energies of the individual C — H bonds in CH_4 are not the same as the average value. Even so, the use of average bond energies makes it possible to estimate the enthalpy changes of certain reactions. Some of the bond energies in Table 9.4 of the textbook are listed here in Table 9.4.

Table 9.4 Bond Energies of Some Common Bonds

Bond	Bond Energy (kJ/mol)
H — H	436
C — H	414
N — H	392
O = O	499
C = O	802
O — H	460
I — Cl	210
N ≡ N	941
N = O	630

We can estimate ΔH for the combustion of methane as follows:

$$CH_4(g) + 2O_2(g) \rightarrow CO_2(g) + 2H_2O(g)$$

$$CH_4 + 2O_2 \xrightarrow{\Sigma BE(\text{reactants})} C + 4H + 4O \xrightarrow{-\Sigma BE(\text{products})} CO_2 + 2H_2O$$

First break the four C — H bonds and two O = O bonds, where Σ BE (reactants) is the sum of the bond energies of bonds broken. Then let the atoms recombine to form the two C = O and four O — H bonds of the products. Σ BE (products) is the sum of the bond energies of all bonds formed from the atoms. Keep in mind that breaking bonds is an endothermic process, and making new bonds is an exothermic process. Therefore

$$\Delta H = \Sigma \text{ BE(reactants)} - \Sigma \text{ BE(products)}$$

$$= [4 \text{ BE (C — H)} + 2 \text{ BE (O = O)}] - [2 \text{ BE (C = O)} + 4 \text{ BE (O — H)}]$$

Inserting values from Table 9.4 gives

$$\Delta H = [4(414) + 2(499)] - [2(802) + 4(460)]$$

$$= 2654 \text{ kJ} - 3444 \text{ kJ}$$

$$= -790 \text{ kJ}$$

If we compare this value with the standard enthalpy of combustion of methane, we get

$$CH_4(g) + 2O_2(g) \rightarrow CO_2(g) + 2H_2O(l) \qquad \Delta H°_{rxn} = -890 \text{ kJ}$$

At first it seems there is a large discrepancy of 100 kJ. However, in our calculation, H_2O is present as a gas! But $\Delta H°$ refers to the standard state of H_2O as a liquid. Therefore, the 100-kJ difference is largely due to the ΔH of vaporization of 2 mol of H_2O, which is 81.4 kJ. Two important conclusions can be drawn here: (1) bond energies are used to estimate enthalpies of reaction of *gas-phase reactions* and (2) enthalpies of reaction calculated from average bond energies are only *approximate*.

EXAMPLE 9.12 Use of Bond Energies to Estimate ΔH_{rxn}

Estimate ΔH_{rxn} for the reaction

$$Cl_2(g) + I_2(g) \rightarrow 2ICl(g)$$

using the bond energies given in Table 9.4 in the text and given that BE(I — Cl) = 210 kJ/mol.

METHOD OF SOLUTION

Recall that

$$\Delta H = \Sigma \text{ BE(reactants)} - \Sigma \text{ BE(products)}$$

$$\Sigma \text{ BE(reactants)} = \text{BE(Cl — Cl)} + \text{BE(I — I)}$$

$$= 243 + 151 = 394 \text{ kJ}$$

and

$$\Sigma \text{ BE(products)} = 2\text{BE(I — Cl)} = 2(210) = 420 \text{ kJ}$$

Substituting yields

$$\Delta H = 394 \text{ kJ} - 420 \text{ kJ}$$

$$= -26 \text{ kJ}$$

TRUE-FALSE QUESTIONS

1. $:\dot{P}:$ is the Lewis symbol for phosphorus.

2. The octet rule is based on the stability of the noble gas elements.

3. The melting point of a solid serves as a rough measure of the strength of attractive forces between ions.

4. The lattice energy of CaO should be much greater than that of LiCl.

5. Satisfactory Lewis structures can be drawn for all molecules.

6. $[\ddot{\text{O}}=\text{N}-\ddot{\text{O}}:]^-$ is called a contributing structure of NO_2^-.

7. The HF molecule has two lone electron pairs.

8. Electronegativity and electron affinity are the same property.

9. The electronegativities of F, Cl, Br, and I are the same.

10. The bond in carbon monoxide is a polar covalent bond.

11. In the SO molecule the S atom is $\delta+$ and the O atom is $\delta-$.

12. The bond in diatomic oxygen is nonpolar (zero percent ionic character).

13. For the compound MgO, the difference in electronegativities (Δ) is 2.3, and therefore, the bond is polar covalent.

14. Diatomic hydrogen, hydroxyl radical, BCl_3, and XeF_4 are all exceptions to the octet rule.

15. The $N \equiv N$ bond dissociation energy is less than the $Cl - Cl$ bond energy.

SELF-TEST A

1. Arrange the following ionic compounds in order of increasing lattice energy: RbI, MgO, $CaBr_2$.

2. Which compound in each pair has the higher melting point?
   a. RbI or KI   b. $MgCl_2$ or $MgBr_2$

3. Draw Lewis dot structures for
   a. IBr
   b. CS_2
   c. PCl_5
   d. SO_2
   e. CO
   f. P_2Cl_4
   g. SO_4^{2-}
   h. PO_4^{3-}

4. Draw contributing structures for SO_3.

5. Which of the following elements is the most electronegative?
   a. Ca   b. P   c. As   d. K   e. Si

6. Which one of the following is the most polar bond?
   a. B — C   b. B — N   c. B — S   d. C — C

7. Which bond has the greatest percent ionic character?
   a. CO   b. SO   c. NaI   d. NaBr

8. Which molecule has a Lewis structure that does not obey the octet rule?
   a. NO   b. PF_3 c. CS_2 d. HCN   e. BF_4^-

9. Which molecule has covalent bonding between atoms?
   a. NaF   b. K_2O c. ICl   d. SrI_2 e. None of these

10. Indicate the molecule in each pair that does not follow the octet rule:
    a. PCl_5 or $AsCl_3$ b. BCl_3 or CCl_4 c. NH_3 or CH_3 d. $SbCl_5$ or NCl_3

11. Consider the following Lewis structures for sulfate ion. Which is the more reasonable structure in terms of formal changes?

a. 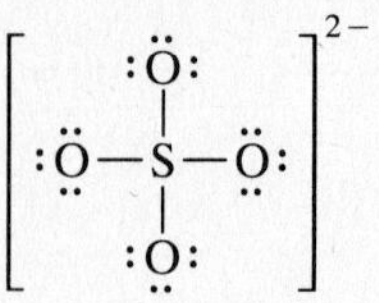 $\left[\begin{array}{c} :\ddot{O}: \\ | \\ :\ddot{O}-S-\ddot{O}: \\ | \\ :\ddot{O}: \end{array}\right]^{2-}$

b. 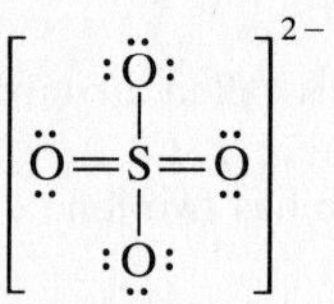 $\left[\begin{array}{c} :\ddot{O}: \\ | \\ \ddot{O}=S=\ddot{O} \\ | \\ :\ddot{O}: \end{array}\right]^{2-}$

12. Assign formal changes to the atoms in the Lewis structures shown:
    a. $:\ddot{N}=N=\ddot{O}:$
    b. $[:\ddot{O}-H]^-$
    c. $[:\ddot{O}-\ddot{C}l:]^-$

13. Write a mathematical expression that would allow you to calculate the lattice energy of $CaCl_2$ using Hess's law.

14. Calculate the $N\equiv N$ bond energy, given that the standard enthalpy of formation of atomic nitrogen is 473 kJ/mol.

15. Estimate the enthalpy change for the following reaction:

    $$N_2(g) + O_2(g) \rightarrow 2NO(g)$$

    Hint: NO has a double bond:

    $BE(N=O) = 630 \text{ kJ}$

SELF-TEST B1

1. Use the Born-Haber cycle to determine the standard enthalpy of formation of sodium bromide, given the following data:

   $\Delta H_{sub}(Na) = 109 \text{ kJ}$

   ionization energy (Na) = 496 kJ

   dissociation energy (Br — Br) = 192 kJ

   electron affinity (Br) = -324 kJ

   lattice energy (NaBr) = 736 kJ

2. The $N\equiv N$ and H — H bond energies in Table 9.4 and standard enthalpy of formation for $NH_3(g)$ are given by

   $$\tfrac{1}{2}N_2(g) + \tfrac{3}{2}H_2(g) \rightarrow NH_3(g) \quad \Delta H^\circ_{rxn} = -46.3 \text{ kJ/mol}$$

   Calculate the average N — H bond energy in ammonia.

SELF-TEST B2

1. Calculate the lattice energy of sodium bromide from the following information:

    $\Delta H_{sub}(Na) = 109$ kJ

    ionization energy (Na) = 496 kJ

    dissociation energy (Br — Br) = 192 kJ

    electron affinity (Br) = -324 kJ

    ΔH_f° (NaBr) = -359 kJ

2. Given the bond energies in Table 9.4, calculate the enthalpy of formation for ammonia, NH_3:

    $\frac{1}{2}N_2(g) + \frac{3}{2}H_2(g) \rightarrow NH_3(g) \quad \Delta H_f^\circ = ?$

ANSWERS

TRUE-FALSE QUESTIONS

1. False. ·P̈· is the symbol.
2. True.
3. True.
4. True.
5. False. Resonance structures, for instance, do not represent real molecules.
6. True.
7. False. Three lone pairs.
8. False. Electron affinity refers to the attraction of a gas-phase atom for an electron. Electronegativity refers to the attraction that an atom in a bond has for the shared-electron pair.
9. False. Electronegativity decreases in going down a group of elements.
10. True.
11. True.
12. True.
13. False. $\Delta > 2.0$ true enough, but that makes the bond *ionic*.
14. True.
15. False. Bond energies of triple bonds are greater than those of single bonds.

SELF-TEST A

1. $RbI > CaBr_2 > MgO$
2. a. KI   b. $MgCl_2$

3. a. :Ï — B̈r: c. [PCl_5 structure]   e. :C≡O: g. $[SO_4]^{2-}$ structure

    b. S̈=C=S̈   d. Ö=S̈—Ö: → :Ö—S̈=Ö   f. [P_2Cl_4 structure]   h. $[PO_4]$ structure

4.

$$\overset{..}{\underset{..}{O}}=S-\overset{..}{\underset{..}{O}}: \leftrightarrow :\overset{..}{\underset{..}{O}}-\overset{\overset{..}{O}:}{\underset{\|}{S}}-\overset{..}{\underset{..}{O}}: \leftrightarrow :\overset{..}{\underset{..}{O}}-S=\overset{..}{\underset{..}{O}}$$

(with $:\overset{..}{O}:$ single-bonded above S in first and third structures)

5. b
6. b
7. d
8. a
9. c
10. a. PCl_5 b. BCl_3 c. CH_3 d. $SbCl_5$
11. b. Because fewer formal charges are present. Particularly, the formal charge on sulfur is zero compared with +2 in a.
12. a. $\overset{..}{N}=\overset{+}{N}=\overset{..}{O}:$ b. $[\overset{..}{\overset{..}{O}}-H]^-$ c. $[\overset{..}{\overset{..}{O}}-Cl]^-$
13. $U = -\Delta H_f^\circ(CaCl_2) + \Delta \dot{H}_{sub}(Ca)$ + first ionization energy (Ca) + second ionization energy (Ca) + bond dissociation energy (Cl_2) + 2 [electron affinity (Cl)]
14. 946 kJ
15. 180 kJ

SELF-TEST B1

1. -359 kJ
2. 391 kJ

SELF-TEST B2

1. 736 kJ
2. -50.7 kJ

Chapter Ten
CHEMICAL BONDING II: MOLECULAR GEOMETRY AND MOLECULAR ORBITALS

- VSEPR Theory
- Polar Molecules
- Valence Bond Theory and Hybrid Orbitals
- Molecular Orbital Theory

VSEPR THEORY

STUDY OBJECTIVES

You should be able to:
1. Predict geometrical shapes of molecules and polyatomic ions.
2. Predict approximate bond angles between atoms within a molecule or polyatomic ion.

Electron-Pair Repulsion. The major features of the geometry of molecules and polyatomic ions can be predicted by applying a simple principle: *The valence-shell electron pairs surrounding an atom are arranged such that they are as far apart as possible.*

According to the valence-shell electron-pair repulsion (VSEPR) theory, a particular molecular geometry results from the orientation of electron pairs around a central atom. For instance, consider the BeH_2 molecule. Drawing its Lewis structure, we count two electron pairs in the valence shell of the Be atom. Two electrons were donated by the Be atom, and one each by the two H atoms:

$$H:Be:H \quad \text{— called a central atom}$$

In order to minimize repulsion, the two electron pairs are located 180° apart. This angle between two bonds to the same atom is called the *bond angle*:

A molecule in which the atoms lie in a straight line is referred to as *linear*.

Table 10.1 Bond Angles and Molecule Shapes

Number of Bonding Electron Pairs about Central Atom	Predicted Bond Angles	Shape of Molecule
2	180°	Linear
3	120°	Trigonal planar
4 (octet)	109.5°	Tetrahedral
5	120°, 90°	Trigonal bipyramid
6	90°	Octahedral

When the central atom of a molecule has no lone pairs in its valence shell, the possible orientations in space around the central atom are determined by the number of bonding pairs of electrons. Table 10.1 gives the shapes of molecules depending on the number of valence electron pairs about the central atom.

Three Electron Pairs. When there are three valence electron pairs around a central atom, they avoid each other best if they align themselves at the corner of a triangle. The central atom sits in the middle of the triangle, and all four atoms lie in the same plane. All bond angles are 120°. Such a molecule is called *trigonal planar*.

Four Electron Pairs. In Chapter 9 we discussed the use of the octet rule to determine the correct formulas of covalent compounds. Lewis structures describe only the bonding within a molecule in a two-dimensional manner. They are not geometrical structures. For instance, the Lewis structure for methane (CH_4) does not tell us whether the molecule is flat (planar) or tetrahedral. When the central atom has four bonding electron pairs (as C does in CH_4), these pairs try to keep as far as possible from each other, while maintaining their distance from the nucleus. The tetrahedral arrangement has less repulsion between the four electron pairs than the flat, two-dimensional geometry. This geometry is illustrated for methane in Figure 10.1. Therefore, four electron pairs around a central atom will be arranged at the corners of a tetrahedron. In methane the H — C — H bond angle is the *tetrahedral angle*, 109.5°.

Five Electron Pairs. Central atoms with expanded octets have five or six electron pairs occupying their valence shells. The geometrical figure defined by the preferred orientation of five electron pairs around the central atom is called a *trigonal bipyramid* (see Table 10.1 in the textbook). The electron pairs that lie around the "equator" are called *equatorial electrons*. The electrons at the "north and south

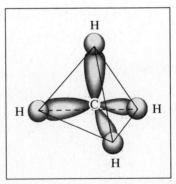

Figure 10.1. The tetrahedral geometry of the methane molecule. The C atom is at the center of a four-sided geometrical figure called a tetrahedron. The four H atoms are found at the four corners. The H—C—H bond angle is the tetrahedral angle, 109.5°.

poles" are called *axial electrons*. The equatorial electron pairs lie in the same plane. In PF_5 the equatorial F — P — F bond angles are 120°. The axial atoms are at right angles to the equatorial plane, and so any bond angle formed by an axial atom, the central atom, and an equatorial atom is 90°:

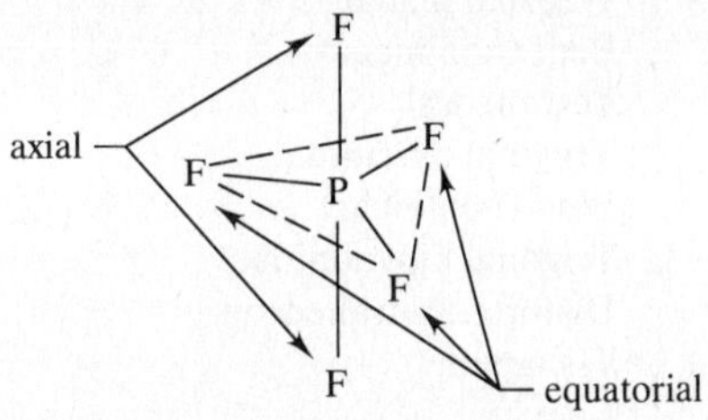

Effect of Lone Electron Pairs. When the central atom contains lone pairs (nonbonding pairs) in addition to bonding pairs, the situation becomes slightly more complicated. The position of the lone electron pairs helps to establish the geometry of a molecule [for instance, in ammonia (NH_3) the nitrogen atom has one lone electron pair]:

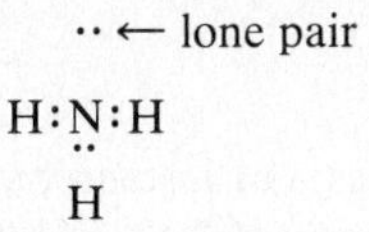

```
    ·· ← lone pair
 H:N:H
    ··
  H
```

There are four electron pairs in the valence shell of the central N atom. Therefore, when the electron pairs are oriented as far from each other as possible, they will be located at the corners of a tetrahedron. But when describing the shape of a molecule, we consider *only the positions of the atoms*. Ammonia is trigonal pyramidal, not tetrahedral. Its geometry is described as a trigonal pyramid because the base of the pyramid bounded by the three H atoms has three sides:

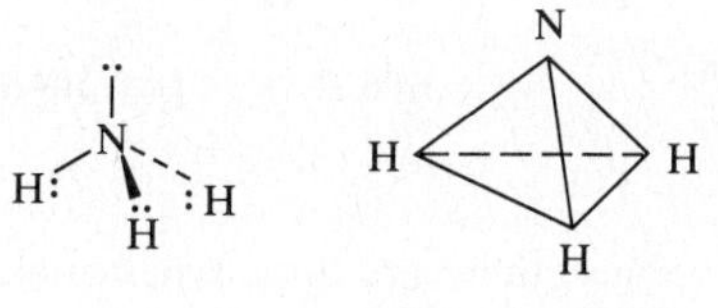

trigonal pyramid

Table 10.2 lists the molecular shapes that result when the central atom has the symbol A, a pair of bonding electrons is B, and a pair of nonbonding electrons is E.

Multiple Bonds. The idea of electron-pair repulsion can be extended to predict geometries of molecules or ions containing multiple bonds by assuming that as far as molecular shape is concerned, electrons in a multiple bond affect the structure in the same way as those in a single bond. Consider SO_3, for example. A contributing structure is

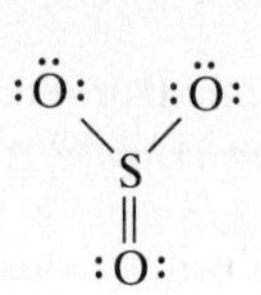

24 valence electrons

Table 10.2 Molecule Shapes as Determined by VSEPR Theory

Molecular Type	Shape of Molecule
AB_2	Linear
AB_3	Trigonal planar
AB_2E	Bent (nonlinear)
AB_4	Tetrahedral
AB_3E	Trigonal pyramidal
AB_2E_2	Bent (nonlinear)
AB_5	Trigonal bipyramidal
AB_4E	Distorted tetrahedron
AB_3E_2	T-shaped
AB_2E_3	Linear
AB_6	Octahedral
AB_5E	Square pyramidal
AB_4E_2	Square planar

There are four electron pairs in the valence shell of the central atom, but remember two of these are used in a double bond. Thus, only three regions of electron density exist around the central atom. The SO_3 molecule has a geometry as if only three regions of electron pairs existed around the central S atom; that is, it is a planar triangle.

Bond Angles. VSEPR also explains qualitatively many observed bond angles. For instance, a true tetrahedral angle, as in CH_4, is 109.5°, but the measured H — N — H bond angle in NH_3 is 106.7°, and in water the H — O — H bond angle is 104.5°:

109.5°   106.7°   104.5°

The deviation from 109.5° arises because there are three types of electron-pair repulsion represented: (1) repulsion between bonding pairs, (2) repulsion between lone pairs, and (3) repulsion between a bonding pair and a lone pair. The force of repulsion is not the same among these three but decreases as follows:

lone pair–lone pair > lone pair–bonding pair > bonding pair–bonding pair

In other words, lone pairs require more space than bonding pairs and tend to push the bonding pairs closer together. Bonding electron pairs are attracted by two nuclei at the same time. These pairs require less space than lone electron pairs.

Thus, the lone pair on the nitrogen in ammonia repels the bonding pairs more strongly than the bonding pairs repel each other. The result is that the H — N — H bond angle is several degrees less than a tetrahedral angle.

In water the oxygen atom has two bonding pairs and two lone pairs. The greater lone pair–bonding pair repulsion causes the two O — H bonds to be pushed in toward each other. The H — O — H angle in water should be less than the H — N — H angle in ammonia.

In general, when nonbonding electrons are present on the central atom, the shape will be close to, but not exactly, as predicted in Table 10.2.

EXAMPLE 10.1   VSEPR Theory

Predict the geometrical shapes (and bond angles) of the following compounds and ions using the VSEPR theory:
a. $SnCl_4$
b. O_3
c. IF_5
d. XeF_2

METHOD OF SOLUTION

a. First determine the number of valence-shell electrons for the central Sn atom. A tin atom has four valence electrons, and each Cl atom contributes one electron which it shares with the Sn atom, and so $4 + 4(1) = 8$. The eight valence-shell electrons are arranged as four electrons pairs around the Sn atom. The Lewis structure is

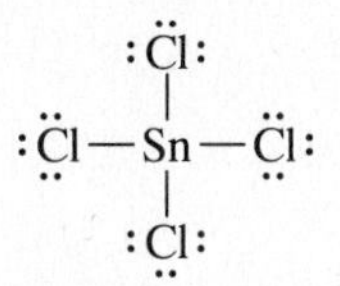

$$:\ddot{C}l\!-\!Sn\!-\!\ddot{C}l:$$

(with $:\ddot{C}l:$ above and $:\ddot{C}l:$ below)

This structure shows that there are no lone pairs about the Sn atom. Thus, the four bonding electron pairs and the Cl atoms as well are oriented at the corners of a tetrahedron. $SnCl_4$ is an AB_4-type molecule and has a tetrahedral structure. The Cl — Sn — Cl bond angles are 109.5°:

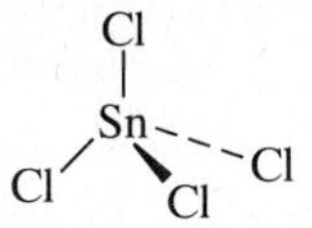

(tetrahedral diagram of Sn with four Cl atoms)

b. Ozone (O_3) has as one contributing structure

(Lewis structure of ozone)

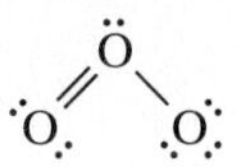

When applying VSEPR theory to a molecule requiring resonance structures, we can use *any one of the contributing structures*. Here we will take the second O atom as the central atom. In VSEPR theory *the double bond is counted as if it were a single bond with one effective electron pair*. The double bond acts as one center of electron density to repel other electron pairs. Therefore the central O atom has the equivalent of three pairs of valence electrons. These three regions of charge are oriented at the corners of a triangle. However, one of these pairs is a lone pair. The molecule is of the type AB_2E. The positions of the three O atoms are described as *nonlinear*. The O — O — O bond angle should be close to 120°. However, the presence of the lone pair compresses the bond angle because the lone pair–bonding pair repulsion is greater than the bonding pair–bonding pair repulsion. The observed angle in ozone is 117°.

c. Now that we have had some practice, we will abbreviate the explanation. To determine the shape of IF_5, count the valence-shell electrons around the central I atom:

from the I atom	7
from the F atoms (5 × 1)	5
total valence-shell electrons	12

The Lewis structure is

```
 F   F
  \ ../
F — I — F
    |
    F
```

valence-shell electron pairs	6
number of lone pairs	1
number of bonding pairs	5
molecular type	AB$_5$E
molecular shape	square pyramid

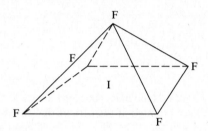

F — I — F bond angles are 90°.

d. To predict the shape of XeF$_2$, first count the valence-shell electrons around Xe:

from the Xe atom	8
from the F atoms (2 × 1)	2
total valence-shell electrons	10

The Lewis structure is

F — Xe — F

valence-shell electron pairs	5
number of lone pairs	3
number of bonding pairs	2
molecular type	AB$_2$E$_3$
molecular shape	linear

POLAR MOLECULES

STUDY OBJECTIVE

You should be able to:
1. Use electronegativity values and molecular geometries to predict whether or not a molecule possesses a dipole moment.

Diatomic Molecules. In the preceding chapter you learned that polar bonds are those in which the centers of positive charge and negative charge do not coincide. This charge separation is called a dipole, and it results from electronegativity differences between two bonded atoms. A molecule with an electric dipole is said to be polar and to possess a dipole moment.

The dipole moment μ is represented by an arrow or vector with the tail at the positive center and the head at the negative center. The length of the arrow represents the magnitude of the dipole moment.

$$\overset{\delta +\ \ \delta -}{H-Cl} \qquad \overset{+\longrightarrow}{H-Cl}$$

Quantitatively the dipole moment is calculated from the equation

$$\mu = Qr$$

where Q is the magnitude of the charge in coulombs from either end of the dipole and r is the charge separation (the distance between the centers of positive and negative charge) in meters. In polar molecules this charge is never as great as one unit of charge. Rather the charge is a partial charge, $\delta +$ or $\delta -$, which signifies a charge of less than one unit. The SI unit for a dipole moment is the coulomb meter. Traditionally, dipole moments have been measured in debye units, where 1 debye (D) = 3.33×10^{-30} C m.

Molecules that possess a dipole moment are called *polar molecules*. Molecules without a dipole moment are called *nonpolar molecules*. All homonuclear diatomic molecules are nonpolar. In general, heteronuclear diatomic molecules are polar.

Polyatomic Molecules. The dipole moment of a molecule containing more than one bond depends both on *bond polarity* and *molecular geometry*. In a polyatomic molecule the arrows representing polar bonds may add together to yield a polar molecule with a resultant dipole moment μ. Conversely, the arrows may cancel each other when added, producing a zero resultant dipole moment. Table 10.3 lists dipole moments of some small molecules in debye units.

Table 10.3 Dipole Moments of Some Molecules

Molecule	μ (D)
H_2O	1.87 D
H_2S	1.10 D
NH_3	1.46 D
CO_2	0
SO_2	1.60 D
BCl_3	0

An important property of the water molecule is its dipole moment of 1.85 D. In the water molecule each bond is polar. The resultant of these bond dipoles yields a dipole moment for the molecule. The center of negative charge lies closer to the O atom than does the center of positive charge:

$$\overset{O}{\underset{H\ \updownarrow\ H}{\nearrow\ \nwarrow}}$$

↙ resultant μ

In CO_2, the two bond dipoles cancel and $\mu = 0$:

$$\overset{\longleftarrow +\ +\longrightarrow}{O=C=O}$$
resultant $\mu = 0$

When the bond dipoles are equal and point in opposite directions, the centers of the positive and negative charge are both on the central atom. Thus, the absence of a molecular dipole in CO_2 suggests that the molecule is linear. Conversely, water has a dipole moment and so cannot be linear.

EXAMPLE 10.2   Polar and Nonpolar Molecules

Predict whether the following molecules are polar or nonpolar:
a. CO
b. H_2CO
c. CCl_4

METHOD OF SOLUTION

The polarity of a diatomic molecule depends on the electronegativity difference of the two bonded atoms.

a. Oxygen is a more electronegative element than carbon. As a result, the $C \equiv O$ bond is polar, with partial positive and negative charges on C and O, respectively:

$$\overset{+\longrightarrow}{C\equiv O}$$

The dipole moment of carbon monoxide has been found to be 0.1 D.

b. The dipolar nature of a polyatomic molecule depends on both the bond polarity and molecular geometry. Formaldehyde is a planar molecule with a polar $C=O$ bond ($\Delta = 1.0$) and two $C-H$ bonds of very low polarity ($\Delta = 0.4$). Neglecting the low polarity $C-H$ bond dipoles, we get

$$\begin{array}{c} H \\ \diagdown \overset{+\longrightarrow}{} \\ C=O \\ \diagup \\ H \end{array}$$

The net effect of these bond dipoles is that the center of positive charge is located approximately on the carbon atom and the center of negative charge lies near the oxygen atom:

$$\begin{array}{c} H \\ \diagdown \ \delta + \ \delta - \\ C=O \\ \diagup \\ H \end{array}$$

Formaldehyde has a measured dipole moment of about 2.5 D.

c. Chlorine is more electronegative than carbon. Therefore CCl_4 has four bond dipoles. However, these polar bonds are arranged in a symmetric tetrahedral fashion about the central carbon atom:

$$\begin{array}{c} Cl \diagdown \diagup Cl \\ C \\ Cl \diagup \diagdown Cl \end{array}$$

In this situation the centers of positive and negative charge are on the carbon atom The bond dipoles have canceled each other, and the molecule as a whole does not possess a dipole moment.

VALENCE BOND THEORY AND HYBRID ORBITALS

STUDY OBJECTIVES

You should be able to:
1. Describe covalent bond formation in terms of overlap of atomic orbitals.
2. Describe the formation of hybrid orbitals from atomic orbitals.
3. Assign a hybridization to the central atom of a given molecule.
4. Describe sigma and pi bond formation.

Covalent Bonds. The *valence bond* theory is one of two quantum mechanical descriptions of chemical bonding currently in use. The other is the molecular orbital theory, which will be discussed in the next section. In the valence bond theory, the atomic orbitals of each bonded atom are essentially the same as in separate isolated atoms. As two separate atoms are brought closer together, their atomic orbitals overlap each other. This overlap produces a region between the two nuclei where the probability of finding an electron is greatly enhanced. The presence of greater electron density between the two atoms tends to attract and hold the nuclei close together:

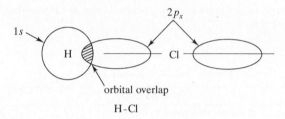

Hybridization. The concept of hybridization is used to account for bonding and geometry in terms of atomic orbitals. Bonding in the linear molecule $BeCl_2$ can illustrate the main features of hybridization. An isolated Be atom has the ground-state electron configuration $1s^2 2s^2$. The orbital diagram for the *valence electrons* is

B [↑↓] [ ][ ][ ]
    2s   2p

In a Be atom the $2p$ orbitals lie quite close in energy to the $2s$ orbital. Orbitals on the same atom that lie close together in energy have an ability to combine with one another, forming what are called *hybrid orbitals*. The presence of two Be — Cl bonds can be accounted for if, before bond formation, a $2s$ electron in a Be atom is promoted into an empty $2p$ orbital:

B [↑] [↑][ ][ ]
    2s   2p

Now there are two unpaired electrons that could participate in two covalent bonds to two chlorine atoms. Since one electron is in an *s* orbital and one is in a *p* orbital, we would expect two types of Be — Cl bonds. However, both Be — Cl bonds are observed to be the same.

This equivalence is explained by hybridization, or "mixing," of the $2s$ and the $2p$ orbitals to create two new equivalent orbitals called *sp* hybrid orbitals:

B [↑][↑] [ ][ ] empty
    2sp   2p

194 / *Chemical Bonding II: Molecular Geometry and Molecular Orbitals*

These two hybrid orbitals point in opposite directions from the Be nucleus. The Be — Cl covalent bonds result from overlap of beryllium *sp* hybrid orbitals that contain one electron each with the half-filled $3p$ orbital on Cl:

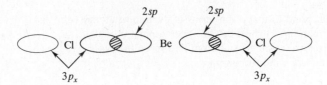

We can summarize the rules of hybridization as follows:
1. Hybridization is a process of mixing orbitals on a single atom in a molecule.
2. Only orbitals of similar energies can be mixed to form hybrid orbitals.
3. The number of hybrid orbitals obtained always equals the number of orbitals mixed together.
4. Hybrid orbitals are identical in shape and are projected out from the nucleus with characteristic orientations in space.

Table 10.4 shows several of the more important hybrid orbitals discussed in the textbook and their characteristic geometrical shapes.

Table 10.4 Hybrid Orbitals and Their Geometries

Designation	Arrangement	Example
sp	Linear (180° angle)	$BeCl_2$
sp^2	Trigonal planar (120° angle)	BF_3
sp^3	Tetrahedral (109.5° angle)	CH_4; NH_4^+
sp^3d	Trigonal bipyramidal (90°, 120° angles)	PCl_5
sp^3d^2	Octahedral (90° angle)	SF_6; $SbCl_6^-$

Multiple Bonds. Double and triple bonds can also be understood in terms of overlap of atomic orbitals. The C = C double bond in the ethylene molecule (see Section 10.6 and Figure 10.16 of the textbook), consists of one sigma (σ) bond and one pi (π) bond:

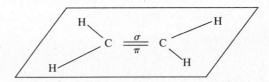

A *sigma bond* is a covalent bond in which the electron density is concentrated in a cylindrical pattern around the line of centers between the two atoms. In the C — C bond, the σ bond results from overlap of sp^2 hybrid orbitals from both of the carbon atoms (Figure 10.16 in the text).

In a *pi bond* the electron density is concentrated above and below the C — C internuclear axis, but not cylindrically as in a sigma bond (Figure 10.2). The pi bond has two regions of electron density, but it

Figure 10.2. Pi bond with electron density located above and below the internuclear axis.

---

should be kept in mind that it is only *one* bond. The π bond in ethylene results from sideways overlap of the unhybridized $2p_z$ atomic orbitals of the two carbon atoms.

EXAMPLE 10.3 Valence Bond Theory

Using unhybridized atomic orbitals, predict the bond angles in H_2O, H_2S, and H_2Se.

METHOD OF SOLUTION

The orbital diagram of the valence electrons in a O atom is

O [↑↓] [↑↓][↑][↑]
    2s $2p_x$ $2p_y$ $2p_z$

Overlap of the two half-filled $2p$ orbitals of oxygen with the half-filled $1s$ orbitals of the two H atoms results in covalent bond formation. The directional properties of two p orbitals (say, the p_y and p_z) are such that the orbitals project out from the O atom at right angles to each other:

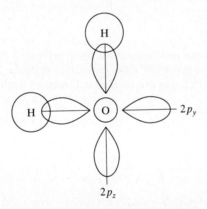

The H — O — H bond angle should be 90°. Actually the measured value is 104°.

The same structure would be expected for H_2S and H_2Se as that predicted for H_2O. In these cases the agreement between experiment and theory is found to be quite good. The H — S — H angle for H_2S is observed to be 92.2°, and for H_2Se it is 91.0°.

EXAMPLE 10.4 Hybrid Orbitals

Using orbital diagrams, outline the steps used to describe the formation of a set of sp^2-hybridized orbitals on a carbon atom.

196 / *Chemical Bonding II: Molecular Geometry and Molecular Orbitals*

METHOD OF SOLUTION

The ground-state orbital diagram for the *valence electrons* of a carbon atom is

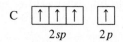

First promote one $2s$ electron to the empty $2p$ orbital. Mixing or hybridization of the $2s$ orbital with two of the $2p$ orbitals creates three equivalent sp^2 hybrid orbitals. These hybrid orbitals point out from the carbon atom to the corners of a triangle. One electron remains in an unhybridized $2p^2$ atomic orbital:

C [↑][↑][↑] [↑]
     $2sp$ $2p$

EXAMPLE 10.5 Hybrid Orbitals

Describe the hybridization of the central atom in
a. BeH_2
b. PCl_5

METHOD OF SOLUTION

First use VSEPR theory to predict a geometry; then match the geometry to the corresponding hybridization.
a. According to VSEPR theory BeH_2 is a linear molecule:

H — Be — H

Therefore, two equivalent orbitals must project out from the central Be atom in opposite directions. The type of hybrid orbitals that point 180° from each other is the sp hybrid type.
b. According to VSEPR theory, PCl_5 has the geometrical shape of a trigonal bipyramid:

```
       Cl Cl
        |/
  Cl — P
       |  ↘Cl
       Cl
```

The type of hybridization that gives five equivalent orbitals that point to the corners of a trigonal bipyramid is sp^3d (see Table 10.5 of the textbook).

EXAMPLE 10.6 Sigma and Pi Bonds

Account for the bonding in H_2CO, which is a triangular molecule.
a. What is the hybridization of the carbon atom?
b. What are the approximate bond angles about the carbon atom?
c. How many sigma and pi bonds are there in the molecule?

METHOD OF SOLUTION

a. The central carbon atom of H_2CO must be using sp^2 hybrid orbitals, since the three hybrid orbitals of this type point to the corners of a triangle. There is also one electron in an unhybridized $2p$ atomic orbital:

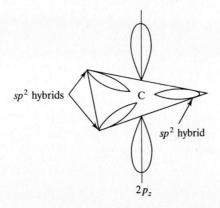

b. Overlap of the half-filled $1s$ orbital of a H atom with the half-filled sp^2 hybrid orbitals leads to formation of a C — H sigma bond. Two C — H bonds are formed in this way. The H — C — H bond angle should be 120°.

c. Oxygen atoms have the valence electron configuration

O [↑↓] [↑↓][↑][↑]
    $2s$ $2p_x\ 2p_y\ 2p_z$

Overlap of one half-filled $2p$ orbital with the remaining sp^2 hybrid yields a C — O sigma bond. The two C — H sigma bonds and the C — O sigma bond are shown below:

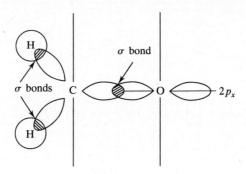

The sideways overlap of the carbon $2p_z$ orbital with the $2p_z$ orbital of the oxygen atom produces a C — O pi bond:

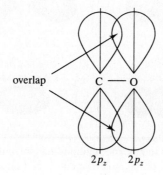

The C — O sigma and pi bonds constitute the C = O double bond:

$$\underset{H}{\overset{H}{\diagdown}}C\overset{\sigma}{\underset{\pi}{=\!=}}O$$

The complete molecule consists of three sigma bonds and one pi bond.

MOLECULAR ORBITAL THEORY

STUDY OBJECTIVES

You should be able to:
1. Describe the formation of bonding and antibonding molecular orbitals.
2. Use molecular orbital theory to predict for diatomic molecules the electron configuration, bond order, magnetic properties, and relative bond length.
3. Describe how delocalized molecular orbits are formed.

Molecular Orbitals. Until now we have considered covalent bonds in terms of the *valence bond theory*. In that theory the electrons in a *molecule* occupy atomic orbitals of the individual *atoms*. Covalent bonds are visualized in terms of the overlap of two half-filled atomic orbitals. This overlap increases the electron density between both atomic centers.

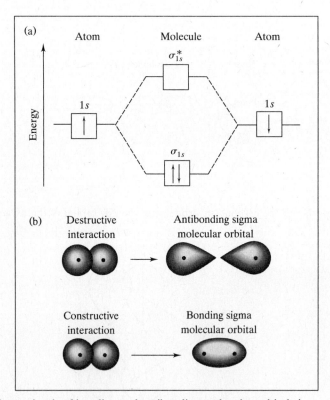

Figure 10.3. (*a*) Energy levels of bonding and antibonding molecular orbitals in the H_2 molecule. The two electrons of the molecule enter the bonding molecular orbital. (*b*) The formation of sigma antibonding and bonding molecular orbitals from the combination of atomic orbitals. In the bonding molecular orbital, there is a buildup of electron density *between* the nuclei.

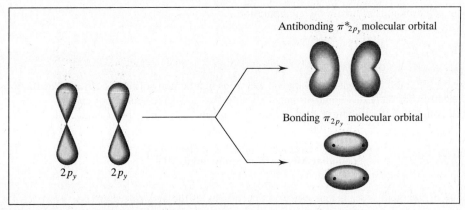

Figure 10.4. Two $2p_y$ atomic orbitals combine to form two π_{2p} molecular orbitals, one bonding and one antibonding.

The *molecular orbital model* has a different approach. In this model the available atomic orbitals are combined to form orbitals that belong to the *entire molecule*. Like atomic orbitals, these molecular orbitals have different shapes, sizes, and energies. Once the molecular orbitals have been derived, electrons are allotted to them in a manner analogous to allotting electrons to the orbitals of atoms.

According to the molecular orbital (MO) theory, the result of the interaction of two $1s$ orbitals is the formation of two molecular orbitals. One of these is a *bonding molecular orbital*, which has a lower energy than the original $1s$ orbitals. The other molecular orbital is an *antibonding orbital*, which is of higher energy than the original $1s$ orbitals. It is the greater stability of the electrons in the bonding MO that results in covalent bond formation. In the bonding MO, electron density is concentrated between the two atoms. On the other hand, the antibonding MO actually has a node (that is, a zero value) in the electron density midway between the two nuclei. There is no contribution to bonding when electrons enter antibonding MOs. The formation of these two MOs is shown in Figure 10.3.

The antibonding and bonding MOs that result from combining $1s$ orbitals are examples of *sigma* (σ) *molecular orbitals*. The electron density of σ MOs is distributed symmetrically about the line of centers between the two atoms. The sigma bonding MO and sigma antibonding MO resulting from the combination of $1s$ atomic orbitals are designated σ_{1s} and σ_{1s}^*, respectively. Sigma MOs also result from the combination of certain other atomic orbitals. For instance, the $2p_x$ orbitals on different atoms combine to form σ_{2p} bonding and σ_{2p}^* antibonding orbitals, as shown in Figure 10.24 in the text.

Two $2p_y$ orbitals and two $2p_z$ orbitals on different atoms must approach each other sideways, as shown in Figure 10.16 of the text. In the resulting molecular orbital, the electron density is concentrated above and below the line joining the two bonded atoms, rather than around the line of centers. Such a molecular orbital is called a *pi molecular orbital*. The symbol π_{2p_z} stands for a bonding pi orbital formed by the combination of two $2p_z$ atomic orbitals. The pi antibonding orbital is designated $\pi_{2p_z}^*$.

Also, the combination of two $2p_y$ atomic orbitals from different atoms forms π_{2p_y} bonding and $\pi_{2p_y}^*$ antibonding orbitals, as shown in Figure 10.4.

Molecular Orbital Configurations. For homonuclear diatomic molecules any two similar atomic orbitals can merge to form two MOs, one bonding and one antibonding orbital.

For H_2, the σ_{1s} and σ_{1s}^* MOs form as the separate atoms are brought closer together, as shown in Figure 10.3. The H_2 molecule has two electrons which are positioned in the lowest-energy orbital, the σ_{1s}. With both electrons in the bonding MO, the molecule is stable. We write the ground-state MO configuration

$$H_2(\sigma_{1s})^2$$

where the superscript 2 means that two electrons occupy the orbital.

The molecular orbital theory predicts He_2 to be unstable. A diatomic helium molecule would have four electrons. Using the same diagram as for H_2, we can see that the third and fourth electrons would enter the σ_{1s}^* orbital, resulting in the configuration

$$He_2(\sigma_{1s})^2(\sigma_{1s}^*)^2$$

In this case the stability gained by having two electrons in a bonding orbital is canceled by the presence of two electrons in the antibonding orbitals. A stable He_2 molecule should not exist.

When comparing the stabilities of molecules, we use the concept of *bond order*:

$$\text{bond order} = \frac{1}{2}\left(\begin{array}{c}\text{number of electrons} \\ \text{in bonding MOs}\end{array} - \begin{array}{c}\text{number of electrons} \\ \text{in antibonding MOs}\end{array}\right)$$

The value of the bond order is a measure of the stability of the molecule. For H_2,

$$\text{bond order} = \tfrac{1}{2}(2 - 0) = 1$$

and for He_2,

$$\text{bond order} = \tfrac{1}{2}(2 - 2) = 0$$

A bond order of 1 represents a single covalent bond, and an order of 0 means the molecule is not stable. Bond orders of 2 and 3 result for molecules with double bonds and triple bonds, respectively.

For most homonuclear diatomic molecules containing atoms of second-period elements, an approximate order of molecular energy levels is

$$\sigma_{1s} < \sigma_{1s}^* < \sigma_{2s} < \sigma_{2s}^* < \pi_{2p_y} = \pi_{2p_z} < \sigma_{2p_x} < \pi_{2p_y}^* = \pi_{2p_z}^* < \sigma_{2p_x}^*$$

The electron configurations of diatomic molecules of the second-period elements and some of their known properties are summarized in Table 10.6 in the textbook. The table shows the bond order of these molecules as predicted by MO theory and the corresponding bond energies and bond lengths. The number of unpaired electrons also correlates with the magnetic properties of molecules, something that the valence bond theory does not do.

A number of rules that apply to molecular orbitals and the stability of molecules are summarized below:
1. The number of molecular orbitals formed is always equal to the number of atomic orbitals combined.
2. The more stable the bonding molecular orbital, the less stable the corresponding antibonding molecular orbital.
3. In a stable molecule, the number of electrons in bonding molecular orbitals is always greater than that in antibonding molecular orbitals.
4. As with atomic orbitals, each molecular orbital can accommodate up to two electrons with opposite spins, in accordance with the Pauli exclusion principle.
5. When electrons are added to molecular orbitals having the same energy, the most stable arrangement is that predicted by Hund's rule. That is, as far as possible, electrons occupy these orbitals singly, and with parallel spins, rather than in pairs.
6. The number of electrons in the molecular orbitals is equal to the sum of all the electrons on the atoms.

Delocalized Molecular Orbitals. In the discussion of Lewis structures we saw that some molecules could not be adequately described by a single structure. In these cases contributing structures were drawn, and the molecule was described by a resonance hybrid. In molecular orbital theory these cases are handled by *delocalized MOs*. In examples discussed so far, all MOs have been localized between two atoms.

Molecular orbitals that are distributed among more than two atoms are called delocalized MOs. For instance, in the nitrite ion, which has a bent structure, the N atom is sp^2 hybridized and has an unhybridized $2p_z$ atomic orbital:

N [↑↓][↑][↑] [↑]
   $2sp^2$ $2p_z$

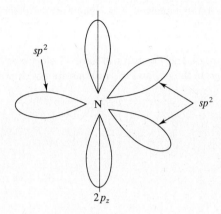

The oxygen atoms also have half-filled $2p$ orbitals:

O [↑↓] [↑↓][↑][↑]
   $2s$ $2p_x\ 2p_y\ 2p_z$

Sigma bonds result when half-filled nitrogen sp^2 orbitals overlap with half-filled $2p$ orbitals from oxygen:

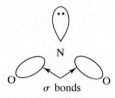

σ bonds

Thus, in NO_2^- the unhybridized p orbital of the N atom can overlap with the p orbitals of *both* O atoms to form a delocalized π orbital. This orbital extends over all three nuclei:

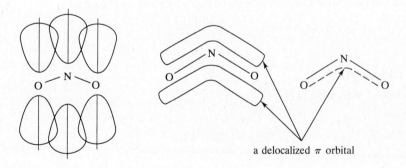

a delocalized π orbital

EXAMPLE 10.7 Molecular Orbital Diagram

Represent bonding in the O_2^- ion by means of a molecular orbital diagram. What is the bond order?

METHOD OF SOLUTION

The O_2 molecule has 16 electrons. The O_2^- ion will have 17 electrons. The energy-level diagram is shown below. The four electrons in the σ_{1s} and σ_{1s}^* orbitals are not shown because they are not involved in bonding:

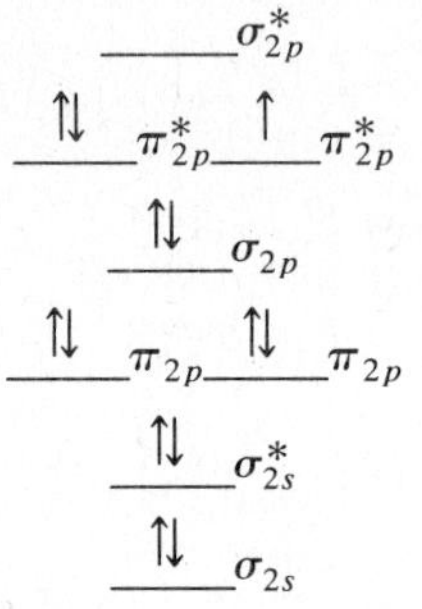

$$
\begin{array}{c}
\underline{}\ \sigma_{2p}^* \\
\underline{\uparrow\downarrow}\ \pi_{2p}^*\quad \underline{\uparrow}\ \pi_{2p}^* \\
\underline{\uparrow\downarrow}\ \sigma_{2p} \\
\underline{\uparrow\downarrow}\ \pi_{2p}\quad \underline{\uparrow\downarrow}\ \pi_{2p} \\
\underline{\uparrow\downarrow}\ \sigma_{2s}^* \\
\underline{\uparrow\downarrow}\ \sigma_{2s}
\end{array}
$$

The electron configuration is

$$O_2^- \quad (\sigma_{1s})^2(\sigma_{1s}^*)^2(\sigma_{2s})^2(\sigma_{2s}^*)^2(\pi_{2p_y})^2(\pi_{2p_z})^2(\sigma_{2p_x})^2(\pi_{2p_y}^*)^2(\pi_{2p_z}^*)^1.$$

$$\text{bond order} = \frac{1}{2}\left(\begin{array}{c}\text{number of electrons}\\ \text{in bonding MOs}\end{array} - \begin{array}{c}\text{number of electrons}\\ \text{in antibonding MOs}\end{array}\right)$$

$$= \frac{1}{2}(8-5) = 1.5$$

EXAMPLE 10.8 Paramagnetism and Diamagnetism

Which of the following species are paramagnetic and which are diamagnetic?
a. N_2
b. N_2^+
c. B_2

METHOD OF SOLUTION

a. The electron configuration of N_2 is

$$N_2 \quad (\sigma_{1s})^2(\sigma_{1s}^*)^2(\sigma_{2s})^2(\sigma_{2s}^*)^2(\pi_{2p_y})^2(\pi_{2p_z})^2(\sigma_{2p_x})^2$$

All orbitals contain pairs of electrons. Because of the lack of unpaired electrons, N_2 is diamagnetic.
b. N_2^+ will have one less electron than N_2. Therefore it has one unpaired electron and is thus paramagnetic.

c. The electron configuration of B_2 is

$$B_2 \quad (\sigma_{1s})^2(\sigma_{1s}^*)^2(\sigma_{2s})^2(\sigma_{2s}^*)^2(\pi_{2p_y})^1(\pi_{2p_z})^1$$

Because of the unpaired electrons, B_2 is paramagnetic.

TRUE-FALSE QUESTIONS

1. In ammonia (NH_3) the nitrogen atom has four pairs of valence-shell electrons which orient themselves at the corners of a tetrahedron. But the geometrical shape of NH_3 is called pyramidal.

2. According to the VSEPR theory, all "electron pair–electron pair" repulsions are equal.

3. The VSEPR theory predicts that XeF_2 is linear.

4. In N_2 the sigma bond could result from the overlap of the $2p_x$ orbital of one N atom with the $2p_x$ orbital of the other N atom.

5. The reason that carbon atoms can form four bonds rather than only two is explained by the mixing of orbitals to form hybrid orbitals.

6. All C atoms in compounds are sp^3 hybridized.

7. The triple bond in N_2 results from one N — N sigma bond and one N — N pi bond.

8. A gas-phase molecule of $HgCl_2$ is linear; therefore, Hg is employing sp hybrid orbitals.

9. BCl_3 has polar bonds but has no dipole moment.

10. For polar molecules, the charge Q on an atom is expressed as $\delta+$ or $\delta-$ and is always less than one electron charge.

11. For each bonding molecular orbital there is a corresponding antibonding MO.

12. The bond order in O_2 is 1.

13. The electron configuration of He_2^+ is $(\sigma_{1s})^2(\sigma_{1s}^*)^1$.

14. N_2 should be paramagnetic according to MO theory.

SELF-TEST A

1. Use VSEPR theory to predict the shapes of the following molecules:
   a. BrF_5 b. HCN   c. BF_3 d. SO_2 e. SCl_2

2. What type of hybrid orbital is used by the central atom of each of the molecules in Problem 1?

3. Use VSEPR theory to predict the shapes of the following polyatomic ions:
   a. BeF_3^- b. AsF_4^- c. ClF_4^- d. NO_3^- e. SO_4^{2-}

4. What type of hybridization is used by the central atom of each of the ions in Problem 3?

5. Which of the following molecules would be tetrahedral?
   a. SO_2 b. SiH_4 c. SF_4 d. BCl_3 e. XeF_4

6. Predict the approximate bond angles in
   a. $GeCl_2$ b. IF_4^- c. $TeCl_4$

7. CCl_4 is a perfect tetrahedron, but $AsCl_4^-$ is a distorted tetrahedron. Explain.

8. $BeCl_2$ and $TeCl_2$ are both covalent molecules, and yet $BeCl_2$ is linear while $TeCl_2$ is nonlinear (bent). Explain.

9. What types of hybrid orbitals can be formed by elements of the third period that cannot be formed by elements of the second period?

10. When we describe the formation of hybrid orbitals on a central atom, must all the available atomic orbitals enter into hybridization?

11. Which molecule should have the largest dipole moment, HBr or HI?

12. Which one of the following molecules has a dipole moment?
    a. CCl_4 b. H_2S c. CO_2 d. BCl_3 e. Cl_2

13. Which of the following molecules have no dipole moment?
    a. ClF   b. BCl_3 c. $BeCl_2$ d. Cl_2O e. H_2CO

14. Choose the best answer. For the water molecule:
    a. The bonds are polar, and the molecule is nonpolar.
    b. The bonds are nonpolar, and the molecule is polar.
    c. The bonds are polar, and the molecule is polar.
    d. The bonds are nonpolar, and the molecule is nonpolar.

15. Hydrogen peroxide (H_2O_2) has a dipole moment of 2.1 D. Which of the bonds in H_2O_2 are polar? Is the molecule linear?

16. Draw a molecular orbital energy-level diagram and determine the bond order for each of the following:
    a. H_2^+ b. HHe   c. He_2^+

17. The compound calcium carbide (CaC_2) contains the acetylide ion C_2^{2-}.
    a. Write molecular orbital electron configurations for C_2 and C_2^{2-}.
    b. Compare their bond order.

18. The bond distance in N_2 is 109 pm, and in N_2^+ it is 112 pm. Explain why the bond distances differ in this way.

19. Represent the bonding in NO_2^-, NO_3^-, and SO_3 using dashed lines to depict delocalized π bonds.

SELF-TEST B1

1. Predict the shape of the $BeCl_2$ molecule.

2. Predict the geometry of the NF_3 molecule.

3. Experimental evidence shows that carbon dioxide has no dipole moment. What does this suggest about its molecular shape?

4. According to VSEPR theory, methane (CH_4) is a tetrahedral molecule. What is the hybridization state of the carbon orbitals?

SELF-TEST B2

1. $BeCl_2$ is a linear molecule. How many electron pairs does the central atom have in its valence shell?

2. NF_3 has a trigonal pyramidal geometry. How many electron pairs are in the valence shell of the central atom?

3. Given that carbon dioxide is a linear molecule, predict whether it is polar or nonpolar.

4. The carbon atom in methane (CH_4) is sp^3 hybridized. What is the geometry of CH_4?

ANSWERS

TRUE-FALSE QUESTIONS

1. True.
2. False. Lone pair–lone pair repulsion > lone pair–bonding pair repulsion > bonding pair–bonding pair repulsion.
3. True.
4. True.
5. True.
6. False. Carbon can participate in sp, sp^2, and sp^3 hybridization.
7. False. One sigma and two pi bonds.
8. True.
9. True.
10. True.
11. True.
12. False. Bond order = 2.
13. True.
14. False. There are no unpaired electrons in N_2.

SELF-TEST A

1. a. Square pyramid   b. Linear   c. Triangular   d. Bent   e. Bent
2. a. sp^3d^2 b. sp c. sp^2 d. sp^2 e. sp^2
3. a. Triangular   b. Distorted tetrahedral   c. Square planar   d. Triangular   e. Tetrahedral
4. a. sp^2 b. sp^3d c. sp^3d^2 d. sp^2 e. sp^3
5. b
6. a. 120°   b. 90°   c. 90° and 120°
7. CCl_4 has four electron pairs about the central carbon atom, while $AsCl_4^-$ has five pairs about the As atom with only four Cl atoms attached. The extra pair of unshared electrons prevents the formation of a tetrahedron by $AsCl_4^-$.
8. There are two electron pairs in the valence shell of Be in $BeCl_2$, but there are four pairs in the valence shell of Te in $TeCl_2$. These four pairs are at the corners of a tetrahedron. When two chlorine atoms bond to a Te atom via two of the pairs, the Cl — Te — Cl bond angle is 109.5°.
9. sp^3d, sp^3d^2
10. No
11. HBr
12. b
13. b and c
14. c
15. The O — H bonds are polar. Nonlinear.

16.

	H_2^+	HHe	He_2^+
σ_{2s}^*		↑	↑
σ_{1s}	↑	↑↓	↑↓
Bond order	0.5	0.5	0.5

17.

	C_2		C_2^{2-}	
σ_{2p}^*	——		——	
π_{2p}^*	——		——	
σ_{2p}	——		↑↓	
π_{2p}	↑↓	↑↓	↑↓	↑↓
π_{2s}^*	↑↓		↑↓	
σ_{2s}	↑↓		↑↓	
Bond order	2		3	

18. The bond order in N_2 is 3, while in N_2^+ it is only 2.5. The greater the bond order, the shorter the bond distance.

19.

$[O-N-O]^-$ $\begin{bmatrix}O-N\begin{smallmatrix}O\\O\end{smallmatrix}\end{bmatrix}^-$ $O=S\begin{smallmatrix}O\\O\end{smallmatrix}$

SELF-TEST B1

1. Linear
2. Tetrahedral
3. It must be symmetrical, which for CO_2 would be linear. The bond dipoles point in opposite directions.
4. sp^3 hybrid orbitals point to the corners of a tetrahedron.

SELF-TEST B2

1. Two electron pairs
2. Four electron pairs
3. Nonpolar
4. Tetrahedral

Chapter Eleven
INTERMOLECULAR FORCES AND LIQUIDS AND SOLIDS

- States of Matter
- Intermolecular Forces
- The Liquid State
- The Solid State
- Phase Changes

STATES OF MATTER

STUDY OBJECTIVE

You should be able to:
1. Describe the characteristic properties of solids, liquids, and gases.

Kinetic Molecular Theory of Liquids and Solids. Matter can exist in three states, or phases: gas, liquid, and solid. Each of these states will be described from a molecular viewpoint in terms of four characteristics: distance between molecules, attractive forces between molecules, the motion of molecules, and the orderliness of the arrangement of molecules (see Figure 11.1).

In Chapter 5 you learned that a gas can be pictured as a collection of molecules that are far apart, are in constant random motion, and exert almost no forces on each other. The kinetic molecular theory was developed to explain the behavior of an ideal gas.

Liquids and solids are quite different from gases. Gases have low density, have high compressibility, and completely fill a container. The condensed states have relatively high density, are almost incom-

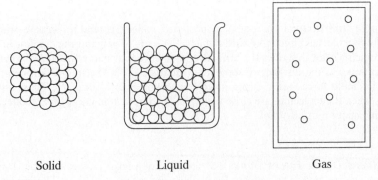

Solid　　　Liquid　　　Gas

Figure 11.1. A molecular view of solids, liquids, and gases. Can you use the kinetic molecular theory to explain the characteristic properties given in Table 11.1?

pressible, and have definite volumes. These properties indicate that the "molecules" of solids and liquids are close together and are held by strong intermolecular forces.

One difference between solids and liquids is that liquids are fluids, but solids are not. In liquids, molecules can move past each other even though they cannot get very far from each other. In solids, molecules are fixed in position, and at most can only vibrate about that position. Many solids are characterized by long-range order of their constituent "molecules." This order gives rise to crystal structures. Table 11.1 summarizes some of the properties of the three states of matter.

Table 11.1 Characteristic Properties of States of Matter

Gas	Liquid	Solid
Assumes the volume and shape of its container	Has a definite volume but assumes the shape of its container	Has a definite volume and shape
Is a fluid (flows readily)	Is a fluid	Is not a fluid
Very compressible	Only slightly compressible	Virtually incompressible
Low density	High density	High density
Molecules far apart	Molecules close together	Molecules close together

INTERMOLECULAR FORCES

STUDY OBJECTIVES

You should be able to:
1. Describe the different types of intermolecular attractive forces.
2. Relate the strength of dispersion forces to molecular mass.

van der Waals Forces. The forces that hold atoms together such as covalent bonds exist within molecules. The forces that hold individual molecules close together as in a solid or a liquid are called *intermolecular forces* (*IMFs*). Intermolecular forces can be grouped for convenience into *van der Waals forces, ion-dipole forces, and hydrogen bonds*. The van der Waals forces cause a gas to deviate from ideal gas behavior and are also responsible for gases condensing to form liquids. They are attractive forces resulting from dipole-dipole interactions, dipole-induced dipole interactions, and dispersion forces. We shall discuss each of these interactions briefly.

Dipole-Dipole Interactions. In a dipolar substance, molecules tend to become oriented with the positive end of one molecule directed toward the negative ends of neighboring dipoles. In Chapter 9 you learned that Coulomb's law describes the attractive force between charged particles, where the force is inversely proportional to the distance of separation squared, $1/r^2$. The interaction energy due to an attractive dipole-dipole force varies as the inverse sixth power of r, or $1/r^6$. Thus, this force is significant only when the molecules are close together. The interaction energy for a collection of dipolar molecules is only about -1.5 kJ/mol.

Dipole-Induced Dipole Interactions. The presence of a polar molecule in the vicinity of another molecule (usually nonpolar) has the effect of polarizing the second molecule. The induced dipole can then interact with the dipole moment of the first molecule, and the two molecules are attracted to each other. Again, the interaction energy is proportional to $1/r^6$. The energy of attraction depends on how easily the molecule is polarized and is approximately -1.0 kJ/mol for a collection of polar and nonpolar molecules.

Dispersion Forces. Nonpolar molecules such as O_2, SF_6, and even the noble gases He, Ne, Ar, Kr, and Xe all show deviations from the ideal gas equation and all condense at low temperatures. We can ask: What kind of attractive forces exist between nonpolar molecules? In 1930, Fritz London proposed that although these molecules have no permanent dipole moments, their electron clouds are fluctuating. In a helium atom, for instance, the electrons occupy a $1s$ orbital, which has a spherical shape. The electrons are in constant motion, and if for an instant they should both move to the same side of the nucleus, an instantaneous dipole will exist. This dipole will polarize a neighboring molecule, and the two dipoles will tend to stick together. The attractive interactions caused by instantaneous dipoles are known as *dispersion forces*.

Another important intermolecular force is the repulsive force between like charges on neighboring molecules. In a liquid, if one of the molecules rotates through 360° relative to another molecule, you might expect that the net force of attraction between the two molecules will average to zero. This does not occur because at the same distance of separation r, the attractive forces are stronger than the repulsive forces. The attractive energy is proportional to $1/r^6$ while the repulsion energy is proportional to $1/r^{12}$. Therefore, at the usual distance of molecular separation in a liquid (300 pm), the attractive force dominates over the repulsive force.

Generalizations concerning van der Waals Forces. All polar molecules exhibit both dipole-dipole attractions and dispersion forces, whereas nonpolar molecules exhibit only dispersion forces. The strength of dispersion forces depends on the polarizability of the molecule and can be as large as or larger than dipole-dipole forces. *Polarizability* is the tendency of an electron cloud to be distorted by the presence of an electrical charge such as that of an ion or the partial charge of a dipole. In general, the polarizability increases as the total number of electrons in a molecule increases. Since molecular mass and number of electrons are related, the polarizability of molecules and the strength of dispersion forces increase with increasing molecular mass.

The strength of intermolecular forces can be summarized as follows:

1. For molecules of approximately equal molecular mass, the IMF increases as polarity increases.
2. For molecules of approximately equal polarity, the IMF increases with increasing molecular mass.

Ion-Dipole Forces. Hydration of ions, which we discussed in Chapter 6, is a good example of ion-dipole interactions. When an ionic compound such as NaBr dissolves, the ions and water molecules are attracted to each other. The source of this attraction is the electrostatic attraction of opposite charges. Recall that water molecules are polar. That is, they have a negative end and a positive end. The Na^+ ions attract the negative end of the water molecule, and the Br^- ions attract the positive end of the water molecule. As the charge of an ion increases, it attracts polar molecules more strongly. Thus Mg^{2+} ions attract water molecules more strongly than Na^+ ions. Al^{3+} ions attract water molecules even more strongly than Mg^{2+}.

Hydrogen Bonding. An additional type of intermolecular force is necessary to explain certain properties of ammonia, water, and hydrogen fluoride. In water, for instance, the attractions are more than just the attractions of one dipole for another. Each hydrogen atom with its partial positive charge is attracted to one of the lone electron pairs of an oxygen atom of a neighboring molecule. This is usually represented as

$$H-\ddot{\underset{|}{\overset{}{O}}}:----\overset{\delta+}{H}-\overset{\overset{H}{|}}{\underset{}{\ddot{O}}}:^{\delta-}$$
$$\phantom{H-\ddot{O}:----}H$$

and is called a hydrogen bond. Hydrogen bonding is limited to compounds containing nitrogen, oxygen,

210 / *Intermolecular Forces and Liquids and Solids*

and fluorine. The requirements for hydrogen bond formation are as follows:

1. The element that is covalently linked to hydrogen must be sufficiently electronegative to attract bonding electrons and leave the hydrogen atom with a significant δ + charge.
2. The electronegative atom, bound to hydrogen by the hydrogen bond, must have a lone pair of electrons.
3. The small size of the hydrogen atom allows it to approach nitrogen, oxygen, and fluorine atoms in neighboring molecules very closely. It is significant that hydrogen bonding is limited to these three elements of the second period. Both sulfur and chlorine are highly electronegative but do not form hydrogen bonds. These third-period atoms are apparently too large and do not present a highly localized nonbonding electron pair for the H atom to be attracted to. Hydrogen bonds are the strongest of the intermolecular forces, with energies of the order 10–40 kJ/mol. Van der Waals interactions correspond to between 2 and 20 kJ/mol.

EXAMPLE 11.1   Types of Intermolecular Forces

Indicate all the different types of intermolecular forces that exist in each of the following substances:
a. $CCl_4(l)$
b. $HBr(l)$
c. $CH_3OH(l)$

METHOD OF SOLUTION

First you must use the concepts of bond polarity and molecular structure to decide whether the molecule is polar or not.
a. CCl_4 is nonpolar. *Answer:* The only type of IMFs are dispersion forces.
b. HBr is a polar molecule. *Answer:* The types of IMFs are dipole-dipole and dispersion forces. There is no hydrogen bonding in HBr. The Br atom is not electronegative enough.
c. CH_3OH is polar and has a hydrogen atom bound to an oxygen atom. *Answer:* Dipole-dipole attractions, dispersion forces, and H bonds exist in liquid methanol.

EXAMPLE 11.2   Types of Intermolecular Forces

Which of the following substances should have the strongest intermolecular attractive forces: N_2, Ar, F_2, or Cl_2?

METHOD OF SOLUTION

Note that none of these molecules is polar and that there is no chance for H bonding to occur. The only intermolecular forces existing in these molecules then are dispersion forces. Dispersion forces increase as the polarizability of the molecule increases, while polarizability increases with molecular mass. *Answer:* The most polarizable molecule will be Cl_2 (70.9 amu).

EXAMPLE 11.3   Intermolecular Forces

The dipole moment μ in HCl is 1.03 D, and in HCN it is 2.99 D. Which one should have the higher boiling point?

METHOD OF SOLUTION

The larger the dipole moment, the stronger the IMF. The stronger the IMF, the higher the temperature needed to provide molecules of the liquid with enough kinetic energy to overcome these attractive forces. The boiling point (b.p.) of HCN should be greater than that of HCl. Observed values are b.p. of HCl = $-85°C$ and b.p. of HCN = $26°C$.

THE LIQUID STATE

STUDY OBJECTIVES

You should be able to:
1. Describe the physical properties called surface tension and viscosity.
2. Compare liquids to solids and gases with respect to the physical properties: density, compressibility, and surface tension.
3. List six unusual properties of water.

Liquid Density. With respect to density, liquids are more like solids than like gases. Both solids and liquids are approximately 1000 times more dense than gases at STP. In general, liquids are 5 to 10% less dense than solids. A notable exception is water; liquid water is 1.09 times more dense than ice.

Liquid Compressibility. Another property of liquids that resembles that of solids is compressibility. Liquids are roughly 10 times more compressible than solids but do not even approach gases, which are 100,000 times more compressible than solids. The differences in compressibility between liquids and solids suggest that liquids have more open space. Apparently in liquids there are holes between atoms. These holes are largely absent in solids.

Surface Tension. A property of a liquid that has no direct counterpart in solids or gases is the surface tension. Surface tension is what makes water bead up on a freshly waxed surface and what makes soap bubbles round. Surface tension is a force that tends to minimize the surface area of a drop of liquid. Energy is required to expand the surface of a liquid, and the surface tension is the amount of energy required to increase the surface area of a liquid by a unit area. Liquids in which strong intermolecular forces exist exhibit high surface tensions.

Water beads up on a freshly waxed surface because its cohesive forces are stronger than its adhesive forces. *Cohesive forces* are the intermolecular forces between like molecules in the drop. *Adhesive forces* are intermolecular attractions between unlike molecules, such as between water and wax. The strong cohesive forces tend to maintain the drop.

When water comes in contact with glass, *adhesion* is stronger than cohesion. Water is pulled against the glass surface, and we say that water "wets" the surface. Wetting results in the spreading of a thin film of water on the glass surface.

Viscosity. One characteristic of liquids that we have all observed is related to how freely they flow. Water pours much more freely than motor oil, and motor oil more readily than glycerol, for instance. The unique pouring characteristics of each liquid are the result of its resistance to flow. *Viscosity* is a measure of resistance to flow. Liquids whose molecules have strong intermolecular forces have greater viscosities than those liquids that have weaker intermolecular forces. Table 11.4 in the textbook lists viscosities of some common liquids.

As the temperature increases, the viscosity of a liquid decreases. Again the property is a result of the molecular behavior of liquids. At high temperatures the average kinetic energy of molecules is increased. Therefore, more molecules have enough energy to overcome intermolecular forces at higher temperatures.

Properties of Water. Given its abundance and familiarity, we often fail to realize just how unusual water actually is. Some of the unusual properties of water are as follows:

1. Water has a considerably greater surface tension than most other liquids.
2. At 4.184 J/g · °C, the specific heat of water is one of the highest of all substances (see Table 6.1 in the textbook).

3. The heats of fusion and vaporization are 6.0 and 40.8 kJ/mol, respectively. H_2O is unusual because 40.8 kJ is a huge vaporization energy. Compare values in Table 11.7 in the textbook.
4. Boiling point. The boiling point of water is about 200°C higher than might be reasonably expected. Boiling point tends to be related to the molar mass. Compare the boiling point of water with that of other low molar mass liquids in Table 11.2.

Table 11.2 Boiling Points of Some Small Molecules

Liquid	Molar Mass (g/mol)	Boiling Point (°C)
H_2O	18	100
CH_4	16	−159
NH_3	17	−33

5. Density versus temperature. As liquids cool, they become more and more dense. As water is cooled from 100°C down to 4°C, it does indeed get more dense. But from 4°C (actually 3.98°C) down to 0°C just the opposite happens: It gets less dense (see Table 11.3).

Table 11.3 Density of H_2O Near 0°C

Temperature (°C)	Water Density (g/cm^3)
10.0	0.9997
5.0	0.99999
3.98	1.00000
2.0	0.99997
0	0.9998

6. Density of liquid versus density of solid. There are over six million known chemical compounds. All, except a dozen or so, have a solid state that is more dense than the liquid. Water, of course, is one of the exceptions. The density of solid water at 0°C is 0.917 g/cm^3, which compares with 0.9998 g/cm^3 for the liquid. Because of its lower density, ice floats on water.

EXAMPLE 11.4 Adhesive and Cohesive Forces

In a glass tube containing a liquid, a curved surface is often noted on the liquid. This curved surface is called a meniscus. For a water-glass interface the water forms a concave upward curve:

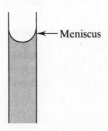

Explain the shape of the water-glass meniscus.

METHOD OF SOLUTION

Water molecules are attracted to a clean glass surface. The shape of the meniscus results from the adhesive forces between glass and water being stronger than the cohesive forces in water. The water molecules spread out over the glass and pull other water molecules after them. This lifts the water along the glass-water interface and produces a concave upward surface.

EXAMPLE 11.5 Intermolecular Forces in Liquids

The viscosities of ether and water are given in Table 11.4 in the textbook as 0.000233 and 0.00101 N s/m^2. Discuss their relative values in terms of the following molecular structures:

$$\underset{\text{water}}{\overset{O}{\underset{H\quad H}{\diagup\diagdown}}} \qquad \underset{\text{ethyl ether}}{\overset{CH_3\quad O\quad CH_3}{\underset{CH_2 CH_2}{\diagdown\diagup\diagdown\diagup}}}$$

METHOD OF SOLUTION

Molecules with strong intermolecular forces have greater viscosities than those that have weak intermolecular forces. Here both molecules are dipolar; water because of the O — H bonds and ether because of the polar C — O bond. However, water has a higher viscosity than ether because of its ability to form hydrogen bonds between molecules.

THE SOLID STATE

STUDY OBJECTIVES

You should be able to:
1. Calculate the spacing between planes in a crystal using Bragg's equation.
2. Determine the number of atoms per unit cell given the type of unit cell.
3. Predict whether the type of crystal formed by a substance is ionic, covalent, molecular, or metallic.

X-Ray Diffraction of Crystals. Crystal structures are analyzed experimentally by the technique of X-ray diffraction. Diffraction is a wave property. When an X-ray beam encounters a single crystal, the beam is scattered from the crystal at only a few angles, rather than randomly. The observed angle θ depends on the spacing d between layers or planes of atoms in the crystal, the wavelength (λ) of the X rays, and the reflection order ($n = 1, 2, 3, \ldots$). These quantities are related by the Bragg equation:

$$n\lambda = 2d \sin \theta$$

The spacing (d) between atomic planes in crystals can be determined by using X rays of known λ and by measuring θ.

Crystal Structure. The essential information obtained from X-ray diffraction studies relates to the geometric form and dimensions of the *unit cell*. The unit cell is the smallest unit which, when repeated over and over again, generates the entire crystal. Three types of unit cell are the simple cubic cell (scc), the body-centered cubic cell (bcc), and the face-centered cubic cell (fcc). These types are illustrated in Figure 11.2.

In a simple cubic cell, particles are located only at the corners of each unit cell. In a body-centered cubic cell, particles are located at the center of the cell as well as at the corners. In a face-centered cubic cell, particles are found at the center of each of the six faces of the cell as well as at the corners. In many calculations involving properties of a crystal, it is important to know how many atoms or ions are contained in each unit cell. Atoms at the corners of unit cells are shared by neighboring unit cells. For a cubic cell, each corner atom is shared by eight unit cells and is counted as 1/8 particle for each unit cell.

214 / *Intermolecular Forces and Liquids and Solids*

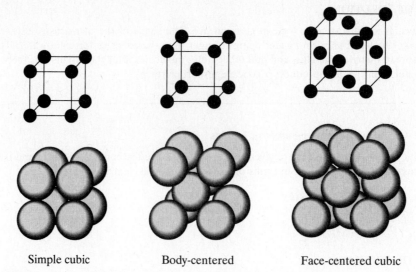

Figure 11.2. Three types of cubic cells. Notice that a "body-centered" sphere lies entirely within the unit cell. Only 1/2 of a "face-centered" sphere lies within the cell. Only 1/8 of a "corner" sphere lies within the cell.

A face-centered atom is shared by two unit cells and is counted as 1/2 of an atom for each unit cell. An atom located at the center belongs wholly to that unit cell.

The packing efficiency is the percentage of the cell space occupied by spheres. The empty space is the interstices, or "holes," between spheres. There are many substances whose atomic arrangement can be pictured as a result of packing together identical spheres so as to achieve maximum density. Many metallic elements and many molecular crystals display these "closest packed" structures. The text describes these two "closest packed" structures in Section 11.4. Alternating layers of spheres in ABABAB... fashion yield a hexagonal close-packed (hcp) structure, while alternating layers of spheres in the ABCABC... fashion generate the cubic close-packed (ccp) structure. In both structures each sphere has a coordination number of 12, which is the maximum coordination number. For both structures the packing efficiency is 74%.

Types of Crystals. In a solid, atoms, molecules, or ions occupy specific positions called lattice points. Crystalline solids are characterized by a regular three-dimensional arrangement of lattice points. In *ionic solids* the lattice points are occupied by positive and negative ions. In all ionic compounds there are continuous three-dimensional networks of alternating positive and negative ions held together by strong electrostatic forces (ionic bonds). Most ionic crystals possess high lattice energies, high melting points, and high boiling points. In ionic solids there are no discrete molecules as are found in molecular or covalent substances.

In *covalent crystals* the lattice points are occupied by atoms that are held by a network of covalent bonds. Diamond, graphite, and quartz are well-known examples. Materials of this type have high melting points and are extremely hard because of the large number of covalent bonds that have to be broken to melt or break up the crystal. The entire crystal can be thought of as one giant molecule.

Covalent compounds form crystals in which the lattice positions are occupied by molecules. Such solids are called *molecular crystals*. Molecular crystals are soft and have low melting points. These properties are the result of the relatively weak intermolecular forces (van der Waals forces and hydrogen bonding) that hold the molecules in the crystal.

Crystals of polar compounds are held together by dipole-dipole forces and dispersion forces. Only dispersion forces occur in the lattices of nonpolar compounds. As a rule polar compounds melt at higher temperatures than nonpolar compounds of comparable molecular mass.

Metallic crystals are quite strong. Most transition metals have high melting points and densities. In metals the array of lattice points is occupied by positive ions. The outer electrons of the metal atoms are loosely held and move freely from ion to ion throughout the metallic crystal. This mobility of electrons in metals accounts for one of the characteristic properties of metals, namely the ability to conduct electricity.

EXAMPLE 11.6 Using the Bragg Equation

When X rays with a wavelength of 0.154 nm are diffracted by a crystal of metallic copper, the angle corresponding to the first-order diffraction ($n = 1$) is found to be 37.06°. What is the spacing between planes of Cu atoms that gives rise to this diffraction angle?

METHOD OF SOLUTION

The spacing between Cu atoms corresponds to d in the Bragg equation, which we rearrange to solve for d:

$$d = \frac{n\lambda}{2 \sin \theta}$$

CALCULATION

For a first-order diffraction $n = 1$. The sine of 37.06° can be determined with your calculator.

$$d = \frac{0.154 \text{ nm}}{2 \sin 37.06°} = \frac{0.154 \text{ nm}}{2(0.603)}$$

$$= 0.128 \text{ nm } (128 \text{ pm})$$

EXAMPLE 11.7 Number of Atoms per Unit Cell

If atoms of a solid occupy a face-centered cubic lattice, how many atoms are there per unit cell?

METHOD OF SOLUTION

In a face-centered cubic cell there are atoms at each of the eight corners, and there is one in each of the six faces.

CALCULATION

$$8 \text{ corners } (\tfrac{1}{8} \text{ atom per corner}) + 6 \text{ faces } (\tfrac{1}{2} \text{ atom per face}) = 4 \text{ atoms per unit cell}$$

EXAMPLE 11.8 Dimensions of a Unit Cell

Potassium crystallizes in a body-centered cubic lattice and has a density of 0.856 g/cm^3 at 25°C.
a. How many atoms are there per unit cell?
b. What is the length of an edge of the cell?

METHOD OF SOLUTION

a. A body-centered cubic structure has one K atom in the center and eight other K atoms, one at each corner of the cube. The corner atoms, however, are shared by 8 adjoining cells. Each corner atom contributes 1/8 of an atom to the unit cell. The total number of K atoms per unit cell is

$$1 + \tfrac{1}{8}(8) = 2 \text{ atoms}$$

b. Ideally the crystal of potassium is made up of a large number of unit cells repeated over and over again. Thus, the density of the unit cell will be the same as the density of metallic K. Recall that density is an intensive property:

$$\text{density} = \frac{\text{mass}}{\text{volume}}$$

$$\text{density (unit cell)} = \frac{\text{mass of 2 K atoms}}{a^3}$$

where a is the length of the side of the unit cell.

CALCULATION

The mass of a K atom is

$$\frac{39.1 \text{ g}}{\text{mol}} \times \frac{1 \text{ mol}}{6.02 \times 10^{23} \text{ atoms}} = 6.50 \times 10^{-23} \text{ g/atom}$$

Substituting into the density equation yields

$$0.856 \text{ g/cm}^3 = \frac{2(6.50 \times 10^{-23} \text{ g/atom})}{a^3}$$

Rearranging yields

$$a^3 = 1.52 \times 10^{-22} \text{ cm}^3$$

$$a = 5.34 \times 10^{-8} \text{ cm} = 534 \text{ pm}$$

EXAMPLE 11.9 Atomic Radius of a Potassium Atom

Determine the radius of a K atom for the metallic potassium crystal in Example 11.8.

METHOD OF SOLUTION

From Figure 11.23 in the text, it is clear that we need the diagonal distance of the body-centered cubic cell because only along the diagonal do the atoms actually touch. The diagonal c will equal $4r$, where r is the atomic radius. From the figure we see that

$$c = 4r \quad \text{and} \quad c = \sqrt{3}\, a$$

Therefore

$$4r = \sqrt{3}\, a$$

where a is the length of an edge. Solving for r, we get

$$r = \frac{a\sqrt{3}}{4} = 534 \text{ pm} \times \frac{\sqrt{3}}{4}$$

$$= 231 \text{ pm}$$

EXAMPLE 11.10 Types of Crystals

What type of force must be overcome in order to melt crystals of the following substances?
a. Mg
b. Cl_2
c. $MgCl_2$
d. SO_2
e. Si

METHOD OF SOLUTION

a. Magnesium is a metal and so it forms metallic crystals. Since the mobile electrons are shared between positive ions, the forces between Mg atoms could be described as covalent bonds, in this case called metallic bonds. The melting point is 1105°C.
b. Cl_2 forms molecular crystals that are held together by intermolecular forces, more specifically dispersion forces. The melting point is -101°C.
c. $MgCl_2$ is an ionic compound. When it melts, Mg^{2+} ions and Cl^- ions break away from their lattice positions. To melt magnesium chloride, ionic bonds must be broken. The melting point is 1412°C.
d. SO_2 is a polar molecule. Thus, dipole-dipole attractions and dispersion forces must be overcome. The melting point is -73°C.
e. Si crystallizes in a diamond structure with each silicon atom bound to four others by covalent bonds. It is a covalent crystal, and covalent bonds must be broken in order for it to melt. The melting point is 1410°C.

PHASE CHANGES

STUDY OBJECTIVES

You should be able to:
1. Describe the dynamic equilibrium between a liquid and its vapor state and list the factors that affect the vapor pressure of a liquid.
2. Describe the significance of the critical point.
3. Define the molar heats of fusion, vaporization, and sublimation.
4. Using a phase diagram, predict what changes a substance will undergo as it is heated or cooled or subjected to pressure.
5. Use the Clausius-Clapeyron equation to relate vapor pressure, temperature, and the heat of vaporization of a liquid.

Liquid-Vapor Equilibrium. When a liquid substance is placed in a closed container, it will not be long before molecules of this substance can be found in the gas phase above the liquid. Vaporization is the process in which liquids become gases. When a portion of a liquid evaporates, the gaseous molecules exert a pressure called the *vapor pressure*. The term *vapor* is often applied to the gaseous state of a substance that is normally a liquid or solid at the temperature of interest.

In a closed container the vapor pressure does not just continually increase; rather a state of equilibrium is reached in which the vapor pressure becomes constant and the amount of liquid remains constant. This equilibrium vapor pressure results from two opposing processes. The opposing process to vaporization is condensation. As the concentration of the molecules in the vapor phase increases, some will strike the liquid surface and condense:

$$\text{liquid} \underset{\text{condensation}}{\overset{\text{vaporization}}{\rightleftharpoons}} \text{vapor}$$

When the two rates become equal, the vapor pressure remains constant and is called the *equilibrium vapor pressure*, or just vapor pressure. Note that in the state of equilibrium, while the amounts of vapor and liquid do not change, there is considerable activity on the molecular level. Thus evaporation and condensation are constantly occurring, but *at the same rates*. Such an equilibrium state is referred to as a *dynamic equilibrium*.

Vapor pressure is a function of the temperature (see Table 11.4). As temperature increases, the evaporation rate increases, and so more molecules exist in the vapor phase at higher temperatures.

Table 11.4 Vapor Pressure of Water

Temperature (°C)	Pressure (mm Hg)
20	17.54
30	31.82
40	55.32
50	92.51
80	355.1
86.5	460.0
100	760.0

In a container open to the atmosphere, boiling of a liquid will occur when the temperature is raised high enough. In order for a bubble to form, the vapor inside the bubble must be able to push back the atmosphere. This will not occur until a temperature is reached at which the vapor pressure is greater than the atmospheric pressure. Therefore, the *boiling point* of a liquid is the temperature at which the vapor pressure is equal to the atmospheric pressure. Since the boiling point of a liquid depends on the atmospheric pressure, and the atmospheric pressure varies daily, the boiling point is not a constant. A *normal boiling point* is defined for purposes of comparing liquid substances. The normal boiling point is the boiling temperature when the external pressure is 1 atm. Figure 11.3 shows the vapor pressures and normal boiling points of three liquids.

The energy required to vaporize 1 mol of a liquid is called the *molar heat of vaporization* (ΔH_{vap}). The value of ΔH_{vap} is directly proportional to the strength of intermolecular forces. Liquids with relatively high heats of vaporization have low vapor pressures and high normal boiling points.

The quantity ΔH_{vap} can be determined experimentally. As Figure 10.39 of the text shows, the vapor pressure of a liquid increases with increasing temperature. The quantitative relationship between the vapor pressure (P) and the absolute temperature (T) is given by the *Clausius-Clapeyron equation*:

$$\ln P = -\frac{\Delta H_{vap}}{RT} + C$$

where ln is the natural logarithm, R is the ideal gas constant in units of $JK^{-1}\,mol^{-1}$, and C is a constant.

A useful form of this equation is

$$\ln\left(\frac{P_1}{P_2}\right) = \frac{\Delta H_{vap}}{R}\left(\frac{T_1 - T_2}{T_1 T_2}\right)$$

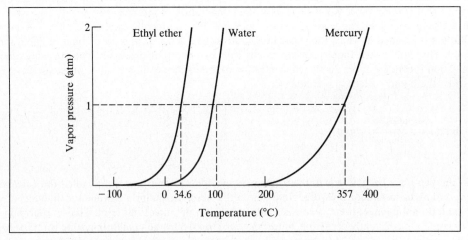

Figure 11.3. The effect of temperature on the vapor pressure of three liquids. Notice that the normal boiling points for the liquids are the temperatures at which their vapor pressures are equal to 1 atm.

The ΔH_{vap} can be calculated if the vapor pressures P_1 and P_2 are measured at two temperatures T_1 and T_2. Alternatively, if ΔH_{vap} is already known and one vapor pressure P_1 is known at temperature T_1, you can calculate the vapor pressure P_2 at some new temperature T_2. See Example 11.11.

Boiling Point as a Measure of IMFs.
The boiling point and the heat of vaporization are convenient measures of the strength of intermolecular forces. In the liquid, molecules are very close together and are strongly influenced by intermolecular forces. In the gas phase, molecules are widely spaced, move rapidly, and have enough energy to overcome intermolecular forces. During vaporization, molecules are completely separated from one another.

The heat of vaporization is the energy necessary to overcome intermolecular forces. The boiling point reflects the kinetic energy that liquid molecules must have to overcome intermolecular forces and escape into the gas phase.

Critical Temperature and Pressure.
When a liquid is heated in a closed container, the vapor pressure increases, but boiling does not occur. The vapor cannot escape, and the vapor pressure continually rises. Eventually a temperature is reached at which the meniscus between liquid and vapor disappears. This temperature is called the critical temperature. The *critical temperature* is the highest temperature at which the substance can exist as a liquid. The *critical pressure* is the lowest pressure that will liquefy a gas at the critical temperature. Critical temperatures and pressures are listed in Table 11.8 in the textbook.

As we saw previously, intermolecular forces are dominant in liquids and effectively determine many of their properties. The critical temperature and pressure are no exceptions. Substances with high critical temperatures have strong intermolecular forces of attraction.

Liquid-Solid Equilibrium.
The temperature at which the solid and liquid are in dynamic equilibrium at 1 atm pressure is called the *normal melting point* or freezing point. Melting is also called fusion. During melting the average distance between molecules is increased slightly as evidenced by the approximately 10% decrease in density:

$$\text{solid} \xrightleftharpoons[\text{freezing}]{\text{fusion}} \text{liquid}$$

The energy required to melt 1 mol of a solid is called the *molar heat of fusion*, ΔH_{fus}. Upon freezing, the substance will evolve the same amount of energy. The ΔH_{fus} is always much less than ΔH_{vap} because vaporization separates molecules completely from each other and so requires more energy than fusion.

Heating Curves.
A heating curve is a convenient way to summarize the solid-liquid-gas transitions for a compound. The heating curve of water is shown in Figure 11.4. It is the result of an experiment in which a given amount of ice (e.g., 1 mol) at some initial temperature below 0°C is slowly heated at *a constant rate*.

The curve shows that the temperature of ice increases on heating (line 1) until the melting point is reached. No rise in temperature occurs while ice is melting (line 2); as long as some ice remains, the temperature stays at 0°C. The length of line 2 is a measure of the heat necessary to melt 1 mol of ice, which is ΔH_{fus}. Along line 3 the temperature of liquid water increases from 0°C to 100°C. The slope of the line Δt per joule of heat depends on the specific heat of liquid water. The greater slope for line 1 than for line 3 means that less heat is required to raise the temperature of ice than of water. This is evidenced by the difference in specific heats of ice and liquid water. The specific heat of ice is 2.09 J/g·°C versus 4.18 J/g·°C for liquid H_2O.

At 100°C the liquid begins to boil, and the heat added is used to bring about vaporization. The temperature does not rise until all the liquid has been transformed to gas. The length of line 4 is the heat required to vaporize 1 mol of liquid. Line 4 will always be longer than line 2 because $\Delta H_{vap} > \Delta H_{fus}$. Line 5 corresponds to the heating of steam. Again the slope of line 5 depends on the specific heat of steam, which is about 1.98 J/g·°C.

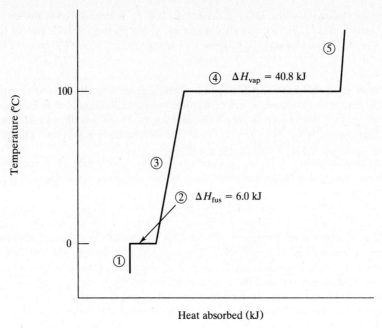

Figure 11.4. The heating curve for water.

Solid-Vapor Equilibrium. Sublimation is the evaporation of a solid directly into the vapor phase. Dry ice and iodine are substances that sublime readily. Ice also sublimes to some extent. As with liquids, the vapor pressure of a solid increases as the temperature increases. The direct conversion of solid to vapor is equivalent to melting the solid first and then vaporizing the liquid. From Hess's law, we obtain

$$\Delta H_{sub} = \Delta H_{fus} + \Delta H_{vap}$$

Phase Diagrams. From discussions in preceding sections we can see that the phase in which a substance exists depends on its temperature and pressure. In addition, two phases may exist in equilibrium at certain temperatures and pressures. Information about the stable phases for a specific compound is summarized by a phase diagram such as that shown in Figure 11.5.

A phase diagram is a graph of pressure versus temperature. The diagram is divided into three regions, one for the solid phase, one for the liquid phase, and one for the gas phase. The line between

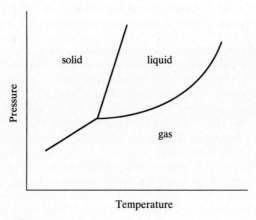

Figure 11.5. A typical phase diagram.

the solid and liquid regions is made up of points of P and T at which the solid and liquid phases are in equilibrium. The line between the solid and gas regions is made up of points of P and T at which solid and gas are in equilibrium. And the line between the liquid and gas regions gives temperatures and pressures at which liquid and gas are in equilibrium. Since all three lines intersect at the *triple point*, this one point describes the conditions under which gas, liquid, and solid are all in equilibrium.

The phase diagram shows at a glance several properties of a subtance: melting point, boiling point, and triple point. If a point (P and T) describing a system falls in the solid region, the subtance exists as a solid. If the point falls on a line such as that between liquid and gas regions, the substance exists as liquid and vapor in equilibrium.

EXAMPLE 11.11 Clausius-Clapeyron Equation

The vapor pressure of water is 55.32 mm Hg at 40.0°C and 92.51 mm Hg at 50.0°C. (See Table 5.1 in the text.) Calculate the molar heat of vaporization of water.

METHOD OF SOLUTION

The Clausius-Clapeyron equation relates the vapor pressure of a liquid to its molar heat of vaporization:

$$\ln\left(\frac{P_1}{P_2}\right) = \frac{\Delta H_{vap}}{R}\left(\frac{T_1 - T_2}{T_1 T_2}\right)$$

CALCULATION

Substituting the value given above yields

$$\ln\left(\frac{55.32 \text{ mm Hg}}{92.51 \text{ mm Hg}}\right) = \frac{\Delta H_{vap}}{8.314 \text{ J K}^{-1} \text{ mol}^{-1}}\left(\frac{313 \text{ K} - 323 \text{ K}}{(313 \text{ K})(323 \text{ K})}\right)$$

$$\ln 0.5980 = \frac{\Delta H_{vap}}{8.314 \text{ J K}^{-1} \text{ mol}^{-1}}\left(\frac{-10 \text{ K}}{1.01 \times 10^5 \text{ K}^2}\right)$$

Taking the logarithm and rearranging yields

$$\frac{-0.5142(8.314 \text{ J K}^{-1} \text{ mol}^{-1})(1.01 \times 10^5 \text{ K})}{-10 \text{ K}} = \Delta H_{vap}$$

$$\Delta H_{vap} = 43{,}000 \text{ J/mol}$$

COMMENT

Compare this result to the ΔH_{vap} listed in Table 11.7 of the text (40,790 J/mol). The observed difference is real. The ΔH_{vap} is less at the boiling point than it is at a lower temperature. Over the entire temperature range of a liquid, ΔH_{vap} is close to, but not really, a constant.

EXAMPLE 11.12 Vapor Pressure and Boiling Point

Using Figure 11.42 in the textbook, estimate the boiling point of water in a pressure cooker at 2.0 atm pressure.

METHOD OF SOLUTION

The boiling point of water at 2.0 atm pressure is the temperature at which the vapor pressure of water is 2.0 atm. According to the figure, this should be approximately 120°C.

EXAMPLE 11.13 Heat of Vaporization

How much heat is evolved when 1.0 g of steam condenses at 100°C?

METHOD OF SOLUTION

The heat of vaporization of water at 100°C is 40.8 kJ/mol. Condensation is the reverse of vaporization, so it releases the same amount of heat:

$$H_2O(g) \rightarrow H_2O(l) \quad \Delta H = -40.8 \text{ kJ/mol}$$

CALCULATION

The heat evolved when 1.0 g of water vapor condenses is

$$q = (\text{mol } H_2O) \times \Delta H_{\text{cond}}$$

$$= 1.0 \text{ g} \times \frac{1 \text{ mol}}{18.0 \text{ g}} \times \frac{-40.8 \text{ kJ}}{1 \text{ mol}}$$

$$= -2.27 \text{ kJ}$$

COMMENT

This is a lot of heat and shows us why burns from steam can be very serious.

EXAMPLE 11.14 Heat of Fusion

The heat of fusion of aluminum is 10.7 kJ/mol. How much energy is required to melt 1 ton of Al at the melting point, 660°C.

METHOD OF SOLUTION

The heat of fusion is the energy required to melt 1 mol of a substance. Therefore we need to convert 2000 lb of Al to moles.

CALCULATION

$$q = (\text{mol } H_2O) \times \Delta H_{\text{fus}}$$

$$= 2000 \text{ lb} \times \frac{454 \text{ g}}{1 \text{ lb}} \times \frac{1 \text{ mol Al}}{27.0 \text{ g Al}} \times \frac{10.7 \text{ kJ}}{1 \text{ mol Al}} = 360 \times 10^3 \text{ kJ (360 MJ)}$$

EXAMPLE 11.15 The Critical Temperature

Discuss the possibilities of liquefying oxygen and carbon dioxide at 25°C by increasing the pressure.

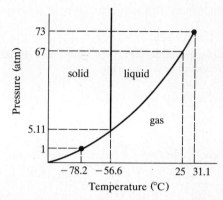

Figure 11.6. The phase diagram for CO_2.

METHOD OF SOLUTION

Above its critical temperature a substance cannot be liquefied by increasing the pressure. Substances having critical temperatures above 25°C can be liquefied at 25°C by application of sufficient pressure. With a critical temperature of 31°C, CO_2 can be liquefied at 25°C with application of enough pressure. The pressure required can be read off the phase diagram for CO_2 (Figure 11.6). A pressure of 67 atm is required to liquefy CO_2 at 25°C. Oxygen has a critical temperature of -119°C; therefore, at a temperature above -119°C, oxygen cannot be liquefied no matter how much pressure is applied. To liquefy O_2, its temperature must be lowered below -119°C and pressure applied.

TRUE-FALSE QUESTIONS

1. PH_3 should exhibit hydrogen bonding because of the lone electron pair on the P atom.

2. Dipole-dipole attractions are the strongest type of intermolecular force.

3. In general, a solid is 5–10% more dense than the liquid of the same compound.

4. Solids are more compressible than liquids.

5. Water beads up on wax surface because its cohesive forces are greater than its adhesive forces.

6. Most chemical substances are like water in that their solid phase will float on their liquid phase.

7. In cubic cells a corner atom is shared by four unit cells.

8. In cubic cells a face-centered atom is shared by two unit cells.

9. Glass is a substance in which the ions occupy regular lattice points.

10. For benzene (C_6H_6) $\Delta H_{vap} = 31.0$ kJ/mol and $\Delta H_{fus} = 13.1$ kJ/mol; therefore, $\Delta H_{sub} = 44.1$ kJ/mol.

11. The critical temperature is the lowest temperature at which a substance can exist as a liquid.

12. The triple point corresponds to a temperature and pressure at which the solid, liquid, and gas phases of a substance are all in equilibrium.

224 / Intermolecular Forces and Liquids and Solids

13. The boiling point of a compound varies as the atmospheric pressure varies.

14. The specific heat of water is 4.184 J/g·°C, and for aluminum it is 0.89 J/g·°C. Therefore, when 100 kJ of heat is absorbed by equal masses of both water and Al, the temperature change for Al will be about 4.7 times greater than Δt for water.

SELF-TEST A

1. Identify the types of intermolecular forces for each of the following substances:
   a. H_2O_2 b. H_2S c. SF_6 d. NOCl   e. NH_3

2. The enthalpies of vaporization of H_2S and H_2Se are 18.7 and 23.2 kJ/mol, respectively. Why is the ΔH_{vap} of water (40.6 kJ/mol) so much higher than the values for H_2S and H_2Se?

3. Which member of each pair has the stronger intermolecular forces of attraction?
   a. CO_2 or SO_2 b. F_2 or Br_2 c. H_2O or H_2S d. HCl or C_8H_{18}

4. In the laboratory, glassware is considered clean when water wets the glass in a continuous film. If droplets of water stand on the glass surface, the glass is considered dirty. Discuss the basis of this cleanliness test.

5. The viscosities of liquids generally decrease with increasing temperature. Water has the following viscosities, in units of N s/m^2: 0.0018 at 0°C; 0.0010 at 20°C; 0.0005 at 55°C; and 0.0003 at 100°C. Interpret this trend on a molecular basis.

6. The meniscus for mercury in a glass tube is concave downward. Explain.

7. At what angle would a first-order reflection be observed in the diffraction of 0.090 nm X rays by a set of crystal planes for which d = 0.500 nm?

8. Calcium oxide, like NaCl, crystallizes in a face-centered cubic cell.
   a. How many Ca^{2+} and O^{2-} ions are in each unit cell?
   b. Given the ionic radius of Ca^{2+} (99 pm) and O^{2-} (140 pm), what is the length of an edge of the unit cell?
   c. Calculate the density of CaO. See Figure 11.26 in the textbook.

9. From the following data calculate Avogadro's number. Potassium crystallizes in a body-centered cubic lattice with an edge of 530 pm. The density of K is 0.86 g/cm^3.

10. Aluminum crystallizes in a face-centered cubic lattice with the length of an edge equal to 405 pm. Assume that a face-centered atom touches each of the surrounding corner atoms. Calculate the length of a diagonal of a face.

11. Pick the substance in each pair that has the higher melting point.
    a. ICl or KCl   d. SiO_2 (quartz) or SiF_4
    b. K or Fe   e. CO_2 or SiO_2
    c. Fe or Cl_2

12. Distinguish between the boiling point of a liquid and the normal boiling point of a liquid.

13. Why does your skin feel cool when water evaporates from it? Why will you feel warmer on a more humid summer day than on a less humid day?

14. a. Use the following data to draw a rough phase diagram for methane: triple point −183.0°C; normal melting point −182.5°C; normal boiling point −161.5°C; critical point −83.0°C; critical pressure 45.6 atm.
    b. What phase will methane exist in at −160°C and 2.0 atm pressure?
    c. At a constant temperature of −182°C the pressure on a sample of methane is gradually

increased from 1.0 mm Hg to 1.0 atm. What phases would be observed, and in what order would they occur?

15. Given the following data for N_2:
    Normal melting point $-210°C$ Specific heat of liquid 2.0 J/g·°C
    Normal boiling point $-196°C$ Specific heat of gas 1.0 J/g·°C
    Heat of fusion 25 J/g   Specific heat of solid 1.6 J/g·°C
    Heat of vaporization 200 J/g
    How much energy in kilojoules is required to convert 1000 g of N_2 at $-206°C$ to N_2 gas at 20°C?

16. In the following pairs indicate which member would have the higher heat of vaporization.
    a. He or Kr   b. CH_3OH or CH_3OCH_3

17. Which one of the following compounds will exhibit hydrogen bonding?
    a. CH_4 b. HBr   c. CH_3OH d. CCl_4

18. The vapor pressure of mercury is 17.3 mm Hg at 200°C. What is its vapor pressure at 300°C if its molar heat of vaporization is 59.0 kJ/mol?

SELF-TEST B1

1. Nickel crystallizes in a face-centered cubic lattice with an edge length of 352 pm. Calculate the density of Ni.

2. Copper crystallizes in a face-centered cubic lattice. The density of copper is 8.96 g/cm^3. What is the length of the edge of a unit cell?

3. The molar heats of fusion and vaporization of potassium are 2.4 and 79.1 kJ/mol, respectively. Estimate the molar heat of sublimation of potassium metal.

4. The heat of vaporization of iodine is 41.7 kJ/mol at the boiling point of iodine (456 K). How much heat is required to vaporize 20.0 g of I_2?

5. In a diffraction experiment, X rays of wavelength 0.154 nm are reflected from a gold crystal. The first-order reflection is at 22.2°. What is the distance between the planes of gold atoms in picometers?

6. The vapor pressure of liquid potassium is 1.00 torr at 341°C and 10.0 torr at 443°C. Calculate the molar heat of vaporization of potassium.

7. The vapor pressure of ethanol is 100 mm Hg at 34.9°C and 400 mm Hg at 63.5°C. What is the molar heat of vaporization of ethanol?

SELF-TEST B2

1. Nickel crystallizes in a face-centered cubic lattice with an edge length of 352 pm. Given that the density of Ni is 8.94 g/cm^3, calculate Avogadro's number.

2. Copper crystallizes in a cubic system and the edge of the unit cell is 361 pm. The density of copper is 8.96 g/cm^3. How many atoms are contained in one unit cell? What type of cubic cell does Cu form?

3. The molar heats of fusion and vaporization of potassium are 2.4 and 81.5 kJ/mol, respectively. Estimate the molar heat of vaporization of liquid potassium.

4. Measurements show that to vaporize 20.0 g of iodine at 456 K requires 3290 J of heat. Calculate the molar heat of vaporization of I_2.

226 / Intermolecular Forces and Liquids and Solids

5. The distance between the planes of gold atoms in a gold crystal is 204 pm. In a diffraction experiment the first-order reflection was observed at 22.2°. What was the wavelength of the X rays?

6. The vapor pressure of liquid potassium is 10.0 torr at 443°C. The heat of vaporization is 82.5 kJ/mol. Calculate the temperature where the vapor pressure is 760 torr. This is the boiling point of liquid potassium.

7. The vapor pressure of ethanol is 400 mm Hg at 63.5°C. Its molar heat of vaporization is 41.7 kJ/mol. Calculate the vapor pressure of ethanol at 34.9°C.

ANSWERS

TRUE-FALSE QUESTIONS

1. False. There is no electronegativity difference between P and H, so the H atom is not positive.
2. False. H bonds are the strongest.
3. True.
4. False. Liquids are more compressible than solids.
5. True.
6. False. Water is a very exceptional substance in this regard.
7. False. Eight unit cells.
8. True.
9. False. Glass is amorphous.
10. True.
11. False. The critical temperature is the *highest* temperature at which a substance can exist as a liquid.
12. True.
13. True.
14. True.

SELF-TEST A

1. a. Dipole-dipole, dispersion forces, and hydrogen bonding
   b. Dipole-dipole and dispersion forces
   c. Dispersion forces
   d. Dipole-dipole and dispersion forces
   e. Dipole-dipole, dispersion forces, and hydrogen bonding
2. H_2O molecules are attracted to each other by hydrogen bonding, whereas H_2S and H_2Se are not.
3. a. SO_2 b. Br_2 c. H_2O d. C_8H_{18}
4. Grease and water do not mix; thus if grease remains on the glass surface, water will bead up. When the surface is clean, the cohesive forces between water and glass cause water to spread out in a thin film and "wet" the glass.
5. Viscosity results from intermolecular forces. As the temperature of water is raised, more and more H bonds are broken and the viscosity decreases.
6. The cohesive forces between Hg atoms are greater than the adhesives forces between glass and mercury.
7. $\theta = 5.2°$
8. a. 4 b. 478 pm c. 3.41 g/cm^3
9. 6.10×10^{23}
10. 573 pm
11. a. KCl b. Fe c. Fe d. SiO_2 e. SiO_2
12. The bubble formation that signifies boiling cannot occur until the vapor pressure is great enough to push back the atmosphere. Thus the boiling point varies with atmospheric pressure. The normal boiling point of a liquid is the temperature at which its vapor pressure is 1.0 atm.

13. Energy is required for water to evaporate. When water evaporates from your skin, the heat of vaporization is absorbed from the skin, thus cooling it. You feel warmer on a humid day because the condensation of water vapor on your skin impedes the evaporation (cooling) process.
14. a.

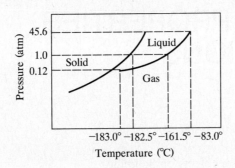

    b. Liquid   c. Gas, then liquid
15. 436 kJ
16. a. Kr   b. CH_3OH
17. c
18. 237 mm Hg

SELF-TEST B1

1. 8.94 g/cm^3
2. 361 pm
3. 81.5 kJ
4. 3.29 kJ
5. 204 pm
6. 82.5 kJ/mol
7. $\Delta H_{vap} = 41.7$ kJ

SELF-TEST B2

1. 6.02×10^{23}
2. Four atoms, face-centered cubic
3. 79.1 kJ
4. 41.7 kJ
5. 0.154 nm
6. 1052 K
7. 100 mm Hg

Chapter Twelve
PHYSICAL PROPERTIES OF SOLUTIONS

- The Solution Process
- Concentration Units
- Temperature and Pressure Effects on Solubility
- Colligative Properties

THE SOLUTION PROCESS

STUDY OBJECTIVES

You should be able to:
1. Describe the role of intermolecular forces in the solution process.
2. Given a particular substance, predict what solvent will be suitable.
3. Describe the exothermic and endothermic steps that occur during the solution process.

Some Definitions. The main subject of this chapter concerns liquid solutions. When one substance is dissolved in another, particles of the solute disperse uniformly throughout the solvent. A *solution* is a homogeneous mixture of two or more substances. The component in greater quantity is called the *solvent*, and the component in lesser amount is called the *solute*. An additional property of the solvent is that it will retain its state of matter in the solution. In contrast, the solute in some cases will not retain its physical state. For example, in a 5% sugar solution in water, the solvent is water. Water is present in a larger amount, and both water and the solution are liquids. Sugar, a solid, is the solute. Several terms are used to describe the degree to which a solute will dissolve in a solvent. Three possibilities occur when a solute and a solvent are brought together:

1. When two liquids are completely soluble in each other in all proportions, they are said to be *miscible*. For example, ethanol and water are miscible.
2. The solute may dissolve partially, reaching a maximum concentration. The solution is then said to be saturated, and the concentration of the solute is called the *solubility*. For example, the solubility of NaCl is 6.1 mol per liter of water.
3. The solute may not dissolve to a measurable extent. If the solute is a solid, it is said to be *insoluble*. If it is a liquid, and the liquids do not mix, they are said to be *immiscible*. Oil and water, for example, are immiscible.

The most general solubility rule is that "like dissolves like." In this rule, the term *like* refers to polarity. "Like dissolves like" means that substances of like polarity will mix to form solutions. And substances of different polarity will be immiscible, or will tend only slightly to form solutions. The rule predicts that two polar substances will form solutions and two nonpolar substances will form solutions, but a polar substance and a nonpolar substance will tend not to mix. Thus, water and oil, being polar and nonpolar substances, respectively, are immiscible. Water and ethanol, both being polar, are miscible. Oil will dissolve in carbon tetrachloride because both substances are nonpolar.

The terms *electrolytes* and *nonelectrolytes*, which were discussed in Chapter 3, are applied to solutes in aqueous solutions. Ionic solids are known as electrolytes. An *electrolyte* is a solute that gives an electrically conducting solution when dissolved in water. The ions supplied by the solute conduct the current through the solution. A *nonelectrolyte* is a substance that on dissolving in water does not yield a solution capable of conducting an electrical current. *Most molecular substances are nonelectrolytes.*

The Solution Process. Dissolving is a process that takes place at the molecular level and can be discussed in molecular terms. When one substance dissolves in another, the particles of the solute disperse uniformly throughout the solvent. The solute particles occupy positions that are normally taken by solvent molecules. The ease with which a solute particle may replace a solvent molecule depends on the relative strengths of three types of interactions:

- Solvent-solvent interaction
- Solute-solute interaction
- Solvent-solute interaction

Imagine the solution process as taking place in three steps, as shown in Figure 12.1. Step 1 is the separation of solvent molecules. Step 2 is the separation of solute molecules. These steps require an input of energy to overcome attractive intermolecular force. Step 3 is the mixing of solvent and solute molecules; it may be exothermic or endothermic. According to Hess's law (see Section 6.6 of the text), the heat of solution is given by

$$\Delta H_{soln} = \Delta H_1 + \Delta H_2 + \Delta H_3$$

The solute will be soluble in the solvent if the solute-solvent attraction is stronger than the solvent-solvent attraction and solute-solute attraction. Such a solution process is exothermic. Only a relatively small amount of the solute will be dissolved if the solute-solvent interaction is weaker than the solvent-solvent and solute-solute interaction; then the solution process will be endothermic.

Here we see that the solution process is assisted by an exothermic heat of solution. Yet there are a number of soluble compounds with endothermic heats of solution. In addition every substance is

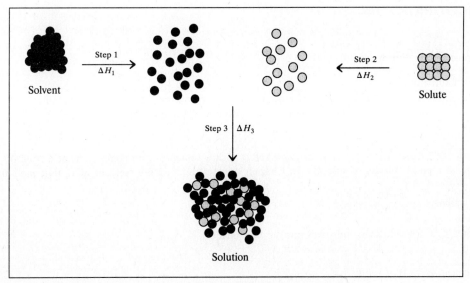

Figure 12.1. A molecular view of the solution process. Think of the solute molecules and solvent molecules first being spread apart, and then being mixed together. The strength of forces holding solvent molecules together and the forces between solvent and solute molecules in the solution are important in determining the solubility.

230 / Physical Properties of Solutions

somewhat soluble no matter what the value of ΔH_{soln}. It turns out that all chemical processes are governed by two factors. The first of these is the energy factor.

Disorder or randomness is the other factor that must be considered. Processes that increase randomness or disorder are also favored. In the pure state, the solvent and solute possess a fair degree of order. Here we mean the ordered arrangement of atoms, molecules, or ions in a three-dimensional crystal. Where order is high, randomness is low.

The order in a crystal is lost when the solute dissolves and its molecules are dispersed in the solvent. The solution process is accompanied by an increase in disorder or randomness. It is the increase in disorder of the system that favors the solubility of any substance.

EXAMPLE 12.1 Solubility

Which would be a better solvent for molecular $I_2(s)$: CCl_4 or H_2O?

METHOD

Using the like-dissolves-like rule, first we identify I_2 as a nonpolar molecule. Therefore it will be more soluble in the nonpolar solvent CCl_4 than in the polar solvent H_2O.

EXAMPLE 12.2 Solubility

In which solvent will NaBr be more soluble, benzene or water?

METHOD

Benzene is a nonpolar solvent and water is a polar solvent. *Answer:* Since NaBr is an ionic compound, it cannot dissolve in nonpolar benzene but is very soluble in water.

CONCENTRATION UNITS

STUDY OBJECTIVES

You should be able to:
1. Calculate concentrations in units of molarity, mass percent, mole fraction, and molality.
2. Convert concentrations expressed in one unit into any of the other units.

Concentration. The term *concentration* refers to how much of one component of a solution is present in a given amount of solution. In Chapter 4 we introduced the concentration unit *molarity*:

$$\text{molarity} = \frac{\text{moles of solute}}{\text{liters of solution}}$$

Three new concentration units will now be introduced:

The percent by mass of solute

$$\text{percent by mass of solute} = \frac{\text{mass of solute}}{\text{mass solute + mass solvent}} \times 100\%$$

$$= \frac{\text{mass of solute}}{\text{mass of soln}} \times 100\%$$

The mole fraction of a component of the solution

$$\text{mole fraction of component A} = X_A = \frac{\text{moles of A}}{\text{sum of moles of all components}}$$

The molality

$$\text{molality} = \frac{\text{moles of solute}}{\text{mass of solvent (kg)}}$$

EXAMPLE 12.3   Concentration Units

The dehydrated form of Epsom salt is magnesium sulfate.
a. What is the percent $MgSO_4$ by mass in a solution made from 16.0 g $MgSO_4$ and 100 mL of H_2O at 25°C? The density of water at 25°C is 0.997 g/mL.
b. What is the mole fraction of each component?
c. Calculate the molality of the solution.

METHOD OF SOLUTION FOR a

Write the equation for percent by mass:

$$\text{percent } MgSO_4 = \frac{\text{mass } MgSO_4}{\text{mass } MgSO_4 + \text{mass water}} \times 100\%$$

CALCULATION FOR a

$$\text{mass } H_2O = 100 \text{ mL} \times \frac{0.997 \text{ g}}{1 \text{ mL}} = 99.7 \text{ g } H_2O$$

$$\text{percent } MgSO_4 = \frac{16.0 \text{ g}}{16.0 \text{ g} + 99.7 \text{ g}} \times 100\% = \frac{16.0 \text{ g}}{115.7 \text{ g}} \times 100\%$$

$$\text{percent } MgSO_4 = 13.8\%$$

METHOD OF SOLUTION FOR b

Write the formula for the mole fraction of $MgSO_4$:

$$X_{MgSO_4} = \frac{\text{mol } MgSO_4}{\text{mol } MgSO_4 + \text{mol } H_2O}$$

CALCULATION FOR b

Convert the masses in part a into moles to substitute into the equation:

$$16.0 \text{ g } MgSO_4 \times \frac{1 \text{ mol } MgSO_4}{120.4 \text{ g } MgSO_4} = 0.133 \text{ mol } MgSO_4$$

$$99.7 \text{ g } H_2O \times \frac{1 \text{ mol } H_2O}{18.0 \text{ g } H_2O} = 5.54 \text{ mol } H_2O$$

The mole fraction of $MgSO_4$ is

$$X_{MgSO_4} = \frac{0.133 \text{ mol}}{0.133 \text{ mol} + 5.54 \text{ mol}} = 0.0235$$

232 / Physical Properties of Solutions

The mole fraction of H_2O is

$$X_{H_2O} = \frac{5.54 \text{ mol}}{5.67 \text{ mol}} = 0.977$$

Notice that the sum of the two mole fractions is 1.000. The sum of the mole fractions of all solution components is always 1.00.

METHOD OF SOLUTION FOR c

Write the formula for molality:

$$\text{molality} = \frac{\text{mol MgSO}_4}{\text{kg H}_2\text{O}}$$

Substitute into this equation the quantities previously calculated in parts a and b.

CALCULATION FOR c

$$\text{molality} = \frac{0.133 \text{ mol MgSO}_4}{99.7 \text{ g}} \times \frac{10^3 \text{ g}}{1 \text{ kg}} = 1.33 \text{ m}$$

EXAMPLE 12.4 Molality and Molarity of the Same Solution

Concentrated hydrochloric acid is 36.5% HCl by mass. Its density is 1.18 g/mL. Calculate:
a. The molality
b. The molarity of concentrated HCl

METHOD OF SOLUTION FOR a

Write the formula for molality:

$$\text{molality} = \frac{\text{mol HCl}}{\text{kg H}_2\text{O}}$$

Next try to find the number of moles of HCl per kilogram of H_2O. We take 100 g of solution and determine how many moles of HCl and how many kilograms of the solvent it contains. A solution that is 36.5% HCl by mass corresponds to 36.5 g HCl/100 g solution. Since 100 g of solution contains 36.5 g HCl, the difference 100 g − 36.5 g must equal the mass of water, which is 63.5 g H_2O. We have two ratios:

$$\frac{36.5 \text{ g HCl}}{100 \text{ g soln}} \quad \text{and} \quad \frac{36.5 \text{ g HCl}}{63.5 \text{ g H}_2\text{O}}$$

CALCULATION FOR a

The moles of HCl is given by

$$\text{mol HCl} = 36.5 \text{ g HCl} \times \frac{1 \text{ mol HCl}}{36.5 \text{ g HCl}} = 1.00 \text{ mol HCl}$$

The kilograms of water is given by

$$\text{kg H}_2\text{O} = 63.5 \text{ g H}_2\text{O} \times \frac{1 \text{ kg}}{10^3 \text{ g}} = 0.0635 \text{ kg H}_2\text{O}$$

Then we calculate molality:

$$\text{molality} = \frac{1.00 \text{ mol HCl}}{0.0635 \text{ kg H}_2\text{O}} = 15.7 \text{ m}$$

METHOD OF SOLUTION FOR b

Write the formula for molarity:

$$\text{molarity} = \frac{\text{moles HCl}}{\text{liters soln}}$$

Take 100 g of solution and determine how many moles of HCl and how many liters of solution are present:

$$\frac{36.5 \text{ g HCl}}{100 \text{ g soln}}$$

CALCULATION FOR b

We know that 100 g of solution contains 36.5 g HCl, which is 1.00 mol HCl. Convert 100 g of solution to volume of solution by using the density:

$$\text{volume of soln} = (100 \text{ g soln}) \times \frac{1 \text{ mL}}{1.18 \text{ g soln}} \times \frac{10^{-3} \text{ L}}{1 \text{ mL}} = 0.0847 \text{ L}$$

$$\text{molarity} = \frac{1.00 \text{ mol HCl}}{0.0847 \text{ L soln}} = 11.8 \text{ } M$$

TEMPERATURE AND PRESSURE EFFECTS ON SOLUBILITY

STUDY OBJECTIVES

You should be able to:
1. Discuss the effects of temperature on the solubility of salts in water.
2. Discuss the effects of temperature and pressure on the solubilities of gases.

Temperature Effect on the Solubility of Solids. Temperature changes strongly affect the solubility of most solid solutes. The effects of temperature on the solubilities of some common salts is shown in Figure 12.8 in the text. In most cases the solubility of a solid in water increases with increasing temperature. However, this is not always true. Table 12.1 lists several solids whose solubility *decreases* with increasing temperature. In general, it is not practical to predict just how the temperature will affect the solubility of a given substance. Temperature effects must be determined experimentally.

Table 12.1 Compounds Whose Solubility Decreases with Increasing Temperature

$CaSO_4$	$Ca(OH)_2$
$Na_2SO_4 \cdot 10H_2O$	$Ca(C_2H_3O_2)_2$
$Ce_2(SO_4)_3$	Li_2SO_4

Pressure and the Solubility of Gases. The solubility of gases in liquids is directly proportional to the gas pressure. *Henry's law* relates gas concentration c, in moles per liter, to the gas pressure p in atmospheres:

$$c = kP$$

where k is a constant for each gas and has units of mol/L · atm. The greater the value of k, the greater the solubility of the gas. A few of Henry's law constants for gases dissolved in water at 25°C are listed in Table 12.2.

Table 12.2 Some Henry's Law Constants

Gas	k (mol/L · atm)
O_2	1.28×10^{-3}
N_2	6.5×10^{-4}
CO_2	3.38×10^{-2}

Gas Solubility and Temperature. A common observation is that a glass of cold water when warmed to room temperature shows the formation of many small air bubbles. This phenomenon results in part from the decreased solubility of gases in water with increasing temperature.

The solubility of an unreactive gaseous solute is due to intermolecular attractive forces between solute molecules and solvent molecules. As the temperature of such a solution is increased, more solute molecules attain sufficient kinetic energy to break away from these attractive forces and enter the gas phase, and so the solubility decreases.

EXAMPLE 12.5 Henry's Law

What is the concentration of O_2 at 25°C in water that is saturated with *air* at an atmospheric pressure of 645 mm Hg? Assume the mole fraction of oxygen in air is 0.209.

METHOD OF SOLUTION

Henry's law states that the concentration of dissolved O_2 (C_{O_2}) is proportional to its partial pressure (P_{O_2}) in atmospheres:

$$C_{O_2} = kP_{O_2}$$

The partial pressure of O_2 in air is found by using Dalton's law of partial pressures:

$$P_{O_2} = X_{O_2} P_T = 0.209\ (645 \text{ mm Hg}) \times \frac{1 \text{ atm}}{760 \text{ mm Hg}}$$

$$= 0.177 \text{ atm}$$

Taking k from Table 12.2, the concentration of dissolved oxygen is

$$C_{O_2} = kP_{O_2} = (1.28 \times 10^{-3} \text{ mol/L} \cdot \text{atm})(0.177 \text{ atm})$$

$$= 2.27 \times 10^{-4} \text{ mol/L}$$

COLLIGATIVE PROPERTIES

STUDY OBJECTIVES

You should be able to:
1. Calculate for a given solution the magnitude of the colligative properties: vapor-pressure lowering, boiling-point elevation, freezing-point depression, and osmotic pressure.
2. Determine the molar mass of a solute from freezing-point depression and osmotic pressure measurements.
3. Calculate the van't Hoff factor (i) for electrolyte solutions

Colligative Properties. The properties of solutions that depend on the number of solute particles in solution are called *colligative properties*. The four colligative properties of interest here are as follows:

1. The vapor-pressure lowering ΔP is given by

$$\Delta P = X_2 P_1^\circ$$

where X_2 is the mole fraction of the solute, and P_1° is the vapor pressure of the pure solvent. Recall that the vapor pressure of a liquid is the pressure exerted by a vapor in equilibrium with its liquid phase. For a solution containing a nonvolatile solute the vapor pressure due to the solvent is less than it is for the pure solvent. The amount of lowering of the vapor pressure can be seen to depend on X_2, the mole fraction of the solute.
2. The boiling-point elevation. The boiling point of a solution is higher than that of pure solvent because the vapor pressure of a solution is always less than the vapor pressure of pure solvent. Thus, a solution must be hotter than a pure solvent if both vapor pressures are to be 1 atm. Figure 12.2 shows the effect that lowering the vapor pressure has on the boiling point. The boiling-point

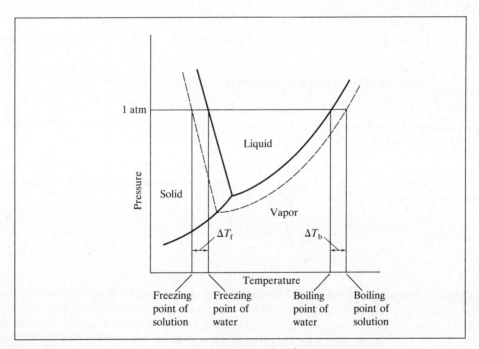

Figure 12.2. Phrase diagram of an aqueous solution, showing the vapor-pressure lowering of a solution, and its effect in raising the boiling point and lowering the freezing point. The dashed curves pertain to the solution, and the solid curves refer to pure solvent.

elevation ΔT_b of a solution of molality m is given by

$$\Delta T_b = K_b m$$

where K_b is a constant called the molal boiling-point elevation constant and has the units °C/m. There is a K_b for each solvent, and Table 12.1 in the text lists K_b values for six solvents. The magnitude of ΔT_b, the increase in boiling point of the solution over the boiling point of the pure solvent, is proportional to the solute, concentration m. The preceding equation is independent of the solute, but it requires that the solute be nonvolatile.

3. The freezing-point depression ΔT_f is given by

$$\Delta T_f = K_f m$$

where K_f is the molal freezing-point depression constant and has the units °C/m. As is the case of K_b, K_f is different for each solvent, and its value must be found in a table such as Table 12.2 in the textbook. The equation applies to all solutes, even volatile ones, because freezing points are rather insensitive to vapor pressure.

4. The osmotic pressure π is given by

$$\pi = MRT$$

where M is the molar concentration, R is the ideal gas constant (0.0821 L · atm/K · mol), and T is the absolute temperature. Osmosis is the net flow of solvent molecules through a semipermeable membrane from a more dilute solution to a more concentrated solution. Note that in this process the solvent flows from a region of greater solvent concentration to a region of lesser solvent concentration. The osmotic pressure of a solution is the pressure required to stop osmosis.

Molar Mass Determination. Colligative properties provide a means to determine the molar mass of the solute. For instance, the freezing-point depression equation is

$$\Delta T_f = K_f m = K_f \times \frac{\text{mol solute}}{\text{kg solvent}}$$

Since ΔT_f can be measured and K_f is known, the molality m of the solution can be calculated. Recall that

$$\text{molality} = \frac{\text{mol solute}}{\text{kg solvent}} = \frac{\text{mass solute/molar mass of solute}}{\text{kg solvent}}$$

When a known mass of solute is dissolved in a known mass of solvent and you already know the molality of the solution, this leaves the *molar mass* of the solute as the only unknown quantity.

Freezing-point depression is more sensitive to the number of moles of solute than the boiling-point elevation because for the same solvent K_f is larger than K_b. The most sensitive method for the determination of the molar mass of a solute is to measure the osmotic pressure. The equation for osmotic pressure then becomes useful in the following form:

$$\pi = MRT = \frac{(\text{g solute/molar mass of solute})RT}{\text{L soln}}$$

Colligative Properties of Electrolyte Solutions. For the same solute concentrations solutions of electrolytes always have greater colligative properties than those of nonelectrolyte solutions. For a 0.1 m methanol solution, ΔT_f is the same as the calculated value. For 0.1 m NaCl, ΔT_f is 1.9 times greater than calculated. And for Na_2SO_4, ΔT_f is 2.7 times greater than the calculated value. Table 12.3 compares values of $\Delta T_f(\text{obs})/\Delta T_f(\text{calc})$ for several solutes. For solutions of electrolytes the ratio is always greater than 1. Since colligative properties depend on the number of solute particles, the most likely explanation of this behavior is that NaCl and Na_2SO_4 dissociate into ions. This gives a greater concentration of solute particles. Therefore, the effective concentration of solute particles in 0.10 m

Table 12.3 van't Hoff Factor i for Several Solutes

Solution	ΔT_f	ΔT_f(calc)	$\Delta T_f/\Delta T_f$(calc)
0.10 m CH$_3$OH	0.186°C	0.186°C	1.0
0.10 m NaCl	0.353°C	0.186°C	1.9
0.10 m Na$_2$SO$_4$	0.502°C	0.186°C	2.7

NaCl is essentially 2×0.10 m, and in Na$_2$SO$_4$ it is almost 3×0.10 m. The freezing-point depression depends on the total concentration of all solute particles. Each ion acts as an independent particle.

The ratio $\Delta T_f/\Delta T_f$(calc) is called the van't Hoff factor i. For 1:1 electrolytes, such as NaCl, KCl, and MgSO$_4$, $i \sim 2$; for 2:1 electrolytes, such as MgCl$_2$ and K$_2$SO$_4$, i is a little less than 3. The deviation from a whole number results from the temporary formation of some ion pairs such as "NaCl" in the solution.

EXAMPLE 12.6 Boiling-Point Elevation

What is the boiling point of an "antifreeze/coolant" solution made from a 50-50 mixture (by volume) of ethylene glycol, C$_2$H$_6$O$_2$ (density 1.11 g/mL), and water?

METHOD OF SOLUTION

The boiling point depends on the molality of the 50-50 mixture. $\Delta T = K_b m$. For simplicity assume 100 mL of the solution. Then the mass of 50 mL (50 vol%) of H$_2$O is 50 g. The mass of 50 mL (50 vol%) of ethylene glycol is 55.5 g.

CALCULATION

$$m = \frac{\text{mol C}_2\text{H}_6\text{O}_2}{\text{kg H}_2\text{O}}$$

$$\text{mol C}_2\text{H}_6\text{O}_2 = 55.5 \text{ g} \times \frac{1 \text{ mol}}{62.0 \text{ g}} = 0.895 \text{ mol}$$

$$m = \frac{0.895 \text{ mol C}_2\text{H}_6\text{O}_2}{0.050 \text{ kg}} = 17.9 \text{ mol/kg} = 17.9 m$$

$$\Delta T_b = K_b m = (0.52 \text{ °C}/m) \times 17.9 \ m = 9.3\text{°C}$$

Therefore the boiling point of the solution is 9.3°C above the normal boiling point of water: b.p. = 109.3°C.

EXAMPLE 12.7 Vapor-Pressure Lowering

Calculate the vapor pressure P of an aqueous solution at 30°C made from 100 g of sucrose (C$_{12}$H$_{22}$O$_{11}$) and 100 g of water. The vapor pressure of water at 30°C is 31.8 mm Hg.

METHOD

Sucrose is a nonvolatile solute, so the vapor pressure over the solution will be due to H$_2$O molecules. The problem can be worked in two ways.

a. The vapor-pressure lowering is proportional to the mole fraction of sucrose, X_2, and $P°$, the vapor pressure of pure water at 30°C:

$$\Delta P = X_2 P_1°$$

First let's calculate the mole fraction of sucrose:

$$X_2 = \frac{n_2}{n_1 + n_2}$$

$$n_2 = 100 \text{ g} \times \frac{1 \text{ mol}}{342 \text{ g}} = 0.292 \text{ mol}$$

$$n_1 = 100 \text{ g} \times \frac{1 \text{ mol}}{18.0 \text{ g}} = 5.55 \text{ mol}$$

$$X_2 = \frac{0.292}{0.292 + 5.55} = \frac{0.292}{5.84}$$

$$= 0.050$$

$$\Delta P = (0.050)(31.8 \text{ mm Hg}) = 1.6 \text{ mm Hg}$$

$$P_1 = 31.8 - 1.6$$

$$= 30.2 \text{ mm Hg}$$

b. Alternatively Raoult's law can be used:

$$P_1 = X_1 P_1^\circ$$

where P_1° and P_1 are the vapor pressures of pure solvent and of the solvent in solution, respectively. From part a, $X_2 = 0.050$. Therefore $X_1 = 1.00 - X_2 = 0.95$, and

$$P_1 = (0.95)(31.8 \text{ mm Hg})$$

$$= 30.2 \text{ mm Hg}$$

EXAMPLE 12.8 Finding Molecular Mass of a Solute

Benzene has a normal freezing point of 5.51°C. The addition of 1.25 g of an unknown compound to 85.0 g of benzene produces a solution with a freezing point of 4.52°C. What is the molecular mass of the unknown compound?

METHOD OF SOLUTION

We need to find the number of moles in 1.25 g of unknown compound X. The freezing-point depression is proportional to the number of moles of X per kilogram of solvent.

CALCULATION

The freezing-point depression is

$$\Delta T_f = 5.51°C - 4.52°C$$

$$= 0.99°C$$

We can calculate the molality of the solution because ΔT_f is given above:

$$m = \frac{\Delta T_f}{K_f}$$

where K_f is the freezing-point depression constant for benzene (Table 12.2 in the textbook):

$$m = \frac{0.99°C}{5.12°C/m} = 0.19 \ m$$

This means that there are 0.19 mol X/kg benzene. The number of moles of X in 0.085 kg of the solvent benzene (the given amount) is

$$0.085 \text{ kg} \times \frac{0.19 \text{ mol } X}{1 \text{ kg benzene}} = 1.6 \times 10^{-2} \text{ mol } X$$

Therefore 1.25 g X = 1.6×10^{-2} mol X, and

$$\text{molar mass of } X = \frac{1.25 \text{ g}}{1.6 \times 10^{-2} \text{ mol}} = 77.4 \text{ g/mol}$$

EXAMPLE 12.9 Colligative Properties of Electrolytes and Nonelectrolytes

List the following aqueous solutions in the order of increasing boiling points: 0.10 m glucose ($C_6H_{12}O_6$); 0.10 m $Ca(NO_3)_2$; 0.10 m NaCl.

METHOD OF SOLUTION

Ethanol is a nonelectrolyte, whereas NaCl and $Ca(NO_3)_2$ are electrolytes. NaCl dissociates into two ions per formula unit, and $Ca(NO_3)_2$ dissociates into three ions per formula unit. The effective molalities are approximately

glucose	0.10 m
NaCl	$\cong$ 0.20 m
$Ca(NO_3)_2$	$\cong$ 0.30 m

Therefore, the boiling-point elevation and the boiling point are greatest for a solution of $Ca(NO_3)_2$, second highest for NaCl, and lowest for glucose.

COMMENT

If you included the van't Hoff factors, the effective molalities would be affected slightly, but not enough to change the predicted results.

TRUE-FALSE QUESTIONS

1. Oil and water do not form solutions and are said to be miscible.

2. The hydration of ions in water is an exothermic process.

3. The heats of solution of ionic solutes are always endothermic.

4. Molality refers to the number of moles of solute per kilogram of solution.

5. To convert a given molarity to molality, you must know the density of the solution.

6. Two hundred fifty milliliters of 0.5 M NaCl solution contains 0.5 mol NaCl.

7. Increasing the temperature of a solution raises the solubility of all solid solutes.

8. The solubility of a gas such as helium in water can be increased by raising the pressure of helium.

9. The solubility of CO_2 in water can be raised by heating the solution.

10. The vapor pressure of water is greater over pure water than over a salt solution at the same temperature.

11. The boiling point of a 1.0 M sucrose solution would be 0.52°C, given $K_b = 0.52°C/m$.

12. The freezing point of a 1.0 M NaCl solution would be $-3.72°C$, given $K_f = 1.86°C/m$.

SELF-TEST A

1. For each of the following pairs predict which substance will be more soluble in water:
   a. $HCl(g)$ or $H_2(g)$
   b. CCl_4 or $AlCl_3$
   c. CH_3Cl or CCl_4
   d. CH_3OH or CH_3-O-CH_3

2. For each of the following pairs predict which substance will be more soluble in $CCL_4(l)$:
   a. H_2O or oil
   b. C_6H_6 (benzene) or CH_3OH
   c. I_2 or NaCl

3. Calculate the molality of a solution prepared by dissolving 28.0 g of urea, $CO(NH_2)_2$, in 450 g of H_2O.

4. How many grams of KI must be dissolved in 300 g of water to make a 0.500 m solution?

5. What is the molarity of a 3.0% hydrogen peroxide (H_2O_2) aqueous solution? The density is essentially 1.0 g/cm^3.

6. Concentrated sulfuric acid is 96% H_2SO_4 by weight. Its density is 1.83 g/mL. Calculate the molarity of concentrated H_2SO_4.

7. What are the mole fractions of methanol and water in a solution that contains 40.0 g CH_3OH and 40.0 g H_2O?

8. An aqueous solution contains 167 g $CuSO_4$ in 820 mL of solution. Calculate the following:
   a. Molarity
   b. Percent by mass
   c. Mole fraction
   d. Molality of the solution
   The density of the solution is 1.195 g/mL.

9. The Henry's law constant for CO is 9.73×10^{-4} mol/L · atm at 25°C. What is the concentration of dissolved CO in water if the partial pressure of CO in the air is 0.015 mm Hg?

10. How many grams of isopropyl alcohol, C_3H_7OH, should be added to 1.0 L of water to give a solution that will not freeze above $-16°C$.

11. What is the boiling point of an aqueous solution of a nonvolatile solute that freezes at $-3.0°C$?

12. When 48 g of glucose (a nonelectrolyte) is dissolved in 500 g of H_2O, the solution has a freezing point of $-0.94°C$. What is the molar mass of glucose?

13. 7.85 g of a compound with an unknown formula is dissolved in 300 g of benzene. The freezing point of the solution is 2.10°C below that of pure benzene. What is the molar mass of the compound?

14. The dart poison in root extracts used by the Peruvian Indians is called curare. It is a nonelectrolyte. The osmotic pressure at 20°C of an aqueous solution containing 0.200 g of curare in 100 mL is 56.2 mm Hg. Calculate the molar mass of curare.

15. What is the concentration of NaCl solution with an osmotic pressure of 10 atm of 25°C?

16. The walls of red blood cells are semipermeable membranes, and the solution of NaCl within those walls exerts an osmotic pressure of 7.82 atm at 37°C. What concentration of NaCl must a *surrounding* solution have so that this pressure is balanced and cell rupture (hemolysis) is prevented?

17. Arrange the following aqueous solutions in order of increasing freezing points, lowest to highest:
    0.100 m ethanol
    0.050 m Ca(NO$_3$)$_2$
    0.100 m NaBr
    0.050 m HCl

18. Calculate the value of i for a 1:1 electrolyte if a 1.0m aqueous solution of the electrolyte freezes at -3.28°C.

GENERAL PROBLEM

19. In the course of research a chemist isolates a new compound. An elemental analysis gives the following: C, 50.7%; H, 4.25%; O, 45.1%. If 5.01 g of the compound is dissolved in 100 g of water, it produces a solution with a freezing point of -0.65°C. What is the molecular formula of the compound?

SELF-TEST B1

1. For each of the following pairs, predict which substance will be more soluble in water:
   a. NaCl(s) or I$_2$(s)
   b. CH$_4$(g) or NH$_3$(g)
   c. CH$_3$OH(l) or C$_6$H$_6$ (benzene)

2. Calculate the molality of a 2.33 M NaCl solution given that its density is 1.089 g/mL.

3. Calculate the molality of a 10% AgNO$_3$ solution.

4. Calculate the freezing point of an aqueous solution that boils at 100.5°C.

5. Benzene melts at 5.50°C. When 2.11 g of naphthalene (C$_{10}$H$_8$) is added to 100 g of benzene, the solution freezes at 4.65°C. Calculate the freezing-point depression constant for benzene.

6. What is the freezing point of a solution made from 1.00 g of C$_6$H$_{12}$O$_6$ (glucose) and 100.0 g of H$_2$O?

SELF-TEST B2

1. For each of the following pairs, predict which substance will be more soluble in CCl$_4$(l):
   a. NaCl(s) or I$_2$(s)
   b. CH$_4$(g) or NH$_3$(g)
   c. CH$_3$OH(l) or C$_6$H$_6$ (benzene)

2. Calculate the molarity of a 2.44 m NaCl solution given that its density is 1.089 g/mL.

3. Calculate the percent AgNO$_3$ by mass in a 0.65 m AgNO$_3$ solution.

4. Calculate the approximate osmotic pressure at 25°C of a solution that has a freezing point of -18°C.

242 / *Physical Properties of Solutions*

5. Calculate the molar mass of naphthalene given that a solution of 2.11 g of naphthalene in 100 g of benzene has a freezing-point depression of 0.85°C.

6. A solution contains 1.00 g of a compound (a nonelectrolyte) dissolved in 100.0 g of water. The freezing point of the solution is −0.103°C. What is the molar mass of the compound?

ANSWERS

TRUE-FALSE QUESTIONS

1. False. Oil and water do not mix. Thus they are called *immiscible*.
2. True.
3. False. The heats of solution of some salts are exothermic.
4. False. Molality is defined as the number of moles of solute per kilogram of *solvent*.
5. True.
6. False. Such a solution will contain 0.125 mol NaCl.
7. False. Several solutes exhibit less solubility at higher temperature.
8. True.
9. False. The solubility of gases decreases with increasing temperature.
10. True.
11. False. ΔT_b is 0.52°C, but T_b would be 100.52°C.
12. True.

SELF-TEST A

1. a. HCl b. $AlCl_3$ c. CH_3Cl d. CH_3OH
2. a. oil b. C_6H_6 c. I_2
3. 1.03 m urea
4. 24.9 g KI
5. 0.88 M H_2O_2
6. 18 M
7. $X_{CH_3OH} = 0.360$; $X_{H_2O} = 0.640$
8. a. 1.28 M
   b. 17.0% $CuSO_4$, 83.0% H_2O
   c. $X_{CuSO_4} = 0.023$, $X_{H_2O} = 0.977$
   d. 1.29 m
9. 1.9×10^{-8} M CO
10. 520 g
11. 100.84°C
12. 190 g/mol
13. 63.8 g/mol
14. 650 g/mol
15. 0.20 M NaCl
16. 0.15 M NaCl
17. NaBr; $Ca(NO_3)_2$; HCl = ethanol
18. $i = 1.76$
19. $C_6H_6O_4$

SELF-TEST B1

1. a. NaCl b. NH_3 c. CH_3OH
2. 2.44 m
3. 0.65 m
4. −1.8°C

5. 5.15°C/m
6. −0.103°C

SELF-TEST B2

1. a. I_2 b. CH_4 c. C_6H_6
2. 2.32 M
3. 9.9%
4. 23 atm
5. 127 g/mol
6. 181 g

Chapter Thirteen
CHEMICAL KINETICS

- Reaction Rate
- The Rate Laws
- Temperature Dependence of Reaction Rates
- Reaction Mechanisms
- Catalysis

REACTION RATE

STUDY OBJECTIVES

You should be able to:
1. Express the rate of a given reaction in terms of the change in concentration with time of a reactant or a product.
2. Define the order of reaction.

The Rate of Reaction. Chemical kinetics is the study of the rates of chemical reactions. The purposes of kinetic studies are to find the factors that affect reaction rates and determine the reaction mechanism. Knowledge of the factors that affect reaction rates enables us to control rates. Finding the reaction mechanism means we can identify the intermediate steps by which reactants are converted into products.

A rate is a change in some quantity with time. The rate of population growth is the change in population per change in time. The change in a quantity such as population is always equal to the difference, population (after) minus population (before). The symbol for "the change in" is the Greek Δ.

$$\Delta(\text{population}) = \text{population}_{\text{final}} - \text{population}_{\text{initial}}$$

The change in time is some appropriate time interval, Δt, where

$$\Delta t = t_{\text{final}} - t_{\text{initial}}$$

The rate of population change is

$$\text{rate} = \frac{\Delta(\text{population})}{\Delta t}$$

As a chemical reaction proceeds, it is the concentrations of reactants and products that change with time. As the reaction $A + B \rightarrow C$ progresses, the concentration of C increases. The rate is expressed as the change in the concentration of C during the time interval Δt:

$$\text{rate} = \frac{\Delta[C]}{\Delta t}$$

For a specific reaction we need to take into account the stoichiometry, that is, we need the balanced equation. For example, let's express the rate of the following reaction in terms of the concentrations of the individual reactants and products:

$$2NO(g) + O_2(g) \rightarrow 2NO_2(g)$$

These concentrations can be analyzed or monitored as a function of time. Notice from the balanced equation that the concentration of NO will decrease twice as fast as that of O_2. The convention used for writing the rate of reaction is

$$\text{rate} = -\frac{\Delta[NO]}{2\Delta t} = -\frac{\Delta[O_2]}{\Delta t} = \frac{\Delta[NO_2]}{2\Delta t}$$

Division of each concentration change by the coefficient from the balanced equation makes all of these rates equal. Notice also a negative sign is inserted before terms involving reactants. Originally $\Delta[NO]$ was negative because the concentration of NO *decreases* with time. Since a negative times a negative makes a positive, this makes the rate of reaction a positive quantity.

For a general equation

$$aA + bB \rightarrow cC$$

the rate can be expressed in terms of any reactant or product:

$$\text{rate} = -\frac{\Delta[A]}{a\Delta t} = -\frac{\Delta[B]}{b\Delta t} = \frac{\Delta[C]}{c\Delta t}$$

Effect of Concentration. The rate of a reaction is proportional to the reactant concentrations. For the reaction $NO + \tfrac{1}{2}O_2 \rightarrow NO_2$, the rate is proportional to the concentrations of NO and O_2:

$$\text{rate} \propto [NO]^x[O_2]^y$$

where x and y are exponents.

The *rate law* or *rate equation* for the reaction is

$$\text{rate} = k[NO]^x[O_2]^y$$

The proportionality constant k is called the *rate constant*. The value of k depends on the reaction and is a constant at constant temperature.

The exponents x and y determine how strongly the concentration affects the rate. The exponent x is called the *order* with respect to NO, and y is the order with respect to O_2. The sum $x + y$ is the overall order. *The values of x and y must be determined from experiment and cannot be derived by any other means.* We will discuss how to determine the order of reaction in the next section. For now we will just use the results. Experiments show that $x = 2$ and $y = 1$ for the NO reaction with O_2. Thus, the rate law for this reaction is

$$\text{rate} = k[NO]^2[O_2]$$

This reaction is second order in nitric oxide and first order in oxygen. It is third order overall.

The fact that the reaction is first order in O_2 means that the rate is directly proportional to the O_2 concentration. If $[O_2]$ doubles or triples, the rate will double or triple also. We can show this mathematically. Consider two experiments. In experiment 1 the concentration of O_2 is c. In experiment 2 the concentration of O_2 is doubled from c to $2c$. If the concentration of NO is the same in both experiments, it will have no effect on the rate. Use of the rate law allows us to write the ratio of the two rates:

$$\frac{\text{rate (expt 2)}}{\text{rate (expt 1)}} = \frac{k[NO]^2(2c)}{k[NO]^2(c)} = 2$$

As discussed, we see that doubling the concentration of a reactant that is first order should cause the rate to double.

If the concentration of O_2 is held constant in two experiments and the concentration of NO doubles (from c to $2c$), the rate law predicts that the rate will quadruple:

$$\frac{\text{rate (expt 2)}}{\text{rate (expt 1)}} = \frac{k[O_2](2c)^2}{k[O_2](c)^2} = 2^2 = 4$$

The fact that the reaction is second order in NO means that the rate is proportional to the square of the concentration of NO. Doubling or tripling of [NO] causes the rate to increase four- or ninefold, respectively.

In general, if the concentration of one reactant doubles while the other reactant concentration is unchanged *and* the rate is:

1. *unchanged*, the order of the reaction is *zero order* with respect to the changing reactant.
2. *doubled*, the order of the reaction is *first order* with respect to the changing reactant.
3. *quadrupled*, the order of the reaction is *second order* with respect to the changing reactant.

EXAMPLE 13.1 Expressing the Rate of Reaction

Write expressions for the rate of the following reaction in terms of each of the reactants and products:

$$2N_2O_5(g) \rightarrow 4NO_2(g) + O_2(g)$$

METHOD OF SOLUTION

Recall that the rate is defined as the change in concentration of a reactant or product with time. Each "change-in-concentration" term is divided by the corresponding stoichiometric coefficient. Terms involving reactants are preceded by a minus sign.

Answer: $\quad \text{rate} = -\dfrac{\Delta[N_2O_5]}{2\,\Delta t} = \dfrac{\Delta[NO_2]}{4\,\Delta t} = \dfrac{\Delta[O_2]}{\Delta t}$

EXAMPLE 13.2 Rate of Reaction

Oxygen gas is formed by the decomposition of nitric oxide:

$$2NO(g) \rightarrow O_2(g) + N_2(g)$$

If the rate of formation of O_2 is 0.054 M/s, what is the rate of change of NO concentration?

METHOD OF SOLUTION

From the stoichiometry of the reaction, 2 mol of NO react for each mole of O_2 that forms. Therefore,

$$\text{rate} = -\frac{\Delta[NO]}{2\,\Delta t} = \frac{\Delta[O_2]}{\Delta t}$$

CALCULATION

$$\frac{\Delta[NO]}{\Delta t} = \frac{-2\Delta[O_2]}{\Delta t} = -2(0.054\ M/s)$$

$$= -0.108\ M/s$$

EXAMPLE 13.3 Concentration Effect on the Rate

The reaction A + 2B → products was found to have the rate law rate = $k[A][B]^3$. By what factor will the rate of reaction increase if the concentration of B is increased from c to $3c$ while the concentration of A is held constant?

METHOD OF SOLUTION

Write a ratio of the rate law expressions for the two different concentrations of B:

$$\frac{\text{rate (expt 2)}}{\text{rate (expt 1)}} = \frac{k[A][3c]^3}{k[A][c]^3} = \frac{3^3 c^3}{c^3}$$

$$= 27$$

Answer: The rate will increase 27-fold when [B] is increased 3-fold.

THE RATE LAWS

STUDY OBJECTIVES

You should be able to:
1. Apply the isolation method to derive the rate law for a reaction.
2. Determine for first- and second-order reactions the concentration of a reactant at any time after the reaction has started or the time required for a given fraction of the sample to react.
3. Apply rate law equations and graphs to determine whether a reaction is first order or second order.

The Isolation Method. One procedure used to determine the rate law for a reaction involves the isolation method. In this method the concentration of all but one reactant is fixed, and the rate of reaction is measured as a function of the concentration of the one reactant whose concentration is varied. Any variation in the rate is due to the variation of this reactant's concentration. In practice the experimenter observes the dependence of the initial rate on the concentration of the reactant.

To determine the order with respect to A in the chemical reaction

2A + B → C

the initial rate would be measured in several experiments in which the concentration of A is varied and the concentration of B is held constant. To determine the order with respect to B, the concentration of A must be held constant and the concentration of B is varied in several experiments. The application of this method is illustrated in Example 13.4.

First-Order Reactions. One of the most widely encountered kinetic forms is the first-order rate equation. In this case the exponent of [A] in the rate law is 1:

A → products

$$\text{rate} = -\frac{\Delta[A]}{\Delta t} = k[A]$$

For a first-order reaction the unit of the rate constant is reciprocal time, $1/t$. Convenient units are s^{-1}, h^{-1}, etc.

The equation that relates the concentration of A remaining to the time interval t is

$$\ln\frac{[A]_0}{[A]} = kt$$

This is a very useful equation called the integrated first-order equation. Here $[A]_0$ is the concentration of A at time zero, and $[A]$ is the concentration of A at time t, which is the time interval since the reaction began. The rate constant k is first order. The concentration $[A]$ decreases as the time increases. This equation allows the calculation of the rate constant k when $[A]_0$ is known, and $[A]$ is measured at time t. Also, once k is known, $[A]$ can be calculated for any future time. In terms of common logarithms the above equation is

$$\log\frac{[A]_0}{[A]} = \frac{kt}{2.303}$$

The factor 2.303 converts common logs (log) to natural logs (ln). We prefer the use of natural logs.

To determine whether a reaction is first order, we rearranged the first-order equation into the form

$$\ln[A] = -kt + \ln[A]_0$$

corresponding to the linear equation

$$y = mx + b$$

Here m is the slope of the line and b is the intercept on the y axis. Comparing the last two equations, we can equate y and x to experimental quantities:

$$y = \ln[A] \quad \text{and} \quad x = t$$

Therefore, the intercept $b = \ln[A]_0$, and the slope of the line $m = -k$. Thus a plot of $\ln[A]$ versus t for a first-order reaction gives a straight line with a slope of $-k$, as shown in Figure 13.9 of the text. If a plot of $\ln[A]$ versus t yields a curved line rather than a straight line, the reaction is not a first-order reaction. This graphical procedure is the method used by most chemists to determine whether or not a given reaction is first order.

The half-life of a reaction, $t_{1/2}$, is a useful concept. For a first-order reaction, the half-life is given by

$$t_{1/2} = \frac{0.693}{k}$$

where 0.693 is a constant and k is the rate constant. Knowledge of the half-life allows the calculation of the rate constant k. The half-life of a reaction is the time required for the concentration of a reactant to decrease to half of the initial value. After one half-life, the ratio $[A]/[A]_0$ is equal to 0.5.

If the reaction continues, then $[A]$ will drop by 1/2 again during the second half-life period as shown in Figure 13.1. After two half-life periods the fraction of the original concentration of A remaining, $[A]/[A]_0$, will be 1/2 of the concentration remaining after the first half-life, so $[A]/[A]_0 = 0.5 \times \frac{1}{2} = 0.25$.

Second-Order Reactions. In a second-order reaction, the rate is proportional either (1) to the square of the concentration of one reactant,

$$A \rightarrow \text{product}$$

$$\text{rate} = -\frac{\Delta[A]}{\Delta t} = k[A]^2$$

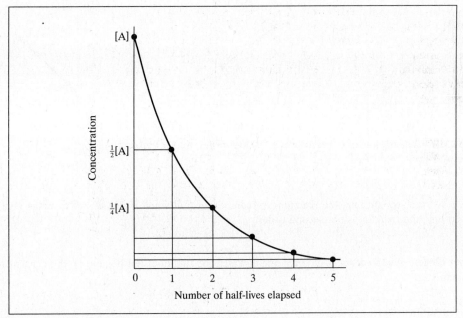

Figure 13.1. A plot of [A] versus time for a first-order reaction is curved (exponential). Over each half-life period, [A] drops in half.

or (2) to the product of the concentrations of two reactants, each raised to the first power,

A + B → product

$$\text{rate} = -\frac{\Delta[A]}{\Delta t} = k[A][B]$$

This reaction is first order in A and first order in B, and so it is second order overall. For a second-order reaction, the rate constant has units $1/\text{molarity} \times \text{time}$, or $1/M \cdot s$, or more clearly, $M^{-1}\, s^{-1}$.

An important equation called the integrated second-order equation is

$$\frac{1}{[A]} = \frac{1}{[A]_0} + kt$$

This equation allows the calculation of the concentration of A at any time t after the reaction has begun. Alternatively, if $[A]_0$, $[A]$, and t are known, the rate constant can be calculated.

The half-life for a second-order reaction is given by

$$t_{1/2} = \frac{1}{k[A]_0}$$

Here we see that the half-life is inversely proportional to the initial concentration. This situation is different from that for a first-order reaction, where $t_{1/2}$ is independent of $[A]_0$.

The second-order equation can also be plotted as a straight line:

$$y = mx + b$$

$$\frac{1}{[A]} = kt + \frac{1}{[A]_0}$$

where $y = 1/[A]$ and $x = t$. This gives the slope $m = k$, as shown in Figure 13.2(c). If a plot of $1/[A]$

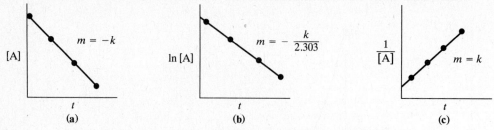

Figure 13.2. Linear plots of (a) zero-order, (b) first-order, and (c) second-order reactions.

versus t yields a straight line, the reaction is second order, but when a plot of $1/[A]$ versus t yields a curved line, the reaction is not second order.

Zero Order. A zero-order reaction is one in which the rate does not depend on the concentrations of reactants:

$$A \rightarrow \text{products}$$

$$\text{rate} = \frac{-[A]}{\Delta t} = k[A]^0 = k$$

A plot of [A] versus time is a straight line. See Figure 13.2(a).

Summary. The integrated rate equations and graphical methods allow us to distinguish between the various overall orders of reaction.

1. The reaction is first order when a plot of ln[A] versus t is a straight line [Figure 13.2(b)], and plots of [A] versus t and $1/[A]$ versus t are curved.
2. The reaction is second order when a plot of $1/[A]$ versus t is a straight line [Figure 13.2(c)] and plots of [A] versus t and ln[A] versus t are curved.
3. The half-life equations also provide a way to distinguish between first- and second-order reactions. The half-life of a first-order reaction is independent of starting concentration, whereas the half-life of a second-order reaction is inversely proportional to the initial concentration.

EXAMPLE 13.4 Finding the Rate Law

The following rate data were collected for the reaction

$$2NO + 2H_2 \rightarrow N_2 + 2H_2O$$

Experiment	$[NO]_0$ (M)	$[H_2]_0$ (M)	$\Delta[N_2]/\Delta t$ (M/h)
1	0.60	0.15	0.076
2	0.60	0.30	0.15
3	0.60	0.60	0.30
4	1.20	0.60	1.21

a. Determine the rate law.
b. Calculate the rate constant.

METHOD OF SOLUTION

a. We want to determine the exponents in the equation

$$\text{rate} = k[NO]^x[H_2]^y$$

To determine the order with respect to H_2, first find two experiments in which [NO] is held constant. This can be done by comparing the data of experiments 1 and 2. When the concentration of H_2 is doubled, the reaction rate doubles. Thus, the reaction is first order in H_2. When the NO concentration is doubled (experiments 3 and 4), the reaction rate quadruples. Thus the reaction is second order in NO. *Answer:* The rate law is

$$\text{rate} = k[NO]^2[H_2]$$

b. Rearrange the rate law from part a:

$$k = \frac{\text{rate}}{[NO]^2[H_2]}$$

Then substitute data from any one of the experiments. Using experiment 1,

$$k = \frac{0.076 \; M/h}{[0.60 \; M]^2[0.15 \; M]} = 1.4/M^2 \; h$$

COMMENT

You should get the same value of k from all four experiments. Note that the units of k are those of a third-order rate constant.

EXAMPLE 13.5 First-Order Reaction

Methyl isocyanide undergoes a first-order isomerization reaction to form methyl cyanide:

$$CH_3NC(g) \rightarrow CH_3CN(g)$$

The reaction was studied at 199°C. The initial concentration of CH_3NC was 0.0258 mol/L, and after 11.4 min analysis showed the concentration of product was 1.30×10^{-3} mol/L.
a. What is the first-order rate constant?
b. What is the half-life of methyl isocyanide?
c. How long will it take for 90% of the CH_3NC to react?

METHOD OF SOLUTION FOR a

This problem illustrates the use of the integrated first-order rate equation

$$\ln \frac{[CH_3NC]_0}{[CH_3NC]} = kt$$

where k is the rate constant and $[CH_3NC]$ is the reactant concentration at time t. The initial concentration is $[CH_3NC]_0 = 0.0258 \; M$. After 11.4 min the product concentration is $1.30 \times 10^{-3} \; M$. This implies that the concentration of CH_3NC remaining must be $0.0258 - 0.0013 = 0.0245 \; M$.

CALCULATION FOR a

Substitution gives

$$\ln \frac{0.0258\ M}{0.0245\ M} = k(11.4\ \text{min})$$

$$k = \frac{\ln 1.05}{11.4\ \text{min}}$$

$$= 4.54 \times 10^{-3}\ \text{min}^{-1}$$

METHOD OF SOLUTION FOR b

For a first-order reaction the half-life equation is

$$t_{1/2} = \frac{0.693}{k}$$

CALCULATION FOR b

$$t_{1/2} = \frac{0.693}{4.54 \times 10^{-3}\ \text{min}^{-1}} = 153\ \text{min}$$

METHOD OF SOLUTION FOR c

If 90% of the initial CH_3NC is consumed, then 10% remains, and therefore,

$$[CH_3NC] = 0.10[CH_3NC]_0$$

CALCULATION FOR c

Substitution into the rate equation gives

$$\ln \frac{[CH_3NC]_0}{0.10[CH_3NC]_0} = (4.54 \times 10^{-3}\ \text{min}^{-1})t$$

Solving for t, we get

$$t = \frac{\ln 10}{4.54 \times 10^{-3}\ \text{min}^{-1}} = \frac{2.303}{4.54 \times 10^{-3}\ \text{min}^{-1}}$$

$$= 507\ \text{min}$$

EXAMPLE 13.6 First-Order Reaction

At 230°C the rate constant for methyl isocyanide isomerization is $9.25 \times 10^{-4}\ \text{s}^{-1}$:

$$CH_3NC \rightarrow CH_3CN$$

a. What fraction of the original isocyanide will remain after 60.0 min?
b. What is the half-life of methyl isocyanide at this temperature?

METHOD OF SOLUTION FOR a

a. Again applying the first-order equation,

$$\ln \frac{[CH_3NC]_0}{[CH_3NC]} = kt$$

The fraction of methyl isocyanide remaining is given by $[CH_3NC]/[CH_3NC]_0$.

CALCULATION FOR a

$$\ln \frac{[CH_3NC]_0}{[CH_3NC]} = -\ln \frac{[CH_3NC]}{[CH_3NC]_0} = kt$$

$$-\ln \frac{[CH_3NC]}{[CH_3NC]_0} = 9.25 \times 10^{-4}\ s^{-1}\ (60\ min) \times \frac{60\ s}{1\ min} = -1.45$$

$$\frac{[CH_3NC]}{[CH_3NC]_0} = 10^{-1.45}$$

$$= 0.035$$

CALCULATION FOR b

The half-life is

$$t_{1/2} = \frac{0.693}{9.25 \times 10^{-4}\ s^{-1}}$$

$$= 749\ s\ or\ 12.4\ min$$

COMMENT

Note the much shorter half-life at 230°C than at 199°C (Example 13.5).

TEMPERATURE DEPENDENCE OF REACTION RATES

STUDY OBJECTIVES

You should be able to:
1. Calculate the activation energy for a reaction when given rate constants at several different temperatures.
2. Describe the concept of activation energy and show how it explains the variation of reaction rate with temperature.
3. Describe a reaction energy profile, including the activation energy and energy change of reaction.

The temperature of a reaction system is an important variable because of its strong effect on reaction rates. As a general rule, reaction rates double with a 10°C rise in temperature. Thus, reaction rates increase exponentially with temperature. In general, the rate equation is rate = $k[A]^x[B]^y$. Since the concentrations [A] and [B] are unaffected by temperature, it is the rate constant that changes with temperature. In 1889, S. Arrhenius found that a plot of the natural logarithm of the rate constant ($\ln k$) versus the reciprocal of the absolute temperature $1/T$ gave a straight line. See Figure 13.14 in the textbook. Arrhenius identified the slope of the line as being related to an energy term

$$m = -\frac{E_a}{R}$$

where R is the ideal gas constant and E_a is the activation energy. The logarithmic form of the Arrhenius equation is

$$\ln k = -\frac{E_a}{RT} + \ln A$$

where ln A is the intercept. Both the Arrhenius A and E_a are constants for a particular reaction. Taking

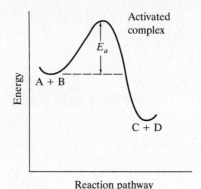

Figure 13.3. Molecules of A and B with average energies must first acquire an energy E_a before they can react.

the antilog of both sides gives the Arrhenius equation:

$$k = Ae^{-E_a/RT}$$

The Meaning of A and E_a. The activation energy E_a is a concept developed by Arrhenius. He divided molecules of reactant into those of average energy and those called activated molecules. Arrhenius's activated molecules are those with enough energy to react. Thus all molecules must first acquire energy in order to react. The high-energy intermediate species resulting from molecular collisions is called the *activated complex*. The activated complex dissociates into the products. Figure 13.3 shows that the *activation energy* is the energy that average reactant molecules must acquire to become activated molecules.

The quantity A can be related to the frequency of molecular collisions. The collision frequency is important because molecules must come into contact if they are to react with one another. This is the basis of the collision theory of chemical reactions.

From the Arrhenius equation we can point out that *reactions for which E_a is large will be much slower than those for which E_a is small*. As E_a increases, the negative exponent increases, and so k decreases.

A convenient equation that can be used to calculate the activation energy is

$$\ln\left(\frac{k_1}{k_2}\right) = \frac{E_a}{R}\left(\frac{T_1 - T_2}{T_1 T_2}\right)$$

where k_1 is the rate constant at temperature T_1, and k_2 is the rate constant at temperature T_2. Use of this equation requires that rate constants k_1 and k_2 be measured at two temperatures T_1 and T_2, respectively.

EXAMPLE 13.7 Calculating Activation Energy

For the reaction

$$NO + O_3 \rightarrow NO_2 + O_2$$

the following rate constants have been obtained:

Temperature, °C	$k\ (M^{-1}\ s^{-1})$
10.0	9.3×10^6
30.0	1.25×10^7

Calculate the activation energy for this reaction.

METHOD OF SOLUTION

Here we are given two rate constants corresponding to two temperatures. The activation energy (E_a) can be calculated by substituting into the equation given earlier that shows the temperature dependence of the rate constant:

$$\ln\left(\frac{k_1}{k_2}\right) = \frac{E_a}{R}\left(\frac{T_1 - T_2}{T_1 T_2}\right)$$

Let

$k_1 = 9.3 \times 10^6\ M^{-1}\ s^{-1}$ at $T_1 = 273 + 10.0 = 283$ K

$k_2 = 1.25 \times 10^7\ M^{-1}\ s^{-1}$ at $T_2 = 273 + 30.0 = 303$ K

Recall $R = 8.31$ J/mol · K. Substitute into the preceding equation to solve for E_a:

$$\ln\left(\frac{9.3 \times 10^6\ M^{-1}\ s^{-1}}{1.25 \times 10^7\ M^{-1}\ s^{-1}}\right) = \frac{E_a}{8.31\ \text{J/mol·K}}\left(\frac{283\ \text{K} - 303\ \text{K}}{(283\ \text{K})(303\ \text{K})}\right)$$

Solving for E_a yields

$$(\ln 0.744)(8.31\ \text{J/mol·K}) = E_a\left(\frac{-20\ \text{K}}{85{,}749\ \text{K}^2}\right)$$

$$(-0.296)(8.31\ \text{J/mol·K}) = E_a(-2.33 \times 10^{-4}\ \text{K}^{-1})$$

$$E_a = 1.06 \times 10^4\ \text{J/mol} \quad \text{(or 10.6 kJ/mol)}$$

EXAMPLE 13.8 Reaction Energy Profile

Draw a reaction energy profile for the following endothermic reaction:

$$2HI(g) \rightarrow H_2(g) + I_2(g) \qquad \Delta H^\circ_{rxn} = 12.5\ \text{kJ}$$

The activation energy $E_a = 185$ kJ/mol. Label the activation energy and the activated complex. What is the activation energy for the reverse reaction?

METHOD OF SOLUTION

The reaction is endothermic so the products are at a higher level than the reactants.

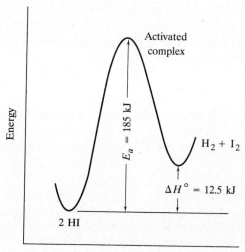

Answer: The activation energy for the reverse reaction is 185 kJ − 12.5 kJ = 173 kJ.

REACTION MECHANISMS

STUDY OBJECTIVES

You should be able to:
1. Describe what is meant by a reaction mechanism and distinguish between elementary steps and balanced chemical reactions.
2. Derive the rate law for a reaction from its proposed mechanism.

Elementary Steps. The purpose of a reaction mechanism is to indicate how the reactants are converted into products. A reaction mechanism consists of a set of so-called *elementary steps*. A mechanism indicates just what molecules must collide with each other and in what sequence. Each elementary step marks the progress of the conversion of molecules of reactants into products. A mechanism points out intermediates that are formed and consumed along the way, as well as which steps are fast and which are slow. This information is not supplied by the balanced chemical equation, because it is intended only to indicate the number of moles of one reactant that are consumed per mole of the other reactant.

Elementary steps have a property called *molecularity*, which pertains to the number of molecules that must collide in a single step. An elementary step that involves three molecules is called *termolecular*. A step that involves two molecules is *bimolecular*, and a step in which one molecule decomposes or rearranges is called *unimolecular*. Thus the equation

$$A \rightarrow B$$

represents a unimolecular reaction in which a single molecule of A reacts to form a single molecule of B. The following equation represents a bimolecular reaction:

$$A + B \rightarrow C$$

that is, one in which a molecule of A collides with a molecule of B and as a result a molecule of C is formed.

Reaction Order for Elementary Steps. The rate of a unimolecular step

$$A \rightarrow B$$

will depend on the number of molecules of A per unit volume of the container. Thus unimolecular reactions follow first-order rate laws:

$$\text{rate} = k[A]$$

The rate of a bimolecular step will depend on the number of molecules of A per unit volume times the number of molecules of B per unit volume:

$$\text{rate} = k[A][B]$$

Therefore, bimolecular steps follow second-order rate laws.

This can be understood also by realizing that A and B must collide with each other if they are to react, and so the rate of a bimolecular process depends on the rate of collisions of A and B. Doubling the concentration of A will double the probability of collision between A and B. Similarly, doubling the concentration of B will also double the rate of collisions of A and B. See Figure 13.12 in the textbook. Reasoning further along these lines, we conclude that termolecular reactions are third order.

Rate-Determining Step. It often turns out that one of the elementary steps in a mechanism is much slower than all the rest. This step determines the overall rate of reaction much as the slowest person ahead of you in a cafeteria line determines how fast you and others move through the line. The slowest step in a sequence of elementary steps is called the *rate-determining step*. This allows us to say that the overall law predicted by a mechanism will be the one corresponding to the rate-determining step.

Testing a Mechanism. The sequence of events in a kinetic study of a reaction that leads to a proposed mechanism is as follows:

1. Determine the rate law experimentally.
2. Propose a mechanism for the reaction.
3. Test the mechanism.

To test a proposed mechanism, assume one step to be the rate-determining step. Then establish the rate law for that step. This gives the rate law predicted by the mechanism. If the mechanism is adequate, then when the predicted rate law is compared with the experimental rate law, the two will match. On the other hand, if the predicted and experimental rate laws do not match, the mechanism has been proved inadequate. Only those mechanisms consistent with all the data can be considered adequate.

EXAMPLE 13.9 Reaction Mechanism

The rate law for the following reaction has been experimentally determined to be third order:

$$2NO(g) + O_2(g) \rightarrow 2NO_2(g) \qquad \text{rate} = k[NO]^2[O_2]$$

Which of the two proposed mechanisms that follow is more satisfactory?

a. $NO + NO \xrightarrow{k_1} N_2O_2$ slow   b. $NO + NO \underset{k_{-1}}{\overset{k_1}{\rightleftharpoons}} N_2O_2$ fast

$N_2O_2 + O_2 \xrightarrow{k_2} 2NO_2$ fast $N_2O_2 + O_2 \xrightarrow{k_2} 2NO_2$ slow

METHOD OF SOLUTION

Testing mechanism a first, we note that the first step is the rate-determining step. The rate law for this bimolecular step is

$$\text{rate} = k_1[NO]^2 \qquad (I)$$

This mechanism predicts a rate law that is second order in NO concentration and zero order in O_2. By comparison, the experimental rate law is second order in NO and first order in O_2. These do not match, and so mechanism a is not satisfactory.

Testing mechanism b, we find that the rate-determining step is the second step. Since it is bimolecular, its rate law should be

$$\text{rate} = k_2[N_2O_2][O_2] \qquad (II)$$

It is not possible to compare this predicted rate law directly with the experimental rate law because of the unique term, which is the concentration of N_2O_2. In this mechanism, N_2O_2 is an *intermediate*. Being formed in step 1 and consumed in step 2, its concentration is always small and usually not measurable.

The way around this situation is to find a mathematical substitution for $[N_2O_2]$. Note that step 1 is reversible and equilibrium can be established. This means that the rates of the forward and reverse reactions are equal:

$$k_1[NO]^2 = k_{-1}[N_2O_2]$$

where the rate constant is k_1 for the forward reaction and k_{-1} for the reverse:

Rearranging terms gives us

$$[N_2O_2] = \frac{k_1}{k_{-1}}[NO]^2$$

We now have an expression for $[N_2O_2]$, which we can substitute into equation II. This yields

$$\text{rate} = \frac{k_2 k_1}{k_{-1}}[NO]^2[O_2]$$

Whenever a reaction step is reversible, we can use the equality of the forward and reverse rates to form an expression to substitute for the concentration of an intermediate.

Now we compare this rate law with the experimental rate law, which is

$$\text{rate} = k[NO]^2[O_2]$$

We see that this mechanism predicts that the reaction will be second order in NO and first order in O_2, just as observed. Also, the collection of constants $k_2 k_1 / k_{-1}$ will equal the rate constant:

$$k = \frac{k_2 k_1}{k_{-1}}$$

Thus, the second mechanism predicts a rate law that matches the experimental order of reaction.

CATALYSIS

STUDY OBJECTIVES

You should be able to:
1. Describe in general how a catalyst increases the rate of a reaction without being consumed.
2. Describe the essentials of heterogeneous, homogeneous, and enzyme catalysis.

Mechanism of a Catalyzed Reaction. A *catalyst* is a substance that increases the rate of a chemical reaction without being consumed in the reaction. For this reason a catalyst does not appear in the stoichiometric equation. Chromium(III) oxide, Cr_2O_3, is the catalyst present in catalytic converters in American automobiles. This compound accelerates the reaction of carbon monoxide with oxygen. By converting CO to CO_2 in the exhaust stream, CO emissions are reduced:

$$CO(g) + \tfrac{1}{2}O_2(g) \xrightarrow{Cr_2O_3} CO_2(g)$$

The lifetime of catalytic converters is quite long; the Cr_2O_3 and other catalysts do not need to be replaced.

All catalysts accelerate reaction rates by providing a new reaction pathway (mechanism) that has a lower activation energy. See Figure 13.17 in the textbook. In the catalytic converter both CO and O_2 are chemically adsorbed on the catalyst's surface. Adsorption of O_2 by Cr_2O_3 weakens the O — O bond enough so that oxygen atoms can react with adsorbed CO. The reaction path involving a weakened O — O bond has a significantly lower activation energy than that of the reaction occurring purely in the gas phase. Recall that as the activation energy is lowered, the rate constant for a reaction increases. After CO_2 is formed, it desorbs from the surface, leaving the Cr_2O_3 catalyst chemically unaltered. The catalyst is not consumed in the reaction. Catalysts are regenerated in one of the last steps of the mechanism. The textbook describes three types of catalysts, which we will review here.

Heterogeneous Catalysis. In heterogeneous catalysis the reactants are in one phase and the catalyst is in another phase. The catalyst is usually a solid, and the reactants are gases or liquids. Ordinarily the site of the reaction is the surface of the solid catalyst. Many industrially important reactions involve gases and are catalyzed by solid surfaces. For example, the Haber process for the synthesis of ammonia is catalyzed by iron plus a few percent of the oxides of potassium and aluminum. The famous Ziegler-Natta catalyst for polymerizing ethylene gas (C_2H_4) into polyethylene polymer contains triethylaluminum and titanium or vanadium salts.

Homogeneous Catalysis. In homogeneous catalysis the reactants, products, and catalysts are all in the same phase, which is usually the gas or the liquid phase. Many reactions are catalyzed by acids. The decomposition of formic acid, HCO_2H, is an example:

$$HCO_2H \rightarrow H_2O + CO$$

Formic acid is normally stable and lasts a long time on the shelf. However, when sulfuric acid is added, bubbles of carbon monoxide gas can be observed immediately. The hydrogen ions from the sulfuric acid initiate a new reaction path. The mechanism is

$$HCO_2H + H^+ \rightarrow HCO_2H_2^+$$

$$HCO_2H_2^+ \rightarrow H_2O + HCO^+$$

$$\underline{HCO^+ \rightarrow CO + H^+}$$

$$HCO_2H \rightarrow CO + H_2O$$

Note that the H^+ ion is consumed in the first step, but, as for all catalysts, H^+ is regenerated. The net reaction is the sum of the three steps. The intermediate $HCO_2H_2^+$ cancels out, as does the catalyst.

Enzyme Catalysis. Enzymes catalyze nearly all chemical reactions that occur in living organisms. All enzymes are protein molecules with molecular masses well over 10,000 amu. All enzyme molecules are highly specific. That is, they can only affect the reaction rates of a few specific reactant molecules. Typically, an enzyme catalyzes a single reaction or a group of closely related reactions. The molecule on which an enzyme acts is called a *substrate*.

The simplest mechanism which explains enzyme activity and is consistent with these trends is one in which the substrate S and the enzyme E form an enzyme-substrate complex, ES. This complex has a lower energy than the activated complex without the enzyme (Figure 13.25 of the text). The enzyme-substrate complex can either dissociate back into E and S or break apart into the products P and the regenerated enzyme E. The simplest mechanism has the two steps shown below:

$$E + S \rightleftharpoons ES$$

$$ES \rightarrow E + P$$

EXAMPLE 13.10 Intermediates and Catalysts

Given the following mechanism for the decomposition of ozone in the stratosphere, identify the intermediate and the catalyst:

$$O_3 + Cl \rightarrow O_2 + ClO$$

$$ClO + O \rightarrow Cl + O_2$$

METHOD OF SOLUTION

Adding the two steps gives the overall reaction

$$O_3 + O \rightarrow 2O_2$$

An intermediate is formed in an early step and removed in a later step and does not appear in the overall equation; therefore ClO is the intermediate. A catalyst is consumed in an early step and is later regenerated. Atomic chlorine, Cl, is the catalyst.

EXAMPLE 13.11 Effect of a Catalyst

The gas-phase decomposition of nitrous oxide has an activation energy of 245 kJ/mol:

$$2N_2O \rightarrow 2N_2 + O_2$$

The rate of this reaction can be accelerated by bringing the N_2O into contact with the surface of a metal such as platinum. The activation energy for the heterogeneous catalysis is 136 kJ/mol. Assuming that the A factor in the Arrhenius equation remains essentially unchanged, how many times faster is the rate of the catalyzed reaction than the uncatalyzed reaction at 500°C?

METHOD OF SOLUTION

The rate constant k is related to the activation energy by the Arrhenius equation

$$k = Ae^{-E_a/RT}$$

The ratio k_{cat}/k can be found by setting up a ratio of Arrhenius equations:

$$\frac{k_{cat}}{k} = \frac{A_{cat}e^{-E_{cat}/RT}}{Ae^{-E_{uncat}/RT}}$$

CALCULATION

From the given data, we obtain $A_{cat}/A = 1.0$ and $T = 273 + 500 = 773$ K. Then substituting in the activation energies, we get

$$\frac{k_{cat}}{k} = \frac{e^{-136,000/(8.31)(773)}}{e^{-245,000/(8.31)(773)}}$$

$$= \frac{e^{-21.17}}{e^{-38.14}} = e^{16.97}$$

$$= 2.3 \times 10^7$$

COMMENT

In this case the platinum catalyst accelerates the rate of reaction some 23 million times. This example shows an unusually large effect.

TRUE-FALSE QUESTIONS

1. For the hypothetical reaction $A + 2B \rightarrow 3C$, the rate of appearance of C, $\Delta[C]/\Delta t$, may also be expressed as

$$\frac{\Delta[C]}{\Delta t} = \frac{3\Delta[B]}{2\Delta t}$$

2. If the concentration of A is tripled in a reaction that is second order in A, the rate of reaction will increase sixfold.

3. In the rate law for a reaction, k is called the rate constant and its units depend on the order of reaction.

4. In a first-order reaction the half-life depends on the initial concentration of reactant.

5. For a first-order reaction, a plot of [A] versus time is curved.

6. For a second-order reaction the rate constant can be determined from the slope of a plot of ln[A] versus t.

7. The rates of some reactions are not affected by the concentration of a reactant.

8. If a reaction has a very large rate constant, it will be likely to have a small activation energy.

Problems 9–12 refer to the following mechanism:

$A + B \rightarrow C + D$ slow

$D + A \rightarrow C + E$ fast

9. The second step is the rate-determining step.

10. The first step is a bimolecular elementary step.

11. The rate law for the first step is rate = k[A][B].

12. The species D is an intermediate.

13. A substance that decreases the activation energy of a reaction but does not appear in the overall chemical equation is called a catalyst.

14. Solid MnO_2, when added to an aqueous hydrogen peroxide solution, causes H_2O_2 to decompose more rapidly. This is an example of homogeneous catalysis.

15. The molecule with which an enzyme binds is called the substrate.

SELF-TEST A

1. The following experimental data were obtained for the reaction $2A + B \rightarrow$ products. What is the rate law for this reaction?

Experiment	Initial Concentration, M		Rate (M/s)
	$[A]_0$	$[B]_0$	
1	0.40	0.20	5.6×10^{-3}
2	0.80	0.20	5.5×10^{-3}
3	0.40	0.40	22.3×10^{-3}

2. For the reaction

   $$30CH_3OH + B_{10}H_{14} \rightarrow 10B(CH_3)_3 + 22H_2$$

   express the rate in terms of the change in concentration with time for each of the reactants and each of the products.

3. The hydrolysis of sucrose ($C_{12}H_{22}O_{11}$) yields the simple sugars glucose ($C_6H_{12}O_6$) and fructose ($C_6H_{12}O_6$), which just happen to be isomers:

$$C_{12}H_{22}O_{11} + H_2O \rightarrow C_6H_{12}O_6 + C_6H_{12}O_6$$

The rate follows the rate equation

$$\text{rate} = k[C_{12}H_{22}O_{11}]$$

At 27°C the rate constant is 2.1×10^{-6}/s.
a. Starting with the sucrose solution with a concentration of 0.10 M at 27°C, what would the concentration of sucrose be 24 h later? (The solution is kept at 27°C).
b. What is the half-life of sucrose at 27°C?

4. The following data were obtained on the rate of disintegration of a pesticide in soil at 30°C:

Time (days)	11	60	93
Pesticide remaining (%)	96	80	71

a. What is the order of reaction with respect to pesticide concentration?
b. What is the value of the rate constant?

5. a. It takes 30.0 min for the concentration of a reactant in a second-order reaction to drop from 0.40 M to 0.30 M. What is the value of the rate constant for this reaction?
b. How long will it take for the concentration to drop from 0.40 M to 0.20 M?

6. The reaction $A + 2B \rightarrow C + D$ has for its experimental rate law $r = k[A]^2[B]$. By what factor will the rate increase if the concentration of A is doubled and the concentration of B is tripled?

7. The rate of the reaction of hydrogen with iodine,

$$H_2 + I_2 \rightarrow 2HI$$

has rate constants of $1.41 \times 10^{-5}/M$ s at 393°C and $1.40 \times 10^{-4}/M$ s at 443°C. Calculate the activation energy for this reaction.

8. At 300 K the rate constant is $1.5 \times 10^{-5}/M$ s for the reaction

$$2NOCl \rightarrow 2NO + Cl_2$$

The activation energy is 90.2 kJ/mol. Calculate the value of the rate constant at 310 K. By what factor did the rate constant increase?

9. The activation energy for the reaction

$$CO + NO_2 \rightarrow CO_2 + NO$$

is 116 kJ/mol. How many times greater is the rate constant for this reaction at 250°C than it is at 200°C?

10. If the reaction $2HI \rightarrow H_2 + I_2$ has an activation energy of 190 kJ/mol and a $\Delta H = 10$ kJ, what is the activation energy for the reverse reaction?

$$H_2 + I_2 \rightarrow 2HI$$

11. The reaction between nitrite ion and oxygen gas is

    $2NO_2^- + O_2 \rightarrow 2NO_3^-$

    and proceeds at a rate that is first order in $[NO_2^-]$. A mechanism has been proposed:

    $2NO_2^- + O_2 \rightarrow 2NO_4^-$ slow

    $NO_4^- + NO_2^- \rightarrow 2NO_3^-$ fast

    Assume that NO_4^- is a very reactive intermediate and that the first step is the rate-determining step. Show that this mechanism is consistent with the experimental rate law.

12. The rate law for the substitution of NH_3 for H_2O in the reaction of a complex ion

    $Ni(H_2O)_6^{2+}(aq) + NH_3(aq) \rightarrow Ni(H_2O)_5(NH_3)^{2+}(aq) + H_2O(l)$

    is first order in $Ni(H_2O)_6^{2+}$ and zero order in NH_3:

    $\text{rate} = k\left[Ni(H_2O)_6^{2+}\right]$

    Show that the following mechanism is consistent with the experimental rate law:

    $Ni(H_2O)_6^{2+} \rightarrow Ni(H_2O)_5^{2+} + H_2O$ slow

    $Ni(H_2O)_5^{2+} + NH_3 \rightarrow Ni(H_2O)_5(NH_3)^{2+}$ fast

13. The rate law for the net reaction

    $2NO + O_2 \rightarrow 2NO_2$

    is rate = $k[NO_2]^2[O_2]$. Could the mechanism of this reaction be a single termolecular process?

14. The rate law for the net reaction

    $H_2(g) + I_2(g) \rightarrow 2HI(g)$

    is rate = $k[H_2][I_2]$. A possible mechanism involves a bimolecular elementary step:

    $H_2 + I_2 \rightarrow 2HI$

    A second possibility has also been proposed:

    $I_2 \rightarrow 2I$ fast

    $2I + H_2 \rightarrow 2HI$ slow

    Show that both mechanisms are consistent with the experimental rate law.

15. The following statements are sometimes made with reference to catalysts. Explain the fallacy in each one of them.
    a. A catalyst is a substance that accelerates the rate of a chemical reaction but does not take part in the reaction.
    b. A catalyst may increase the rate of a reaction going in one direction without increasing the rate of the reaction going in the reverse direction.

264 / Chemical Kinetics

16. a. Identify the catalyst in the mechanism below.
    b. What is the overall reaction?
    c. Is this reaction an example of homogeneous or heterogeneous catalysis?

$$H_2O_2(aq) + Br_2(aq) \rightarrow 2H^+(aq) + O_2(aq) + 2Br^-(aq)$$

$$H_2O_2(aq) + 2H^+(aq) + 2Br^-(aq) \rightarrow Br_2(aq) + H_2O(l)$$

GENERAL PROBLEM

17. Cooking an egg involves the denaturation of a protein called albumin from a soluble globular protein to an insoluble fibrous protein. This reaction has an activation energy of roughly 75 kJ/mol. The time required to achieve a certain degree of denaturation is inversely proportional to the rate constant for the process. At sea level, water boils at 100°C, and it takes 3 min to properly "boil an egg." Calculate how long it would take to cook the usual "three-minute egg" on top of Pikes's Peak at 445 m elevation on a day when the atmospheric pressure is 460 mm Hg. *Hint:* First find the boiling temperature of water. See Table 11.4.

SELF-TEST B1

1. The rate law for the reaction 2A + B → C was found to be rate = $k[A][B]^2$. If the concentration of B is tripled, by how many times will the reaction rate increase?

2. A certain first-order reaction A → B is 40% complete (40% of the reactant is used up) in 75 s. What is the rate constant and the half-life of this reaction?

3. The rate constant for a first-order reaction is 4.60×10^{-4} s^{-1} at 350°C. If the activation energy is 92.0 kJ/mol, calculate the rate constant at 320°C.

4. The activation energy for the reaction $CH_3CO \rightarrow CH_3 + CO$ is 71.0 kJ/mol. What is the ratio of the rate constants k_2/k_1 at 172°C and 86°C?

5. Tritium is a radioactive isotope of hydrogen with a half-life of 12.5 years. If you initially had 5.0×10^{-6} mol of tritium, calculate its decay rate in atoms per minute.

6. Thiosulfate ion is oxidized by iodine in aqueous solution according to the equation

$$2S_2O_3^{2-}(aq) + I_2(aq) \rightarrow 2S_4O_6^{2-}(aq) + 2I^-(aq)$$

If 0.025 mol of $S_2O_3^{2-}$ is consumed in 0.50 L solution per minute:
a. Calculate the rate of removal of $S_2O_3^{2-}$ in M/s.
b. What is the rate of removal of I_2?

7. The decomposition of hydrogen peroxide,

$$2H_2O_2 \rightarrow 2H_2O + O_2$$

has an activation energy of 75 kJ/mol. In the presence of a catalyst the activation energy is of 60 kJ/mol. Calculate the ratio of the rate constants at 25°C for the reaction, k_{cat}/k_{uncat}. Assume that the frequency factor is the same in both cases.

8. The reaction of nitric oxide and chlorine,

$$2NO(g) + Cl_2(g) \rightarrow 2NOCl(g)$$

has been proposed to proceed by the following two-step mechanism:

$$NO(g) + Cl_2(g) \rightarrow NOCl_2(g)$$

$$NO(g) + NOCl_2(g) \rightarrow 2NOCl(g)$$

What rate law is predicted if the first step is the rate-determining step?

SELF-TEST B2

1. Use the following data to determine the rate law for the reaction $2A + B \rightarrow C$:

Experiment	$[A]_0$	$[B]_0$	Rate (M/s)
1	0.25	0.10	0.012
2	0.25	0.20	0.048
3	0.50	0.10	0.024

2. The half-life of a certain first-order reaction is 102 s. What percent of the initial concentration of reactant will remain after 75 s?

3. Calculate the activation energy for a reaction given that the rate constant is 4.60×10^{-4} s^{-1} at 350°C and 1.87×10^{-4} s^{-1} at 320°C.

4. The rate constant for the reaction $CH_3CO \rightarrow CH_3 + CO$ is 100 times larger at 172°C than at 86°C. Calculate the activation energy for this reaction.

5. A sample of tritium (an isotope of hydrogen) contains 5.0×10^{-6} moles of tritium and it decays at a rate of 3.2×10^{11} atoms per minute. Calculate the decay constant k and the half-life.

6. Thiosulfate ion is oxidized by iodine in aqueous solution according to the equation

    $$2S_2O_3^{2-}(aq) + I_2(aq) \rightarrow 2S_4O_6^{2-}(aq) + 2I^-(aq)$$

    If 4.2×10^{-4} mol of I_2 is consumed in 1.0 L of the solution every second, what is the rate of consumption of $S_2O_3^{2-}$?

7. The rate of decomposition of hydrogen peroxide,

    $$2H_2O_2 \rightarrow 2H_2O + O_2$$

    is increased 430 times by the presence of a certain catalyst at 25°C. By how many kilojoules is the activation energy lowered by the catalyst?

8. The reaction of nitric oxide and chlorine,

    $$2NO(g) + Cl_2(g) \rightarrow 2NOCl(g)$$

    has been proposed to proceed by the following two-step mechanism:

    $$NO(g) + Cl_2(g) \rightarrow NOCl_2(g)$$

    $$NO(g) + NOCl_2(g) \rightarrow 2NOCl(g)$$

    What rate law is predicted if the second step is the rate-determining step?

ANSWERS

TRUE-FALSE QUESTIONS

1. True.
2. False. The rate will increase ninefold.
3. True.
4. False. The half-life of a first-order reaction is independent of concentrations.
5. True.
6. False. You must plot $1/[A]$ versus t.
7. True.
8. True.
9. False. The first step is the rate-determining step; it is the slow step.
10. True.
11. True.
12. True.
13. True.
14. False. It is heterogeneous catalysis.
15. True.

SELF-TEST A

1. Rate = $k[B]^2$
2. Rate = $-\dfrac{1}{30}\dfrac{\Delta[CH_3OH]}{\Delta t} = -\dfrac{\Delta[B_{10}H_{14}]}{\Delta t} = \dfrac{1}{22}\dfrac{\Delta[H_2]}{\Delta t} = \dfrac{1}{10}\dfrac{\Delta[B(OCH_3)_3]}{\Delta t}$
3. a. [Sucrose] = 0.083 M b. $t_{1/2} = 3.3 \times 10^5$ s
4. a. First order   b. 3.7×10^{-3} per day
5. a. $2.76 \times 10^{-2}/M \cdot$ min   b. 90 min
6. Twelvefold
7. 182 kJ/mol
8. $k = 4.8 \times 10^{-5}/M$ s; $k_2/k_1 = 3.21$
9. 16.8
10. 180 kJ/mol
11. The rate law for the rate-determining step is

    rate = $k[NO_2^-][O_2]$

    The concentration of dissolved O_2 is a constant according to Henry's law (Chapter 12). Therefore $k[O_2] = K$ (a constant) and rate = $K[NO_2^-]$.
12. The rate law for the rate-determining step is

    rate = $k\left[Ni(H_2O)_6^{2+}\right]$

    This rate law matches the experimental rate law.
13. Yes
14. The rate law for the rate-determining step is

    rate = $k_2[I]^2[H_2]$

    Obtaining a mathematical substitution for [I] from

    $k_1[I_2] = k_{-1}[I]^2$

yields

$$\text{rate} = \frac{k_2 k_1}{k_{-1}}[H_2][I_2]$$

This predicted rate law matches the experimental rate law given in the problem.
15. a. The catalyst reacts in one of the first few steps but then is regenerated in the last step.
    b. If E_a is lowered for the forward reaction, it must also be lowered for the reverse.
16. a. Br_2 b. $2H_2O \rightarrow 2H_2O + O_2$ c. Homogeneous
17. 7.4 min

SELF-TEST B1

1. 9 times
2. $k = 6.8 \times 10^{-3}$ s^{-1}, $t_{1/2} = 102$ s
3. 1.87×10^{-4} s^{-1}
4. $k_2/k_1 = 100$
5. 3.2×10^{11} atoms/min
6. a. 8.3×10^{-4} M/s   b. 4.2×10^{-4} mol I_2/L · s
7. 430
8. Rate = $k[NO][Cl_2]$

SELF-TEST B2

1. Rate = $k[A][B]^2$
2. 60%
3. 92 kJ/mol
4. 71.0 kJ/mol
5. $k = 1.2 \times 10^{-7}$ min^{-1}; $t_{1/2} = 12.14$ years
6. 8.4×10^{-4} mol $S_2O_3^{2-}$/L · s
7. 15 kJ
8. Rate = $k[NO]^2[Cl_2]$

Chapter Fourteen
CHEMICAL EQUILIBRIUM

- The Equilibrium State
- What K_c Tells Us
- Factors Affecting Equilibrium

THE EQUILIBRIUM STATE

STUDY OBJECTIVES

You should be able to:
1. Write the equilibrium constant expressions for homogeneous and heterogeneous chemical reactions, given balanced chemical equations.
2. Determine an equilibrium constant value, given either equilibrium concentrations, or initial concentrations and the equilibrium concentration of at least one component.
3. Given K_c for a reaction, calculate the value of K_c for the same reaction when it is balanced by using some multiple of the coefficients in the given equation. Also, find K_c for the reverse reaction, and convert K_c to K_p for reactions involving gases.

The Equilibrium Constant. A state of chemical equilibrium exists when the concentrations of reactants and products are observed to remain constant with time. When a mixture of SO_2 and O_2, for instance, is introduced into a closed vessel at a temperature of 700 K, a reaction that produces SO_3 occurs:

$$2SO_2 + O_2 \rightarrow 2SO_3$$

When a specific concentration of SO_3 is reached, no additional SO_3 is formed even though some SO_2 and O_2 remain. From that time on, the concentrations of SO_2, O_2, and SO_3 stay constant, and we say the system has reached chemical equilibrium.

The constant concentrations are the result of a reversible chemical reaction. In the reverse reaction some SO_3 decomposes back into SO_2 and O_2:

$$2SO_3 \rightarrow 2SO_2 + O_2$$

When the rates of forward and reverse reactions are the same, no net chemical change occurs and a state of chemical equilibrium exists. The equilibrium state is referred to as *dynamic* because of the continual conversions of reactants into products and products into reactants at the molecular level. A reversible reaction is represented by the opposing arrows in the chemical equation

$$2SO_2 + O_2 \rightleftharpoons 2SO_3$$

In quantitative terms, the equilibrium state is described by the *equilibrium constant expression*, which is also called the mass action expression:

$$K_c = \frac{[SO_3]^2}{[SO_2]^2[O_2]}$$

Here the brackets indicate concentration in moles per liter (molarity). In this expression, note that the concentration of the product is in the numerator and the concentrations of the reactants are in the denominator. Also, the concentration of each component is raised to a power equal to its coefficient in the balanced equation. The value of this expression is called the *equilibrium constant*.

The equilibrium constant value for a reaction can be determined by measuring the concentrations of all components at equilibrium and substituting these values into the K_c expression. The equilibrium concentrations of SO_2, O_2, and SO_3 are related to each other by the equilibrium constant expression, which at 700 K has the following value:

$$K_c = \frac{[SO_3]^2}{[SO_2]^2[O_2]} = 4.3 \times 10^6$$

The Form of K_c and the Equilibrium Equation. The equilibrium constant expression and its value depend on how the equation is balanced. Often an equation can be balanced with more than one set of coefficients, as shown below for instance:

$$2SO_2 + O_2 \rightleftharpoons 2SO_3$$

$$SO_2 + \tfrac{1}{2}O_2 \rightleftharpoons SO_3$$

How does this affect the equilibrium constant? For the latter equation, the equilibrium constant expression is written

$$K_c' = \frac{[SO_3]}{[SO_2][O_2]^{1/2}}$$

where the prime is just to help us keep track of the constant to which we are referring. Note that K_c' is the square root of K_c:

$$K_c' = \frac{[SO_3]}{[SO_2][O_2]^{1/2}} = \sqrt{\frac{[SO_3]^2}{[SO_2]^2[O_2]}} = \sqrt{K_c}$$

Therefore its value is

$$K_c' = \sqrt{4.3 \times 10^6} = 2.1 \times 10^3$$

The equilibrium constant expression for the reaction written in the reverse direction,

$$2SO_3 \rightleftharpoons 2SO_2 + O_2$$

is

$$K_c'' = \frac{[SO_2]^2[O_2]}{[SO_3]^2}$$

By inspection you can tell that it is the reciprocal of K_c for the forward reaction:

$$K_c'' = \frac{[SO_2]^2[O_2]}{[SO_3]^2} = \frac{1}{K_c}$$

Therefore the value of K_c'' is

$$K_c'' = \frac{1}{4.3 \times 10^6} = 2.3 \times 10^{-7}$$

Always use the K_c expression and value that are consistent with the way in which the balanced equation is written.

K_p and K_c. For reactions involving gases, the equilibrium constant expression can be written in terms of the partial pressures, P_i, of each gaseous component. For the equation

$$2SO_2 + O_2 \rightleftharpoons 2SO_3$$

$$K_p = \frac{P_{SO_3}^2}{P_{SO_2}^2 P_{O_2}}$$

The pressure, in atmospheres, of each gas at equilibrium is raised to a power corresponding to its coefficient in the balanced equation. The subscript p in K_p indicates that the equilibrium constant has a value that was calculated by using equilibrium pressures in atmospheres, rather than in units of moles per liter.

The values of K_p and K_c are related. Recall that for an ideal gas $PV = nRT$, and so the pressure of an ideal gas is proportional to its concentration:

$$P = \frac{n}{V}RT$$

where n/V is moles per liter. Substitution of $(n/V)RT$ for the pressure of each gas gives the equation relating K_p and K_c that was derived in the text:

$$K_p = K_c(RT)^{\Delta n}$$

where R is the ideal gas constant 0.0821 L · atm/K · mol and Δn is the change in the number of moles of gas when going from reactants to products. For the preceding reaction $\Delta n = -1$, and at 700 K, K_p is given by

$$K_p = K_c(RT)^{-1} = \frac{K_c}{RT} = \frac{4.3 \times 10^6}{(0.0821)(700)} = 7.5 \times 10^4$$

In general, $K_p \neq K_c$ except in the case when $\Delta n = 0$.

Heterogeneous Equilibria. Whenever a reaction involves reactants and products that exist in different phases, it is called a *heterogeneous reaction*, and in a closed container a *heterogeneous equilibrium* will result. For example, when steam is brought into contact with charcoal in a closed container, the following equilibrium is established:

$$C(s) + H_2O(g) \rightleftharpoons H_2(g) + CO(g)$$

The equilibrium constant expression for this reaction will *not* be the same as for a homogeneous reaction.

The usual expression for the constant would be

$$K = \frac{[H_2][CO]}{[C][H_2O]}$$

However, the concentration of a *pure* solid is itself a constant and is not changed by the addition to or removal of some of that solid. The concentration (mol/L) of a solid such as charcoal depends only on its density and not on its amount. Remember that solids have their own volume and do not fill their containers as gases do. This means that the equilibrium constant expression can be written

$$K_c = K[C] = \frac{[H_2][CO]}{[H_2O]}$$

where the constant concentration of the pure solid [C] has been incorporated into the equilibrium constant itself. Thus the expression $[CO][H_2]/[H_2O]$ does not contain the solid. In general, concentrations of solids and pure liquids do not appear in equilibrium constant expressions. In this reaction, equilibrium will be maintained as long as some $C(s)$ is present. The *amount* of solid carbon present does not affect the point of equilibrium.

Multiple Equilibria. When the product molecules of one equilibrium reaction become reactants in a second equilibrium process, we have an example of multiple equilibria. In such a case the overall reaction is the sum of the two individual reactions.

For instance, consider the reactions at 700°C,

$$NO_2 \rightleftharpoons NO + \tfrac{1}{2}O_2 \qquad K_2 = 0.012$$

followed by

$$SO_2 + \tfrac{1}{2}O_2 \rightleftharpoons SO_3 \qquad K_1 = 20$$

The overall reaction is

$$NO_2 + SO_2 \rightleftharpoons NO + SO_3 \qquad K_c = ?$$

The equilibrium constant value for the overall reaction is given by *the product of the equilibrium constants of the individual reactions*:

$$K_c = K_1 K_2 = (0.012)(20) = 0.24$$

Summary of Rules for Writing Equilibrium Constant Expressions

- The concentrations of the reacting species in the solution phase are expressed in mol/L. In the gaseous phase, the concentrations can be expressed in mol/L or in atmospheres. The constant K_c is related to K_p by a simple equation.
- The concentrations of pure solids, pure liquids, and solvents are constants and do not appear in equilibrium constant expressions.
- The equilibrium constant (K_c or K_p) is treated as a dimensionless quantity.
- In quoting a value for the equilibrium constant, we must specify the balanced equation and the temperature.
- If a reaction can be expressed as the sum of two or more reactions, the equilibrium constant for the overall reaction is given by the product of the equilibrium constants of the individual reactions.

EXAMPLE 14.1 The Equilibrium Constant Expression

Write the equilibrium constant expressions for the following reversible reactions:
a. $4NH_3(g) + 5O_2(g) \rightleftharpoons 4NO(g) + 6H_2O(g)$
b. $BaO(s) + CO_2(g) \rightleftharpoons BaCO_3(s)$

METHOD OF SOLUTION

Remember that the equilibrium constant expression has the concentrations of the products in the numerator and those of the reactants in the denominator. Raise each concentration to a power equal to the coefficient of that substance in the balanced equation. Include concentration terms for gaseous components only; leave out the concentrations of pure liquids and solids as they are not included in the equilibrium constant expression.

Answer:

a. $K_c = \dfrac{[NO]^4[H_2O]^6}{[NH_3]^4[O_2]^5}$ b. $K_c = \dfrac{1}{[CO_2]}$

EXAMPLE 14.2 The Equilibrium Constant Value

Ammonia is synthesized from hydrogen and nitrogen according to the equation

$$3H_2 + N_2 \rightleftharpoons 2NH_3$$

An equilibrium mixture at a given temperature was analyzed and the following concentrations were found: 0.31 mol N_2/L; 0.90 mol H_2/L; and 1.4 mol NH_3/L. What is the equilibrium constant value?

METHOD OF SOLUTION

When the concentrations of each component of a chemical system at equilibrium are known, the value of K_c can be determined readily by substituting these concentrations into the K_c expression:

$$K_c = \frac{[NH_3]^2}{[H_2]^3[N_2]}$$

CALCULATION

$$K_c = \frac{(1.4)^2}{(0.90)^3(0.31)}$$

$$= 8.7$$

EXAMPLE 14.3 The Equilibrium Constant Value

When 3.0 mol of I_2 and 4.0 mol of Br_2 are placed in a 2.0-L reactor at 150°C, the following reaction occurs until equilibrium is reached:

$$I_2(g) + Br_2(g) \rightleftharpoons 2IBr(g)$$

Chemical analysis then shows that the reactor contains 3.2 mol of IBr. What is the value of the equilibrium constant K_c for the reaction?

METHOD OF SOLUTION

We need to know the equilibrium concentrations of I_2, Br_2, and IBr. These, when substituted into the K_c expression, will give the value of K_c. First find the number of moles of I_2 and Br_2 at equilibrium. Initially, the *amounts* of each component in the reactor are 3.0 mol of I_2, 4.0 mol of Br_2, and zero moles of IBr. The number of moles of I_2 remaining at equilibrium is given by the *initial number of moles of I_2 minus the moles of I_2 reacted*. This is also true for Br_2. The information that 3.2 mol of IBr are formed tells us that 1.6 mol of I_2 and 1.6 mol of Br_2 must have reacted. We know this because the balanced equation states that 2 mol of IBr are formed when 1 mol of I_2 and 1 mol of Br_2 react.

CALCULATION

moles I_2 at equilibrium = 3.0 mol − 1.6 mol = 1.4 mol I_2

moles Br_2 at equilibrium = 4.0 mol − 1.6 mol = 2.4 mol Br_2

Substituting the equilibrium concentrations into the equilibrium constant expression yields

$$K_c = \frac{[IBr]^2}{[I_2][Br_2]} = \frac{(3.2 \text{ mol}/2.0 \text{ L})^2}{(1.4 \text{ mol}/2.0 \text{ L})(2.4 \text{ mol}/2.0 \text{ L})}$$

$$= 3.0$$

COMMENT

The following table containing the given information would have helped us to keep track of the initial and equilibrium (final) amounts of each component:

Amount	I_2	+	Br_2	⇌	2IBr
Initial	3.0 mol		4.0 mol		0
Change	—		—		—
Equilibrium	—		—		3.2 mol

If you look down the IBr column, the change in IBr must be +3.2 mol. If IBr increases by 3.2 mol, then Br_2 and I_2 both change by −1.6 mol and the table becomes

Amount	I_2	+	Br_2	⇌	2IBr
Initial	3.0 mol		4.0 mol		0
Change	−1.6 mol		−1.6 mol		3.2 mol
Equilibrium	—		—		3.2 mol

The equilibrium amounts are

Amount	I_2	+	Br_2	⇌	2IBr
Initial	3.0 mol		4.0 mol		0
Change	−1.6 mol		−1.6 mol		3.2 mol
Equilibrium	1.4 mol		2.4 mol		3.2 mol

Keep in mind that the equilibrium concentration equals the initial concentration plus the change in concentration due to reaction to reach equilibrium. The change in concentration is positive for a product and negative for a reactant. Notice also that only the *change in concentration* follows the laws of stoichiometry. When 2 mol IBr are formed, then 1 mol of I_2 must have reacted. This is always true. On the other hand, the equilibrium concentrations of I_2 and Br_2 are not equal, and it is not true that mol IBr = 2 mol I_2. Remember that many sets of concentrations of I_2, Br_2, and IBr can equal 3.0 when substituted into the K_c expression.

EXAMPLE 14.4 The Equilibrium Constant Value

The equilibrium constant for the reaction

$$H_2(g) + I_2(g) \rightleftharpoons 2HI(g)$$

at 400°C is

$$K_c = \frac{[HI]^2}{[H_2][I_2]} = 64$$

What is the equilibrium constant value for the reverse reaction?

$$2HI(g) \rightleftharpoons H_2(g) + I_2(g)$$

METHOD OF SOLUTION

Look to see if the reactions are related in some way. Note that the equilibrium constant for the second reaction is the reciprocal of the equilibrium expression for the forward reaction:

$$K'_c = \frac{[H_2][I_2]}{[HI]^2}$$

CALCULATION

$$K'_c = \frac{1}{K_c} = \frac{1}{64} = 0.016$$

EXAMPLE 14.5 K_p **and Partial Pressures at Equilibrium**

At 400°C, $K_c = 64$ for the reaction

$$H_2(g) + I_2(g) = 2HI(g)$$

a. What is the value of K_p for this reaction?
b. If, at equilibrium, the partial pressures of H_2 and I_2 in a container are 0.20 atm and 0.50 atm, respectively, what is the partial pressure of HI in the mixture?

METHOD OF SOLUTION

a. The equation relating K_p to K_c is

$$K_p = K_c(RT)^{\Delta n}$$

Here the change in the number of moles of gas, Δn, is

$$\Delta n = 2 \text{ mol HI} - 1 \text{ mol } H_2 - 1 \text{ mol } I_2 = 0$$

Since $\Delta n = 0$, K_p and K_c are the same:

$$K_p = K_c(RT)^0 = K_c$$

$$= K_c = 64$$

b. Writing the equilibrium constant expression

$$K_p = \frac{P^2_{HI}}{P_{H_2} P_{I_2}} = 64$$

and substituting the given pressures

$$\frac{P^2_{HI}}{(0.20)(0.50)} = 64$$

the partial pressure of HI is

$$P_{HI} = \sqrt{(0.20)(0.50)(64)}$$

$$= 2.5 \text{ atm}$$

EXAMPLE 14.6 K_p and K_c Values

What are the values of K_p and K_c at 1000°C for the reaction

$$CaCO_3(s) \rightleftharpoons CaO(s) + CO_2(g)$$

if the pressure of CO_2 in equilibrium with $CaCO_3$ and CaO is 3.87 atm?

METHOD OF SOLUTION

Enough information is given to find K_p first. Write the K_p expression for this heterogeneous reaction:

$$K_p = P_{CO_2} = 3.87$$

Then to get K_c, rearrange the equation,

$$K_p = K_c(RT)^{\Delta n}$$

where Δn, the change in the number of moles of gas in the reaction, is $+1$.

CALCULATION

$$K_c = \frac{K_p}{(RT)^{\Delta n}} = \frac{3.87}{[(0.0821)(1273)]^1}$$

$$= 0.0370$$

WHAT K_c TELLS US

STUDY OBJECTIVES

You should be able to:
1. Determine whether a reaction is at equilibrium or not, and if it is not, predict the direction in which a net reaction will occur.
2. Predict the relative extents of several reactions given the values of their equilibrium constants.
3. Given K_c and the initial concentrations, calculate the equilibrium concentrations of all components.

Predicting the Direction of Reaction. Equilibrium constants provide useful information about chemical reaction systems. In this section, we will use them to predict the direction a reaction will proceed to establish equilibrium. We will also use them to determine the extent of reaction.

The reaction quotient, Q_c, is a useful aid in predicting whether or not a reaction system is at equilibrium. Take, for instance, the reaction

$$2SO_2 + O_2 \rightleftharpoons 2SO_3$$

The reaction quotient is

$$Q_c = \frac{[SO_3]_0^2}{[SO_2]_0^2[O_2]_0}$$

You will notice that Q has the same algebraic form of the concentrations terms as does K_c. However, the concentrations here are not necessarily equilibrium concentrations. We will call them initial concentrations, $[\ ]_0$. When a set of initial concentrations is substituted into the reaction quotient, Q_c takes on a certain value. In order to predict whether the system is at equilibrium or not, the magnitude of Q_c must be compared with that of K_c.

- When $Q_c = K_c$, the reaction is at equilibrium, and no net reaction will occur.
- When $Q_c > K_c$, the system is not at equilibrium, and a reaction will occur in the reverse direction until $Q_c = K_c$.
- When $Q_c < K_c$, the system is not at equilibrium, and a reaction will occur in the forward direction until $Q_c = K_c$.

Extent of Reaction. The term *extent of reaction* refers to the degree of conversion of reactants to products before chemical equilibrium is reached. The extent of reaction is related to the magnitude of K_c. Since K_c is proportional to the concentrations of products divided by the concentrations of reactants present at equilibrium, then $K_c \gg 1$ for reactions in which more products are present than reactants at equilibrium. Conversely, $K_c \ll 1$ for reactions in which more reactants are present at equilibrium than products. In this case equilibrium is reached before appreciable concentrations of products build up. *The larger the value of K_c, the greater the extent of reaction before equilibrium is reached*. See Example 14.7. In general, we can say:

When $K_c \gg 1$, the forward reaction will go nearly to completion.

When $K_c \ll 1$, the reaction will not go forward to an appreciable extent.

Calculation of Equilibrium Concentrations. Not only can we estimate the extent of reaction from the K_c value, but also the expected concentrations at equilibrium can be calculated from a knowledge of the initial concentrations and the K_c value. See Example 14.9 for this important type of calculation. In these types of problems it will be very helpful to recall that

equilibrium concentration = initial concentration ± the change due to reaction

where the plus sign is used for a product and the minus sign for a reactant.

EXAMPLE 14.7 Predicting Direction of Reaction

At a certain temperature the reaction

$$CO(g) + Cl_2(g) \rightleftharpoons COCl_2(g)$$

has an equilibrium constant $K_c = 13.8$. Is the following mixture an equilibrium mixture? If not, in which direction (forward or reverse) will reaction occur to reach equilibrium?

$[CO]_0 = 2.5\ M \quad [Cl_2]_0 = 1.2\ M \quad [COCl_2]_0 = 5.0\ M$

METHOD OF SOLUTION

Recall that for the system to be at equilibrium $Q_c = K_c$. Substitute the given concentrations into the reaction quotient for the reaction, and determine Q_c.

CALCULATION

$$Q_c = \frac{[COCl_2]_0}{[CO]_0[Cl_2]_0} = \frac{(5.0)}{(2.5)(1.2)} = 1.7$$

Compare Q_c to K_c. Since $Q_c < K_c$, the reaction mixture is not an equilibrium mixture, and a forward reaction will bring the system to equilibrium. *Answer:* Only a forward reaction will cause Q_c to increase until $Q_c = K_c$.

EXAMPLE 14.8 Extent of Reaction

Arrange the following reactions in order of their increasing tendency to proceed toward completion (least extent → greatest extent):
a. $CO + Cl_2 \rightleftharpoons COCl_2$ $K_c = 13.8$
b. $N_2O_4 \rightleftharpoons 2NO_2$ $K_c = 2.1 \times 10^{-4}$
c. $2NOCl \rightleftharpoons 2NO + Cl_2$ $K_c = 4.7 \times 10^{-4}$

METHOD OF SOLUTION

The larger the value of K_c, the more products there are at equilibrium compared to reactants, and the farther a reaction will proceed toward completion (the greater the extent of reaction).
Answer: b < c < a.

EXAMPLE 14.9 Equilibrium Concentrations

At 400°C, $K_c = 64$ for the equilibrium

$$H_2(g) + I_2(g) \rightleftharpoons 2HI(g)$$

If 1.00 mol of H_2 and 2.00 mol of I_2 are introduced into an empty 0.50-L reaction vessel, find the equilibrium concentrations of all components at 400°C.

METHOD OF SOLUTION

The equilibrium constant expression contains the desired concentrations, and we will use it to obtain the answer:

$$K_c = \frac{[HI]^2}{[H_2][I_2]} = 64$$

First we need expressions for the equilibrium concentrations of H_2, I_2, and HI. Begin by tabulating the initial concentrations.

Concentration	H_2	+	I_2	$\rightleftharpoons$	2HI
Initial	1.00 mol/0.50 L		2.00 mol/0.50 L		0
Change	—		—		—
Equilibrium	—		—		—

Since the answer involves three unknowns, we will relate the concentrations to each other by introducing the variable x. Recall that the equilibrium concentration = initial concentration ± change in concentration. Let x be the change in concentration of H_2. That is, let x be the number of moles of H_2 reacting per liter. From the coefficients of the balanced equation we can tell that if the change in H_2 is $-x$, then the change in I_2 must also be $-x$, and the change in HI must be $+2x$. The next step is to complete the table in units of molarity:

Concentration	H_2	+	I_2	$\rightleftharpoons$	2HI
Initial (M)	2.0		4.0		0
Change (M)	$-x$		$-x$		$2x$
Equilibrium (M)	$(2.0 - x)$		$(4.0 - x)$		$2x$

Now substitute the equilibrium concentrations from the table into the K_c expression,

$$K_c = \frac{(2x)^2}{(2.0 - x)(4.0 - x)} = 64$$

and solve for x:

$$\frac{(2x)^2}{x^2 - 6.0x + 8.0} = 64$$

Rearranging, we get

$$4x^2 = 64x^2 - 384x + 512$$

and grouping yields

$$60x^2 - 384x + 512 = 0$$

We will use the general method of solving a quadratic equation of the form

$$ax^2 + bx + c = 0$$

The root x is given by

$$x = \frac{-b \pm \sqrt{b^2 - 4ac}}{2a}$$

In this case, $a = 60$, $b = -384$, and $c = 512$. Therefore,

$$x = \frac{-(-384) \pm \sqrt{(384)^2 - 4(60)(512)}}{2(60)} = \frac{384 \pm \sqrt{2.5 \times 10^4}}{120}$$

$$= \frac{384 \pm 158}{120} = 1.9 \text{ and } 4.5 \text{ mol/L}$$

Recall that x is the number of moles of H_2 (or I_2) reacting per liter. Of the two answers (roots), only 1.9 is reasonable, because the value 4.5 M would mean that more H_2 (or I_2) reacted than was present at the start. This would result in a negative equilibrium concentration, which is physically meaningless. We therefore use the root $x = 1.9\ M$ to calculate the equilibrium concentrations:

$$[H_2] = 2.0 - x = 2.0\ M - 1.9\ M = 0.1\ M$$

$$[I_2] = 4.0 - x = 4.0\ M - 1.9\ M = 2.1\ M$$

$$[HI] = 2x = 2(1.9\ M) = 3.8\ M$$

The results can be checked by plugging these concentrations back into the K_c expression to see if $K_c = 64$:

$$K_c = \frac{[HI]^2}{[H_2][I_2]} = \frac{(3.8)^2}{(0.1)(2.1)} = 68$$

Thus the concentrations we have calculated are correct. The difference between 64 and 68 results from rounding off to maintain the correct number of significant figures. Therefore, our result is correct only to the number of significant figures given in the problem.

FACTORS AFFECTING EQUILIBRIUM

STUDY OBJECTIVES

You should be able to:
1. Predict how the equilibrium concentrations of reactants and products are shifted by changes in concentrations of reactants and products, by pressure changes, and by temperature changes.

Le Chatelier's Principle. When a reaction reaches a state of chemical equilibrium under a particular set of conditions, no further changes in the concentrations of reactants and products occur. If a change is made in the conditions under which the system is at equilibrium, chemical change will occur in such a way as to establish a new equilibrium. The factors that can influence equilibrium are change in concentration, change in pressure (or volume), and change in temperature.

What effect does a change in one of these factors have on the extent of reaction? This question can be answered qualitatively by using *Le Chatelier's principle*: When a stress is applied to a system in a state of dynamic equilibrium, the system will, if possible, shift to a new position of equilibrium in which the stress is partially offset. To interpret this statement, take, for example, the reaction

$$2NO_2(g) \rightleftharpoons N_2O_4(g)$$

First, the reaction must be at equilibrium. Let's add more N_2O_4 as an illustration of a change in concentration. The concentration of N_2O_4 increases, and the equilibrium is disturbed. The system will respond by using up part of the additional N_2O_4. In this case, a net reverse reaction will partially offset the increased N_2O_4 concentration. The net reverse reaction brings the system to a new state of equilibrium. When equilibrium is reestablished, more NO_2 will be present than there was before the N_2O_4 was added. Thus, the position of equilibrium has shifted to the left.

Le Chatelier's principle will predict the direction of the net reaction or "shift in equilibrium" that brings the system to a new equilibrium. In this case the stress of adding more N_2O_4 was partially offset by a net reverse reaction that consumed some of the additional N_2O_4. The key to the use of Le Chatelier's principle is to recognize which net reaction, forward or reverse, will partially offset the change in conditions.

Changes in Volume and Pressure. The pressure of a system of gases in chemical equilibrium can be increased by decreasing the available volume. This change causes the concentration of *all components* to increase. The stress will be partially offset by a net reaction that will lower the total concentration of gas molecules. Consider our previous example reaction,

$$2NO_2(g) \rightleftharpoons N_2O_4(g)$$

When the molecules of both gases are compressed into a smaller volume, their total concentration increases (this is a stress). A net forward reaction (shift to the right) will bring the system to a new state of equilibrium, in which the *total concentration* of all molecules will be lowered somewhat. This partially offsets the initial stress on the system. Notice that when 2 moles of NO_2 react, only 1 mole of N_2O_4 is formed. When equilibrium is reestablished, more N_2O_4 and less NO_2 will be present than before the pressure increase occurred. In this case the equilibrium has shifted to the right. In general, an increase in pressure by decreasing the volume will result in a net reaction that decreases the total concentration of gas molecules. When the total number of moles of gaseous products and of gaseous reactants are equal in the balanced equation, no shift in equilibrium will occur.

Changes in Temperature. If the temperature of a system is changed, a change in the value of K_c occurs. An increase in temperature always shifts the equilibrium in the direction of the endothermic reaction, while a temperature decrease shifts the equilibrium in the direction of the exothermic reaction. See Figure 14.8 in the text. Thus, for endothermic reactions the value of K_c increases with increasing temperature, and for exothermic reactions the value of K_c decreases with increasing temperature. In the case of our example reaction, K_c will decrease as the temperature is increased because the equilibrium will shift in the direction of the endothermic reaction, that is, in the reverse direction:

$$2NO_2(g) \rightleftharpoons N_2O_4(g) \qquad \Delta H^\circ_{rxn} = -58.0 \text{ kJ}$$

We can explain this in terms of Le Chatelier's principle. As heat is added to the system, it represents a stress on the equilibrium. The equilibrium will shift in the direction that will consume some of the added heat. This partially offsets the stress. In this reaction, the equilibrium shifts to the left, and K_c decreases.

Remember, of these three types of changes—concentration, pressure, and temperature—*only* changes in temperature will actually alter the K_c value.

There are two other factors related to chemical reactions that do not affect the position of equilibrium. The first of these is a catalyst. Catalysts speed up the rate at which equilibrium is reached by lowering the activation energy barrier. Of course, this speeds up the reverse reaction as well. The net result is that changes in the concentration of a catalyst will not affect the equilibrium concentrations of reactants and products, nor will they change the value of K_c.

The second of the two factors is the addition of an inert gas to a system at equilibrium. This will cause an increase in the total pressure within the reactor. However, none of the partial pressures of reactants or products are changed, and so the equilibrium is not upset, and no shifting is needed to bring the system back to equilibrium.

EXAMPLE 14.10 Changing Conditions Affecting Equilibrium

For the reaction at equilibrium

$$2NaHCO_3(s) \rightleftharpoons Na_2CO_2(s) + H_2O(g) + CO_2(g) \qquad \Delta H^\circ_{rxn} = 128 \text{ kJ}$$

state the effects (increase, decrease, no change) of the following stresses on the number of moles of sodium carbonate, Na_2CO_3, at equilibrium in a closed container. Note that Na_2CO_3 is a solid (this is a heterogeneous equation); its concentration will remain constant, but its amount can change.
a. Removing $CO_2(g)$
b. Adding $H_2O(g)$
c. Raising the temperature
d. Adding $NaHCO_3(s)$

METHOD OF SOLUTION

Apply Le Chatelier's principle.
a. If CO_2 concentration is lowered, the system will react in such a way as to offset the change. That is, a shift to the right will replace some of the missing CO_2. *Answer:* Increases moles of $Na_2CO_3(s)$.
b. Addition of $H_2O(g)$ exerts a stress on the equilibrium that is partially offset by a shift in the equilibrium to the left (net reverse action). This consumes Na_2CO_3 as well as some of the extra H_2O. *Answer:* Decreases moles of $Na_2CO_3(s)$.

c. An increase in temperature will increase the K_c value of the endothermic reaction. *Answer:* There is a shift to the right, and the extent of reaction is increased and more Na_2CO_3 is formed.
d. The position of a heterogeneous equilibrium does not depend on the amounts of pure solids or liquids present. The same equilibrium is reached whether the system contains 1 g of $NaHCO_3(s)$ or 10 g of $NaHCO_3(s)$. *Answer:* No shift in the equilibrium occurs.

TRUE-FALSE QUESTIONS

1. The K_c expression for the equilibrium

    $2NOCl(g) \rightleftharpoons 2NO(g) + Cl_2(g)$

    is $K_c = [NOCl]^2/[NO]^2[Cl_2]$.

2. If $K_c = 9.5$ for the reaction

    $2NH_3(g) \rightleftharpoons 3H_2(g) + N_2(g)$

    then $K_c = 0.11$ for the reaction

    $N_2(g) + 3H_2(g) \rightleftharpoons 2NH_3(g)$.

3. The K_c expression for the reaction

    $S(s) + O_2(g) \rightleftharpoons SO_2(g)$

    is $K_c = [SO_2]/[S][O_2]$.

4. Consider the equilibrium in a closed reactor,

    $N_2(g) + O_2(g) \rightleftharpoons 2NO(g)$

    If more N_2 is added, the equilibrium will shift to the left.

5. For the reaction

    $N_2(g) + O_2(g) \rightleftharpoons 2NO(g) \quad \Delta H°_{rxn} = 180.7$ kJ

    the equilibrium constant, K_p, will increase as the temperature increases.

6. The following reaction mixture is at equilibrium at a constant temperature:

    $3H_2(g) + N_2(g) \rightleftharpoons 2NH_3(g)$

    If NH_3 is removed, the equilibrium constant will increase.

7. The balanced equation

    $A(g) + 2B(g) \rightleftharpoons 3C(g)$

    implies that at equilibrium, [C] will be three times greater than [A].

8. Consider the reaction at equilibrium

    $A(g) + 2B(g) \rightleftharpoons 3C(g)$

If more B is added, a new equilibrium will be reached. At the new equilibrium, [A] will be greater than [A] in the original equilibrium.

9. The concentration of an ideal gas is related to its pressure by

$$\frac{n}{V} = \frac{RT}{P}$$

10. For the reaction at 25°C

$$Cl_2(g) + Br_2(g) \rightleftharpoons 2BrCl(g) \qquad K_c = 5.0$$

Therefore, the value of K_p is 5.0.

SELF-TEST A

1. Write the equilibrium constant expressions for the following reactions:
   a. $4NH_3(g) + 3O_2(g) \rightleftharpoons 2N_2(g) + 6H_2O(g)$
   b. $2N_2O(g) + 3O_2(g) \rightleftharpoons 2N_2O_4(g)$
   c. $2ClO_2(g) + F_2(g) \rightleftharpoons 2FClO_2(g)$
   d. $H_2(g) + Br_2(l) \rightleftharpoons 2HBr(g)$
   e. $C(s) + CO_2(g) \rightleftharpoons 2CO(g)$
   f. $CuO(s) + H_2(g) \rightleftharpoons Cu(s) + H_2O(g)$

2. The equilibrium constant K_c for the reaction

$$Ni(s) + 4CO(g) \rightleftharpoons Ni(CO)_4(g)$$

is 5.0×10^4 at 25°C. What is value of the equilibrium constant for the reaction?

$$Ni(CO)_4(g) \rightleftharpoons Ni(s) + 4CO(g)$$

3. The decomposition of HI(g) is represented by the equation

$$2HI(g) \rightleftharpoons H_2(g) + I_2(g) \qquad K_c = 64$$

If the equilibrium concentrations of H_2 and I_2 at 400°C are found to be $[H_2] = 4.2 \times 10^{-4}$ M and $[I_2] = 1.9 \times 10^{-3}$ M, what is the equilibrium concentration of HI?

4. For the reaction $CH_4(g) + 2H_2S(g) \rightleftharpoons CS_2(g) + 4H_2(g)$, $K_p = 2.05 \times 10^9$ at 25°C. Calculate K_p and K_c, at this temperature, for

$$2H_2(g) + \tfrac{1}{2}CS_2(g) \rightleftharpoons H_2S(g) + \tfrac{1}{2}CH_4(g)$$

5. A 1.00-L vessel initially contains 0.777 mol of $SO_3(g)$ at 1100 K. What is the value of K_c for the following reaction if 0.520 mol of SO_3 remain at equilibrium?

$$2SO_3(g) \rightleftharpoons 2SO_2(g) + O_2(g)$$

6. Initially a 1.0-L vessel contains 10.0 mol of NO and 6.0 mol of O_2 at a certain temperature. They react until equilibrium is established:

$$2NO(g) + O_2(g) \rightleftharpoons 2NO_2(g)$$

At equilibrium the vessel contains 8.8 mol of NO_2. Determine the value of K_c at this temperature.

7. Given the reaction

$$N_2 + O_2 \rightleftharpoons 2NO \qquad K_c = 2.5 \times 10^{-3} \text{ at } 2130°C$$

Decide whether the following mixture is at equilibrium or if a net forward or reverse reaction will occur: [NO] = 0.005; [O_2] = 0.25; [N_2] = 0.020 mol/L.

8. Hydrogen iodide decomposes according to the equation

$$2HI(g) \rightleftharpoons H_2(g) + I_2(g) \qquad K_c = 0.0156 \text{ at } 400°C$$

A 0.55-mol sample of HI was injected into a 2.0-L reaction vessel held at 400°C. Calculate the concentration of HI at equilibrium.

9. A sample of 2.00 mol of NOCl was placed in a 2.00-L reaction vessel at 400°C. After equilibrium was established, it was found that 24% of the NOCl had dissociated according to the equation

$$2NOCl(g) \rightleftharpoons 2NO(g) + Cl_2(g)$$

Calculate the equilibrium constant K_c for the reaction.

10. For the reaction

$$N_2(g) + O_2(g) \rightleftharpoons 2NO(g) \qquad K_p = 3.80 \times 10^{-4} \text{ at } 2000°C$$

what equilibrium pressures of N_2, O_2, and NO will result when a 10-L reactor vessel is filled with 2.00 atm of N_2 and 0.400 atm of O_2 and the reaction is allowed to come to equilibrium?

11. For the equilibrium

$$N_2O_4(g) \rightleftharpoons 2NO_2(g) \qquad K_c = 0.36 \text{ at } 100°C$$

a sample of 0.25 mol of N_2O_4 is allowed to dissociate and come to equilibrium in a 1.5-L flask at 100°C. What are the equilibrium concentrations of NO_2 and N_2O_4?

12. What changes in the equilibrium composition of the reaction

$$2SO_2(g) + O_2(g) \rightleftharpoons 2SO_3(g) \qquad \Delta H°_{rxn} = -197 \text{ kJ}$$

will occur if it experiences the following stresses?
a. The partial pressure of $SO_3(g)$ is increased.
b. Inert Ar gas is added.
c. The temperature of the system is decreased.
d. The total pressure of the system is increased by reducing the available volume.
e. The partial pressure of $O_2(g)$ is decreased.

13. Assume for the chemical equilibrium

$$PCl_5(g) \rightleftharpoons PCl_3(g) + Cl_2(g) \qquad \Delta H°_{rxn} = 92.9 \text{ kJ}$$

a. What is the effect on K_c of lowering the temperature?
b. What is the effect on the equilibrium concentration of PCl_3 of adding Cl_2?
c. What is the effect on the equilibrium concentrations of compressing the mixture to a smaller volume?
d. What is the effect on the equilibrium pressure of Cl_2 of removing PCl_3?

14. Consider the equilibrium

$$SO_2(g) + NO_2(g) \rightleftharpoons NO(g) + SO_3(g)$$

where $K_c = 85.0$ at 460°C. The following mixture was prepared in a reactor at 460°C:

$[SO_2] = 0.0200\ M$ $\quad$ $[NO_2] = 0.0200\ M$ $\quad$ $[NO] = 0.100\ M$ $\quad$ $[SO_3] = 0.100\ M$

What will the concentrations of the four gases be when equilibrium is reached?

15. Arrange the following reactions in their increasing tendency to proceed toward completion:
    a. $2HF \rightleftharpoons H_2F_2$ $\quad$ $K_c = 1 \times 10^{-13}$
    b. $2H_2 + O_2 \rightleftharpoons 2H_2O$ $\quad$ $K_c = 3 \times 10^{81}$
    c. $2NOCl \rightleftharpoons 2NO + Cl_2$ $\quad$ $K_c = 4.7 \times 10^{-4}$

16. For the decomposition of calcium carbonate

$$CaCO_3(s) \rightleftharpoons CaO(s) + CO_2(g) \qquad \Delta H^\circ_{rxn} = 175\ kJ$$

how will the amount (not concentration) of $CaCO_3(s)$ change with the following stresses?
a. $CO_2(g)$ is removed.
b. $CaO(s)$ is added.
c. The temperature is raised.
d. The volume of the container is decreased.

17. At 2000°C, 5.0×10^{-3} mol of CO_2 is introduced into a 1.0-L container and the following reaction comes to equilibrium:

$$2CO_2(g) \rightleftharpoons 2CO(g) + O_2(g) \qquad K_c = 6.4 \times 10^{-7}$$

a. Calculate the equilibrium concentrations of CO and O_2.
b. What fraction of the CO_2 is decomposed at equilibrium?

18. In a closed container at 1900 K nitrogen reacts with oxygen to yield NO:

$$N_2(g) + O_2(g) \rightleftharpoons 2NO(g) \qquad K_p = 2.3 \times 10^{-4}$$

However, $NO(g)$ quickly reacts with oxygen to produce $NO_2(g)$:

$$2NO(g) + O_2(g) \rightleftharpoons 2NO_2(g) \qquad K_p = 1.3 \times 10^{-4}$$

Calculate K_p and K_c at 1900 K for the net reaction

$$N_2(g) + 2O_2(g) \rightleftharpoons 2NO_2(g)$$

SELF-TEST B1

1. Copper can be extracted from its ores by heating Cu_2S in air:

$$Cu_2S(l) + O_2(g) \rightleftharpoons 2Cu(l) + SO_2(g) \qquad \Delta H^\circ_{rxn} = -250\ kJ$$

Predict the direction of the shift of the equilibrium position in response to each of the following changes in conditions:
a. Adding $O_2(g)$
b. Compressing the vessel volume in half
c. Raising the temperature

2. A sample of nitrosyl bromide is heated to 100°C in a 10.0-L container in order to partially decompose it:

$$2NOBr(g) \rightleftharpoons 2NO(g) + Br_2(g)$$

At equilibrium the container is found to contain 0.0585 mol of NOBr, 0.105 mol of NO, and 0.0524 mol of Br_2. Calculate the value of K_c.

3. For the reaction

$$2HBr(g) \rightleftharpoons H_2(g) + Br_2(g)$$

$K_p = 1.4 \times 10^{-5}$ at 700 K. What is the value of K_p for the following reaction at the same temperature?
   a. $H_2(g) + Br_2(g) \rightleftharpoons 2HBr(g)$
   b. $HBr(g) \rightleftharpoons \frac{1}{2}H_2(g) + \frac{1}{2}Br_2(g)$

4. The reaction

$$PCl_5(g) \rightleftharpoons PCl_3(g) + Cl_2(g)$$

has the equilibrium constant value $K_c = 0.24$ at 300°C.
   a. Is the following reaction mixture at equilibrium?

$[PCl_5] = 5.0$ mol/L $[PCl_3] = 2.5$ mol/L $[Cl_2] = 1.9$ mol/L

   b. Predict the direction in which the system will react to reach equilibrium, and calculate the equilibrium concentrations.

5. In the decomposition of ammonium chloride,

$$NH_4Cl(s) \rightleftharpoons NH_3(g) + HCl(g)$$

$K_p = 4.8$ at 427°C. Calculate K_c for this reaction.

SELF-TEST B2

1. Copper can be extracted from its ores by heating Cu_2S in air:

$$Cu_2S(l) + O_2(g) \rightleftharpoons 2Cu(l) + SO_2(g) \qquad \Delta H^\circ_{rxn} = -250 \text{ kJ}$$

   Predict the direction of the shift of the equilibrium position in response to each of the following changes in conditions:
   a. Adding $SO_2(g)$
   b. Adding $Cu(l)$
   c. Lowering the temperature

2. The decomposition of NOBr is represented by the equation

$$2NOBr(g) \rightleftharpoons 2NO(g) + Br_2(g) \qquad K_c = 0.0169$$

   At equilibrium the concentrations of NO and Br_2 are 1.05×10^{-2} M and 5.24×10^{-3} M, respectively. What is the concentration of NOBr?

3. For the reaction

$$H_2(g) + Br_2(g) \rightleftharpoons 2HBr(g)$$

$K_p = 7.1 \times 10^4$ at 700 K. What is the value of K_p for the following reactions at the same temperature?
  a. $2HBr(g) \rightleftharpoons H_2(g) + Br_2(g)$
  b. $\frac{1}{2}H_2(g) + \frac{1}{2}Br_2(g) \rightleftharpoons HBr(g)$

4. The reaction

$$PCl_5(g) \rightleftharpoons PCl_3(g) + Cl_2(g)$$

has the equilibrium constant value $K_c = 0.24$ at 300°C.
  a. Is the following reaction mixture at equilibrium?

$[PCl_5] = 2.6$ mol/L $[PCl_3] = 0.5$ mol/L $[Cl_2] = 0.5$ mol/L

  b. Predict the direction in which the system will react to reach equilibrium, and calculate the equilibrium concentrations.

5. In the decomposition of ammonium chloride,

$$NH_4Cl(s) \rightleftharpoons NH_3(g) + HCl(g)$$

$K_c = 1.5 \times 10^{-3}$ at 427°C. Calculate K_p for this reaction.

ANSWERS

TRUE-FALSE QUESTIONS

1. False. $K_c = [NO]^2[Cl_2]/[NOCl]^2$.
2. True.
3. False. $K_c = [SO_2]/[O_2]$.
4. False. Equilibrium will shift to the right.
5. True.
6. False. Only temperature change will alter the K_c value. At constant temperature the value of K_c is constant.
7. False. The balanced equation says that whenever 3 mol of C are formed, 1 mol of A must react. Actually there are many possible ratios of [C] to [A] that satisfy the K_c value.
8. False. [A] will decrease.
9. False. $n/V = P/RT$.
10. True.

SELF-TEST A

1. a. $K_c = \dfrac{[N_2]^2[H_2O]^6}{[NH_3]^4[O_2]^3}$ b. $K_c = \dfrac{[N_2O_4]^2}{[N_2O]^2[O_2]^3}$ c. $K_c = \dfrac{[FClO_2]^2}{[ClO_2]^2[F_2]}$

   d. $K_c = \dfrac{[HBr]^2}{[H_2]}$ e. $K_c = \dfrac{[CO]^2}{[CO_2]}$ f. $K_c = \dfrac{[H_2O]}{[H_2]}$

2. 2.0×10^{-5}
3. $[HI] = 1.1 \times 10^{-4}\ M$
4. $K_p = 2.0 \times 10^{-5}$; $K_c = 8.2 \times 10^{-7}$
5. $K_c = 0.031$
6. $K_c = 34$
7. Net reverse reaction
8. $[HI] = 0.22\ M$
9. $K_c = 0.012$
10. $P_{NO} = 0.017$ atm; $P_{N_2} = 1.99$ atm; $P_{O_2} = 0.39$ atm

11. $[NO_2] = 0.17\ M$; $[N_2O_4] = 0.080\ M$
12. a. The partial pressures of SO_2 and O_2 will increase.
    b. No changes in equilibrium partial pressures.
    c. The partial pressures of SO_2 and O_2 decrease while that of SO_3 increases.
    d. The partial pressures of SO_2 and O_2 decrease while that of SO_3 increases.
    e. The partial pressure of SO_2 increases and that of SO_3 decreases.
13. a. K_c will decrease
    b. $[PCl_3]$ decreases
    c. $[PCl_5]$ increases; $[PCl_3]$ and $[Cl_2]$ decrease
    d. Cl_2 pressure increases
14. $[SO_2] = [NO_2] = 0.0118\ M$; $[NO] = [SO_3] = 0.108\ M$
15. a. Least extent of reaction
    b. Greatest extent of reaction
    c. Intermediate of the three examples
16. a. Decreases   b. No change   c. Decreases   d. Increases
17. a. $[O_2] = 1.5 \times 10^{-4}\ M$; $[CO] = 3.0 \times 10^{-4}\ M$
    b. 6.1%
18. $K_p = K_c = 3.0 \times 10^{-8}$

SELF-TEST B1

1. a. Right   b. No shift   c. Left
2. 0.0169
3. a. 7.1×10^4 b. 3.7×10^{-3}
4. a. No
   b. Reverse, $[PCl_5] = 6.0\ M$, $[PCl_3] = 1.5\ M$, $[Cl_2] = 0.93\ M$
5. $K_c = 1.5 \times 10^{-3}$

SELF-TEST B2

1. a. Left   b. No shift   c. Right
2. $5.85 \times 10^{-3}\ M$
3. a. 1.4×10^{-5} b. 266
4. a. No
   b. Forward, $[PCl_5] = 2.35\ M$, $[PCl_3] = 0.75\ M$, $[Cl_2] = 0.75\ M$
5. $K_p = 4.8$

Chapter Fifteen
ACIDS AND BASES: GENERAL PROPERTIES

- Definitions of Acids and Bases
- The Autoionization of Water and the pH Scale
- Strengths of Acids and Bases
- Molecular Structure and Strengths of Acids
- Lewis Acids and Bases

DEFINITIONS OF ACIDS AND BASES

STUDY OBJECTIVES

You should be able to:
1. Write equations according to the Brønsted theory that show the processes occurring when acids and bases dissolve in water.
2. Describe the state of the proton in aqueous solution.

The Proton in Aqueous Solution. Acid-base reactions are an important type of chemical reaction. In Chapter 3 you learned that S. Arrhenius proposed that an acid is a substance that produces H^+ ions (protons) in aqueous solution, and a base is a substance that produces OH^- ions (hydroxide) in aqueous solution. Recall that protons, like other ions, are hydrated in solution. This means that each proton is bonded to one or more water molecules. The hydrated nature of the proton in aqueous solution is usually emphasized by writing its formula as H_3O^+, which is called the hydronium ion. In practice, both $H^+(aq)$ and $H_3O^+(aq)$ are used to represent the hydrated proton.

Brønsted Acids and Bases. J. N. Brønsted proposed a more general theory of acid-base reactions than that of Arrhenius. According to Brønsted's theory, an *acid* is defined as a substance that is able to donate a proton, and a *base* is a substance capable of accepting a proton, in a chemical reaction. For example, when HBr gas dissolves in water, it can be viewed as donating a proton to the solvent. Thus HBr is a Brønsted acid:

$$HBr(aq) + H_2O(l) \rightarrow H_3O^+(aq) + Br^-(aq)$$
$$\text{acid} \qquad\quad \text{base}$$

The water molecule is the proton acceptor; thus water is a Brønsted base in this reaction. All acid-base reactions, according to Brønsted, involve the transfer of a proton. Acceptance of a proton by H_2O forms the hydronium ion, H_3O^+.

In the above example, HBr and Br^- make up a conjugate acid-base pair. The addition of a proton (H^+) to a base (Br^-) gives its conjugate acid (HBr). Removing a proton (H^+) from an acid (HBr) gives its conjugate base (Br^-). A conjugate acid-base pair consists of two species that differ from each other by the presence or absence of a proton. Thus H_3O^+ and H_2O are also a conjugate acid-base pair.

When ammonia dissolves in water, another proton transfer reaction takes place:

$$NH_3(aq) + H_2O(l) \rightleftharpoons NH_4^+(aq) + OH^-(aq)$$

In this reaction water is the acid and ammonia is the base. Substances that can behave as an acid in one proton transfer reaction and as a base in another are called *amphoteric*. Thus water is an amphoteric substance. The conjugate acid-base pairs are

$$NH_4^+ - NH_3 \quad \text{and} \quad H_2O - OH^-$$

Such pairs are usually labeled as shown below the following equation:

$$\underset{base_1}{NH_3(aq)} + \underset{acid_2}{H_2O(l)} \rightleftharpoons \underset{acid_1}{NH_4^+(aq)} + \underset{base_2}{OH^-(aq)}$$

The subscript 1 designates one conjugate pair and the subscript 2 the other pair.

The double arrow indicates that the reaction is reversible. In the reverse reaction OH^- acts as a base and NH_4^+ acts as an acid.

EXAMPLE 15.1 Conjugate Acid-Base Pair

Write the formula of the conjugate base of H_2SO_4.

METHOD OF SOLUTION

A conjugate base differs from its conjugate acid by the lack of a proton:

$$\underset{\text{acid}}{H_2SO_4} \quad \underset{\text{conjugate base}}{HSO_4^-}$$

EXAMPLE 15.2 Conjugate Acid-Base Pairs

Consider the reaction

$$HSO_4^-(aq) + HCO_3^-(aq) \rightleftharpoons SO_4^{2-}(aq) + H_2CO_3(aq)$$

a. Identify the acids and bases for the forward and reverse reactions.
b. Identify the conjugate acid-base pairs.

METHOD OF SOLUTION

a. In the forward reaction, HSO_4^- is the proton donor, which makes it an acid. The proton acceptor HCO_3^- is a base. In the reverse reaction the proton donor (acid) is H_2CO_3, and SO_4^{2-} is the proton acceptor (base).
b. $\underset{acid_1}{HSO_4^-} + \underset{base_2}{HCO_3^-} \rightleftharpoons \underset{base_1}{SO_4^{2-}} + \underset{acid_2}{H_2CO_3}$
the subscripts 1 and 2 designate the two conjugate acid-base pairs.

EXAMPLE 15.3 Brønsted Acid Theory

An advantage of the Brønsted theory is that acid-base reactions can be described in nonaqueous solvents. Indicate whether each of the following would be an acid or a base in a solvent that is liquid

(glacial) acetic acid, CH_3COOH:
a. CH_3COOH
b. H_2SO_4
c. H_2O

METHOD OF SOLUTION
a. By analogy to water, an acetic acid molecule can donate a proton to another acetic acid molecule:

$$CH_3COOH + CH_3COOH \rightarrow CH_3COOH_2^+ + CH_3COO^-$$
$$\text{acid}_1 \qquad \text{base}_2 \qquad \text{acid}_2 \qquad \text{base}_1$$

Thus acetic acid is amphoteric.

b. The stronger of the two acids will be the proton donor:

$$H_2SO_4 + CH_3COOH \rightarrow HSO_4^- + CH_3COOH_2^+$$
$$\text{acid}_1 \qquad \text{base}_2 \qquad \text{base}_1 \qquad \text{acid}_2$$

c. Here we observe the same reaction as when acetic acid is dissolved in water:

$$H_2O + CH_3COOH \rightarrow H_3O^+ + CH_3COO^-$$
$$\text{base}_1 \qquad \text{acid}_2 \qquad \text{acid}_1 \qquad \text{base}_2$$

THE AUTOIONIZATION OF WATER AND THE pH SCALE

STUDY OBJECTIVES

You should be able to:
1. Write the ion product constant for the autoionization of water, and use this constant to relate $[H^+]$ and $[OH^-]$ in aqueous solutions.
2. Explain what is meant by the pH scale, and calculate pH from a knowledge of $[H^+]$ or $[OH^-]$.
3. Carry out numerical calculations involving the relationships among $[H^+]$, $[OH^-]$, pH, and pOH.

Autoionization and the Ion Product of Water. Pure water is itself a very weak electrolyte and ionizes according to the equation

$$H_2O \rightleftharpoons H^+(aq) + OH^-(aq)$$

It is written this way to be consistent with the Arrhenius theory of acids and bases. According to the Brønsted theory, the reaction is viewed as a proton transfer from one water molecule to another:

$$H_2O(l) + H_2O(l) \rightleftharpoons H_3O^+(aq) + OH^-(aq)$$
$$\text{acid}_1 \qquad \text{base}_2 \qquad \text{acid}_2 \qquad \text{base}_1$$

Since water can act as both an acid and a base, it is amphoteric.

This reaction is reversible and equilibrium is established. In pure water at 25°C, $[H^+] = [OH^-] = 1.0 \times 10^{-7}$ M. Note, this is a very small concentration and tells us that very few H_2O molecules are ionized. In pure water, $[H_2O] = 55.6$ M. Taking the ratio $[H^+]/[H_2O] = 1.0 \times 10^{-7}$ $M/55.6$ M, we see that only one water molecule out of 550 million is ionized.

At equilibrium, the product of the hydrated proton (abbreviated as H^+) and hydroxide ion concentrations equals a constant called the ion product constant for water, K_w:

$$K_w = [H^+][OH^-] = 1.0 \times 10^{-14}$$

The ion product tells us that the H^+ concentration times the OH^- concentration is a constant. When an acid is added to water, the $[H^+]$ increases. Thus the $[OH^-]$ must decrease in order for K_w to remain constant. In acid solutions $[H^+] > [OH^-]$. Similarly, when a base is added to water and $[OH^-]$ increases, then $[H^+]$ must decrease. Acidic, basic, and neutral solutions are characterized by the following conditions:

neutral $[H^+] = [OH^-]$
acidic $[H^+] > [OH^-]$
basic $[H^+] < [OH^-]$

In all cases, $[H^+][OH^-] = 1.0 \times 10^{-14}$ at 25°C. Therefore, if the value of $[H^+]$ is known, the value of $[OH^-]$ can be calculated, and vice versa.

The pH Scale. The concentration of $H^+(aq)$ in a solution can be expressed in terms of the pH scale. The pH of a solution is defined as the negative logarithm of the hydrogen ion concentration:

$$pH = -\log[H^+]$$

Recall that the logarithm of a number is the power to which 10 must be raised in order to equal the number. For example, the logarithm of 100 is 2.0 because raising 10 to the second power gives 100:

$$100 = 10^2$$

$$\log 100 = 2$$

$$x = 10^{\log x}$$

Therefore, if $x = 10^3$, then $\log x = 3$, and if $x = 10^{-3}$, then $\log x = -3$.

For a neutral solution $[H^+] = 1 \times 10^{-7}\ M$, and so

$$pH = -\log(1.0 \times 10^{-7}) = -(\log 1.0 + \log 10^{-7})$$

The log of 1 is zero, and the log of 10^{-7} is -7. Therefore, the pH of a neutral solution is 7.0:

$$pH = -(0 - 7.0) = 7.0$$

Likewise for an acidic solution, if

$$[H^+] = 1 \times 10^{-5}\ M$$

then

$$pH = -\log(1 \times 10^{-5}) = -(0 - 5.0) = 5.0$$

All acidic solutions have a pH below 7.0.

When $[H^+]$ is not an exact power of 10, we must evaluate the log of the decimal part as well. Take a basic solution where $[H^+] = 2.0 \times 10^{-9}\ M$, for instance:

$$pH = -\log(2.0 \times 10^{-9}) = -(\log 2.0 + \log 10^{-9})$$

The log of 2.0 is 0.301; therefore

$$pH = -(0.30 - 9.00) = 8.70$$

Note that all basic solutions have a pH above 7.0. The pH values corresponding to selected sets of $H^+(aq)$ and $OH^-(aq)$ concentration are given in Table 15.1.

Table 15.1 Relationship of pH and pOH in Aqueous Solutions

$[H^+]$	$[OH^-]$	pH	pOH	Nature of Solution
10^0	10^{-14}	0	14	acidic
10^{-1}	10^{-13}	1	13	acidic
10^{-2}	10^{-12}	2	12	acidic
10^{-3}	10^{-11}	3	11	acidic
10^{-6}	10^{-8}	6	8	acidic
10^{-7}	10^{-7}	7	7	neutral
10^{-8}	10^{-6}	8	6	basic
10^{-11}	10^{-3}	11	3	basic
10^{-12}	10^{-2}	12	2	basic
10^{-13}	10^{-1}	13	1	basic
10^{-14}	10^0	14	0	basic

The pOH Scale. A scale just like the pH scale has been devised for the hydroxide ion concentration, where

$$pOH = -\log[OH^-]$$

The pH and pOH are related to each other through the K_w expression:

$$[H^+][OH^-] = K_w = 1 \times 10^{-14}$$

Taking the log of both sides yields

$$\log[H^+] + \log[OH^-] = -14$$

Multiplying both sides by -1 yields

$$-\log[H^+] + (-\log[OH^-]) = 14$$

$$pH + pOH = 14$$

The sum of the pH and pOH values of any solution is always 14 at 25°C. You can see this also in Table 15.1. It is important to notice that acidic solutions have pH values below 7.0. Thus, highly acidic solutions have *low* pH values, not high values. Solutions with pH values *above* 7.0 are *basic*.

It is also important to notice that a change in pH of one unit corresponds to a 10-fold change in $[H^+]$. As H^+ drops from 10^{-8} to 10^{-9}, for instance, the pH changes from 8 to 9. A change of 2.0 pH units corresponds to a 100-fold change in H^+ ion concentration. Never say "A pH of 2 is twice as acidic as a pH of 4." It is really 100 times more acidic!

Changing pH Units to $[H^+]$. Given the pH, how do we calculate the $[H^+]$? Recall

$$pH = -\log[H^+]$$

$$\log[H^+] = -pH$$

Taking the antilog of both sides, we get

$$\text{antilog}(\log[H^+]) = \text{antilog}(-pH)$$

$$[H^+] = 10^{-pH}$$

Any electronic calculator with exponential (antilogarithmic) functions 10^x or e^x will easily make the calculation of H^+ ion concentrations from pH values. Just enter $-$pH for x and push 10^x.

EXAMPLE 15.4 Using K_w

The H^+ ion concentration in a certain solution is 5.0×10^{-5} M. What is the OH^- ion concentration?

METHOD OF SOLUTION

The ion product of water is applicable to all aqueous solutions:

$$K_w = [H^+][OH^-] = 1 \times 10^{-14}$$

When $[H^+]$ is known, we can solve for $[OH^-]$.

CALCULATION

$$[OH^-] = \frac{1 \times 10^{-14}}{[H^+]} = \frac{1 \times 10^{-14}}{5 \times 10^{-5}}$$

$$[OH^-] = 2.0 \times 10^{-10} \, M$$

EXAMPLE 15.5 Comparing pH Values

The pH of many cola-type soft drinks is about 3.0. How many times greater is the H^+ concentration in these drinks than in neutral water?

METHOD OF SOLUTION

First write out the H^+ ion concentrations in the cola drink and in neutral water:

$[H^+]_{cola} = 1.0 \times 10^{-3}$ M, and $[H^+]_{neut} = 1.0 \times 10^{-7}$ M. Then the ratio is

$$\frac{[H^+]_{cola}}{[H^+]_{neut}} = \frac{1.0 \times 10^{-3} \, M}{1.0 \times 10^{-7} \, M} = 10^4 = 10{,}000$$

COMMENT

Here is another approach. Since a change of 1.0 pH unit corresponds to a 10-fold change in H^+ concentration, a change of 4.0 pH units corresponds to $10 \times 10 \times 10 \times 10 = 10^4$ or a 10,000-fold increase in H^+ concentration.

EXAMPLE 15.6 pH and pOH

The OH^- ion concentration in a certain ammonia solution is 7.2×10^{-4} M. What is the pOH and pH?

METHOD OF SOLUTION

The pOH is the negative logarithm of the OH^- ion concentration:

$$\text{pOH} = -\log[OH^-]$$

CALCULATION

$$\text{pOH} = -\log(7.2 \times 10^{-4}) = -(-3.14)$$

$$= 3.14$$

The pH and pOH are related by

$$pH + pOH = 14.00$$

$$pH = 14.00 - pOH = 14.00 - 3.14$$

$$= 10.86$$

EXAMPLE 15.7 H^+ Ion Concentration from pH

What is the H^+ ion concentration in a solution with a pOH of 3.9?

METHOD OF SOLUTION

Recall that pH + pOH = 14. With the pOH given, the pH can be calculated, and then the H^+ ion concentration can be determined.

CALCULATION

$$pH = 14.0 - pOH = 10.0$$

Since

$$pH = -\log[H^+]$$

$$-pH = \log[H^+]$$

and taking the antilog of both sides

$$10^{-pH} = [H^+]$$

$$[H^+] = 10^{-10.1} = 7.9 \times 10^{-11} \, M$$

STRENGTHS OF ACIDS AND BASES

STUDY OBJECTIVES

You should be able to:
1. Define the terms *strong* and *weak* as they are applied to acids and bases.
2. Explain the relationship between the relative strength of an acid and the strength of its conjugate base.
3. Calculate the pH of given solutions of strong acids and strong bases.

Strong Acid and Bases. The stronger the acid, the more completely it ionizes in water, producing $H_3O^+(aq)$ and an anion. HBr is a strong acid; in solution it ionizes 100%:

$$HBr(aq) + H_2O(l) \xrightarrow{100\%} H_3O^+(aq) + Br^-(aq)$$

At equilibrium, the concentration of HBr molecules is zero because all have dissociated. The four most common strong acids used in chemical laboratories are HCl, HNO_3, H_2SO_4, and $HClO_4$.

Like the strong acids, the strongest bases ionize 100% in aqueous solution. KOH, a commonly used base, is an example. It is purchased as a white solid in pellet form. It dissolves readily in water to give a solution of K^+ and OH^- ions:

$$KOH(aq) \xrightarrow{100\%} K^+(aq) + OH^-(aq)$$

A solution of 0.1 M KOH is made by dissolving 0.1 mol of KOH in 1 L of solution. In this solution, $[K^+]$ is 0.1 M, $[OH^-]$ is 0.1 M, and [KOH] is essentially zero. Of the strong bases, only NaOH and KOH are commonly used in the laboratory. RbOH, $Mg(OH)_2$, and $Ca(OH)_2$ are also strong bases.

Weak Acids and Bases. Most acids and bases are categorized as weak because they ionize only partially in water. Hydrofluoric acid is a typical weak acid in water:

$$HF(aq) + H_2O(l) \xrightleftharpoons{2.6\%} H_3O^+(aq) + F^-(aq)$$

The ionization reaction is reversible, so at equilibrium the concentration of undissociated HF is much greater than the concentrations of $H^+(aq)$ and $F^-(aq)$.

Quantitatively the extent of ionization is expressed in terms of percent ionization, which is the percent of the total concentration of the acid (or base) that is in the ionic form.

$$\text{percent ionization} = \frac{\text{amount ionized (mol/L)}}{\text{initial concentration (mol/L)}} \times 100$$

For example, a 1.0 M HF solution has the following concentrations: $[H^+] = [F^-] = 0.026$ M and $[HF] = 0.974$ M. Thus, the ionic form amounts to only 2.6% of the initial concentration of HF. We would say that HF is only 2.6% ionized in a 1.0 M HF solution. HF then is a weak acid, unlike HCl, HBr, and HI, which are strong acids. A number of weak acids and their conjugate bases are listed in Table 15.3 of the text. Figure 15.1 compares the relative concentrations of HA, H^+, and A^- in solutions of strong and weak acids, where A^- is a general symbol for an anion.

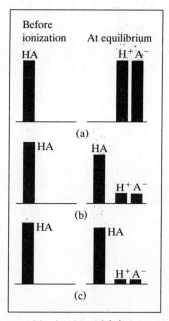

Figure 15.1. Illustration of the extent of ionization of (*a*) a strong acid that ionizes 100%, (*b*) a weak acid that ionizes a few percent, and (*c*) a very weak acid. A^- represents the anion.

Ammonia is a weak base. In a 0.01 M solution of the weak base ammonia, only 1.3% of the ammonia molecules are ionized:

$$NH_3(aq) + H_2O(l) \xrightleftharpoons{1.3\%} NH_4^+(aq) + OH^-(aq)$$

Strong versus Concentrated. Sometimes confusion arises in the use of these two terms. In chemistry *strong* does not mean highly concentrated. As we have pointed out, strength is related to the *percent ionization* of an acid or a base. In general, strength refers to the extent of reaction before equilibrium is reached. *Concentration*, on the other hand, refers to *how much* of a solute, such as an acid or a base, is dissolved in a certain amount of solvent.

Strength of Conjugate Bases. The strength of an acid is related to the strength of its conjugate base. Take, for instance, the dissociation of the weak acid HF:

$$HF(aq) + H_2O \xrightleftharpoons{2.6\%} H_3O^+(aq) + F^-(aq)$$

Here we see that HF is a proton donor and H_2O is a proton acceptor. The conjugate base of HF is F^-. By comparison, the strong acid HCl dissociates 100% to form the conjugate base Cl^-:

$$HCl(aq) + H_2O(l) \xrightarrow{100\%} H_3O^+(aq) + Cl^-(aq)$$

In the case of the fluoride ion, the reverse reaction occurs to a greater extent than for Cl^-. Fluoride is a much stronger base than Cl^-. It is always true that strong acids have relatively weak conjugate bases and weak acids have relatively strong conjugate bases. *Conjugate bases of weak acids are stronger than the conjugate bases of strong acids.*

Table 15.3 in the text lists several important conjugate acid-base pairs, according to their relative strengths. At this point, it is worth remembering that ions such as CH_3COO^-, NO_2^-, and CN^- are bases and will react with water, as any base would, by accepting a proton:

$$CH_3COO^-(aq) + H_2O(l) \rightleftharpoons CH_3COOH(aq) + OH^-(aq)$$

Solutions containing these ions may be basic, with $[OH^-] > [H^+]$.

The pH of Strong Acid and Base Solutions. The pH of a strong acid solution depends on the hydrogen ion concentration. When hydrochloric acid is added to water, two reactions occur that produce $H^+(aq)$:

$$H_2O \rightleftharpoons H^+ + OH^-$$

$$HCl \rightarrow H^+ + Cl^-$$

In calculating the concentration of $H^+(aq)$, we can usually neglect the autoionization of water because it occurs only to a slight extent compared to the ionization of the strong acid. For example, in a 0.0052 M HCl solution, $[H^+] = [Cl^-] = 0.0052$ M because of the complete ionization of HCl. The pH is given by

$$pH = -\log(5.2 \times 10^{-3})$$

$$= 2.28$$

The Leveling Effect. One way to compare the strengths of acids is to choose a reference base such as water and measure the extent to which the acid donates a proton to the base. Choosing water as the reference base causes a problem because H_3O^+ is the strongest acid that can exist in aqueous solution. Acids stronger than H_3O^+ react with water to produce H_3O^+. Thus HCl, which is a stronger acid than

H_3O^+, reacts with water completely to form H_3O^+ and Cl^-:

$$HCl(aq) + H_2O(l) \rightarrow H_3O^+(aq) + Cl^-(aq)$$

If proton transfer to the solvent is essentially complete for several acids of different strengths, no distinction in acid strength may be made between them. Water is said to have a *leveling effect* on acids stronger than H_3O^+. On the other hand, acids that are weaker than H_3O^+ are not leveled by water. For instance, CH_3COOH, HNO_2, and HCN show a wide variation in their percent ionization.

EXAMPLE 15.8 Relative Conjugate Base Strength

The strengths of the following acids increase in the order $HCN < HF < HNO_3$. Arrange the conjugate bases of these acids in order of increasing base strength.

METHOD OF SOLUTION

Recall that as the strengths of acids increase, the strengths of their conjugate bases decrease. The acid strength increases from left to right:

$HCN < HF < HNO_3$

increasing acid strength →

Answer: Therefore, the strengths of their conjugate bases decrease from left to right.

$CN^- > F^- > NO_3^-$

decreasing base strength →

← increasing base strength

EXAMPLE 15.9 pH of Strong Acid versus a Weak Acid

Compare the pH of 1.0 M HCl with that of 1.0 M HF, given that HF is only 2.6% dissociated.

METHOD OF SOLUTION

Hydrochloric acid is a strong acid; therefore, 1.0 M HCl contains 1.0 M H^+ and 1.0 M Cl^-:

$$pH = -\log[H^+] = -\log(1.0)$$
$$= 0.0$$

Hydrofluoric acid is a weak acid. If it is 2.6% dissociated, then the concentration of H^+ is given by

$$[H^+] = 0.026[HF]_0 = 0.026[1.0\ M] = 0.026\ M$$

and

$$pH = -\log[H^+] = -\log(0.026)$$
$$= 1.58$$

COMMENT

The pH of a strong acid solution is always 1-2 pH units lower than the pH of a weak acid of the same concentration.

MOLECULAR STRUCTURE AND STRENGTHS OF ACIDS

STUDY OBJECTIVES

You should be able to:
1. Predict the relative strengths of binary acids.
2. Predict the relative strengths of oxacids.

Binary Acids. Two factors are related to the acid strengths of nonmetal hydrides. These are the *polarity* of the X — H bond (X stands for a nonmetal atom) and the *strength of the bond* between the nonmetal atom and the hydrogen atom. The polarity of a X — H bond increases with the electronegativity of the nonmetal atom. The greater the electronegativity of the nonmetal atom, the more it withdraws electrons from the hydrogen atom. This facilitates the release of H^+ as a proton. Thus *acid strength increases as the electronegativity of the nonmetal atom increases*.

The strength of the X — H bond decreases as the atomic radius of X increases. A hydrogen ion is more easily broken away from a larger atom than a smaller one. This happens because the electron cloud in a large atom is more diffuse. The smaller the nonmetal atom, the more dense the electron cloud. This greater electron density results in a greater attraction for the proton. Thus *acid strength increases as the atomic radius of the nonmetal atom increases*.

When comparing acid strengths of hydrides in a group versus those within a period, there is a reversal of the relative importance of these two factors (electronegativity and atomic radius). Within a series of hydrides of elements in a group the acid strength increases with increasing atomic radius of the nonmetal. For example, note the following hydrides of the Group 6A elements:

$H_2S < H_2Se < H_2Te$

increasing acid strength →

increasing nonmetal radius →

← increasing electronegativity of X

In this series, the electronegativity of X varies only slightly. This factor is outweighed by the more significant increases in atomic radius, as shown below:

	H_2S	H_2Se	H_2Te
Radius of X (pm)	104	117	137
Electronegativity of X	2.5	2.4	2.1

The importance of bond polarity in determining acid strength outweighs atomic radius considerations when going across a row of the periodic table. For instance, the hydrides of the elements phosphorus, sulfur, and chlorine show increasing acid strength:

$PH_3 < H_2S < HCl$

increasing acid strength →

increasing electronegativity of X →

← increasing nonmetal radius

	PH_3	H_2S	HCl
Radius of X (pm)	110	104	99
Electronegativity of X	2.1	2.5	3.0

The smaller decrease in atomic radius is insignificant in comparison to the larger change in electronegativity. The trend in acid strength results from significant increases in bond polarity when going across the row.

Oxoacids. For oxoacids with the same structure, but with different central atoms, whose elements are in the same group, acid strength increases with increasing electronegativity of the central atom. Thus, acid strength increases in the following series where the central atom is a halogen element:

$$HOI < HOBr < HOCl$$

In each molecule, the O — H bond strength is approximately the same. In this series the ability of the halogen atom to withdraw electron density from the O — H bond increases with increasing electronegativity. As the O — H bond in a series of acids becomes more polar, its tendency to become ionized increases, and the acid strength increases.

For oxoacids that have the same central atom, but differing numbers of attached oxygen atoms, the acid strength increases with increasing oxidation number of the central atom. Thus, HNO_3 is stronger than HNO_2, for example:

$$\begin{array}{cc} O & \\ \| & \\ O-N-O-H & O-N-O-H \end{array}$$

The oxygen atoms draw electrons away from the nitrogen atom, making it more positive. The more positive the N atom, the more effective it is in withdrawing electrons from the O — H bond. This increases the polarity of the O — H bond.

EXAMPLE 15.10 Relative Acid Strength

Which member of each of the following pairs is the stronger acid?
a. HCl or HBr
b. HCl or H_2S

METHOD OF SOLUTION

a. These acids are nonmetal hydrides of elements within Group 6A of the periodic table. Variations in atomic radius are more important within a group than electronegativity variations. The stronger acid is the one with the greater radius of its nonmetal atom. *Answer:* HBr is the stronger acid.
b. These acids are nonmetal hydrides of elements within the third row of the periodic table. The electronegativity changes more significantly going across a row than does the atomic radius. *Answer:* HCl, with its more polar bond, is the stronger acid.

EXAMPLE 15.11 Relative Acid Strength

Which member of each of the following pairs is the stronger acid?
a. $HClO_3$ or $HBrO_3$
b. H_3PO_3 or H_3PO_4

METHOD OF SOLUTION

a. These two oxoacids have the same structure:

$$\begin{array}{cc} O & O \\ \| & \| \\ O-Cl-O-H & O-Br-O-H \end{array}$$

The electronegativity of Cl is greater than that of Br. The greater the electronegativity of the central atom, the more electrons of the O — H bond are drawn away from the H atom, and the more readily the hydrogen ion is released. *Answer:* $HClO_3$ is stronger.

b. *Answer:* For oxoacids of the same element, the acid strength increases as the number of O atoms increases. H_3PO_4 is the stronger acid because of the greater number of oxygen atoms bonded to the phosphorus atom.

LEWIS ACIDS AND BASES

STUDY OBJECTIVES

You should be able to:
1. Define acids and bases according to the Lewis system.
2. Identify Lewis acids and bases in given chemical equations.

Electron-Pair Donors and Acceptors. It shouldn't be too surprising that the chemist who described the chemical bond as a "pair of electrons lying between two atomic centers" (G. N. Lewis) would notice that unshared electron pairs are central to the acid-base reaction. For H^+ to be transferred from an acid to a base, the base must possess at least one unshared electron pair:

$$H_2O + :NH_3 \rightleftharpoons [OH^- - H^+ \rightarrow :NH_3] \rightleftharpoons OH^- + H:NH_3^+$$

| electron-pair donor | H^+ leaves H_2O | proton accepts electron pair |

A *Lewis base* is an electron-pair donor, and a *Lewis acid* is an electron-pair acceptor. The ammonia molecule is the Lewis base, and the proton is the acid. Note that a new covalent bond is formed by the donation of the electron pair. Recall that a covalent bond involves the sharing of an electron pair by two atoms. When both electrons of the shared pair are donated by the same atom, the bond is called a *coordinate covalent bond* or a *dative bond*. An acid-base reaction in the Lewis system is donation of an electron pair to form a new dative bond between the acid and the base.

The Lewis system is the most general of the three acid-base systems because in this system acid-base reactions can occur even without a solvent, which is necessary in the Brønsted system. The reaction of $CaO(s)$ with $CO_2(g)$ is

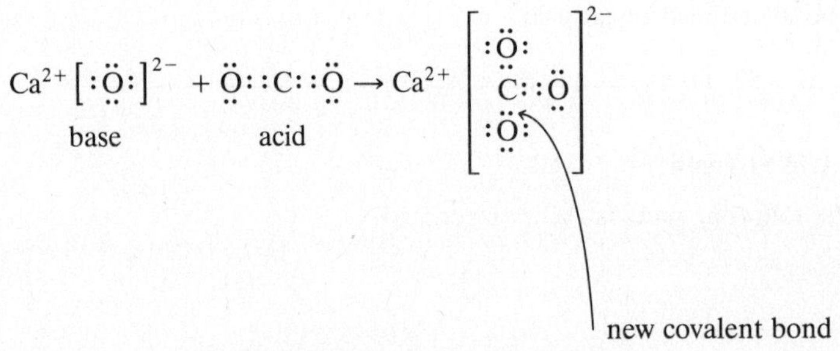

$$Ca^{2+}[:\ddot{O}:]^{2-} + \ddot{O}::C::\ddot{O} \rightarrow Ca^{2+}\left[\begin{array}{c} :\ddot{O}: \\ C::\ddot{O} \\ :\ddot{O}: \end{array}\right]^{2-}$$

base   acid

new covalent bond

EXAMPLE 15.12 Lewis Acids and Bases

Identify the Lewis acids and bases in each of the following reactions:
a. $Ag^+(aq) + Cl^-(aq) \rightarrow AgCl(s)$
b. $Hg^{2+}(aq) + 4I^-(aq) \rightarrow HgI_4^{2-}(aq)$
c. $BF_3(g) + NF_3(g) \rightarrow F_3N - BF_3(s)$
d. $SO_2(g) + H_2O(l) \rightarrow H_2SO_3(aq)$

METHOD OF SOLUTION

Lewis acids are species that accept electron pairs. In each of the above reactions the first reactant species is the Lewis acid, and the second reactant is the Lewis base (electron-pair donor). The Lewis acids in a and b are capable of accepting electron pairs because they are positive ions and have previously lost electrons. In c the boron atom has an incomplete octet. And in d the sulfur in SO_2 is "oxidized," that is, its valence electrons are shifted toward the more electronegative oxygen atoms. In all four reactions new coordinate covalent bonds are formed.

TRUE-FALSE QUESTIONS

1. In the following reaction HSO_3^- and HCO_3^- are a conjugate acid-base pair:

   $$HSO_3^- + HCO_3^- \rightarrow SO_3^{2-} + H_2CO_3$$

2. The conjugate acid of CO_3^{2-} is H_2CO_3.

3. HCO_3^- can be an acid or a base.

4. In aqueous solutions at 25°C the product $[H^+] \times [OH^-]$ is always equal to 1×10^{-14}.

5. When the pH of a solution is 2.0, the pOH is 16.0.

6. A solution with a pH of 6.0 has a H^+ concentration that is 100 times greater than that of a solution of pH 8.0.

7. The OH^- ion concentration in 0.10 M $Ca(OH)_2$ is 0.20 M.

8. If a 0.10 M solution of a weak acid (HA) is 1.0% ionized, the H^+ ion concentration is 0.001 M.

9. The conjugate base of a strong acid is a strong base.

10. The bond polarity in HF is greater than in HCl; therefore, HF is a stronger acid than HCl.

11. HNO_2 is a stronger acid than HNO_3.

12. In the reaction

    $$HNO_2(aq) + NH_3(aq) \rightarrow NH_4^+(aq) + NO_2^-(aq)$$

    the Lewis base is ammonia. Ammonia is also a Brønsted base.

SELF-TEST A

1. Identify the conjugate bases of the following acids:
   a. CH_3COOH b. H_2S c. HSO_3^- d. $HClO$

2. Identify the conjugate acid-base pairs in each of the following reactions:
   a. $NH_3 + H_2CO_3 \rightleftharpoons NH_4^+ + HCO_3^-$
   b. $CO_3^{2-} + HSO_3^- \rightleftharpoons HCO_3^- + SO_3^{2-}$
   c. $HCl + HSO_3^- \rightleftharpoons Cl^- + H_2SO_3$
   d. $HCO_3^- + HPO_4^{2-} \rightleftharpoons H_2CO_3 + PO_4^{3-}$

3. What is the $H^+(aq)$ concentration in each of the following?
   a. Neutral water   b. 1 M HCl   c. 0.01 M HNO_3

4. What is the $OH^-(aq)$ concentration in each of the following?
   a. Neutral water   b. 0.01 M NaOH   c. 0.01 M $Sr(OH)_2$

5. What is the $H^+(aq)$ concentration in 1×10^{-3} M NaOH?

6. What is the $OH^-(aq)$ concentration in 2.0 M HCl?

7. What is the pH of the following solutions?
   a. 0.015 M HCl   b. 0.015 M NaOH

8. What is the pOH of the following solutions?
   a. 0.30 M $Ca(OH)_2$ b. 2.0×10^{-3} M $HClO_4$

9. The pH of solution A is 2.0, and the pOH of solution B is 10.0. How many times greater is the $H^+(aq)$ concentration in solution A than in solution B?

10. The pH of a solution is 4.45. What is the H^+ ion concentration?

11. Classify each of the following as a weak or strong acid:
    a. H_2SO_4
    b. CH_3COOH
    c. HClO
    d. HCl
    e. HNO_2
    f. $HClO_4$

12. Use Table 15.3 of the textbook to arrange the following anions in order of increasing base strength.
    HSO_4^- CH_3COO^- Cl^- NO_2^-

13. Which member of each of the following pairs is the stronger acid?
    a. H_2CO_3 or H_2SiO_3
    b. H_3AsO_4 or H_3AsO_3
    c. H_3PO_2 or H_3PO_3
    d. $HClO_2$ or $HClO_3$
    e. H_3PO_4 or H_3AsO_4

14. Which member of each of the following pairs is the stronger acid?
    a. HI or HBr   b. H_2O or H_2S c. H_2Te or HI   d. H_2Se or AsH_3

15. Choose the Lewis acids and bases that could be used to form the following compounds:
    a. HCl   b. H_3O^+ c. BCl_3NH_3 d. $Al(OH)_4^-$ e. MgF_2

16. What is the pH of a 0.0055 M HA (weak acid) solution that is 8.2% ionized?

17. Predict the direction in which the equilibrium will lie for the following reaction:

$$C_6H_5COO^- + HF \rightleftharpoons C_6H_5COOH + F^-$$

18. Acid strength increases in the following series: weakest—HCN, HF, to HSO_4^-—strongest. Which of the following is the strongest base?
    a. SO_4^{2-} b. F^- c. CN^- d. HSO_4^-

SELF-TEST B1

1. Calculate the pH of a 0.0051 M HCl solution.

2. What is the H^+ ion concentration in a solution with a pOH of 10.0?

3. Calculate the pOH of a 0.125 M NaOH solution.

4. Use Table 15.3 in the text to arrange the following acids in order of increasing acid strength: HCl, HF, HNO_2.

5. The pH of solution A is 2.0 and the pH of solution B is 4.0. How many times greater is the $H^+(aq)$ concentration in solution A than in solution B?

6. Which of the following acids is the stronger: HNO_2 or HNO_3?

SELF-TEST B2

1. Calculate the H^+ ion concentration of an acid solution with a pH of 2.29.

2. What is the OH^- ion concentration in a solution with a pH of 4.0?

3. What is the concentration of OH^- in a NaOH solution which has a pOH of 0.90?

4. Use Table 15.3 in the text to arrange the following bases in order of increasing base strength: NO_2^-, Cl^-, and F^-.

5. The pH of solution B is 4.0. The $H^+(aq)$ concentration in solution A is 100 times greater than in solution B. What is the pH of solution A?

6. Predict which of the following is the stronger base: NO_2^- or NO_3^-.

ANSWERS

TRUE-FALSE QUESTIONS

1. False. The conjugate acid-base pairs have formulas that differ only by one proton.
2. False. The conjugate acid is HCO_3^-.
3. True.
4. True.
5. False. The pOH = 12.
6. True.
7. True.
8. True.
9. False. The conjugate base of a strong acid is a weak base.
10. False. The bond polarity is not the only influence. The smaller size of the F^- ion as compared with Cl^- means that it is more difficult to break the H — F bond than the H — Cl bond.
11. False. HNO_3 is stronger.
12. True.

SELF-TEST A

1. a. CH_3COO^- b. HS^- c. SO_3^{2-} d. ClO^-

2. a. $NH_3 + H_2CO_3 \rightleftharpoons NH_4^+ + HCO_3^-$
   base₁ acid₂ acid₁ base₂

   b. $CO_3^{2-} + HSO_3^- \rightleftharpoons HCO_3^- + SO_3^{2-}$
   base₁ acid₂ acid₁ base₂

   c. $HCl + HSO_3^- \rightleftharpoons Cl^- + H_2SO_3$
   acid₁ base₂ base₁ acid₂

   d. $HCO_3^- + HPO_4^{2-} \rightleftharpoons H_2CO_3 + PO_4^{3-}$
   base₂ acid₁ acid₂ base₁

3. a. $1 \times 10^{-7} M$ b. $1\ M$ c. $0.01\ M$
4. a. $1 \times 10^{-7}\ M$ b. $0.01\ M$ c. $0.02\ M$
5. $[H^+] = 1 \times 10^{-11}\ M$
6. $[OH^-] = 5 \times 10^{-15}\ M$
7. a. pH = 1.82 b. pH = 12.18
8. a. pOH = 0.22 b. pOH = 11.3
9. 100
10. $[H^+] = 3.55 \times 10^{-5}\ M$
11. a. Strong b. Weak c. Weak d. Strong e. Weak f. Strong
12. $Cl^- < HSO_4^- < NO_2^- < CH_3COO^-$
13. a. H_2CO_3 b. H_3AsO_4 c. H_3PO_3 d. $HClO_3$ e. H_3PO_4
14. a. HI b. H_2S c. HI d. H_2Se
15. a. $H^+ + Cl^- \rightarrow HCl$ d. $Al^{3+} + 4OH^- \rightarrow Al(OH)_4^-$
    b. $H^+ + H_2O \rightarrow H_3O^+$ e. $Mg^{2+} + 2F^- \rightarrow MgF_2$
    c. $BCl_3 + NH_3 \rightarrow BCl_3NH_3$
16. 3.34
17. To the right
18. c

SELF-TEST B1

1. pH = 2.29
2. $1 \times 10^{-4}\ M$
3. pOH = 0.90
4. $HF < HNO_2 < HCl$
5. 100 times
6. HNO_3

SELF-TEST B2

1. $0.0051\ M\ H^+$
2. $1 \times 10^{-10}\ M$
3. $0.125\ M\ OH^-$
4. $Cl^- < NO_2^- < F^-$
5. pH = 2
6. NO_2^-

Chapter Sixteen
ACID-BASE EQUILIBRIA

- Weak Acids and Weak Bases
- Diprotic and Polyprotic Acids
- Hydrolysis and Acid-Base Properties of Salts
- Buffer Solutions
- Titration Curves and Indicators

WEAK ACIDS AND WEAK BASES

STUDY OBJECTIVES

You should be able to:
1. Write the *acid ionization constant* for any weak acid and *base ionization constant* for any weak base.
2. Calculate the concentrations of H^+, A^-, and undissociated weak acid HA, given K_a.
3. For a weak base B in water, calculate the concentrations of OH^-, H^+, BH^+, and B, given K_a.

Ionization Constants. Weak acids and weak bases are ionized only to a small extent in aqueous solution. In both cases the ionization reaction is reversible and equilibrium is established. For instance, hydrocyanic acid ionizes in water as follows:

$$HCN(aq) \rightleftharpoons H^+(aq) + CN^-(aq)$$

The equilibrium constant for this reaction is called the *acid ionization constant*, K_a. The expression for this constant is

$$K_a = \frac{[H^+][CN^-]}{[HCN]} = 4.9 \times 10^{-10}$$

The value of K_a is experimentally determined, and Table 16.1 in the text and Table 16.1 below list K_a values for a number of weak acids.

Table 16.1 Values of K_a for Some Monoprotic Acids

Formula	Name	K_a Value
HSO_4^-	Hydrogen sulfate ion	1.1×10^{-2}
HF	Hydrofluoric acid	7.1×10^{-4}
HNO_2	Nitrous acid	4.5×10^{-4}
HCOOH	Formic acid	1.7×10^{-4}
CH_3COOH	Acetic acid	1.8×10^{-5}
HSO_3^-	Hydrogen sulfite ion	6.3×10^{-8}
HCN	Hydrocyanic acid	4.9×10^{-10}

Methylamine is an example of a weak base. It ionizes in water as shown in the following equation:

$$CH_3NH_2(aq) + H_2O(l) \rightleftharpoons CH_3NH_3^+(aq) + OH^-(aq)$$

The equilibrium constant for this reaction is called the *base ionization constant*, K_b:

$$K_b = \frac{[CH_3NH_3^+][OH^-]}{[CH_3NH_2]} = 4.4 \times 10^{-4}$$

Table 16.2 in the text and Table 16.2 below list ionization constant values for a number of weak bases. Note that the concentration of water is not shown in the K_b expression. The concentration of water is not measurably affected by the reaction and is essentially a constant.

Table 16.2 Values of K_b for Several Weak Bases

Formula	Name	K_b Value
CH_3NH_2	Methylamine	4.4×10^{-4}
CN^-	Cyanide ion	2.0×10^{-5}
NH_3	Ammonia	1.8×10^{-5}
C_5H_5N	Pyridine	1.7×10^{-9}

Acid and Base Strength. The K_a value is the quantitative measure of acid strength. The larger the K_a, the greater the extent of the ionization reaction and the stronger the acid. Therefore, acetic acid with $K_a = 1.8 \times 10^{-5}$ is a stronger acid than hydrocyanic acid with $K_a = 4.9 \times 10^{-10}$. The acids listed in Table 16.1 in this chapter are arranged in order of decreasing acid strength. The K_a values for strong acids are never tabulated. Strong acids are essentially 100% ionized. Thus, the concentration of the nonionized HA are much too small to detect and measure accurately.

The percent ionization introduced in the previous chapter is another way of comparing the strengths of acids. The percent ionization is defined as follows:

$$\text{percent ionization} = \frac{\text{moles of acid ionized per liter of solution}}{\text{initial concentration of acid}} \times 100\%$$

Examples follow that illustrate the problem-solving techniques used to calculate $[H^+]$, $[OH^-]$, pH, and percent ionization of weak acids and bases.

EXAMPLE 16.1 Percent Ionization of a Weak Acid

a. Calculate the concentrations of H^+, F^-, and HF in a 0.31 M HF (hydrofluoric acid) solution.
b. What is the percent ionization?

METHOD OF SOLUTION

a. First write the ionization reaction and the acid ionization constant expression for hydrofluoric acid. The value for K_a can be found in Table 16.1:

$$HF(aq) \rightleftharpoons H^+(aq) + F^-(aq)$$

$$K_a = \frac{[H^+][F^-]}{[HF]} = 7.1 \times 10^{-4}$$

The equilibrium concentrations of all species must be written in terms of a single unknown. Let x equal the moles of HF ionized per liter of solution. This means that x is the molarity of H^+ at equilibrium because one H^+ ion is formed for each HF molecule that ionizes. Also $[H^+] = [F^-] = x$ because one F^- ion is formed for each hydrogen ion. It is helpful to tabulate the initial and equilibrium concentrations of all species. The initial concentrations are those that existed before any ionization of HF occurred. The change in concentration is x. When x moles of HF have ionized, x moles of H^+ and x moles of F^- are formed. The equilibrium concentrations are found as usual by adding algebraically the change in concentration to the initial concentration:

Concentration	HF $\rightleftharpoons$	H^+	+ F^-
Initial (M)	0.31	0	0
Change (M)	$-x$	$+x$	$+x$
Equilibrium (M)	$(0.31 - x)$	x	x

CALCULATION

The value of x can be determined by substituting the equilibrium concentrations into the K_a expression:

$$K_a = \frac{(x)(x)}{(0.31 - x)} = 7.1 \times 10^{-4}$$

The small magnitude of K_a indicates that very little hydrofluoric acid actually ionizes. After all, it is a weak acid. Therefore, the value of x is much less than 0.31, and $0.31 - x \cong 0.31\ M$. This approximation simplifies the equation and allows us to avoid solving a quadratic equation:

$$K_a = \frac{x^2}{0.31} = 7.1 \times 10^{-4}$$

Solving for x,

$$x = \sqrt{(0.31)(7.1 \times 10^{-4})} = 1.5 \times 10^{-2}\ M$$

Was our approximation valid? As a rule, if x is equal to or less than 5% of the initial concentration $[HF]_0$, then $0.31 - x \cong 0.31\ M$. To test the validity of the approximation, divide x by the initial concentration of hydrofluoric acid, $[HF]_0$:

$$\frac{x}{[HF]_0} \times 100\% = \frac{1.5 \times 10^{-2}}{0.31} \times 100\% = 4.8\%$$

Since x is *less than* 5% of the initial amount of HF, our approximate solution is close enough to the accurate result to be valid. Knowing x, we can utilize the bottom row in the table to answer part a of the problem:

$$[H^+] = [F^-] = 1.5 \times 10^{-2}\ M$$

$$[HF] = 0.31\ M - 0.015\ M = 0.30\ M$$

b. The percent ionization was already calculated above:

$$\text{percent ionization} = \frac{x}{[HF]_0} \times 100\% = 4.8\%$$

COMMENT

In this example we have followed three basic steps common to all problems of this type.
1. Write the acid dissociation equilibrium and express the equilibrium concentration of all species in terms of an unknown (x), where x equals the concentration of $H^+(aq)$.
2. Substitute these concentrations into the K_a expression and look up the value of K_a in a table.
3. Solve for x, and calculate all equilibrium concentrations, the pH, and percent ionization.

EXAMPLE 16.2 pH of a Weak Base Solution

What is the pH of a 0.010 M C_5H_5N (pyridine) solution?

METHOD OF SOLUTION

Pyridine is a weak base. Write the equation for the reaction of pyridine in water and look up its K_b value in Table 16.2:

$$C_5H_5N + H_2O \rightleftharpoons C_5H_5NH^+ + OH^-$$

Let x equal the moles of C_5H_5N ionized per liter to reach equilibrium. Note, when x moles per liter of C_5H_5N are ionized, x moles per liter of OH^- and x moles per liter of $C_5H_5N^+$ are formed. These changes are listed in the following table:

Concentration	H_2O +	C_5H_5N	$\rightleftharpoons$	$C_5H_5NH^+$	+	OH^-
Initial (M)		0.010		0		0
Change (M)		$-x$		$+x$		$+x$
Equilibrium (M)		$(0.010 - x)$		x		x

CALCULATION

Substitute the equilibrium concentrations into the base constant expression:

$$K_b = \frac{[C_5H_5NH^+][OH^-]}{[C_5H_5N]} = \frac{(x)(x)}{(0.010 - x)} = 1.7 \times 10^{-9}$$

Make the approximation $0.010 - x \cong 0.010$ M:

$$\frac{x^2}{0.010} = 1.7 \times 10^{-9}$$

$$x = \sqrt{(0.010)(1.7 \times 10^{-9})}$$

$$= 4.1 \times 10^{-6} \, M$$

Testing the assumption that the approximation is reasonable yields

$$\frac{x}{[C_5H_5N]_0} \times 100\% = \frac{4.1 \times 10^{-6}}{0.010} \times 100\% = 0.041\%$$

Therefore x is less than 5% of the initial concentration of pyridine, and the assumption is valid. At equilibrium

$$[OH^-] = x = 4.1 \times 10^{-6} \, M$$

To find the pH, first calculate the pOH:

$$pOH = -\log(4.1 \times 10^{-6}) = 5.38$$

Since

$$pOH + pH = 14.00$$

then

$$pH = 14.00 - 5.38$$
$$= 8.62$$

DIPROTIC AND POLYPROTIC ACIDS

STUDY OBJECTIVES

You should be able to:
1. Write ionic equations for the various stages of ionization of diprotic and polyprotic acids, and predict the relative extent of each stage.
2. Calculate the concentration of the related species present in a solution of a given polyprotic acid.

Stages of Ionization. The acids listed in Table 16.1 of the text are monoprotic acids; that is, they produce one proton per acid molecule. Acetic acid contains four hydrogen atoms per molecule, but it is still monoprotic because only one of them is ionized in solution. The H atoms bonded to the C atom do not ionize because the electronegativity difference between C and H is not great enough to produce sufficient positive charge on the hydrogen atoms:

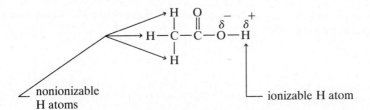

Several acids that contain more than one ionizable H atom were mentioned in Chapter 3. Acids that contain two ionizable H atoms are called *diprotic acids*. Those with three or more ionizable H atoms are called *polyprotic acids*. For example, sulfuric acid, H_2SO_4, is a diprotic acid, while phosphoric acid, H_3PO_4, with its three ionizable H atoms, is a polyprotic acid. The ionization reactions occur stepwise; the conjugate base in the first step becomes the acid in the second step of ionization. For sulfurous acid:

$$H_2SO_3 \rightleftharpoons H^+ + HSO_3^- \qquad K_{a1} = 1.3 \times 10^{-2}$$

$$HSO_3^- \rightleftharpoons H^+ + SO_3^{2-} \qquad K_{a2} = 6.3 \times 10^{-8}$$

This example is typical in that $K_{a1} \gg K_{a2}$. As a rule of thumb the first stage has a K_a about 10^7 times greater than the second stage. It is always more difficult to remove a H^+ ion from a negative ion than from a neutral molecule. Table 16.3 of the text lists acid ionization constants for several polyprotic acids.

Here are some important observations concerning polyprotic acids:

1. Keep in mind that the predominant species in solution come from the first ionization step.
2. Since all reactants and products (ions and nonionized acid molecules) exist in the same solution, they are in equilibrium with each other. At equilibrium, each species can have only one concentration.
3. The concentration of all species in solution must be consistent with both K_{a1} and K_{a2}.

EXAMPLE 16.3 Solution Containing a Diprotic Acid

Calculate the concentrations of H_2SO_3, HSO_3^-, H^+, and SO_3^{2-} in a 1.0 M H_2SO_3 solution.

METHOD OF SOLUTION

First write the equations for the stepwise ionization of H_2SO_3. Include K_a values:

$$H_2SO_3 \rightleftharpoons H^+ + HSO_3^- \qquad K_{a1} = 1.3 \times 10^{-2}$$

$$HSO_3^- \rightleftharpoons H^+ + SO_3^{2-} \qquad K_{a2} = 6.3 \times 10^{-8}$$

CALCULATION

Based on the fact that $K_{a1} \gg K_{a2}$, we assume that essentially all the H^+ comes from step 1. Let x be the concentration of H_2SO_3 ionized to reach equilibrium:

Concentration	H_2SO_3	$\rightleftharpoons$	H^+	+	HSO_3^-
Initial (M)	1.00		0		0
Change (M)	$-x$		x		x
Equilibrium (M)	$(1.00 - x)$		x		x

$$K_{a1} = \frac{[H^+][HSO_3^-]}{[H_2SO_3]} = \frac{(x)(x)}{1.0 - x} = 1.3 \times 10^{-2}$$

$$\frac{x^2}{1.0 - x} = 1.3 \times 10^{-2}$$

For this acid K_{a1} is quite large, and the percent ionization will be greater than 5%. Rearranging the preceding equation into quadratic form yields

$$ax^2 + bx + c = 0$$

$$x^2 + (1.3 \times 10^{-2})x - 1.3 \times 10^{-2} = 0$$

where

$$a = 1 \qquad b = 1.3 \times 10^{-2} \qquad c = -1.3 \times 10^{-2}$$

Substituting into the quadratic formula,

$$x = \frac{-b \pm \sqrt{(b^2 - 4ac)}}{2a} = \frac{-1.3 \times 10^{-2} \pm \sqrt{(1.3 \times 10^{-2})^2 - (4)(1)(-1.3 \times 10^{-2})}}{(2)(1)}$$

yields

$$x = 0.11 \, M$$

Therefore, at equilibrium

$$[H^+] = [HSO_3^-] = 0.11 \, M$$

and

$$[H_2SO_3] = 1.00 - 0.11 = 0.89 \, M$$

which corresponds to 11% ionization. To determine the concentration of SO_3^{2-}, we must consider the second stage of ionization:

$$HSO_3^- \rightleftharpoons H^+ + SO_3^{2-}$$

Let y be the equilibrium concentration of SO_3^{2-}, and note that $x \gg y$:

Concentration	HSO_3^-	$\rightleftharpoons$	H^+	+	SO_3^{2-}
Initial (M)	0.11		0.11		0
Change (M)	$-y$		$+y$		$+y$
Equilibrium (M)	$(0.11 - y)$		$(0.11 + y)$		$+y$

$$K_{a2} = \frac{[H^+][SO_3^{2-}]}{[HSO_3^-]} = \frac{(0.11 + y)(y)}{(0.11 - y)} = 6.3 \times 10^{-8}$$

Since K_{a2} is very small, as for a typical weak acid, we can assume that less than 5% of the HSO_3^- dissociates. Therefore

$$0.11 \pm y \cong 0.11$$

With this assumption

$$K_{a2} = \frac{(0.11)(y)}{(0.11)} = y = [SO_3^{2-}]$$

The concentration of SO_3^{2-} is equal to K_{a2}:

$$[SO_3^{2-}] = 6.3 \times 10^{-8} \, M$$

Checking the validity of the assumption,

$$0.11 - (6.3 \times 10^{-8}) \cong 0.11 \, M$$

or

$$\frac{6.3 \times 10^{-8}}{0.11} \times 100\% = 5.7 \times 10^{-5}\% \text{ ionization}$$

COMMENT

Because $K_{a1} \gg K_{a2}$ the species produced in step 1 will be present in much greater concentration than those from step 2. This can be seen by comparing the percent ionization of each step (11% vs. 0.000057%). Also, since H_2SO_3 is a weak acid, the concentration of undissociated H_2SO_3 should be greater than that of all ionic species. Thus we expect the following order of concentrations:

$$[H_2SO_3] > [H^+] \cong [HSO_3^-] > [SO_3^{2-}]$$

HYDROLYSIS AND ACID-BASE PROPERTIES OF SALTS

STUDY OBJECTIVES

You should be able to:
1. Predict which ions of a given salt will hydrolyze, and predict whether a solution will be acidic, basic, or neutral.
2. Calculate the pH of a given salt solution.

Hydrolysis. A salt is an ionic compound formed by the reaction of an acid and a base. In dealing with salts, it will be useful to recall that they are *strong* electrolytes and are *completely dissociated* in water. The term *salt hydrolysis* refers to the reaction of a cation or an anion or both with water, which splits the water molecule into two parts. Salt hydrolysis usually affects the pH of a solution.

During hydrolysis a cation will attach itself to the hydroxide part of the water molecule, causing a proton to be released. Cations that undergo hydrolysis produce acidic solutions:

$$M^+ + HOH \rightleftharpoons H^+ + MOH$$

Certain metal ions such as Al^{3+}, Cr^{3+}, Fe^{3+}, Be^{2+}, and Bi^{3+} are good examples of acidic cations. NH_4^+ is also an acidic cation.

Anions that undergo hydrolysis produce basic solutions. An anion that causes hydrolysis acts as a Brønsted base; it accepts a proton from water and releases a OH^- ion:

$$A^- + HOH \rightleftharpoons A + OH^-$$

Anion Hydrolysis. Anions that cause hydrolysis are Brønsted bases because they can accept a proton from water.

$$A^- + H_2O \rightleftharpoons HA + OH^-$$

Not all anions cause hydrolysis. *You must be able to recognize those that do and those that don't.* Quite simply, those anions with the ability to undergo hydrolysis are conjugate bases of weak acids. The conjugate base of any weak acid in Table 16.1 of the text is a weak base and will raise the pH of an aqueous solution. The extent of hydrolysis is measured quantitatively by the base ionization constant, K_b.

In contrast, the conjugate bases of strong acids have no tendency to accept protons from H_2O. They do not cause hydrolysis. They do not have acid-base properties.

The K_b Value of a Conjugate Base. The fluoride ion is the conjugate base of the weak acid hydrofluoric acid (HF). As a weak base it can accept a proton from water. The resulting OH^- ions make the solution basic. The force that makes this reaction possible is the tendency of the weak acid HF to remain undissociated:

$$F^- + H_2O \rightleftharpoons HF + OH^- \qquad K_b = \frac{[HF][OH^-]}{[F^-]}$$

The base dissociation constant K_b for F^- is related to the K_a for the conjugate acid HF by

$$K_a K_b = K_w$$

or after rearranging,

$$K_b = \frac{K_w}{K_a}$$

From this equation you can see that *the weaker the acid, the stronger its conjugate base*, and vice versa. Thus the conjugate bases of strong acids are extremely weak bases with no measurable acid-base properties. Three realms of base strength can be distinguished:

1. The conjugate bases of strong acids ($K_a > 0.1$) are too weak to cause measurable hydrolysis. Thus Cl^-, Br^-, and NO_3^- ions do not hydrolyze in aqueous solutions:

$$Cl^- + H_2O \not\rightarrow HCl + OH^-$$

2. The conjugate bases of acids with a K_a between 10^{-1} and 10^{-5} are weak bases. Thus F^-, NO_2^-, $HCOO^-$ (formate), and CH_3COO^- (acetate) ions hydrolyze to a small extent and produce low

concentrations of OH$^-$ ions in solution:

$$NO_2^- + H_2O \rightleftharpoons NHO_2 + OH^-$$

3. The conjugate bases of very weak acids ($K_a < 10^{-5}$) are moderately strong bases. Thus solutions containing CN$^-$, HS$^-$, HPO$_4^{2-}$, and CO$_3^{2-}$ ions can be quite basic.

Cation Hydrolysis. It is important to know which metal ions cause hydrolysis and which do not. Those that cause hydrolysis are *small, highly charged* metal ions such as Al^{3+}, Cr^{3+}, Fe^{3+}, Bi^{3+}, and Be^{2+}. These can split H$_2$O molecules, yielding acidic solutions. In the case of the hydrated Al^{3+} ion the reaction is

$$Al(H_2O)_6^{3+}(aq) + H_2O(l) \rightleftharpoons Al(OH)(H_2O)_5^{2+}(aq) + H_3O^+(aq)$$

The singly charged metal ions associated with strong bases do not cause hydrolysis. The commonly encountered ions Na$^+$, K$^+$, Ca^{2+}, and Mg^{2+} do not have a great enough positive charge to attract a OH$^-$ from a H$_2$O molecule and bond to it. Recall that NaOH and KOH are strong bases. That is, they ionize completely in solution. These ions, K$^+$ and OH$^-$, for instance, have no tendency to stay together in solution:

$$KOH(aq) \rightarrow K^+(aq) + OH^-(aq)$$

The hydrolysis reaction that would bond an OH$^-$ to a K$^+$ ion does not occur!

$$K^+(aq) + H_2O(l) \not\rightarrow KOH(aq) + H^+(aq)$$

Another acidic cation that you will need to be familiar with is the ammonium ion (NH$_4^+$). Solutions of ammonium ion are acidic because NH$_4^+$ is a Brønsted acid. Recall that NH$_4^+$ is the conjugate acid of the weak base NH$_3$,

$$NH_4^+ + H_2O \rightleftharpoons NH_3 + H_3O^+$$

or simply

$$NH_4^+ \rightleftharpoons NH_3 + H^+ \qquad K_a = \frac{[NH_3][H^+]}{[NH_4^+]}$$

where $K_a = K_w/K_b$. Here K_b is the base ionization constant of NH$_3$, the conjugate of NH$_4^+$:

$$K_a = \frac{K_w}{K_b}$$

$$= \frac{1.0 \times 10^{-14}}{1.8 \times 10^{-5}} = 5.6 \times 10^{-10}$$

The hydrolysis of NH$_4^+$ is the same as the ionization of the acid NH$_4^+$.

Table 16.3 lists the various anions and cations we have discussed according to their acid-base properties. You should review the material above and in the textbook to see why each ion has the property described.

Table 16.3 Acid-Base Properties of Ions

	Acidic	Basic	Neutral
Cations	Al^{3+}, Cr^{3+}, Fe^{3+}, Bi^{3+}, Be^{2+}, NH$_4^+$		Na$^+$, K$^+$, Mg^{2+}, Ca^{2+}
Anions	HSO$_4^-$	CH$_3$COO$^-$, F$^-$, NO$_2^-$ all conjugate bases of weak acids	Cl$^-$, Br$^-$, NO$_3^-$ all conjugate bases of strong acids

Salts That Produce Neutral Solutions. When a salt is dissolved in water, the solution will acquire a pH consistent with the acid-base properties of the cation and anion making up the salt. In general, salts containing alkali metal ions or alkaline earth metal ions (except Be^{2+}) and the conjugate bases of strong acids do not undergo hydrolysis; their solutions are neutral. Consequently, a solution of KCl, for example, is neutral. A K^+ ion does not accept a OH^- ion from H_2O, and a Cl^- ion does not accept a H^+ ion from H_2O.

Salts That Produce Basic Solutions. Salts producing a basic pH are salts of weak acids and strong bases. These salts have a neutral cation and a basic anion. For example, NaF dissociates into Na^+ and F^- ions. The Na^+ ion has no acidic or basic properties, but the fluoride ion is a conjugate base of a weak acid and therefore is a basic anion:

$$F^- + H_2O \rightleftharpoons HF + OH^-$$

$$\begin{array}{c} \text{NaF (a salt)} \\ \diagup \quad \diagdown \\ \text{NaOH} \quad \text{HF} \\ \text{strong} \quad \text{weak} \\ \text{base} \quad \text{acid} \end{array}$$

Salts That Produce Acidic Solutions. Salts producing acidic solutions are salts of strong acids and weak bases. Such salts contain cations that hydrolyze water and anions that cannot (the neutral anions). Such a salt has a cation that is a conjugate acid of a weak base. Thus solutions of these salts are acidic. The salt NH_4Cl is a strong electrolyte and dissociates into ammonium and chloride ions when dissolved in water. The NH_4^+ ion is a conjugate acid of the weak base NH_3 and releases a H^+ ion into solution:

$$NH_4Cl(aq) \rightarrow NH_4^+(aq) + Cl^-(aq) \quad \text{(strong electrolyte)}$$

$$NH_4^+ \rightleftharpoons NH_3 + H^+ \quad K_a = K_w/K_b = 5.6 \times 10^{-10}$$

The Cl^- ion is neutral, having no acidic or basic properties. NH_4Cl is a salt of a strong acid (HCl) and a weak base (NH_3).

Solutions of certain salts such as $AlCl_3$ and $FeCl_3$ are also acidic. These contain small, highly charged cations and anions that are conjugate bases of strong acids.

Salts in Which Both the Anion and Cation Hydrolyze. Salts derived from a weak acid and a weak base contain cations and anions that will hydrolyze. Whether a solution containing such a salt is acidic, basic, or neutral depends on the relative strengths of the acidic and basic anions. For example, a solution of ammonium fluoride is acidic because NH_4^+ is a stronger acid than F^- is a base:

$$NH_4F(aq) \rightarrow NH_4^+(aq) + F^-(aq) \quad \text{(strong electrolyte)}$$

$$NH_4^+ \rightleftharpoons NH_3 + H^+ \quad K_a = 5.6 \times 10^{-10}$$

$$F^- + H_2O \rightleftharpoons HF + OH^- \quad K_b = 1.4 \times 10^{-11}$$

NH_4F is a salt of a weak acid and a weak base. On the other hand, another such salt, NH_4CN, is a basic salt because CN^- ion is a stronger base than NH_4^+ is an acid.

The pH of Salt Solutions. The pH of a salt solution is calculated just as it is for any weak acid or base. See Example 16.6 for an illustration of this important calculation.

EXAMPLE 16.4 Hydrolysis of Ions

Write net ionic equations to show which of the following ions hydrolyze in aqueous solution:
a. NO_3^-
b. NO_2^-
c. NH_4^+

METHOD OF SOLUTION

Strong acids have extremely weak conjugate bases which have no affinity for H^+ ions. They have neutral properties. Weak acids have stronger conjugate bases with significant affinity for a H^+ ion. The conjugate bases of weak acids have measurable basic properties.

a. The nitrate ion is the conjugate base of a strong acid, HNO_3:

$$HNO_3 \rightarrow H^+ + NO_3^-$$

Therefore NO_3^- has no acidic or basic properties and has no tendency to react with water to form HNO_3:

$$NO_3^- + H_2O \not\rightarrow HNO_3 + OH^-$$

b. The nitrite ion is the conjugate base of a weak acid, HNO_2:

$$HNO_2 \rightleftharpoons H^+ + NO_2^-$$

Therefore NO_2^- is a stronger base than NO_3^- and will accept a proton from water to form HNO_2:

$$NO_2^- + H_2O \rightleftharpoons HNO_2 + OH^-$$

c. The ammonium ion is the conjugate acid of a weak base:

$$NH_3 + H_2O \rightarrow NH_4^+ + OH^-$$

Therefore it behaves as an acid by donating a proton to water:

$$NH_4^+ + H_2O \rightleftharpoons NH_3 + H_3O^+$$

EXAMPLE 16.5 Acid-Base Properties of Ions

Predict whether the following aqueous solutions will be acidic, basic, or neutral:
a. KI
b. NH_4I
c. CH_3COOK

METHOD OF SOLUTION

a. What we must decide is whether either of the ions of the salt undergoes hydrolysis. KI is a salt of a strong acid (HI) and a strong base (KOH). Neither K^+ nor I^- has acidic or basic properties. Thus a solution containing KI remains neutral.

b. NH_4I is a salt of a strong acid (HI) and a weak base (NH_3). As explained in part a, iodine does not hydrolyze. Since the NH_4^+ ion is the conjugate acid of a weak base, it will donate protons to water:

$$NH_4^+ + H_2O \rightleftharpoons NH_3 + H_3O^+$$

Thus a solution containing NH_4I will be acidic.

c. CH_3COOK is a salt of a weak acid and a strong base. The K^+ ion will not hydrolyze, but CH_3COO^- (acetate ion) is the conjugate base of a weak acid, acetic acid. Thus acetate ions will accept protons from water and form acetic acid and a OH^- ion:

$$CH_3COO^- + H_2O \rightleftharpoons CH_3COOH + OH^-$$

Solutions containing CH_3COOK will be basic.

EXAMPLE 16.6 pH of a Solution Containing a Salt

Calculate the pH of 0.25 M $C_6H_7O_6Na$ (sodium ascorbate) solution.

METHOD OF SOLUTION

First decide what ions undergo hydrolysis. Then use the appropriate ionization constant to calculate the equilibrium concentrations. Sodium ascorbate is a strong electrolyte:

$$C_6H_7O_6Na(aq) \rightarrow C_6H_7O_6^-(aq) + Na^+(aq)$$

The ascorbate ion is a conjugate base of a weak acid (ascorbic acid, Table 16.1 in the text), and so it undergoes hydrolysis. Na^+ does not and is neutral:

$$C_6H_7O_6^-(aq) + H_2O \rightleftharpoons C_6H_8O_6 + OH^-$$

It is important to use the correct equilibrium constant value. Here we need the K_b for the ascorbate ion. From Table 16.1 in the text, we get

$$K_b = \frac{K_w}{K_a} = \frac{1.0 \times 10^{-14}}{8.0 \times 10^{-5}} = 1.3 \times 10^{-10}$$

CALCULATION

The OH^- ion concentration in a solution that is 0.25 M $C_6H_7O_6^-$ is obtained from the equilibrium constant expression

$$K_b = \frac{[C_6H_8O_6][OH^-]}{[C_6H_7O_6^-]}$$

Let x be moles of OH^- formed per liter of solution:

Concentration	H_2O +	$C_6H_7O_6^-$	$\rightleftharpoons$	$C_6H_8O_6$ +	OH^-
Initial (M)		0.25		0	0
Change (M)		$-x$		$+x$	$+x$
Equilibrium (M)		$(0.25 - x)$		x	x

Substitute into the equilibrium constant expression:

$$K_b = \frac{(x)(x)}{0.25 - x} = 1.3 \times 10^{-10}$$

Apply the approximation that $0.25 - x \cong 0.25$ M:

$$\frac{x^2}{0.25} = 1.3 \times 10^{-10}$$

$$x = \sqrt{(0.25)(1.3 \times 10^{-10})} = 5.7 \times 10^{-6} \, M = [OH^-]$$

$$pOH = -\log[OH^-] = 5.24$$

The pH is $14.0 - 5.24 = 8.76$.

BUFFER SOLUTIONS

STUDY OBJECTIVES

You should be able to:
1. Describe the effect of common ions on the percent ionization of weak acids and bases.
2. Calculate the pH of a buffer solution, given the concentrations of weak acid or base and their salts.
3. Determine the pH of a buffer solution after the addition of small amounts of strong acid base.
4. Describe how to prepare a buffer solution with a specific pH.

The Common Ion Effect. A 1.0 M HF solution has equal concentrations of hydrogen ion and fluoride ion; $[H^+] = [F^-] = 2.6 \times 10^{-2}$ M. Upon dissolving enough sodium fluoride (NaF) to bring the F^- ion concentration up to 1.0 M, the hydrogen ion concentration will fall to 7.1×10^{-4} M. The effect of addition of NaF to a HF solution is an example of the common ion effect. In this equilibrium, the fluoride ion is the common ion.

The equilibrium in the original hydrofluoric acid solution was

$$HF(aq) \rightleftharpoons H^+(aq) + F^-(aq)$$

Then sodium fluoride, a strong electrolyte, was added. This caused an increase in F^- concentration:

$$NaF(aq) \rightarrow Na^+(aq) + F^-(aq)$$

According to Le Chatelier's principle, the addition of F^- ions will shift the weak acid equilibrium to the left. This consumes some of the F^- ions and some $H^+(aq)$ and lowers the percent ionization. In effect, the percent ionization of a weak acid, HA, is repressed by the addition to the solution of its conjugate base, F^-. The shift in equilibrium caused by the addition of an ion common to one of the products of the original equilibrium reaction is called the *common ion effect*.

Buffers. Any solution like the preceding one, which contains both a weak acid (HF) and its conjugate base (F^-), has the ability to neutralize small amounts of either a strong acid or a strong base, with very little change in its own pH. Such a solution is called a *buffer* because it resists significant changes in the pH. A buffer must contain an acid to react with any OH^- ions that may be added and a base to react with any added H^+ ions. A solution containing (1) a weak acid or weak base and (2) its salt is a buffer solution.

When a small amount of a strong acid is added to the buffer, it is neutralized by the weak base. Also, when a small amount of a strong base is added to a buffer, it is neutralized by the weak acid. In the hydrofluoric acid–fluoride buffer, for example, the following neutralization reactions take place:

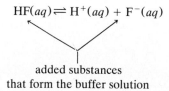

added substances
that form the buffer solution

On addition of a small quantity of strong base, the neutralization reaction is

$$OH^- + HF \rightarrow H_2O + F^-$$

On addition of a small quantity of strong acid, the neutralization reaction is

$$H^+ + F^- \rightarrow HF$$

In order to be most effective, the amounts of weak acid and weak base used to prepare the buffer must be considerably greater than the amounts of strong acid or strong base that may be added later.

Buffer capacity refers to the amount of acid or base that can be neutralized by a buffer solution before its pH is appreciably affected.

A buffer has equal capacity to neutralize either acid or base if the concentrations of weak acid and weak base are large and equal to each other. Each weak acid–conjugate base buffer has a characteristic pH range in which it is most effective. For a HF-F$^-$ buffer

$$K_a = \frac{[H^+][F^-]}{[HF]}$$

When HF and its conjugate base are present in equal concentrations, the H$^+$ ion concentration is

$$[H^+] = K_a$$

In terms of logarithms,

$$\log[H^+] = \log K_a$$

Changing signs

$$-\log[H^+] = -\log K_a$$

and substituting yield

$$pH = pK_a$$

Here the pK_a is defined in the same way as the pH:

$$pK_a = -\log K_a$$

A buffer has equal effectiveness against added acid and added base when its pH = pK_a.

There is a rather narrow pH range, called the *buffer range*, over which a buffer is most effective. The effective buffer range is

$$pH = pK_a \pm 1.0$$

For HF, p$K_a = -\log(7.1 \times 10^{-4}) = 3.15$, and its buffer range is pH = 2.15 − 4.15.

The pH of a Buffer. To calculate the pH of a buffer requires that the concentrations of the weak acid (such as HF) and a soluble salt of the weak acid (NaF) be substituted into the ionization constant expression for the weak acid:

$$HF(aq) \rightleftharpoons H^+(aq) + F^-(aq)$$

$$K_a = \frac{[H^+][F^-]}{[HF]}$$

$$[H^+] = \frac{[HF]}{[F^-]} K_a$$

The pH is found from pH = −log[H$^+$]. Alternatively, the Henderson-Hasselbalch equation, which is introduced in the textbook in Section 16.6, can be used to calculate buffer pH:

$$pH = pK_a + \log \frac{[\text{conjugate base}]}{[\text{acid}]}$$

This equation holds only if the acid is less than 5% ionized. This requirement is almost always met because the ionization of the weak acid is lowered by the addition of the conjugate base in accord with the common ion effect.

EXAMPLE 16.7 Conjugate Base to Weak Acid Ratio in a Buffer

What ratio of [F$^-$] to [HF] would you use to make a buffer of pH = 2.85?

METHOD OF SOLUTION

Start with the Henderson-Hasselbalch equation:

$$\text{pH} = \text{p}K_a + \log \frac{[\text{conjugate base}]}{[\text{acid}]}$$

CALCULATION

Substitute in the desired pH and the pK_a value of HF;

$$2.85 = 3.15 + \log \frac{[\text{F}^-]}{[\text{HF}]}$$

$$\log \frac{[\text{F}^-]}{[\text{HF}]} = -0.30$$

$$\frac{[\text{F}^-]}{[\text{HF}]} = 10^{-0.30} = 0.50$$

The ratio of [F$^-$] to [HF] would be 0.50, or 1 to 2.

EXAMPLE 16.8 pH of a Buffer Solution

Calculate the pH of a buffer solution that is 0.25 M HF and 0.50 M NaF.

METHOD OF SOLUTION

The pH of a buffer is given in terms of the Henderson-Hasselbalch equation:

$$\text{pH} = \text{p}K_a + \log \frac{[\text{conjugate base}]}{[\text{acid}]} = \text{pH} = 3.15 + \log \frac{[\text{F}^-]}{[\text{HF}]}$$

[F$^-$] = 0.50 M because NaF is a strong electrolyte and dissociates completely. [HF] = 0.25 M. Substituting into the preceding equation yields

$$\text{pH} = 3.15 + \log \frac{0.50}{0.25} = 3.15 + \log 2.0 = 3.15 + 0.30$$

$$= 3.45$$

COMMENT

Actually [HF] = 0.25 − x. But we can assume that x is less than 5% of [HF]$_0$ because HF is a weak acid and because the addition of the common ion, F$^-$, represses the dissociation of HF. Also, note that this buffer has a greater capacity to neutralize added acid than it does added base because [F$^-$] > [HF].

EXAMPLE 16.9 Percent Ionization in a Buffer Solution

Calculate the percent ionization of HF in the buffer solution of the preceding example.

METHOD OF SOLUTION

$$\text{percent ionization} = \frac{[\text{H}^+]}{[\text{HF}]_0} \times 100\%$$

where

$$[H^+] = 10^{-pH} = 10^{-3.45}$$

$$= 3.55 \times 10^{-4} \, M$$

$$\text{percent ionization} = \frac{3.55 \times 10^{-4} \, M}{0.25} \times 100\% = 0.14\%$$

COMMENT

By comparison, the percent ionization of HF in a 0.25 M HF solution is 5.3%. The common ion effect is responsible for the much lower percent ionization found in the buffer solution.

EXAMPLE 16.10 Adding Strong Acid to a Buffer

Suppose 3.0 mL of 2.0 M HCl is added to 100 mL of the buffer described in Example 16.8. What is the new pH of the buffer after the HCl is neutralized?

METHOD OF SOLUTION

The HCl is neutralized in this buffer by the reaction, which goes to completion:

$$H^+ + F^- \rightarrow HF$$

Additional HF is generated, and some F^- is consumed. The number of moles of H^+ added as HCl is

$$0.003 \, L \, (2.0 \, mol/L) = 0.006 \, mol \, H^+$$

The number of moles of HF originally present in 100 mL of buffer was

$$0.100 \, L \, (0.25 \, M) = 0.025 \, mol \, HF$$

The number of moles of F^- originally present in 100 mL of buffer was

$$0.100 \, L \, (0.50 \, M) = 0.050 \, mol \, F^-$$

CALCULATION

After the added H^+ is neutralized by $H^+ + F^- \rightarrow HF$,

the number of moles of HF is $0.025 + 0.006 = 0.031$ mol

the number of moles of F^- is $0.050 - 0.006 = 0.044$ mol

The new pH can be found as usual from

$$pH = pK_a + \log\frac{[F^-]}{[HF]}$$

$$= 3.15 + \log\frac{(0.044 \, mol/0.103 \, L)}{(0.031 \, mol/0.103 \, L)} = 3.15 + \log 1.42 = 3.15 + 0.15$$

$$= 3.30$$

COMMENT

The pH has dropped only 0.15 unit from 3.45 due to the addition of 3.0 mL of 2.0 M HCl; thus the use of the term *buffer solution*.

TITRATION CURVES AND INDICATORS

STUDY OBJECTIVES

You should be able to:
1. Describe the shape of titration curves for titrations involving a strong acid and a strong base, a weak acid and a strong base, and a strong acid and a weak base.
2. Choose the correct indicator for a particular acid-base titration.

Titration Curves. Acid-base titrations were discussed in Chapter 4. With the introduction to pH in the present chapter, it is informative to follow the pH as a function of the progress of the titration. A graph of pH versus volume of titrant added is called a *titration curve*. Initially, the pH is that of the unknown solution. As titrant is added, the pH becomes that of a partially neutralized solution of unknown plus titrant. The pH at the equivalence point refers to the [H^+] when just enough titrant has been added to completely neutralize the unknown. If more titrant is added after the equivalence point has been reached, the pH assumes a value consistent with the pH of excess titrant.

Titration curves are shown in the textbook. Figure 16.8 of the text shows the *titration of a strong acid by addition of a strong base*. The main features of this curve can be stated briefly. The pH starts out quite low. As base is slowly added, it is neutralized, and the pH is determined by the unreacted acid. Near the equivalence point, the pH begins to rise more rapidly. At the equivalence point the pH changes rapidly (about 5.0 units upon the addition of only two drops of base). Beyond the equivalence point the pH is determined by the amount of excess base that is added.

The pH at the equivalence point of an acid-base titration is the pH of the salt solution that is formed by neutralization. NaCl, $NaNO_3$, NaBr, KCl, and KI are examples of salts that can be formed in titrations of strong acids and strong bases. These salts yield ions that do not cause hydrolysis. Therefore, the equivalence point in a titration of a strong acid with a strong base occurs at pH 7.0.

Figure 16.8 of the text shows titration curves for a strong acid with a weak base and for a weak acid with a strong base. In the *titration of a strong acid with a weak base*, the first part of the curve is the same as in the strong acid versus strong base titration. However, the pH at the equivalence point is below 7.0 because the cation of the salt is the conjugate acid of the weak base used in the titration. Hydrolysis of this salt yields an acidic solution. In the titration of hydrochloric acid with ammonia, the salt produced is NH_4Cl. The Cl^- ion does not hydrolyze, but NH_4^+ does:

$$NH_4^+ + H_2O \rightleftharpoons NH_3 + H_3O^+$$

In the *titration of a weak acid with a strong base*, the initial pH is greater than in the titration of a strong acid. At the equivalence point, the pH is above 7.0 because of salt hydrolysis. The anion of the salt is the conjugate base of the weak acid used in the titration. In the titration of acetic acid with sodium hydroxide the salt produced is CH_3COONa. The Na^+ ion does not hydrolyze, but CH_3COO^- does:

$$CH_3COO^- + H_2O \rightleftharpoons CH_3COOH + OH^-$$

This makes the solution basic at the equivalence point.

Indicators. Indicators are used in the laboratory to reveal the equivalence point of a titration. Ideally, the indicator will change color at the pH corresponding to the equivalence point of a particular titration. This is the *endpoint* of a titration. It is hoped that the color change will coincide with the equivalence point and that any volume difference between the equivalence point and the endpoint will be small.

Indicators are usually weak organic acids that have distinctly different colors in the nonionized (molecular) and ionized forms:

$$\underset{\text{color 1}}{HIn(aq)} \rightleftharpoons H^+(aq) + \underset{\text{color 2}}{In^-(aq)} \qquad K_a = \frac{[H^+][In^-]}{[HIn]}$$

To determine the pH range in which an indicator will change color, we can write the K_a expression in logarithmic form:

$$\mathrm{pH} = \mathrm{p}K_a + \log \frac{[\mathrm{In}^-]}{[\mathrm{HIn}]}$$

The color of an indicator depends on which form predominates. Typically, when $[\mathrm{In}^-]/[\mathrm{HIn}] \geq 10$, the solution will be color 2, and when $[\mathrm{In}^-]/[\mathrm{HIn}] \geq 0.1$, the solution will be color 1. Thus the color change will occur between

$$\mathrm{pH} = \mathrm{p}K_a + \log(10/1) = \mathrm{p}K_a + 1.0$$

and

$$\mathrm{pH} = \mathrm{p}K_a + \log(1/10) = \mathrm{p}K_a - 1.0$$

At the middle point of the pH range over which the color changes, $[\mathrm{HIn}] = [\mathrm{In}^-]$, and $\mathrm{pH} = \mathrm{p}K_a$.

Like any weak acid each HIn has a characteristic $\mathrm{p}K_a$, and so each indicator changes color at a characteristic pH. Table 16.6 of the text lists a number of indicators used in acid-base titration and the pH ranges over which they change color.

The choice of indicator for a particular titration depends on the pH at which the equivalence point is expected to occur. For instance, in the titration of acetic acid by sodium hydroxide, which is discussed in Example 16.16 of the text, the pH at the equivalence point is 8.72. According to Table 16.6 in the text, both cresol red and phenolphthalein change color over ranges that include pH 8.72. Therefore either of these indicators would show the equivalence point of this titration.

EXAMPLE 16.11 Net Ionic Equations

Write the net ionic equations for the neutralization reactions that occur during the following titrations. Predict whether the pH at the equivalence point will be above, below, or equal to 7.0:
a. Titration of HI with NH_3
b. Titration of HI with NaOH
c. Titration of NaOH with HF

METHOD OF SOLUTION

a. Net ionic equations are written to show the predominant species in solution. HI is 100% dissociated, but NH_3 is a weak base. The net ionic equation for neutralization is

$$\mathrm{H}^+ + \mathrm{NH}_3 \rightarrow \mathrm{NH}_4^+$$

The result of neutralization is production of the conjugate acid NH_4^+ of the weak base NH_3. NH_4^+ is an acid. *Answer:* The pH at the equivalence point will be below 7.0.

b. This is a titration of a strong acid and a strong base. The net ionic equation is

$$\mathrm{H}^+ + \mathrm{OH}^- \rightarrow \mathrm{H}_2\mathrm{O}$$

The resulting solution is neutral at the equivalence point. *Answer:* pH 7.0.

c. This titration involves a weak acid and a strong base:

$$\mathrm{HF} + \mathrm{OH}^- \rightarrow \mathrm{H}_2\mathrm{O} + \mathrm{F}^-$$

The reaction produces the conjugate base of the weak acid. *Answer:* The solution will be basic at the equivalence point, pH > 7.0.

EXAMPLE 16.12 Choosing an Indicator

Choose an indicator for the titration of 50 mL of a $0.10M$ HI solution with $0.10M$ NH_3:

$$HI(aq) + NH_3(aq) \rightarrow NH_4I(aq)$$

METHOD OF SOLUTION

First we need to know the pH at the equivalence point. This is the titration of a strong acid with a weak base. The net ionic reaction is

$$H^+(aq) + NH_3(aq) \rightarrow NH_4^+(aq)$$

5.0×10^{-3} mol + 5.0×10^{-3} mol yields 5.0×10^{-3} mol

The product NH_4^+ ion is a weak acid, and so we expect a slightly acidic solution at the equivalence point.

CALCULATION

The concentration of NH_4^+ formed at the equivalence point is 5.0×10^{-3} mol/0.100 L = 5.0×10^{-2} M NH_4^+. The $[H^+]$ at the equivalence point is just the pH of a 0.050 M NH_4^+ solution:

$$NH_4^+(aq) \rightleftharpoons NH_3(aq) + H^+(aq) \qquad K_a = 5.6 \times 10^{-10}$$

Concentration	NH_4^+	$\rightleftharpoons$	H^+	+	NH_3
Initial (M)	0.050		0		0
Change (M)	$-x$		$+x$		$+x$
Equilibrium (M)	$(0.050 - x)$		x		x

$$K_a = \frac{[NH_3][H^+]}{[NH_4^+]} = \frac{x^2}{(0.050 - x)} = \frac{x^2}{0.050} = 5.6 \times 10^{-10}$$

Solving for x yields

$$x = [NH_3] = [H^+] = 5.3 \times 10^{-6} \ M$$

and the pH is 5.28 at the equivalence point. According to Table 16.6 of the text, chlorophenol blue and methyl red are indicators that will change color in the vicinity of pH 5.28.

TRUE-FALSE QUESTIONS

1. Acid strength is measured by the magnitude of the acid ionization constant K_a.

2. In a solution of a weak acid, HA, [HA] > $[H^+]$.

3. In a solution of a weak base, $[OH^-] < [H^+]$.

4. In a solution of a polyprotic acid, the major ionic species in solution will be produced in the first ionization step.

5. All cations and anions can react with water, producing acidic or basic solutions.

6. The common ion effect from adding C_6H_5COONa (sodium benzoate) to a C_6H_5COOH solution (benzoic acid) will cause the percent ionization of benzoic acid to decrease and the pH to rise.

7. A buffer solution can be made from HCl and NaCl.

8. A buffer solution can be made from NaCN and HCN.

9. No matter how much strong acid you add to a buffer solution, it will maintain a constant pH.

10. The Henderson-Hasselbalch equation is just the logarithmic form of the acid ionization constant expression.

11. The equivalence point of all titrations occurs at pH 7.0.

12. Methyl red would be a suitable indicator for a titration that has a pH of 5.0 at the endpoint.

SELF-TEST A

1. What is the H^+ ion concentration in a 0.82 M HOCl solution? $K_a = 3.2 \times 10^{-3}$.

2. Calculate the percent ionization in a 1.0 M solution of nitrous acid, HNO_2. See Table 16.1 for K_a.

3. What is the pH of a 0.090 M CH_3COOH (acetic acid) solution? See Table 16.1 for K_a.

4. A 1.0 M HF solution is only 2.6% ionized. What is the K_a value for HF?

5. Calculate the concentration of H^+, OH^-, F^- ions and undissociated HF molecules in a 0.010 M HF solution.

6. What is the pH of a 2.0 M NH_3 solution?

7. The first and second ionization constants of H_2CO_3 are 4.2×10^{-7} and 4.8×10^{-11}. Calculate the concentrations of H^+ ion, HCO_3^-, CO_3^{2-}, and undissociated H_2CO_3 in a 0.080 M H_2CO_3 solution.

8. A 0.1 M solution of Na_2CO_3 is
   a. Acidic   b. Basic   c. Neutral

9. Classify the solutions of the following salts as acidic, basic, or neutral:
   a. $AlCl_3$
   b. Na_2SO_4
   c. $NaClO_4$
   d. KCN
   e. NH_4Cl
   f. Na_2SO_3

10. Calculate the pH of a 0.30 M CH_3COONa solution. See Table 16.1 for the K_a of CH_3COOH.

11. What is the H^+ ion concentration of a 0.010 M NaCN solution? *Hint:* Use K_a for HCN to calculate K_b for CN^-.

12. Which one of the following aqueous solutions (all 0.010 M) will have the highest pH value?
    a. CH_3COOH b. HCl   c. NaCl   d. KF

13. A buffer solution is prepared by mixing 500 mL of 0.60 M CH_3COOH with 500 mL of a 1.00 M CH_3COONa solution. What is the pH of this solution?

14. A buffer solution is prepared by mixing 300 mL of 0.10 M HNO_2 with 200 mL of 0.40 M $NaNO_2$.
    a. Calculate the pH of the resulting solution.
    b. What is the new pH after 2.0 mL of 2.0 M HCl are added to this buffer?

15. What is the effective buffer range of hydrocyanic acid (HCN)?

16. Which one of the following mixtures is suitable for making a buffer solution with an optimum pH of about 9.2?
    a. $NaC_2H_3O_2$ and $HC_2H_3O_2$
    b. NH_4Cl and NH_3
    c. HF and NaF
    d. HNO_2 and $NaNO_2$
    e. NaCl and HCl

17. A 20.0-mL portion of a solution of 0.0200 M HNO_3 is titrated with 0.010 M KOH.
    a. What is the pH at the equivalence point?
    b. How many milliliters of KOH are required to reach the equivalence point?
    c. What will the pH be after only 10.0 mL of KOH are added?
    d. What will the pH be after 45.0 mL of KOH are added?

18. Consider the titration of 50.0 mL of 0.10 M CH_3COOH with 0.10 M NaOH. What is the pH at the equivalence point?

19. The color of methyl red indicator changes from red to yellow at pH 5.2. What is the K_a for the weak acid form of this indicator?

GENERAL PROBLEMS

20. Calculate the SO_3^{2-} concentration in a 0.10 M H_2SO_3 solution whose pH was adjusted to 10.0 by the addition of NaOH. *Hint:* Consider the overall reaction $H_2SO_3 \rightleftharpoons 2H^+ + SO_3^{2-}$ for which $K = K_{a1} \cdot K_{a2}$.

21. The indicator thymol blue is a weak acid with a K_a of approximately 1×10^{-9}. In 0.1 M HCl the indicator is yellow, and in 0.1 M NaOH it is blue. What color is a solution of pH 7.0 to which thymol blue has been added?

22. Calculate the pH of a solution prepared by mixing 25.0 mL of 0.10 M HCl and 25.0 mL of 0.25 M CH_3COONa?

SELF-TEST B1

1. HA is a weak acid. If a 0.020 M HA solution is 2.5% dissociated, what is the K_a value of the weak acid?

2. Consider a 0.100 M solution of the diprotic acid H_2S. What pH is required to produce a H_2S solution in which the sulfide ion concentration is 5.0×10^{-18} M? *Hint:* Use the overall equilibrium

$$H_2S(aq) \rightleftharpoons 2H^+(aq) + S^{2-}(aq)$$

for which $K = K_{a1} \cdot K_{a2}$.

3. What is the pH of a buffer solution prepared by dissolving 0.15 mol of benzoic acid (C_6H_5COOH) and 0.45 mol of sodium benzoate (C_6H_5COONa) in enough water to make 400 mL of solution.

4. Bromothymol blue is an ordinary acid-base indicator. It has a $K_a = 1.6 \times 10^{-7}$. At what pH does this indicator change color?

5. Given that K_a for formic acid (HCOOH) is 1.7×10^{-4}, find the K_b value for the formate ion, $HCOO^-$.

6. What is the approximate pH to be expected at the equivalence point in the titration of a strong acid with a weak base?

SELF-TEST B2

1. What is the percent dissociation of a weak acid in a 0.020 M HA solution? The K_a of the acid is 1.3×10^{-5}.

2. Calculate the sulfide ion concentration in a 0.10 M H_2S solution to which strong acid has been added to bring the pH to 4.97. *Hint:* H_2S is a diprotic acid. Use K_{a1} and K_{a2} from Table 16.3 of the text.

3. Determine the ratio of benzoic acid to benzoate ion that must be present in a buffer solution if it is to have a pH of 4.67.

4. Bromothymol blue is an ordinary acid-base indicator. It has a $K_a = 1.6 \times 10^{-7}$. Its undissociated form HIn has a yellow color, and its conjugate base In^- is blue. What color would a solution have at pH 5.6?

5. Given that K_b for the formate ion $HCOO^-$ is 5.9×10^{-11}, find the K_a value for formic acid, HCOOH.

6. The pH at the equivalence point in a titration was found to be approximately 9. Describe the category of titration in terms of the acid and base strengths.

ANSWERS

TRUE-FALSE QUESTIONS

1. True.
2. True.
3. False. $[OH^-] > [H^+]$.
4. True.
5. False. The conjugate bases of strong acids do not react with water.
6. True.
7. False. You need a weak acid and its salt. HCl is a strong acid.
8. True.
9. False. All buffers have a buffer capacity.
10. True.
11. False. The pH at the equivalence point depends on the strengths of the acid and base. It will be 7 only for a strong acid–strong base titration.
12. True.

SELF-TEST A

1. $[H^+] = 1.6 \times 10^{-4}$ M
2. 2.1%
3. pH = 2.9
4. $K_a = 6.9 \times 10^{-4}$
5. $[H^+] = [F^-] = 2.3 \times 10^{-3}$ M, [HF] = 7.7×10^{-3} M, $[OH^-] = 4.3 \times 10^{-12}$ M
6. pH = 11.8
7. $[H^+] = [HCO_3^-] = 1.8 \times 10^{-3}$ M
   $[CO_3^{2-}] = 4.8 \times 10^{-11}$ M
   $[H_2CO_3] = 0.080$ M
8. b
9. a. Acidic   b. Neutral   c. Neutral   d. Basic   e. Acidic   f. Basic
10. pH = 9.1
11. $[H^+] = 2.2 \times 10^{-11}$ M
12. d

13. pH = 4.96
14. a. pH = 3.77   b. 3.66
15. pH range 8.3–10.3
16. b
17. a. pH = 7.0   b. 40.0 mL   c. 2.00   d. 10.89
18. 8.72
19. $K_a = 6.3 \times 10^{-6}$
20. $[SO_3^{2-}]/[H_2SO_3] = K/[H^+]^2 \doteq 8.2 \times 10^{10}$. Therefore essentially all the H_2SO_3 has dissociated to SO_3^{2-}, and $[SO_3^{2-}] = 0.10$ M.
21. Yellow
22. pH = 4.92

SELF-TEST B1

1. $K_a = 1.3 \times 10^{-5}$
2. pH = 4.97
3. pH = 4.67
4. pH = 6.80
5. $K_b = 5.9 \times 10^{-11}$
6. pH $\cong$ 8 – 10

SELF-TEST B2

1. 2.5%
2. 5×10^{-18} M
3. 1 : 3
4. Yellow
5. $K_a = 1.7 \times 10^{-4}$
6. Weak acid vs. strong base

Chapter Seventeen
SOLUBILITY EQUILIBRIA

- Solubility and Solubility Product
- Predicting Precipitation Reactions and Separation of Ions by Precipitation
- Factors Affecting Solubility: Common Ion Effect and pH
- Complex Ions and Solubility

SOLUBILITY AND SOLUBILITY PRODUCT

STUDY OBJECTIVES

You should be able to:
1. Write the solubility product expression for an insoluble salt.
2. Calculate the solubility product constant of a salt given its solubility, and vice versa.
3. Predict whether a precipitate will form when two solutions of salts are mixed.

The Solubility Product Constant. The solubility rules were discussed in Chapter 3. In this chapter, solubility is treated quantitatively in terms of equilibrium. In a saturated solution of silver bromide, for example, the solubility equilibrium is

$$AgBr(s) \rightleftharpoons Ag^+(aq) + Br^-(aq)$$

The equilibrium constant for the reaction in which a solid salt dissolves is called the solubility product constant, K_{sp}:

$$K_{sp} = [Ag^+][Br^-]$$

The solubility product expression is always written in the form of the equilibrium constant expression for the solubility reaction. Two other examples are

$$Ca(OH)_2(s) \rightleftharpoons Ca^{2+}(aq) + 2OH^-(aq)$$

$$K_{sp} = [Ca^{2+}][OH^-]^2$$

and

$$Ag_2CrO_4(s) \rightleftharpoons 2Ag^+(aq) + CrO_4^{2-}(aq)$$

$$K_{sp} = [Ag^+]^2[CrO_4^{2-}]$$

The solubility product constant (also called a K_{sp} value) can be calculated by substituting the concentrations of the ions in a *saturated solution* into the solubility product expression. Take AgBr again as an example. Since AgBr is a strong electrolyte, all of the solid AgBr that dissolves is dissociated into

Ag$^+$ and Br$^-$ ions,

$$AgBr(aq) \rightarrow Ag^+(aq) + Br^-(aq)$$

The solubility of AgBr at 25°C is 8.8×10^{-7} moles per liter of solution. Therefore, the ion concentrations in a saturated AgBr solution are

$$[Ag^+] = [Br^-] = 8.8 \times 10^{-7} M$$

The solubility product constant is found by substituting these concentrations into the solubility product:

$$K_{sp} = [Ag^+][Br^-] = (8.8 \times 10^{-7})(8.8 \times 10^{-7})$$

$$= 7.7 \times 10^{-13}$$

This calculation points out the difference between the solubility and the K_{sp} value. These two quantities are *not* the same. Here we see that the solubility is used to determine the ion concentrations and that substitution of these into the solubility product expression gives the solubility product constant (K_{sp} value). Table 17.1 in the text and Table 17.1 in this chapter list solubility product constants for a number of slightly soluble salts, including silver bromide.

Table 17.1 Solubility Products of Some Slightly Soluble Salts at 25°C

Name	Formula	K_{sp}
MX type		
Calcium sulfate	CaSO$_4$	2.4×10^{-5}
Silver chloride	AgCl	1.6×10^{-10}
Silver bromide	AgBr	7.7×10^{-13}
MX$_2$ and M$_2$X Type		
Lead chloride	PbCl$_2$	2.4×10^{-4}
Calcium hydroxide	Ca(OH)$_2$	8.0×10^{-6}
Lead iodide	PbI$_2$	1.4×10^{-8}
Magnesium fluoride	MgF$_2$	6.6×10^{-9}
Magnesium hydroxide	Mg(OH)$_2$	1.2×10^{-11}
Silver chromate	Ag$_2$CrO$_4$	1.1×10^{-12}

The K_{sp} value can be used to calculate the solubility of a compound in moles per liter (the molar solubility) or in grams per liter. For example, the K_{sp} value for CaSO$_4$ is 2.4×10^{-5}. What is the solubility of CaSO$_4$ in mol/L? Start with the solubility equilibrium:

$$CaSO_4(s) \rightleftharpoons Ca^{2+}(aq) + SO_4^{2-}(aq) \qquad K_{sp} = [Ca^{2+}][SO_4^{2-}]$$

Let s equal the solubility in mol/L. At equilibrium, both [Ca^{2+}] and [SO$_4^{2-}$] are equal to s because when s moles of CaSO$_4$ dissolve in 1 L, s moles of Ca^{2+} ions and s moles of SO$_4^{2-}$ ions are produced. Substitute into the K_{sp} expression:

$$K_{sp} = [Ca^{2+}][SO_4^{2-}] = (s)(s) = s^2$$

Since

$$K_{sp} = 2.4 \times 10^{-5}$$

then

$$s^2 = 2.4 \times 10^{-5}$$

Hence the solubility is

$$s = \sqrt{2.4 \times 10^{-5}} = 4.9 \times 10^{-3} \, M$$

It is important to notice that the molar solubility and K_{sp} are not the same. *Molar solubility and K_{sp} are related but are not the same quantity.*

Solubility and K_{sp}. Table 17.1 lists a number of solubility product constants for various salts that will be needed in this chapter. Notice that within a series of salts of the same type of formula (MX, MX$_2$, etc.), *solubility increases as the solubility product constant increases*. Comparing the K_{sp}'s of the three MX salts, one can tell that CaSO$_4$ is the most soluble of the three. Even though we can predict the relative solubilities of salts from their K_{sp} values, remember that the K_{sp} value is *not* the same as the solubility.

EXAMPLE 17.1 Calculating a K_{sp} Value

The solubility of calcium phosphate in water is 8.0×10^{-5} g per 100 mL of solution. What is the solubility product constant for Ca$_3$(PO$_4$)$_2$?

METHOD OF SOLUTION

Write the solubility equilibrium and the K_{sp} expression:

$$Ca_3(PO_4)_2(s) \rightleftharpoons 3Ca^{2+}(aq) + 2PO_4^{3-}(aq)$$

$$K_{sp} = [Ca^{2+}]^3[PO_4^{3-}]^2$$

Calculate s, the molar solubility of Ca$_3$(PO$_4$)$_2$, from the given solubility. The formula mass of Ca$_3$(PO$_4$)$_2$ is 310.2 g/mol:

$$s = \frac{8.0 \times 10^{-5} \, g}{100 \, mL} \times \frac{1 \, mL}{10^{-3} \, L} \times \frac{1 \, mol}{310.2 \, g} = 2.6 \times 10^{-6} \, mol/L$$

The solubility product constant (K_{sp} value) can be calculated by substitution of the ion concentrations in a saturated solution into this expression. When 2.6×10^{-6} mol of Ca$_3$(PO$_4$)$_2$ dissolves in 1 L, the Ca^{2+} concentration is

$$[Ca^{2+}] = 3s = 3(2.6 \times 10^{-6}) = 7.8 \times 10^{-6} \, M$$

The PO$_4^{3-}$ concentration is

$$[PO_4^{3-}] = 2s = 2(2.6 \times 10^{-6}) = 5.2 \times 10^{-6} \, M$$

CALCULATION

Now substitute the ion concentrations into the K_{sp} expression:

$$K_{sp} = (7.8 \times 10^{-6})^3 (5.2 \times 10^{-6})^2$$

$$= (4.7 \times 10^{-16})(2.7 \times 10^{-11})$$

$$= 1.3 \times 10^{-26}$$

EXAMPLE 17.2 Calculating a Molar Solubility

The K_{sp} value for Ag_2CrO_4 is 1.1×10^{-12}. Calculate the molar solubility of silver chromate.

METHOD OF SOLUTION

Write the solubility equilibrium,

$$Ag_2CrO_4(s) \rightleftharpoons 2Ag^+(aq) + CrO_4^{2-}(aq)$$

and the K_{sp} expression,

$$K_{sp} = [Ag^+]^2[CrO_4^{2-}]$$

Let s = molar solubility of Ag_2CrO_4. Whenever s moles of Ag_2CrO_4 dissolve, $2s$ moles of Ag^+ and s moles of CrO_4^{2-} are produced:

$$[Ag^+] = 2s \qquad [CrO_4^{2-}] = s$$

Summarize the changes in concentrations as follows:

Concentration	$Ag_2CrO_4(s) \rightleftharpoons 2Ag^+(aq) +$	$CrO_4^{2-}(aq)$
Initial (M)	0	0
Change (M)	$2s$	s
Equilibrium (M)	$2s$	s

CALCULATION

Substitute these values into the K_{sp} expression:

$$K_{sp} = [2s]^2[s] = 1.1 \times 10^{-12}$$

and solve for s:

$$4s^2(s) = 1.1 \times 10^{-12}$$

$$4s^3 = 1.1 \times 10^{-12}$$

$$s = \sqrt[3]{\frac{1.1 \times 10^{-12}}{4}} = \sqrt[3]{2.75 \times 10^{-13}}$$

$$= 6.5 \times 10^{-5} \, M$$

PREDICTING PRECIPITATION REACTIONS AND SEPARATION OF IONS BY PRECIPITATION

STUDY OBJECTIVES

You should be able to:
1. Predict whether a precipitate will form when two solutions of salts are mixed.
2. Describe how to selectively precipitate one of two ions in a solution.

Criteria for Precipitate Formation. When two solutions containing dissolved salts are mixed, formation of an insoluble compound is always a possibility. From a knowledge of solubility rules (Chapter 3) and solubility products (Table 17.1), you can predict whether a precipitate will form. For a dissolved salt, $MX(aq)$, the ion product (Q) is $Q = [M^+]_0[X^-]_0$, where $[\ ]_0$ stands for the initial concentrations.

Any one of three conditions for the ion product Q may exist after two solutions are mixed:

1. $Q = K_{sp}$. For a saturated solution, the value of Q is equal to K_{sp}. No precipitate will form in this case as no net reaction occurs in a system at equilibrium.
2. $Q < K_{sp}$. In an unsaturated solution the value of Q is less than K_{sp}. No precipitate will form in this case.
3. $Q > K_{sp}$. In a supersaturated solution more of the salt is dissolved than the solubility allows. In this case, an unstable situation exists, and some solute will precipitate from solution until a saturated solution is attained. At this point $Q = K_{sp}$ and equilibrium is reestablished.

Suppose 500 mL of a solution containing $2 \times 10^{-5} M$ Ag^+ is mixed with 500 mL of solution containing 2×10^{-4} M Br^-. Will a precipitate form? First determine the new concentrations of Ag^+ and Br^- in the mixture. The total volume is 1 L, and so accounting for dilution, the ion concentrations in the new solution are just half of what they were in the separate solutions. The initial concentrations before any precipitate forms are

$$[Ag^+]_0 = 1 \times 10^{-5}\ M \qquad [Br^-]_0 = 1.0 \times 10^{-4}\ M$$

The ion product for AgBr is

$$Q = [Ag^+]_0[Br^-]_0$$

To predict whether a precipitate of AgBr will form, calculate the ion product Q and compare it to the K_{sp} value:

$$Q = [Ag^+]_0[Br^-]_0 = (1 \times 10^{-5})(1 \times 10^{-4}) = 1 \times 10^{-9}$$

$$K_{sp} = 7.7 \times 10^{-13}$$

Therefore, $Q > K_{sp}$, and so the mixture corresponds to a supersaturated solution of AgBr, which means that some $AgBr(s)$ will precipitate,

$$Ag^+(aq) + Br^-(aq) \rightarrow AgBr(s)$$

until a new equilibrium is reached, at which point $Q = K_{sp}$.

Criteria for Separation of Ions by Precipitation. Ions in a solution can be separated from each other on the basis of the different solubilities of their salts. For example, when slowly adding a solution containing a OH^- ion to a solution that contains both Ca^{2+} ions and Mg^{2+} ions, we find that $Mg(OH)_2$ precipitates before $Ca(OH)_2$. $Mg(OH)_2$, with a K_{sp} of 1.2×10^{-11}, precipitates first because it is less soluble than $Ca(OH)_2$; whose K_{sp} is 8.0×10^{-6}.

The concentrations of the ions in solution also help determine which ions will precipitate first. Even though $Mg(OH)_2$ is less soluble than $Ca(OH)_2$, $Ca(OH)_2$ would precipitate first when a OH^- ion is added if the concentration of a Ca^{2+} ion was much greater than the concentration of a Mg^{2+} ion. It is best to determine the concentration of a OH^- ion needed to cause a precipitate of both compounds. Since Ca^{2+} and Mg^{2+} ion concentrations will be given, then the concentration of OH^- that must be exceeded to initiate the precipitation of $Ca(OH)_2$ and $Mg(OH)_2$ are, respectively,

$$[OH^-] = \sqrt{\frac{K_{sp}}{[Ca^{2+}]}} \quad \text{and} \quad [OH^-] = \sqrt{\frac{K_{sp}}{[Mg^{2+}]}}$$

Examples 17.4 and 17.5 illustrate the separation of Cl^- and CrO_4^{2-} (chromate) ions.

EXAMPLE 17.3 Predicting Formation of a Precipitate

a. Predict whether or not a precipitate of PbI_2 will form when 200 mL of $0.015 M$ $Pb(NO_3)_2$ and 300 mL of $0.050\ M$ NaI are mixed together, given that $K_{sp}(PbI_2) = 1.4 \times 10^{-8}$.

b. If the answer is yes, what concentrations of $Pb^{2+}(aq)$ and $I^-(aq)$ will exist when equilibrium is reestablished?

METHOD OF SOLUTION FOR a

Recall that $Pb(NO_3)_2$ and NaI are both strong electrolytes. When the solutions are mixed, will the following reaction occur?

$$Pb^{2+}(aq) + 2NO_3^-(aq) + 2Na^+(aq) + 2I^-(aq) \rightarrow PbI_2(s) + 2Na^+(aq) + 2NO_3^-(aq)$$

A precipitate will form only if the ion product exceeds the K_{sp} value: $[Pb^{2+}]_0[I^-]_0^2 > K_{sp}$. When the two solutions are mixed, 500 mL of new solution is formed. Immediately after mixing, the initial ion concentrations would be

$$[Pb^{2+}]_0 = 0.015 M \times \frac{200\ \text{mL}}{500\ \text{mL}} = 6.0 \times 10^{-3}\ M$$

$$[I^-]_0 = 0.050 M \times \frac{300\ \text{mL}}{500\ \text{mL}} = 3.0 \times 10^{-2}\ M$$

Then $Q = [Pb^{2+}]_0[I^-]_0^2 = (6.0 \times 10^{-3})(3.0 \times 10^{-2})^2 = 5.4 \times 10^{-6}$.

Answer: $Q > K_{sp}$, and therefore a precipitate will form.

METHOD OF SOLUTION FOR b

What concentrations of Pb^{2+} and I^- remain in solution? In order to determine how many moles of Pb^{2-} and I^- precipitated from solution, consider the net reaction

$$Pb^{2+}(aq) + 2I^-(aq) \rightarrow PbI_2(s)$$

Look to see if there is a limiting reagent. Then determine the amount of the excess reagent that remains after complete reaction of the limiting reagent. The initial number of moles of the Pb^{2+} ion was $0.015\ M \times 0.200\ L = 0.003$ mol, and of the I^- ion it was $0.050\ M \times 0.300\ L = 0.015$ mol. The Pb^{2+} ion, being in the smaller amount, is the limiting reagent and will react essentially completely. The 0.003 mol of Pb^{2+} will react with 0.006 mol of I^-, leaving 0.009 mol of excess I^- (0.015 − 0.006) dissolved in solution.

CALCULATION

The I^- concentration is

$$[I^-] = 0.009\ \text{mol}/0.50\ L = 0.018\ M$$

The Pb^{2+} concentration is controlled by the PbI_2 solubility equilibrium. In other words, some PbI_2 dissolves by the reverse reaction:

$$PbI_2(s) \rightleftharpoons Pb^{2+} + 2I^-$$

$$K_{sp} = [Pb^{2+}][I^-]^2 = 1.4 \times 10^{-8}$$

$$[Pb^{2+}] = \frac{K_{sp}}{[I^-]^2} = \frac{1.4 \times 10^{-8}}{(0.018)^2} = 4.3 \times 10^{-5}\ M$$

COMMENT

The percentage the of Pb^{2+} ion remaining unprecipitated is $[4.3 \times 10^{-5} \, M(0.5 \, L)/0.003 \, mol] \times 100 = 0.72\%$. This confirms our assumption that essentially all the Pb^{2+} ion precipitated.

EXAMPLE 17.4   Selective Precipitation of an Ion

A solution contains $0.10 \, M \, Cl^-$ and $0.010 \, M \, CrO_4^{2-}$. If $AgNO_3$ solution is added dropwise, which will precipitate first: AgCl or Ag_2CrO_4?

METHOD OF SOLUTION

The solubility product is a number that the product of the ion concentrations can never exceed at equilibrium. First write the two equilibria of interest and their solubility constants:

$$AgCl(s) \rightleftharpoons Ag^+(aq) + Cl^-(aq) \qquad K_{sp} = [Ag^+][Cl^-] = 1.6 \times 10^{-10}$$

$$Ag_2CrO_4(s) \rightleftharpoons 2Ag^+ + CrO_4^{2-} \qquad K_{sp} = [Ag^+]^2[CrO_4^{2-}] = 1.1 \times 10^{-12}$$

CALCULATION

The highest Ag^+ ion concentration possible in a solution of $0.10 \, M \, Cl^-$ is

$$[Ag^+] = \frac{1.6 \times 10^{-10}}{[Cl^-]} = \frac{1.6 \times 10^{-10}}{0.10} = 1.6 \times 10^{-9} \, M$$

The highest Ag^+ ion concentration possible in a solution of $0.010 \, M \, CrO_4^{2-}$ is

$$[Ag^+] = \sqrt{\frac{1.1 \times 10^{-12}}{[CrO_4^{2-}]}} = \sqrt{\frac{1.1 \times 10^{-12}}{0.010}} = 1.0 \times 10^{-5} \, M$$

Therefore silver chloride will precipitate before silver chromate because of the lower Ag^+ ion concentration needed to produce a saturated AgCl solution.

EXAMPLE 17.5   Completeness of Precipitation

In the above example AgCl precipitates before Ag_2CrO_4 as the $Ag^+(aq)$ ion is added dropwise to the solution. What percentage of the Cl^- ion in solution will have precipitated when Ag_2CrO_4 just begins to precipitate?

METHOD OF SOLUTION

As we found above, Ag_2CrO_4 begins to precipitate when $[Ag^+] = 1.0 \times 10^{-5} \, M$.

CALCULATION

The Cl^- ion concentration at this Ag^+ ion concentration is

$$[Cl^-] = \frac{1.6 \times 10^{-10}}{[Ag^+]} = \frac{1.6 \times 10^{-10}}{1.0 \times 10^{-5} \, M} = 1.6 \times 10^{-5} \, M$$

The percentage of Cl⁻ remaining in solution *unprecipitated* is

$$\%\text{Cl}^- = \frac{1.6 \times 10^{-5} \, M}{0.10 \, M} \times 100\%$$

$$= 0.016\%$$

Therefore, the percentage of the Cl⁻ ion precipitated is $100\% - 0.016\% = 99.98\%$.

FACTORS AFFECTING SOLUBILITY: COMMON ION EFFECT AND pH

STUDY OBJECTIVES

You should be able to:
1. Calculate the solubility of a salt in a solution containing an ion common to that salt.
2. Predict the effect of pH on the solubility of a salt.

The Common Ion Effect. The effect of adding an ion common to one already in equilibrium in a solubility reaction is to decrease the solubility of the salt. In the case of AgBr solubility,

$$\text{AgBr}(s) \rightleftharpoons \text{Ag}^+(aq) + \text{Br}^-(aq)$$

The addition of either Ag^+ or Br^- ions will shift the equilibrium to the left, in accord with Le Chatelier's principle, thus decreasing the amount of AgBr dissolved. The Ag^+ ions can be added by pouring in a solution of AgNO_3. Recall that AgNO_3 is very soluble and is a strong electrolyte:

$$\text{AgNO}_3(s) \Rightarrow \text{Ag}^+(aq) + \text{NO}_3^-(aq)$$

The nitrate ion will not interfere with AgBr solubility because it is not a common ion.

Additional Br^- ions could be supplied by adding NaBr, for instance. Sodium bromide is very soluble and is a strong electrolyte:

$$\text{NaBr}(s) \rightarrow \text{Na}^+(aq) + \text{Br}^-(aq)$$

The sodium ion will not affect the solubility of AgBr because it is not a common ion.

Here we see that the addition of an ion that is common to one already in the solubility equilibrium shifts the equilibrium to the left, which *decreases* the solubility. This is equivalent to the case of a weak acid, where the presence of a common ion decreases the percent ionization. Several calculations involving the common ion effect appear at the end of this section.

pH and Solubility. The pH can affect the solubility of a solute in two ways. One of these is through the common ion effect. Consider the solubility equilibrium of an insoluble hydroxide such as Mg(OH)_2 or Ca(OH)_2:

$$\text{Ca(OH)}_2(s) \rightleftharpoons \text{Ca}^{2+}(aq) + 2\text{OH}^-(aq)$$

Upon the addition of NaOH, for instance, the pH of the solution will increase. The equilibrium position will shift to the left because of the added OH^- ion (a common ion). Therefore the solubility of Ca(OH)_2 will decrease proportionately.

The other way that pH can affect solubility is the following. If a salt contains a basic anion such as F^-, CH_3COO^-, or CN^-, the anion will react with added H^+ ion, and the solubility will be pH dependent. Take, for instance, the effect of adding a strong acid on the solubility of silver acetate. To explain this, we need two reaction steps. In the first one silver acetate dissolves, and in the second the acetate ion combines with added $H^+(aq)$ ions:

$$CH_3COOAg(s) \rightleftharpoons CH_3COO^-(aq) + Ag^+(aq)$$

$$H^+(aq) + CH_3COO^-(aq) \rightarrow CH_3COOH(aq)$$

In the presence of H^+, the concentration of acetate ion is lowered by the occurrence of the second reaction. The second reaction occurs essentially completely because it forms a weak acid. According to Le Chatelier's principle, the solubility equilibrium will shift to the right, causing more silver acetate to dissolve.

The solubilities of salts containing anions that do *not* hydrolyze, such as Cl^-, Br^-, I^-, and NO_3^-, are not affected by pH.

EXAMPLE 17.6 Common Ion Effect

What is the solubility of $PbCl_2$ in 0.50 M NaCl solution?

METHOD OF SOLUTION

Write the solubility equilibrium reaction for $PbCl_2$ and the K_{sp} expression. The K_{sp} value is in Table 17.1:

$$PbCl_2(s) \rightleftharpoons Pb^{2+}(aq) + 2Cl^-(aq)$$

$$K_{sp} = [Pb^{2+}][Cl^-]^2 = 2.4 \times 10^{-4}$$

All of the $[Pb^{2+}]$ ion comes from the dissolution reaction. Let s = molar solubility of $PbCl_2$. The only source of Pb^{2+} is $PbCl_2$. And so,

$$[Pb^{2+}] = s$$

Chlorine ions are contributed by $PbCl_2$ and by NaCl (a strong electrolyte),

$$NaCl(aq) \rightarrow Na^+(aq) + Cl^-(aq)$$

and so

$$[Cl^-] = 0.50\ M + 2s$$

A table here would help summarize the changes in concentrations:

Concentration	$PbCl_2(s) \rightleftharpoons$	$Pb^{2+}(aq)$ +	$2Cl^-(aq)$
Initial (M)		0	0.50
Change (M)		s	$2s$
Equilibrium (M)		s	$(0.50 + 2s)$

CALCULATION

Substituting into K_{sp}, we get

$$[Pb^{2+}][Cl^-]^2 = K_{sp}$$

$$(s)(0.50 + 2s)^2 = 2.4 \times 10^{-4}$$

If we assume $0.50 \gg 2s$, then solving the problem is greatly simplified:

$$(s)(0.50)^2 = 2.4 \times 10^{-4}$$

$$s = 9.6 \times 10^{-4} M$$

COMMENT

Note that the assumption $0.50 \gg 2s$ was valid; $0.50 \gg 1.9 \times 10^{-3}$ M.

EXAMPLE 17.7 Solubility of a Metal Hydroxide

What is the solubility of $Pb(OH)_2$ in a buffer solution of pH 8.0? In one of pH 9.0? Given $K_{sp} = 1.2 \times 10^{-15}$.

METHOD OF SOLUTION

Write the solubility equilibrium and the K_{sp} expression for $Pb(OH)_2$:

$$Pb(OH)_2(s) \rightleftharpoons Pb^{2+}(aq) + 2OH^-(aq)$$

$$K_{sp} = [Pb^{2+}][OH^-]^2 = 1.2 \times 10^{-15}$$

Let s equal the molar solubility of $Pb(OH)_2$. The concentration of Pb^{2+} will be equal to s. The concentration of OH^- ion will not be $2[Pb^{2+}]$ because in a buffer solution, the H^+ ion and OH^- ion concentrations are maintained constant by the buffer. At pH 8.0, the pOH is 6.0, and $[OH^-] = 1.0 \times 10^{-6}$ M. In the buffer solution, the $[OH^-]$ ion concentration is *maintained constant* at 1.0×10^{-6} M. The OH^- from dissolution of $Pb(OH)_2$ is neutralized by the buffer. Therefore we can substitute as follows to obtain the solubility at pH 8.

CALCULATION

$$[Pb^{2+}] = s$$

$$[Pb^{2+}] = \frac{K_{sp}}{[OH^-]^2}$$

$$s = \frac{1.2 \times 10^{-15}}{(1.0 \times 10^{-6})^2} = 1.2 \times 10^{-3} M$$

And at pH 9, where $[OH^-] = 1.0 \times 10^{-5}$ M,

$$s = \frac{1.2 \times 10^{-15}}{(1.0 \times 10^{-5})^2} = 1.2 \times 10^{-5} M$$

COMMENT

The lower solubility of $Pb(OH)_2$ at pH 9.0 than at pH 8.0 is due to the common ion effect.

EXAMPLE 17.8 Effect of pH on Solubility

Which of the following salts will be more soluble at acidic pH than in pure water?
a. AgBr
b. $Ba(OH)_2$
c. $MgCO_3$

METHOD OF SOLUTION

First we need to identify the ions present in solutions of these compounds and then identify those with acid-base properties.

a. $AgBr(s) \rightleftharpoons Ag^+ + Br^-$; neither Ag^+ nor Br^- has acid-base properties, and so the solubility of AgBr is not affected by pH.
b. $Ba(OH)_2 \rightleftharpoons Ba^{2+} + 2OH^-$.
   Comparing equilibria in water and in acidic solution, as $[H^+]$ increases, $[OH^-]$ decreases, and the solubility equilibrium will shift to the right, causing the solubility of $Ba(OH)_2$ to increase.
c. Carbonate ion is a base, and in acid solution it combines with $H^+(aq)$:

$$MgCO_3 \rightleftharpoons Mg^{2+} + CO_3^{2-}$$

$$H^+ + CO_3^{2-} \rightarrow HCO_3^-$$

With the increase in $[H^+]$ in acid solution, $[CO_3^{2-}]$ decreases, and the solubility equilibrium shifts to the right, causing more $MgCO_3$ to dissolve.
Answer: b and c.

COMPLEX IONS AND SOLUBILITY

STUDY OBJECTIVES

You should be able to:
1. Write the formation constant expression for a given complex ion.
2. Explain the effect of complex ion formation on the solubility of a salt whose metal ion forms a complex ion in solution.
3. Calculate the solubility of a salt in a solution where the metal ion of the salt forms a complex ion.

The Formation Constant. Many transition metal ions act as Lewis acids in aqueous solution by accepting electron pairs from Lewis bases. For instance, a Ag^+ ion reacts with two CN^- ions,

$$Ag^+(aq) + 2CN^-(aq) \rightleftharpoons Ag(CN)_2^-(aq)$$

to form the complex ion $Ag(CN)_2^-$. An ion made up of the metal ion bonded to one or more molecules or ions is called a *complex ion*. Some examples are $Ag(NH_3)_2^+$, $Ni(H_2O)_6^{2+}$, and $Cu(CN)_4^{2-}$.

A number of complex ions are extremely stable and so have an important effect on whatever chemical species exist in solution. A measure of the tendency of a particular metal ion to form a certain complex ion is given by its formation constant, K_f. The formation constant is the equilibrium constant for the reaction that forms the complex ion. For example,

$$Ag^+ + 2CN^- \rightleftharpoons Ag(CN)_2^-$$

$$K_f = \frac{[Ag(CN)_2^-]}{[Ag^+][CN^-]^2} = 1.0 \times 10^{21}$$

Table 17.3 in the text lists K_f values for several selected complex ions. The more stable the complex ion, the greater the extent of reaction, and the greater the value of K_f.

The formation of a complex ion has a strong effect on the solubility of a metal salt. For example, AgI, which has a very low solubility in water, will dissolve in an aqueous solution of NaCN. The stepwise process is

$$\begin{array}{ll} AgI(s) \rightleftharpoons Ag^+(aq) + I^-(aq) & K_{sp} \\ Ag^+(aq) + 2CN^-(aq) \rightleftharpoons Ag(CN)_2^-(aq) & K_f \\ \hline \text{overall} \quad AgI(s) + 2CN^-(aq) \rightleftharpoons Ag(CN)_2^-(aq) + I^-(aq) & K \end{array}$$

The formation of the complex ion in the second step causes a decrease in [Ag$^+$]. Therefore the first equilibrium shifts to the right according to Le Chatelier's principle. This is why AgI is more soluble in CN$^-$ solution than in pure water. The net equation shows the overall reaction.

The equilibrium constant for the overall reaction is the product of the equilibrium constants for the two steps:

$$K = K_{sp} \times K_f$$

Tables 17.1 and 17.3 of the text give, respectively,

$$K_{sp} = 8.3 \times 10^{-17}$$

$$K_f = 1.0 \times 10^{21}$$

Therefore

$$K = 8.3 \times 10^4$$

The large value of K shows that AgI is very soluble in NaCN solution.

EXAMPLE 17.9 Ratio of Free Metal Ion to Complexed Metal Ion

Calculate the ratio of concentrations of free Ag$^+$ ions to Ag(CN)$_2^-$ complex ions in a NaCN solution in which [CN$^-$] = 0.010 M.

METHOD OF SOLUTION

The relative concentrations of Ag$^+$ and Ag(CN)$_2^-$ ions are controlled by the formation constant for the complex ion. Write the formation reaction and the formation constant expression for Ag(CN)$_2^-$:

$$Ag^+ + 2CN^- \rightleftharpoons Ag(CN)_2^-$$

$$K_f = \frac{[Ag(CN)_2^-]}{[Ag^+][CN^-]^2}$$

Obtain the K_f value from Table 17.3 of the text:

$$K_f = 1.0 \times 10^{21}$$

The ratio [Ag$^+$]/[Ag(CN)$_2^-$] can be obtained by rearranging the K_f expression:

$$\frac{[Ag^+]}{[Ag(CN)_2^-]} = \frac{1}{K_f[CN^-]^2} = \frac{1}{1.0 \times 10^{21}(0.01)^2}$$

$$\text{ratio} = 1.0 \times 10^{-17}$$

COMMENT

The extremely small ratio of uncomplexed to complexed silver ions shows that essentially all Ag^+ ions are complexed.

EXAMPLE 17.10 Concentration of Uncomplexed Metal Ion

Calculate the concentration of the free Ag^+ ions in a solution formed by adding 0.20 mol of $AgNO_3$ to 1 L of 1.0 M CaCN.

METHOD OF SOLUTION

In this solution Ag^+ ions will complex with CN^- ions, and the concentration of Ag^+ will be determined by the equation

$$Ag^+ + 2CN^- \rightleftharpoons Ag(CN)_2^- \qquad K_f = 1.0 \times 10^{21}$$

Since K_f is so large, we expect the Ag^+ to react essentially quantitatively to form 0.20 mol of $Ag(CN)_2^-$. To find the concentration of free Ag^+ at equilibrium, use the equilibrium constant expression

$$K_f = \frac{[Ag(CN)_2^-]}{[Ag^+][CN^-]^2}$$

Then

$$[Ag^+] = \frac{[Ag(CN)_2^-]}{K_f [CN^-]^2}$$

CALCULATION

Substitute any known equilibrium concentrations into the above equation. At equilibrium, $[Ag(CN)_2^-] = 0.20\ M$, and

$$[CN^-] = [CN^-]_0 - 2[Ag(CN)_2^-]$$

$$= 1.0\ M - 0.40\ M = 0.60\ M$$

Therefore

$$[Ag^+] = \frac{(0.20)}{(1.0 \times 10^{21})(0.60)^2}$$

$$= 5 \times 10^{-22}\ M$$

COMMENT

This concentration corresponds to only three Ag^+ ions per 100 mL!

EXAMPLE 17.11 Effect of Complex Formation on Solubility

Calculate the molar solubility of silver bromide in 6.0 M NH_3.

METHOD OF SOLUTION

The solubility of AgBr is determined by two equilibria. The solubility and complex ion equilibria are

$$AgBr(s) \rightleftharpoons Ag^+ + Br^- \qquad K_{sp} = 7.7 \times 10^{-13}$$
$$Ag^+ + 2NH_3 \rightleftharpoons Ag(NH_3)_2^+ \qquad K_f = 1.5 \times 10^7$$

overall $\quad AgBr(s) + 2NH_3 \rightleftharpoons Ag(NH_3)_2^+ + Br^- \qquad K = K_{sp} \times K_f$

The equilibrium constant of this net reaction controls the solubility of AgBr:

$$K = K_{sp} \times K_f = 1.2 \times 10^{-5}$$

$$= \frac{[Ag(NH_3)_2^+][Br^-]}{[NH_3]^2}$$

CALCULATION

Let s equal the solubility of AgBr in 6.0 M NH$_3$:

Concentration	AgBr(s) + 2NH$_3$	$\rightleftharpoons$ Ag(NH$_3$)$_2^+$	+ Br$^-$
Initial (M)		6.0	0	0
Change (M)		$-2s$	$+s$	$+s$
Equilibrium (M)		$(6.0 - 2s)$	s	s

Substitute into the K expression for the overall reaction:

$$K = \frac{(s)(s)}{(6.0 - 2s)^2} = 1.2 \times 10^{-5}$$

The left-hand side is a perfect square. Taking the square root of both sides gives

$$\sqrt{\frac{s^2}{(6.0 - 2s)^2}} = \sqrt{(1.2 \times 10^{-5})}$$

$$\frac{s}{6.0 - 2s} = 3.5 \times 10^{-3}$$

$$s = 2.1 \times 10^{-2} \, M$$

COMMENT

The solubility of AgBr in pure water is 8.8×10^{-7} M. The enhanced solubility in this example is due to the formation of the complex ion Ag(NH$_3$)$_2^+$.

TRUE-FALSE QUESTIONS

1. The solubility product constant expression for Ag$_3$PO$_4$ is $K_{sp} = [3 \, Ag^+]^3[PO_4^{3-}]$.

2. For CaSO$_4$ $K_{sp} = 2.4 \times 10^{-5}$; therefore the solubility of CaSO$_4$ is 2.4×10^{-5} mol/L.

3. For a AgCl solution, when the ion product $[Ag^+][Cl^-] < K_{sp}$, the solution is not saturated.

4. In a solution of CaF$_2$, $2[F^-] = [Ca^{2+}]$.

5. According to Table 17.1, the solubility of AgI is greater than that of AgBr.

6. The solubility of AgI will be greater in a solution containing $AgNO_3$ than it is in pure water.

7. The solubility of $Pb(OH)_2$ in water will be less at pH 7 than at pH 4.

8. The solubility of AgBr is less at pH 7 than at pH 4.

9. In general, the larger the value of K_f, the formation constant, the more stable the complex.

10. The solubility of sulfide compounds in H_2S solution is unaffected by pH.

SELF-TEST A

1. Which of the following is the *least* soluble?
   a. SrF_2 $K_{sp} = 2.8 \times 10^{-9}$
   b. $Zn(OH)_2$ $K_{sp} = 1.8 \times 10^{-14}$
   c. PbI_2 $K_{sp} = 1.4 \times 10^{-8}$
   d. BaF_2 $K_{sp} = 1.7 \times 10^{-6}$

2. At a certain temperature the solubility of barium chromate ($BaCrO_4$) is 1.8×10^{-5} mol/L. What is the K_{sp} value at this temperature?

3. If the solubility of $Fe(OH)_2$ in water is 7.7×10^{-6} mol/L at a certain temperature, what is its K_{sp} value at that temperature?

4. What is the molar solubility of silver phosphate (Ag_3PO_4) in water? $K_{sp} = 1.8 \times 10^{-18}$.

5. AgCl is most soluble in:
   a. 0.1 M $AgNO_3$
   b. 0.2 M NaCl
   c. 2.0 M HCl
   d. Pure water

6. Will a precipitate of $BaSO_4$ form when 400 mL of 0.020 M Na_2SO_4 is added to 700 mL of 0.001 M $BaCl_2$? $K_{sp}(BaSO_4) = 1.1 \times 10^{-10}$.

7. Some municipal water supplies contain Ca^{2+} at a concentration of 3×10^{-3} M. Will a precipitate of $CaSO_4$ form when 0.10 mol of Na_2SO_4 is dissolved in 0.50 L of this water? $K_{sp}(CaSO_4) = 2.4 \times 10^{-5}$.

8. Will a precipitate of MgF_2 form when 600 mL of solution that is 2.0×10^{-4} M in $MgCl_2$ is added to 300 mL of 1.1×10^{-2} M NaF solution? $K_{sp}(MgF_2) = 6.6 \times 10^{-9}$.

9. A solution of NaOH is added dropwise to a solution that is 0.010 M Mn^{2+} and 0.010 M Mg^{2+}.
   a. Which cation precipitates first?
   b. What concentration of OH^- is necessary to begin precipitation?
   c. What is the concentration of the "first cation to precipitate" when the second cation begins to precipitate?
   d. What percentage of the "first cation to precipitate" remains in solution when the second cation begins to precipitate?
   Given $K_{sp}[Mn(OH)_2] = 1.6 \times 10^{-13}$.

10. Which of the following salts should be more soluble in acid solution than in pure water?
    a. $CaCO_3$ b. $MgCl_2$ c. $NaNO_3$ d. LiBr   e. KI

11. Which of the following is the most stable complex ion?
    a. $Zn(NH_2CH_2CH_2NH_2)_2^{2+}$ $K_f = 2.3 \times 10^{10}$
    b. $Ag(NH_3)_2^+$ $K_f = 1.5 \times 10^7$
    c. $Ag(CN)_2^-$ $K_f = 1.0 \times 10^{21}$

12. Calculate the concentration of the free copper ion in a solution made from 1×10^{-2} M $Cu(NO_3)_2$ and 1.0 M NH_3.

13. What is the molar solubility of AgCl in 3.0 M NH_3?

14. What is the molar solubility of AgBr in 0.100 M $Na_2S_2O_3$? Given:

$$Ag^+ + 2S_2O_3^{2-} \rightleftharpoons Ag(S_2O_3)_2^{3-} \qquad K_f = 1.0 \times 10^{13}$$

15. a. Will a precipitate form when 400 mL of 0.20 M K_2CrO_4 is added to 400 mL of 0.10 M $AgNO_3$? $K_{sp}(Ag_2CrO_4) = 1.1 \times 10^{-12}$.
    b. If yes, what is the silver ion concentration left in solution?

16. What is the minimum concentration of aqueous NH_3 required to prevent $AgCl(s)$ from precipitating from 1.0 L of solution prepared from 0.20 mol of $AgNO_3$ and 0.010 mol of NaCl?

SELF-TEST B1

1. If K_{sp} for CH_3COOAg (silver acetate) is 4.0×10^{-3} and K_a for CH_3COOH is 1.8×10^{-5}, calculate the equilibrium constant value for the net reaction

$$CH_3COOAg(s) + H^+(aq) \rightleftharpoons Ag^+(aq) + CH_3COOH(aq)$$

2. If 1.0 mL of 1.0×10^{-3} M $Ba(NO_3)_2$ is added to 99.0 mL of 1×10^{-4} M Na_2CO_3, will $BaCO_3$ precipitate from this solution? $K_{sp}(BaCO_3) = 8.1 \times 10^{-9}$.

3. The solubility of $PbBr_2$ is 0.392 g per 100 mL at 20°C. What is the K_{sp} value for $PbBr_2$?

4. What is the maximum possible concentration of Mg^{2+} ion in a basic solution with pH 9.0? $K_{sp}[Mg(OH)_2] = 1.2 \times 10^{-11}$.

5. What is the molar solubility of Ag_3PO_4 in 0.20 M $AgNO_3$? $K_{sp} = 1.8 \times 10^{-18}$.

SELF-TEST B2

1. The equilibrium constant value for the net reaction

$$CH_3COOAg(s) + H^+(aq) \rightleftharpoons Ag^+(aq) + CH_3COOH(aq)$$

is $K = 220$. Given that K_a for CH_3COOH is 1.8×10^{-5}, calculate the K_{sp} value for CH_3COOAg (silver acetate).

2. Solid $Ba(NO_3)_2$ is slowly dissolved in a solution containing 1.0×10^{-4} M Na_2CO_3. At what Ba^{2+} concentration will a precipitate of $BaCO_3$ just begin to form?

3. For $PbBr_2$, $K_{sp} = 4.9 \times 10^{-5}$. What is the solubility of $PbBr_2$ in moles per liter?

4. Calculate the maximum pH that will just prevent precipitation of $Mg(OH)_2$ from an aqueous solution containing 0.12 M Mg^{2+}.

5. What is the molar solubility of Ag_3PO_4 in 0.20 M Na_3PO_4?

ANSWERS

TRUE-FALSE QUESTIONS

1. False. $K_{sp} = [Ag^+]^3[PO_4^{3-}]$.
2. False. The solubility is $s = \sqrt{K_{sp}}$.
3. True.
4. False. $[F^-] = 2[Ca^{2+}]$.
5. False. The greater K_{sp} value for AgBr indicates it is more soluble.
6. False. Due to the common ion effect, the solubility of AgI will be less in $AgNO_3$ solution than in water.
7. True.
8. False. The solubility of AgBr should not be affected by pH.
9. True.
10. False. The solubilities of sulfides are strongly affected by pH.

SELF-TEST A

1. b
2. $K_{sp} = 3.2 \times 10^{-10}$
3. $K_{sp} = 1.8 \times 10^{-15}$
4. $s = 1.6 \times 10^{-5} M$
5. d
6. Yes
7. Yes
8. No
9. a. Mn^{2+}
   b. $[OH^-] = 4.0 \times 10^{-6} M$
   c. $[Mn^{2+}] = 1.3 \times 10^{-4} M$
   d. 1.3% of Mn^{2+} is unprecipitated
10. a
11. c
12. $[Cu^{2+}] = 2.35 \times 10^{-16} M$
13. $s = 0.13 M$
14. $s = 0.049 M$
15. a. Yes b. $[Ag^+] = 3.8 \times 10^{-6} M$
16. $[NH_3] = 0.91 M$

SELF-TEST B1

1. $K = 220$
2. No
3. $K_{sp} = 4.9 \times 10^{-6}$
4. $0.12\ M\ Mg^{2+}$
5. $s = 2.2 \times 10^{-16} M$

SELF-TEST B2

1. $K_{sp} = 4 \times 10^{-3}$
2. $8.1 \times 10^{-5} M$
3. $s = 1.1 \times 10^{-2} M$
4. pH = 9.0
5. $s = 2.1 \times 10^{-6} M$

Chapter Eighteen
CHEMISTRY IN THE ATMOSPHERE

- Earth's Atmosphere and Its Regions
- Depletion of the Ozone Layer
- The Greenhouse Effect and Acidic Precipitation
- Photochemical Smog
- Indoor Air Pollution

EARTH'S ATMOSPHERE AND ITS REGIONS

STUDY OBJECTIVES

You should be able to:
1. List the major components and several of the minor components of the atmosphere.
2. Describe the approximate altitude and characteristics of four regions of Earth's atmosphere.
3. Describe the chemical processes responsible for light emission by the aurora borealis and aurora australis.

Earth's Atmosphere. Earth's atmosphere is unique in the solar system due to its high oxygen (O_2) content. Molecular oxygen makes up 21% of the atmosphere by volume. Table 18.1 in the text shows that N_2, O_2, and Ar make up over 99.9% of the atmosphere by volume. The other noble gas elements (besides argon) are only present in trace amounts. Associated with the high oxygen content is the formation of ozone (O_3), a molecule that absorbs ultraviolet light and filters it from sunlight. This radiation is known to be harmful to many of the life forms that inhabit Earth's surface.

Nitrogen is the major constituent of the atmosphere. Nitrogen molecules are held together by a triple bond and therefore are extremely stable. Biological and industrial nitrogen fixation converts the elemental form of nitrogen into nitrogen compounds. Lightning plays an important role by initiating the production of about 30 million tons of nitric acid each year. There are three steps in the mechanism of nitric acid production. Lightning provides the energy to break up the N_2 molecule and initiates the first step in the reaction sequence:

$$N_2(g) + O_2(g) \rightarrow 2NO(g)$$

$$2NO(g) + O_2(g) \rightarrow 2NO_2(g)$$

$$2NO_2(g) + H_2O(l) \rightarrow HNO_3(g) + HNO_2(g)$$

The nitrogen in nitric acid is an essential plant nutrient.

Carbon dioxide, the fourth most abundant component, makes up only 0.033% of the air. Carbon dioxide is extremely important to Earth's heat balance even though its relative concentration is low. In this chapter we will review the effects of air pollutants on ozone, on Earth's temperature, and on other phenomena related to the atmosphere.

Regions of the Atmosphere. For convenience the atmosphere is divided into four regions with respect to altitude. The lowest region is the *troposphere*. This region extends up to about 10 km. The temperature decreases with altitude in the troposphere from about 25°C at sea level to −55°C at 10 km. The troposphere contains about 80% of the mass of the atmosphere and virtually all the precipitation, clouds, and water vapor. All the weather occurs in the troposphere. Remember that the atmosphere gradually thins out as the altitude increases. The atmospheric pressure at 10 km is about 230 mm Hg.

Above 10 km we enter the *stratosphere*, which extends upward to 50 km. In the stratosphere the temperature increases with altitude from about −55°C at 10 km to −10°C at 50 km. The atmospheric pressure at 50 km is only 0.5 mm Hg. For the most part there is very little mixing of the contents of the troposphere and the stratosphere.

Above the stratosphere we enter the *mesophere*. As in the troposphere, the temperature decreases with altitude, falling to −90°C at 80 km. From 80 km to 100 km we find the *ionosphere* (also called the thermosphere). In this region ions are produced by high energy solar radiation. In this region, the temperature increases slightly with increasing altitude. The temperatures associated with the different regions of the atmosphere are shown in Figure 13.27 of the text.

Auroral Displays. Chemical processes in the ionosphere produce the phenomena called "northern lights," or aurora borealis, in the Northern Hemisphere and aurora australis in the Southern Hemisphere. In Chapter 7 you learned that excited atoms emit light when electrons jump from a higher to a lower energy state. The text explains that excited O atoms and N_2^+ ions are produced in the ionosphere when protons from solar flares are ejected from the sun and eventually collide with molecules in the ionosphere:

$$N_2 + \text{high energy proton} \rightarrow N_2^{+*} + e^- + \text{the proton}$$

$$O_2 + \text{high energy proton} \rightarrow 2O^* + 2e^- + \text{the proton}$$

The green and red colors seen in the auroral displays are emitted by O* atoms, and the blue and violet colors are emitted by N_2^{+*} ions. The symbols contain an asterisk to indicate excess energy.

During the emission process excited atoms (denoted by an asterisk) drop to the ground state and emit photons ($h\nu$):

$$O^* \rightarrow O + h\nu$$

$$N_2^{+*} \rightarrow O + h\nu$$

The auroral displays occur near Earth's poles because solar protons are attracted there by Earth's magnetic field.

DEPLETION OF THE OZONE LAYER

STUDY OBJECTIVES

You should be able to:
1. Write the steps in the mechanism of ozone formation.
2. Write chemical equations describing the destruction of ozone in the stratosphere.

The Ozone Layer. Ozone is a gas that is present in the atmosphere in a layer or shell that extends around the entire Earth called the "ozone layer." This layer is centered in the stratosphere about 25 km above Earth's surface. Ozone molecules have the ability to absorb some of the ultraviolet (UV) light present in sunlight. It is in this way that the ozone layer protects life on Earth from some of the harmful effects of sunlight. Ultraviolet light causes skin cancer and eye damage in humans and mutations in plants.

The term *ozone layer* is misleading because it may imply that the layer is pure ozone. Actually the concentration of ozone is extremely low. Ozone is spread thinly at a concentration of about 10 ppm at elevations between 15 and 35 km. If all the ozone in the atmosphere was collected into a layer of pure ozone at standard temperature and pressure (STP), it would only be 3 mm thick! The low concentration and unstable nature of ozone combine to make the ozone layer a fragile part of the atmosphere.

Ozone is formed in the ozone layer by a two-step mechanism initiated by the action of UV light on oxygen molecules. In the first step, UV light dissociates molecular oxygen into oxygen atoms. Then an oxygen atom combines with an oxygen molecule, forming ozone:

$$O_2(g) + h\nu \rightarrow 2O(g)$$
$$\underline{O(g) + O_2(g) \rightarrow O_3(g)} \quad \text{(occurs twice)}$$
$$\text{net } 3O_2(g) \rightarrow 2O_3(g)$$

Ozone, being an unstable molecule, tends to undergo reactions that convert it to the more stable O_2 molecule. Ozone absorbs UV light with wavelengths in the range from 200 to 300 nm. When ultraviolet light is absorbed by ozone, the molecule splits into atomic oxygen and O_2:

$$O_3(g) + h\nu \rightarrow O_2(g) + O(g)$$

Atomic oxygen can also destroy ozone:

$$O(g) + O_3(g) \rightarrow 2O_2(g)$$

Depletion of Ozone. The concentration of ozone in the stratosphere depends on a balance between the rate of its production from molecular oxygen and its rate of destruction:

$$3O_2(g) \rightarrow 2O_3(g) \quad \text{production of ozone}$$

$$2O_3(g) \rightarrow 3O_2(g) \quad \text{destruction of ozone}$$

Ozone is said to be in a steady state. This refers to the situation where the rate of ozone production is approximately equal to its rate of destruction. Any process that speeds up the destruction of ozone will cause its stratospheric concentration to decrease.

The depletion of stratospheric ozone has been associated with man-made substances called chlorofluorocarbons, or CFCs for short. The CFCs were developed in the 1930s by chemists who were searching for a new refrigerant to replace blocks of ice and the toxic and corrosive ammonia and sulfur dioxide then in commercial use. Fluorocarbon-12 (CF_2Cl_2) was their answer, and it quickly found application in refrigerators, freezers, and air conditioners. In the 1950s, fluorocarbon-11 ($CFCl_3$) became a basic propellant for the new aerosol can industry, fostering the development of the spray can. Soon, it was also developed as a blowing agent for styrofoam in insulation, cushions, plastic furniture, and sealants. While there are a number of other fluorocarbon compounds, fluorocarbon-11 and fluorocarbon-12 find the widest applications. The advantage of CFCs was that they were nontoxic, noncorrosive, and nonreactive.

There are no known chemical reactions in the troposphere that destroy CFCs. Because of their inertness, CFCs accumulate in the troposphere and gradually diffuse up to the stratosphere. Since ozone filters out UV light, there is more UV light in the ozone layer than below it. In the ozone layer CFCs are bombarded by high energy UV radiation and break apart. This photon-induced dissociation of fluorocarbon molecules (called photolysis) breaks a C — Cl bond and releases a chlorine atom. In the case of chlorofluorocarbon-12,

$$CCl_2F_2 + h\nu \rightarrow CClF_2 + Cl$$

The Cl atom attacks an ozone molecule, converting it to the more stable oxygen molecule and forming an intermediate molecule, chlorine monoxide. Compounding the problem is that once a chlorine atom has destroyed an ozone molecule, then it is regenerated when the intermediate ClO reacts with an O atom. Recall that O atoms are prevalent in the stratosphere:

$$Cl(g) + O_3(g) \rightarrow ClO(g) + O_2(g)$$
$$ClO(g) + O(g) \rightarrow O_2(g) + Cl(g)$$
$$\text{net} \quad \overline{O(g) + O_3(g) \rightarrow 2O_2(g)}$$

Since Cl atoms accelerate the destruction of ozone and are regenerated, they are true catalysts. An increase in the destruction of ozone has the expected effect of lowering ozone concentration in the atmosphere.

The net result of ozone depletion is that UV light, which is usually absorbed by ozone, is no longer sufficiently filtered from sunlight. More UV light can proceed through the atmosphere to Earth's surface. The National Research Council predicts 10,000 additional cases of skin cancer in the United States for each 1% depletion in the ozone layer. This problem will be around for a long time because currently the troposphere is a giant reservoir of CFCs. Estimates are that there are already enough CFCs in the troposphere to cause depletion of ozone for the next 100 years!

Currently intense efforts are underway to develop CFC substitutes. The idea is to maintain the noncorrosive properties but make a molecule that is a little more reactive. HCFC-123 ($C_2HCl_2F_3$), where the H represents hydrogen, is a promising possibility. Introduction of a hydrogen atom into CFC molecules makes them more reactive. Chemists are synthesizing a number of HCFC molecules to try as the first generation of substitutes. The goal is to find molecules that are more easily oxidized in the troposphere, before they reach the stratosphere.

THE GREENHOUSE EFFECT AND ACIDIC PRECIPITATION

STUDY OBJECTIVES

You should be able to:

1. Explain how some gases produce a greenhouse effect.
2. Describe the likely chemical events by which sulfur dioxide leads to acid rain.

The Greenhouse Effect. Each year that passes sees another 9 gigatons (Gtons) of carbon dioxide added to Earth's atmosphere. This is only about 1/1,000,000 of the total weight of the atmosphere. Though relatively small in amount, carbon dioxide has an important effect on Earth's average temperature. Carbon dioxide, water, and a few other gases such as methane are called "greenhouse gases." This means that they act like the glass in a greenhouse. They let sunlight through but do not let heat out. The main components of the air, O_2 and N_2, are not greenhouse gases.

"Greenhouse warming" occurs in the following way. Visible light from the sun passes directly through the main components of the atmosphere: N_2, O_2, CO_2, and H_2O vapor. This radiant energy is absorbed by Earth's oceans and land areas and warms the surface. All warm objects radiate infrared radiation, called IR. The oceans and land areas emit IR, which passes directly through N_2 and O_2 into outer space. The loss of this energy would mean that Earth would be a cooler place. However, $CO_2(g)$ and $H_2O(g)$ absorb IR and in effect slow down the loss of heat from Earth's surface. Thus Earth's surface is many degrees warmer than it would be without $CO_2(g)$ and $H_2O(g)$ in the air. This is the so-called greenhouse effect. Some carbon dioxide has always been present in the atmosphere, and some warming is considered desirable. Since atmospheric carbon dioxide is increasing at the rate of 1 ppm per year, this is expected to lead to an "enhanced greenhouse effect."

Greenhouse gases reduce the heat loss to outer space by absorbing IR. Infrared radiation increases the vibrational energy level of molecules that absorb it. Then these vibrationally excited molecules reemit IR as they drop to the ground state. Statistically half of the IR is aimed downward toward Earth's surface and half is aimed upward toward outer space. The IR aimed back toward the surface eventually causes the temperature of the air to rise.

Scientists have been trying for many years to estimate rates of global warming and its effects. The carbon dioxide content of the atmosphere is expected to double by the year 2050. Some of the predicted effects are:

1. A temperature increase of 3°C to 5°C by that year
2. Melting of glaciers and icecaps that will cause a 2 foot rise in sea level with its accompanying flooding of sea coasts and major cities
3. Widespread changes in climate

Suggestions to lower carbon dioxide emissions center around less dependence on fossil fuels for energy. Carbon dioxide emissions could be reduced by increasing efficiency of automobiles, home heating, and electric power production. Additional reduction could result from replacing existing fossil fuel electric power plants with ones that utilize solar energy and nuclear energy.

Methane, nitrous oxide, and CFCs are also strong absorbers of IR. These gases are present in only trace amounts, but they enhance the greenhouse effect significantly because they absorb IR at wavelengths that water and CO_2 cannot absorb. In a sense they close an open window that would allow some IR to escape the greenhouse.

Acid Rain. The term *acid rain* was coined in 1872 by an English chemist who used it to describe the increasingly acid precipitation that fell on the industrial city of Manchester. In the century since then, acid rain has grown to be an environmental problem of global proportions. Acid rain is rainwater with a pH of less than 5.5. Precipitation in the northeastern part of the United States has an average pH of 4.3 and in some specific storms the pH has been as low as 2.8.

The text points out that "normal" rain is slightly acid because it contains carbonic acid. As rain passes through air containing carbon dioxide, the CO_2 dissolves and forms carbonic acid. Carbon dioxide is an acid anhydride:

$$CO_2(g) + H_2O(l) \rightarrow H_2CO_3(aq)$$

Because carbonic acid is a weak acid, the pH of rain normally will not go below 5.5. The presence of several stronger acids accounts for the lower pH values of acid rain. Acid rain usually contains sulfuric acid (H_2SO_4), sulfurous acid (H_2SO_3), and nitric acid (HNO_3). We will discuss only the sulfur oxoacids.

These acids do not start out in the atmosphere as oxoacids. The text points out that the precursors of the oxoacids are the acidic oxides, SO_2 and SO_3. Sulfur dioxide is produced by the combustion of fossil fuels such as coal and from the processing of sulfide ores at smelters:

$$2ZnS(s) + 3O_2(g) \rightarrow 2ZnO(s) + 2SO_2(g) \quad \text{smelting an ore}$$

Coal and petroleum contain between 1% and 5% S. Combustion of sulfur in fossil fuels yields SO_2:

$$S(s) + O_2(g) \rightarrow SO_2(g) \quad \text{combustion of sulfur}$$

Once in the atmosphere some of the sulfur dioxide is oxidized to sulfur trioxide by reactions that are currently being intensely studied. The effects of light and ozone on this oxidation are mentioned in the text. The net oxidation reaction is

$$2SO_2(g) + O_2(g) \rightarrow 2SO_3(g)$$

In the final step, sulfur dioxide and sulfur trioxide combine with water to form sulfurous acid and sulfuric acid, respectively. Like CO_2, SO_2 and SO_3 are acid anhydrides:

$$SO_2(g) + H_2O(l) \rightarrow H_2SO_3(aq)$$

$$SO_3(g) + H_2O(l) \rightarrow H_2SO_4(aq)$$

Streams and lakes show the most dramatic effects of acid rain. It is known that natural waters with a low pH can kill fish eggs, salamander eggs, and frog eggs. The extent of change in acidity of a lake or stream when under the stress of acid rain is determined mainly by the buffering capacity of the surrounding soil. Watershed soils containing limestone or bicarbonate are alkaline and can resist rapid changes in pH, making them less susceptible to harm. The bedrock in the Rocky Mountains has very little limestone in comparison to the eastern United States. This makes the western states much more susceptible to acid rain than the eastern United States.

Trees are also susceptible to acid rain. Acid rainfall causes damage to leaves and the growing tissues of trees. The needles of firs, spruces, and pines turn yellow and fall off. Among the possible causes are acid rain and ozone from polluted air. These pollutants damage the cell membranes of needles, allowing nutrients to escape.

Acid precipitation can acidify the soil, interfering with nutrient availability. To grow normally trees require adequate supplies of 16 elements. Several of these, notably Ca, Mg, and K, are taken up by tree roots as cations from aqueous solution in the soil. When sulfuric acid is deposited in soil by precipitation, nutrient cations can be leached from the root zone. This can greatly affect the health of forests.

Sulfur dioxide emissions can be reduced by removing SO_2 after combustion but before it leaves the stack and is released into the atmosphere. Powdered limestone ($CaCO_3$) is injected into the hot gases leaving the combustion zone. Heat causes the carbonate to decompose into quicklime (CaO). The quicklime then reacts with SO_2 to form calcium sulfite ($CaSO_3$):

$$CaCO_3(s) \rightarrow CaO(s) + CO_2(g)$$

$$CaO(s) + SO_2(g) \rightarrow CaSO_3(s)$$

About half of the SO_2 is removed by contact with the dry CaO. The remaining SO_2 must be removed by spraying the hot gases with a suspension of quicklime. This process, called "scrubbing," creates huge amounts of calcium sulfite to dispose of. Retrofitting of scrubbers onto established power plants is very expensive and significantly raises the cost of electricity.

PHOTOCHEMICAL SMOG

STUDY OBJECTIVES

You should be able to:
1. Distinguish between primary and secondary pollutants
2. Describe how ozone is formed in polluted air.

Photochemical smog is formed by the reactions of automobile exhaust in the presence of sunlight. Reactions initiated by photons are called photochemical reactions. The text points out that smog begins with certain primary pollutants. These substances may or may not be objectionable by themselves. These are transformed by sunlight or by ordinary chemical reactions into secondary pollutants. Secondary pollutants are involved directly in the buildup of smog. Nitric oxide is a good example of a primary pollutant. It is formed at the high temperatures inside an internal combustion engine when nitrogen and oxygen from air react. Nitric oxide does not build up in the air because it is rapidly converted to nitrogen dioxide.

$$N_2(g) + O_2(g) \rightarrow 2NO(g)$$

$$2NO(g) + O_2(g) \rightarrow 2NO_2(g)$$

Nitrogen dioxide is involved in a chain of reactions that produce ozone. First NO_2 is photochemically decomposed by sunlight:

$$NO_2(g) + h\nu \rightarrow NO(g) + O(g)$$

Oxygen atoms initiate a number of reactions in polluted air. An important one is the formation of ozone:

$$O(g) + O_2(g) \rightarrow O_3(g)$$

Exposure to 0.1 to 1.0 ppm of ozone produces headaches, burning eyes, and irritation to the respiratory passages. Another secondary pollutant is peroxyacetyl nitrate, better known as PAN. PAN literally brings tears to your eyes. It is an example of a lachrymator, a compound that causes burning of the eyes and tears. Onions, as you know, contain a lachrymator.

Unburned hydrocarbons in automobile exhaust also produce secondary air pollutants. The oxidation of unburned hydrocarbons produces various alcohols and organic acids. These can condense to produce an aerosol. Aerosols are liquid droplets dispersed in air. They are objectionable because they reduce visibility and make the air look hazy.

Efforts to control smog are usually focused on reducing the source of primary pollutants. The catalytic converters on automobiles are designed to remove NO, CO, and unburned hydrocarbons from automobile exhaust. The catalyst contains platinum and palladium metals. Nitrogen monoxide, a primary pollutant, is converted back to nitrogen and oxygen by a reaction whose rate is increased by the catalyst in the converter:

$$2NO(g) \rightarrow N_2(g) + O_2(g)$$

Carbon monoxide and unburned hydrocarbons are both oxidized to carbon dioxide by the catalytic converter, as shown by the following equations (C_5H_{12} represents a typical hydrocarbon):

$$2CO(g) + O_2(g) \rightarrow 2CO_2(g)$$

$$C_5H_{12}(g) + 8O_2(g) \rightarrow 5CO_2(g) + 6H_2O(l)$$

INDOOR AIR POLLUTION

STUDY OBJECTIVE

You should be able to:
1. Describe the source of radon and how it gets into buildings.
2. Describe why the main health effects of radon are to the lungs.

Radon Gas. All isotopes of radon are radioactive. Concern about radon in homes began in 1984 when a worker at a nuclear power plant in Pennsylvania found that he was setting off the plant's radiation monitor alarms in the morning upon *arriving* at the plant. The source of the radioactivity was traced to the worker's home, which measured an extremely high level of radioactivity in the basement air. The air in his home contained an isotope of radon. Radon-222 atoms undergo radioactive decay by emitting alpha particles. Radon is always associated with uranium deposits. Some uranium occurs naturally in most soils and rocks in widely varying amounts. Radon-222 has a half-life of 3.8 days. This gives it time to migrate up out of the ground and to enter buildings through cracks in foundations. It can also dissolve in ground water and enter a home via well water.

The element radon has been known since its discovery in 1900. As a member of the noble gas group of elements, radon is a colorless, odorless, and tasteless gas with very little tendency to combine with other elements. Therefore, it tends to stay in the air. Radon atoms in the air can be inhaled into the lungs. If radon atoms decay in the lungs, we get a dose of radiation. In addition, its decay products, particularly polonium-218 and polonium-214, are deposited in the lungs. These isotopes are radioactive and contribute more to the radiation dose than that received from radon itself. These alpha-emitting isotopes are of the greatest health concern because they are solids and become trapped in respiratory passages. Here their radioactivity can damage nearby cells, which over a long period of time leads to lung cancer. In order to better assess the danger to people, health workers are trying to sort out the importance of the three main causes of lung cancer: cigarette smoking, secondary smoke, and radon gas.

TRUE-FALSE QUESTIONS

1. N_2 molecules contain a triple bond and are very stable.

2. Nitrogen fixation refers to the conversion of nitrogen compounds into N_2 molecules.

3. O_2 molecules are produced by photosynthesis.

4. The troposphere is the thinnest layer of Earth's atmosphere and contains most of the mass of the atmosphere.

5. The ozone layer is in the ionosphere.

6. Excited atoms and ions are produced in the ionosphere by collisions of protons and electrons from the sun with N_2 and O_2 molecules.

7. The blue and violet colors observed in many auroras result from excited N_2^+ ions returning to the ground state.

8. The ozone layer is a thin layer of ozone surrounding the entire Earth that is 3 mm thick.

9. CFCs break down in the troposphere, producing chlorine atoms. These atoms drift up to the ozone layer and attack ozone.

10. The Montreal protocol is an international agreement to spy on countries still making CFCs.

11. HCFCs are the first generation of replacements for CFCs.

12. ClO, called chlorine monoxide, acts as an intermediate in the ozone destruction mechanism.

13. So-called greenhouse gases transmit infrared radiation and absorb visible light.

14. Like CO_2, N_2 and O_2 are also "greenhouse gases."

15. For rain to be considered acid rain, its pH must be below 7.0.

16. Both SO_2 and SO_3 are converted to acids by rainwater.

17. Fossil fuels contain sulfur, which when burned forms SO_2.

18. Ozone, nitrogen dioxide, and haze are primary pollutants.

19. The function of catalytic converters is to remove CO_2 and H_2O from automobile exhaust.

20. Radon gas is a compound.

21. Besides radon, its decay products polonium-218 and polonium-214 are radioactive.

22. The principal health effect of radon is that it causes heart and arterial disease.

SELF-TEST A

1. Name the regions of the atmosphere in order of increasing altitude.

2. a. Write out the steps in the mechanism by which ozone is formed in the stratosphere.
   b. What other naturally occurring molecule absorbs UV light in the stratosphere?

3. a. Write out the steps in the mechanism of ozone destruction by chlorine atoms.
   b. Identify the catalyst and the intermediate.

4. What does the abbreviation CFC stand for? Write the formulas of two CFCs.

5. Name two environmental problems related to use of CFCs.

6. Give two sources of methane in the atmosphere.

7. Name four "greenhouse gases" besides carbon dioxide.

8. By what means does Earth lose heat?

9. Write chemical equations that show what happens when acid rain reacts with iron and with limestone.

10. a. List three primary pollutants removed from automobile exhaust by catalytic converters.
    b. Explain briefly the role of each in the formation of smog.

11. The strength of the C — Cl bond in CF_2Cl_2 is 318 kJ/mol. What is the wavelength of a photon that has enough energy to break a C — Cl bond? Is this in the UV or visible region of the spectrum?

ANSWERS

TRUE-FALSE QUESTIONS

1. True.
2. False. It is the conversion of N_2 molecules into nitrogen compounds.
3. True.
4. True.
5. False. The ozone layer is in the stratosphere.
6. True.
7. True.
8. False. The ozone layer has a very low concentration of ozone. Also it is 20 km thick.
9. False. CFCs break up under the influence of UV light in the stratosphere. This produces Cl atoms right in the ozone layer.
10. False. The agreement is to phase out CFC production by the year 2000.
11. True.
12. True.
13. False. Greenhouse gases transmit visible light and absorb IR.
14. False. N_2 and O_2 do not absorb IR.
15. False. Its pH must be below 5.5.
16. True.
17. True.
18. False. They are secondary pollutants.
19. False. It is to remove CO, NO, and unburned hydrocarbons.
20. False. Radon is an element.
21. True.
22. False. Radon causes lung cancer.

SELF-TEST A

1. Troposphere, stratosphere, mesosphere, ionosphere
2. a. $O_2(g) + h\nu \rightarrow 2O(g)$ b. O_2
   $O(g) + O_2(g) \rightarrow O_3(g)$
3. a. $Cl(g) + O_3(g) \rightarrow ClO(g) + O_2(g)$
   $ClO(g) + O(g) \rightarrow O_2(g) + Cl(g)$
   b. The catalyst is atomic Cl, and the intermediate is chlorine monoxide, ClO.

4. CFC stands for chlorofluorocarbon. CF_2Cl_2 and $CFCl_3$.
5. Ozone depletion and global warming
6. Landfills and natural gas leaks
7. CH_4, H_2O, N_2O, CFCs
8. Emission of IR to space
9. $Fe(s) + 2H^+(aq) \rightarrow Fe^{2+}(aq) + H_2(g)$
   $CaCO_3(s) + 2H^+(aq) \rightarrow Ca^{2+}(aq) + CO_2(g) + H_2O(l)$
10. a. Carbon monoxide, nitric oxide, unburned hydrocarbons
    b. CO is toxic but may not be involved in reactions producing smog. NO is oxidized to NO_2. Photolysis of NO_2 leads to ozone and PAN. Oxidation of unburned hydrocarbons produces alcohols and organic acids that eventually form aerosols and haze.
11. The wavelength is 376 nm, which is in the UV region.

Chapter Nineteen
ENTROPY, FREE ENERGY, AND EQUILIBRIUM

- The Second Law of Thermodynamics
- Gibbs Free Energy
- Free Energy and Equilibrium

THE SECOND LAW OF THERMODYNAMICS

STUDY OBJECTIVES

You should be able to:
1. Explain the meaning of the term *spontaneous process*.
2. Predict, for a given process, whether entropy of the system increases or decreases.
3. State the second law of thermodynamics.
4. Calculate the standard entropy change for a given reaction using a table of standard absolute entropies.

Spontaneous Processes. A large and important part of experimental chemistry deals with spontaneous reactions, that is, reactions that take place "without outside influence." One goal of thermodynamics is to gain the ability to predict whether a reaction will take place when a set of given reactants are brought together. Here we want to know what property of a system can be used as a criterion for predicting spontaneous processes. The textbook points out that the sign of ΔH by itself is not an adequate guide to spontaneity because while some spontaneous reactions have been found to be exothermic (ΔH is negative), many others have been found to be endothermic (ΔH is positive).

It is also important to remember that the term *spontaneous* doesn't necessarily mean a fast reaction rate. The term describes reactions that occur without outside influence, but it tells us nothing about how fast or slow the reaction rate is.

Entropy. In addition to the heat absorbed or evolved in a spontaneous process, another factor, called entropy, which is a measure of the disorder of reactants and products, must also be considered. The *entropy* (S) is a state function that increases in value as the disorder or randomness of the system increases. Entropy has the units J/K · mol. Intuitively we consider a system to be "ordered" if it is arranged according to some plan or method. The system is "disordered" when its parts are helter-skelter within the system and their arrangement is random.

Order and disorder in chemical systems are discernible at the molecular level. Crystalline solids are highly ordered, with molecules or ions occupying fixed lattice sites and with the unit cell repeated identically over and over again. Liquids are less ordered than solids because the solid lattice has broken down, and molecules or ions have kinetic energy of translation. Thus the structure of liquids is more random than that of solids.

Gases are more random than liquids. On vaporization, the molar volume increases about 1000-fold, and it is much more difficult to pin down the position of any one molecule. For a given substance, the

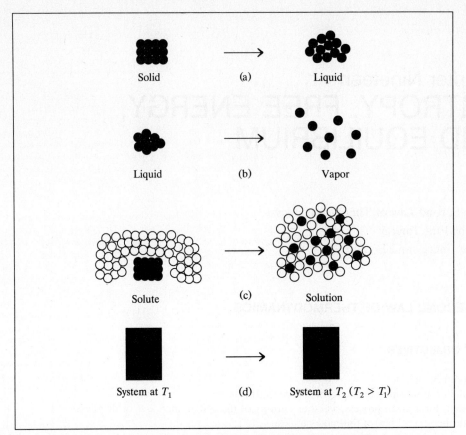

Figure 19.1. Processes that lead to an increase in entropy of the system lead to an increase in randomness. (*a*) Melting, (*b*) vaporization, (*c*) dissolving a solid, (*d*) heating.

molecular order increases:

    gas < liquid < solid

whereas the entropy (disorder) increases in the opposite direction:

    $S_{solid} < S_{liquid} < S_{gas}$

For chemical reactions in which solids or liquids are converted to gases, the entropy change ΔS is positive. It is negative for the condensation of a gas or the freezing of a liquid. In general:

1. If a reaction produces an *increase* in the number of moles of gaseous compounds, ΔS is *positive*.
2. If the total number of moles of gaseous compounds is *decreased*, ΔS is *negative*.
3. If there is *no net change* in the total number of gas molecules, then ΔS is *either a small positive or a small negative number*.

Figure 19.1 illustrates the increases in disorder related to several ordinary processes.

The Second Law. The second law of thermodynamics is concerned with predicting the direction of spontaneous change. It states that the *entropy of the universe increases in a spontaneous change*. The term *universe* as used here refers to a system and all its surroundings. For an isolated system,

$$\Delta S_{univ} = \Delta S_{sys} + \Delta S_{surr} \geq 0$$

where ΔS_{sys} stands for the entropy change of the system and ΔS_{surr} is the entropy change of the surroundings. For spontaneous change,

$$\Delta S_{sys} + \Delta S_{surr} > 0$$

and for a reaction at equilibrium (no change),

$$\Delta S_{sys} + \Delta S_{surr} = 0$$

Entropy Changes in Chemical Reactions. Not only can the sign ΔS for a chemical reaction be estimated, but the actual value of ΔS can also be calculated. The *absolute* value of the entropy S of an element or compound can be determined by very careful experimentation. Appendix 3 in the textbook lists experimental values of the absolute entropy of a number of substances in their standard states at 1 atm and 25°C. Recall that the degree superscript refers to the standard state of the substance. These values are called absolute or sometimes standard entropies.

One can confirm that the entropy of a gas is greater than that of a liquid by comparing the *absolute entropy* of H_2O in the liquid and gas phases:

$$S° \text{ of } H_2O(l) = 69.94 \text{ J/K} \cdot \text{mol}$$

$$S° \text{ of } H_2O(g) = 188.72 \text{ J/K} \cdot \text{mol}$$

For a reaction,

$$aA + bB \rightarrow cC + dD$$

the standard entropy change for the reaction is given by

$$\Delta S°_{rxn} = \Sigma nS° \text{ (products)} - \Sigma mS° \text{ (reactants)}$$

where m and n are stoichiometric coefficients. When applied to the above reaction, we get

$$\Delta S°_{rxn} = [cS°(C) + dS°(D)] - [aS°(A) + bS°(B)]$$

EXAMPLE 19.1 Changes in Entropy

Predict the sign of ΔS_{sys} for each of the following reactions using just the qualitative ideas discussed in this section:
a. $H_2O_2(l) \rightarrow H_2O(l) + \frac{1}{2}O_2(g)$
b. $H^+(aq) + OH^-(aq) \rightarrow H_2O(l)$
c. $CaO(s) + CO_2(g) \rightarrow CaCO_3(s)$

METHOD OF SOLUTION

a. The number of moles of gaseous compounds in the products is greater than in the reactant. Entropy increases during this reaction. *Answer:* The sign of ΔS is positive.
b. Two reactants combine into one product in this reaction. Order is increased and so entropy decreases. *Answer:* The sign of ΔS is negative.
c. The number of gas molecules is decreased. *Answer:* The sign of ΔS is negative.

EXAMPLE 19.2 The Second Law

The entropy change of the system is negative for the precipitation reaction

$$Ag^+(aq) + Cl^-(aq) \rightarrow AgCl(s) \qquad \Delta H° = -65 \text{ kJ}$$

Since ΔS decreases rather than increases in this reaction, why is this reaction spontaneous?

METHOD OF SOLUTION

According to the second law,

$$\Delta S_{sys} + \Delta S_{surr} > 0$$

In order to predict whether a reaction is spontaneous, both ΔS_{sys} and ΔS_{surr} must be considered, not just ΔS_{sys}. Since the entropy change of the system is negative, the only way for the inequality to be true is if ΔS_{surr} is positive and greater in amount than ΔS_{sys}. In this case, this is a reasonable assumption because the reaction is exothermic. This means that heat is liberated to the surroundings, which causes increased thermal motion and disorder of molecules in the surroundings. Thus, the sum of ΔS_{sys} and ΔS_{surr} is greater than zero even though ΔS_{sys} is negative.

EXAMPLE 19.3 Calculation of the Entropy Change for a Reaction

Use absolute entropies to calculate the standard entropy change (ΔS°_{rxn}) for the reaction

$$H_2(g) + \tfrac{1}{2}O_2(g) \rightarrow H_2O(l)$$

METHOD OF SOLUTION

The standard entropy change is given by

$$\Delta S^\circ_{rxn} = S^\circ[H_2O(l)] - [S^\circ(H_2) + \tfrac{1}{2}S^\circ(O_2)]$$

Look up the standard entropy values in Appendix 3 of the text:

$$\Delta S^\circ_{rxn} = 1\ mol\left(\frac{69.9\ J}{K \cdot mol}\right) - \left[1\ mol\left(\frac{130.1\ J}{K \cdot mol}\right) + \frac{1}{2}\ mol\left(\frac{205.0\ J}{K \cdot mol}\right)\right]$$

$$= 69.9\ J/K - 233.5\ J/K$$

$$= -163.6\ J/K$$

COMMENT

This reaction is known to be spontaneous. The value of ΔS°_{rxn} applies only to the system; ΔS_{univ} will be positive.

GIBBS FREE ENERGY

STUDY OBJECTIVES

You should be able to:
1. Calculate free energy changes for chemical reactions, given a table of standard free energies of formation.
2. Predict, using given ΔS° and ΔH° values, at what temperature a reaction will be spontaneous under standard conditions.

Gibbs Free Energy. Quite often it is not possible to calculate ΔS_{surr}. This makes the second law difficult to apply when it is in the form given in the previous section. The Gibbs free energy, expressed in terms of enthalpy and entropy, refers only to the system, yet can be used to predict spontaneity. The Gibbs free energy change (ΔG) for a reaction carried out at constant temperature and pressure is given by

$$\Delta G = \Delta H - T\Delta S$$

where both ΔH and ΔS refer to the system.

The Gibbs free energy of the system will *decrease* ($\Delta G < 0$) in a spontaneous process and will *increase* ($\Delta G > 0$) in a nonspontaneous process. The free energy criteria are summarized as follows:

If ΔG is negative, the forward reaction is spontaneous.

If ΔG is zero, the reaction is at equilibrium.

If ΔG is positive, the forward reaction is nonspontaneous.

Note that the reverse reaction will have a negative ΔG and will be spontaneous.

Any process that occurs spontaneously can be utilized to perform useful work. Burning fuels and the oxidation of food components all provide energy to do work. Thermodynamics tells us that the maximum possible work (w) that can be obtained from a spontaneous process is equal to the Gibbs free energy change:

$$\Delta G = w_{max}$$

In any spontaneous reaction, the free energy decreases until it reaches a minimum value. The state of minimum free energy corresponds to the state of chemical equilibrium.

Calculation of $\Delta G°_{rxn}$. The free energy change for a reaction can be calculated in two ways. When both ΔH and ΔS are known, then $\Delta G = \Delta H - T\Delta S$ will give the free energy change at the temperature T. Also, $\Delta G°_{rxn}$ can be calculated from standard free energies of formation in a manner analogous to the calculation of $\Delta H°$ by using enthalpies of formation of the reactants and products. The standard free energies of formation ($\Delta G°_f$) of selected compounds are tabulated in Appendix 3 of the text. Just as for the standard enthalpies of formation, the free energies of formation of elements in their standard states are equal to zero.

For a general reaction

$$a\text{A} + b\text{B} \rightarrow c\text{C} + d\text{D}$$

The standard free energy change is given by

$$\Delta G°_{rxn} = [c\,\Delta G°_f(\text{C}) + d\,\Delta G°_f(\text{D})] - [a\,\Delta G°_f(\text{A}) + b\,\Delta G°_f(\text{B})]$$

In general,

$$\Delta G°_{rxn} = \Sigma\, n\, \Delta G°_f(\text{products}) - \Sigma\, m\, \Delta G°_f(\text{reactants})$$

where n and m are stoichiometric coefficients.

Temperature and the Free Energy Change. From the equation $\Delta G = \Delta H - T\Delta S$ we can see that temperature too will influence the spontaneity of reaction. If both ΔH and ΔS are positive, then at low temperature, as long as $\Delta H > T\Delta S$, ΔG is positive and the process will be nonspontaneous. However, as temperature increases, the $T\Delta S$ term increases and eventually $\Delta H = T\Delta S$. At this point ΔG is zero. With further T increase, $T\Delta S > \Delta H$, making $\Delta G > 0$, and the reaction becomes spontaneous. Table 19.4 in the textbook summarizes the four possible situations affecting the ΔG of a reaction.

In the above situation, where both $\Delta H°$ and $\Delta S°$ are positive, the temperature at which $\Delta H° = T\Delta S°$, and also at which $\Delta G° = 0$, can be calculated from the equation $T = \Delta H°/\Delta S°$. Above this temperature, the reaction favors the products at equilibrium.

For a phase transition, $\Delta G = 0$ when the two phases coexist in equilibrium. For example, at the boiling point (T_{bp}) the liquid and vapor phases are in equilibrium, and

$$\Delta H_{vap} - T_{bp} \Delta S_{vap} = 0$$

Rearranging gives

$$\Delta S_{vap} = \frac{\Delta H_{vap}}{T_{bp}}$$

This equation allows the calculation of the entropy of vaporization from knowledge of the heat of vaporization and the boiling temperature.

EXAMPLE 19.4 Calculation of the Free Energy Change for a Reaction

Calculate $\Delta G°_{rxn}$ at 25°C for the following reaction using Appendix 3 in the textbook and given that $\Delta G°_f(Fe_2O_3) = -741.0$ kJ/mol:

$$2Al(s) + Fe_2O_3(s) \rightarrow Al_2O_3(s) + 2Fe(s)$$

METHOD OF SOLUTION

$$\Delta G°_{rxn} = \Delta G°_f(Al_2O_3) + 2\Delta G°_f(Fe) - [2\Delta G°_f(Al) + \Delta G°_f(Fe_2O_3)]$$

$$= 1 \text{ mol}\,(-1576.41 \text{ kJ/mol}) + 0 - [0 + 1 \text{ mol}\,(-741.0 \text{ kJ/mol})]$$

$$= -1576.41 \text{ kJ} + 741.0 \text{ kJ} = -835.4 \text{ kJ}$$

EXAMPLE 19.5 Effect of Temperature on ΔG

Hydrate lime, $Ca(OH)_2$, can be re-formed into quicklime, CaO, by heating:

$$Ca(OH)_2(s) \xrightarrow{\Delta} CaO(s) + H_2O(g)$$

At what temperatures is this reaction spontaneous under standard conditions (that is, where H_2O is formed at 1 atm pressure)? Given:

$$\Delta H°_f[Ca(OH)_2] = -986.2 \text{ kJ/mol}$$

$$S°[Ca(OH)_2] = 83.4 \text{ J/K} \cdot \text{mol}$$

METHOD OF SOLUTION

This reaction is nonspontaneous at room temperature. The temperature above which the reaction becomes spontaneous under standard conditions corresponds to $\Delta G° = 0$ and is given by

$$T = \frac{\Delta H°}{\Delta S°}$$

$\Delta H°$ and $\Delta S°$ must be calculated separately:

$$\Delta H° = [\Delta H°_f(CaO(s)) + \Delta H°_f(H_2O(g))] - \{\Delta H°_f[Ca(OH)_2(s)]\}$$

From Appendix 3 in the textbook and given data,

$$\Delta H° = 1\text{ mol}(-635.55\text{ kJ/mol}) + 1\text{ mol}(-241.83\text{ kJ/mol}) - 1\text{ mol}(-986.2\text{ kJ/mol})$$

$$= 108.82\text{ kJ} \quad \text{(or } 1.088 \times 10^5\text{ J)}$$

$$\Delta S° = S°[CaO(s)] + S°[H_2O(g)] - S°[Ca(OH)_2(s)]$$

$$= 1\text{ mol}(39.8\text{ J/K}\cdot\text{mol}) + 1\text{ mol}(188.7\text{ J/K}\cdot\text{mol}) - 1\text{ mol}(83.4\text{ J/K}\cdot\text{mol})$$

$$= +145.1\text{ J/K}$$

$$T = \frac{\Delta H°}{\Delta S°} = \frac{1.088 \times 10^5\text{ J}}{145.1\text{ J/K}} = 750\text{ K}$$

Answer: At temperatures above 750 K the reaction is spontaneous.

EXAMPLE 19.6 Entropy of Fusion

The heat of fusion of water, ΔH_{fus}, at 0°C is 6.02 kJ/mol. What is ΔS_{fus} for 1 mol of H_2O at the melting point?

METHOD OF SOLUTION

$$\Delta S_{fus} = \frac{\Delta H_{fus}}{T_{mp}} = \frac{6.02 \times 10^3\text{ J/mol}}{273\text{ K}}$$

$$= +22.1\text{ J/K}\cdot\text{mol}$$

COMMENT

The increase in entropy upon melting of the solid corresponds to the increase in molecular disorder in the liquid state.

FREE ENERGY AND EQUILIBRIUM

STUDY OBJECTIVES

You should be able to:
1. Calculate ΔG, the free energy change under nonstandard state conditions.
2. Calculate an equilibrium constant from a knowledge of $\Delta G°$, and vice versa.

ΔG and $\Delta G°$. Recall that $\Delta G°$ refers to the standard free energy change. All the values we have calculated so far relate to processes in which the reactants are present in their standard states and are converted to products in their standard states. However, in many cases neither the reactants nor the products are present at standard concentration (1 M) and standard pressure (1 atm). Under nonstandard state conditions, we use the symbol ΔG.

The relationship between ΔG and $\Delta G°$ is

$$\Delta G = \Delta G° + RT \ln Q$$

where R is the gas constant, T is the absolute temperature, and Q is the reaction quotient. For a given reaction at a given temperature the value of $\Delta G°$ is constant, but the value of Q depends on the composition of the reacting mixture; therefore ΔG will depend on Q. To calculate ΔG, first find $\Delta G°$, then calculate Q from the given concentrations of reactants and products, and finally use the preceding equation.

Under special conditions this equation reduces to an extremely important relationship. At equilibrium, $Q = K$, and therefore $\Delta G = 0$. The equation then becomes

$$0 = \Delta G° + RT \ln K$$

or

$$\Delta G° = -RT \ln K$$

This equation relates the equilibrium constant of a reaction to its standard free energy change. Thus, if $\Delta G°$ can be calculated, K can be determined, and vice versa.

Three possible relationships exist between $\Delta G°$ and K, because $\Delta G°$ can be negative, positive, or zero.

1. When $\Delta G°$ is *negative*, $\ln K$ is positive and $K > 1$. The products are favored over reactants at equilibrium. The extent of reaction is large.
2. When $\Delta G°$ is *positive*, $\ln K$ is negative and $K < 1$. The reactants are favored over products at equilibrium. The extent of reaction is small.
3. When $\Delta G° = 0$, $\ln K$ is zero and $K = 1$. The reactants and products are equally favored at equilibrium.

EXAMPLE 19.7 Calculating the Equilibrium Constant

The standard free energy change for the reaction

$$\tfrac{1}{2}N_2(g) + \tfrac{3}{2}H_2(g) \rightleftharpoons NH_3(g)$$

is $\Delta G°_{rxn} = 26.9$ kJ/mol at 700 K. Calculate the equilibrium constant at this temperature.

METHOD OF SOLUTION

The equilibrium constant is related to the standard free energy change by the equation

$$\Delta G°_{rxn} = -RT \ln K_p$$

CALCULATION

Since the gas constant R has units involving joules and the free energy change has units involving kilojoules, we must be careful to use consistent units. In terms of joules, we get

$$26.9 \times 10^3 \text{ J/mol} = -(8.31 \text{ J/mol} \cdot \text{K})(700 \text{ K})\ln K_p$$

$$-4.62 = \ln K_p$$

Taking the antilog of both sides gives

$$e^{-4.62} = K_p$$

Use of a calculator with an e^x key yields

$$K_p = 9.8 \times 10^{-3}$$

EXAMPLE 19.8 ΔG **at Nonstandard State Conditions**

Using data given in the preceding example, calculate ΔG at 700 K if the reaction mixture consists of 30.0 atm of H_2, 20.0 atm of N_2, and 0.50 atm of NH_3.

METHOD OF SOLUTION

Under nonstandard conditions, ΔG is related to the reaction quotient Q by the equation

$$\Delta G = \Delta G^\circ + RT \ln Q_p$$

where

$$Q_p = \frac{P_{NH_3}}{P_{N_2}^{1/2} P_{H_2}^{3/2}} = \frac{(0.50)}{(20.0)^{1/2}(30.0)^{3/2}} = 6.8 \times 10^{-4}$$

From Example 19.7, $\Delta G^\circ = 26.9$ kJ/mol. Substituting yields

$$\Delta G = 26.9 \text{ kJ/mol} + (8.31 \text{ J/K} \cdot \text{mol})(700 \text{ K})\ln(6.8 \times 10^{-4})$$

$$= 26.9 \text{ kJ/mol} - 42{,}400 \text{ J/mol}\left(\frac{1 \text{ kJ}}{10^3 \text{ J}}\right) = 26.9 \text{ kJ/mol} - 42.4 \text{ kJ/mol}$$

$$= -15.5 \text{ kJ/mol}$$

COMMENT

By making the partial pressures of N_2 and H_2 high and that of NH_3 low, the reaction is spontaneous in the forward reaction. This condition corresponds to $Q_p < K_p$, and so the reaction proceeds in the forward direction until $Q_p = K_p$.

TRUE-FALSE QUESTIONS

1. Reactions for which energy must be supplied from outside the system to carry out the reactions are spontaneous processes.

2. Spontaneous processes always occur very rapidly.

3. The entropy of a substance in the gas phase is always greater than the entropy of the same substance in the liquid phase.

4. According to the second law of thermodynamics, if $\Delta S_{sys} > 0$, the process must be spontaneous.

5. A reaction which is spontaneous in one direction is nonspontaneous in the reverse direction.

6. If ΔG is positive, the reaction is spontaneous.

7. ΔG_f° of elements in their standard states is zero.

8. Reactions for which ΔG° is positive and ΔS° is negative will be nonspontaneous at low temperature but will become spontaneous if the temperature is raised sufficiently.

9. ΔS for vaporization of a substance is always positive.

10. When $\Delta G^\circ = 0$, the reaction is at equilibrium.

SELF-TEST A

1. Which of the following processes are spontaneous?
   a. Melting of ice at $-10°C$ and 1 atm pressure
   b. Evaporation of water at $30°C$ when the relative humidity is less than 100%
   c. Water + $NaCl(s) \rightarrow$ salt solution

2. From each pair of substances, choose the one having the larger standard entropy at $25°C$:
   a. $H_2O(l)$ or $H_2O(g)$
   b. $SiO_2(s)$ or $CO_2(g)$
   c. $Ag^+(g)$ or $Ag^+(aq)$
   d. $F_2(g)$ or $Cl_2(g)$
   e. $2Cl(g)$ or $Cl_2(g)$

3. Predict, using the intuitive ideas about entropy, whether ΔS_{sys} will be positive, negative, or essentially zero for each of the following:
   a. $Ca(OH)_2(s) + CO_2(g) \rightarrow CaCO_3(s) + H_2O(g)$
   b. $CuSO_4(s) \rightarrow Cu^{2+}(aq) + SO_4^{2-}(aq)$
   c. $2HCl(g) + Br_2(l) \rightarrow 2HBr(g) + Cl_2(g)$
   d. $SO_2(g) + \frac{1}{2}O_2(g) \rightarrow SO_3(g)$
   e. $Ag^+(aq) + 2CN^-(aq) \rightarrow Ag(CN)_2^-(aq)$

4. At the boiling point, $35°C$, the heat of vaporization of MoF_6 is 25 kJ/mol. Calculate ΔS for the vaporization of MoF_6.

5. Calculate $\Delta G°_{rxn}$ for the following reaction at 298 K:

   $$2H_2(g) + CO(g) \rightleftharpoons CH_3OH(g)$$

   given that $\Delta H° = -90.7$ kJ and $\Delta S° = -221.5$ J/K for this process.

6. For the reaction at 298 K,

   $$Mg(s) + \tfrac{1}{2}O_2(g) \rightarrow MgO(s)$$

   $\Delta H° = -602$ kJ and $\Delta G° = -569$ kJ. Calculate $\Delta S°$.

7. Using Appendix 3 of the text calculate $\Delta G°$ values for the following reactions:
   a. $3CaO(s) + 2Al(s) \rightarrow 3Ca(s) + Al_2O_3(s)$
   b. $ZnO(s) \rightarrow Zn(s) + \tfrac{1}{2}O_2(g)$

8. Consider the following three reactions. Which one will have the greatest equilibrium constant?
   a. $N_2 + O_2 \rightleftharpoons 2NO$
   b. $N_2 + 2O_2 \rightleftharpoons N_2O_4$
   c. $N_2 + \tfrac{1}{2}O_2 \rightleftharpoons N_2O$
   Given:

   $\Delta G°_f(NO) = +86.7$ kJ/mol

   $\Delta G°_f(N_2O_4) = +98.3$ kJ/mol

   $\Delta G°_f(N_2O) = +103.6$ kJ/mol

9. In Chapter 14, we saw that for the reaction

   $$H_2(g) + I_2(g) = 2HI(g)$$

   the equilibrium constant at $400°C$ is $K_p = 64$. Calculate the value of $\Delta G°$ at this temperature.

10. Calculate $\Delta G°$ and K_p at 25°C for the following reaction:

    $NO(g) + \frac{1}{2}O_2(g) = NO_2(g)$

11. The synthesis of $O_2(g)$ is often carried out in general chemistry laboratories by the decomposition of $KClO_3$,

    $KClO_3(s) \rightarrow KCl(s) + \frac{3}{2}O_2(g)$

    for which $\Delta H° = -44.7$ kJ and $\Delta S° = +59.1$ J/K. Is this reaction spontaneous at 25°C under standard conditions?

12. For the reaction

    $N_2 + O_2 \rightarrow 2NO$

    the following are given:

    $\Delta H° = 180.7$ kJ   and $\Delta S° = 24.7$ J/K

    a. Is this reaction spontaneous at 25°C?
    b. Above what temperature will this reaction become spontaneous under standard conditions?

13. For the reaction $2SO_2(g) + O_2(g) \rightleftharpoons 2SO_3(g)$, $K_p = 7.4 \times 10^4$ at 700 K. If, in a reaction vessel at 700 K, we have the following partial pressures, what is ΔG?

    $P_{SO_2} = 1.2$ atm $P_{O_2} = 0.5$ atm $P_{SO_3} = 50$ atm

    Predict the direction of reaction.

SELF-TEST B1

1. The enthalpy of vaporization of mercury is 58.5 kJ/mol and the normal boiling point is 630 K. What is the entropy of vaporization of mercury?

2. Calculate $\Delta G°$ for the following reaction at 298 K:

    $O_3(g) \rightarrow O_2(g) + O(g)$

    Given: $\Delta H° = 106.5$ kJ and $\Delta S° = 127.3$ J/K.

3. The autoionization of water at 25°C has the equilibrium constant

    $2H_2O(l) \rightleftharpoons H_3O^+(aq) + OH^-(aq)$ $K = 1.0 \times 10^{-14}$.

    Calculate the value of $\Delta G°$ for this reaction.

4. Above what temperature does the reaction

    $H_2(g) \rightarrow 2H(g)$

    become spontaneous? Given: $\Delta H° = 436$ kJ and $\Delta S° = 98.6$ J/K.

5. Calculate ΔG for the following reaction at 25°C when the pressure of CO_2 is 0.001 atm.

    $CaCO_3(s) \rightarrow CaO(s) + CO_2(g)$

    Given: $\Delta H° = 177.8$ kJ and $\Delta S° = 160.5$ J/K.

SELF-TEST B2

1. The enthalpy of vaporization (ΔH_{vap}) of mercury is 58.5 kJ/mol, and the entropy of vaporization is 92.6 J/K · mol. Calculate the normal boiling point of mercury.

2. Calculate $\Delta S°$ for the reaction at 298 K:

   $$O_3(g) \rightarrow O_2(g) + O(g)$$

   Given: $\Delta H° = 106.5$ kJ and $\Delta G° = 68.6$ kJ.

3. The standard free energy change for the autoionization of water at 25°C is $\Delta G° = 79.8$ kJ:

   $$2H_2O(l) \rightleftharpoons H_3O^+(aq) + OH^-(aq)$$

   What is the equilibrium constant value for this process?

4. The following reaction is nonspontaneous at low temperature:

   $$H_2(g) \rightarrow 2H(g)$$

   However, the reaction will proceed spontaneously above 4400 K. If $\Delta H° = 436$ kJ, is the standard entropy change a positive or negative quantity?

5. At 25°C, what pressure of CO_2 is in equilibrium with $CaCO_3$ and CaO?

   $$CaCO_3(s) \rightleftharpoons CaO(s) + CO_2(g) \qquad \Delta G° = 130.0 \text{ kJ}$$

ANSWERS

TRUE-FALSE QUESTIONS

1. False. The statement describes a nonspontaneous process.
2. False. The rates of chemical reactions vary widely. A spontaneous reaction can be very slow.
3. True.
4. False. The process is spontaneous when $\Delta S_{univ} > 0$, not ΔS_{sys}.
5. True.
6. False. ΔG is negative for a spontaneous reaction.
7. True.
8. False. The reaction will not become spontaneous at high temperature; $T\Delta S$ is negative.
9. True.
10. False. When the reaction is at equilibrium, $\Delta G = 0$; not $\Delta G°$.

SELF-TEST A

1. b and c
2. a. $H_2O(g)$ b. $CO_2(g)$ c. $Ag^+(g)$ d. $Cl_2(g)$ e. $2Cl(g)$
3. a. Essentially zero   b. Positive   c. Positive   d. Negative   e. Negative
4. $\Delta S = 81$ J/mol · K
5. $\Delta G° = -24{,}700$ J
6. $\Delta S° = -110$ J/K
7. a. $\Delta G° = 236$ kJ   b. $\Delta G° = 318.2$ kJ
8. b
9. $\Delta G° = -23.2$ kJ
10. $\Delta G° = 34.85$ kJ; $K_p = 1.29 \times 10^6$

11. Yes, $\Delta G° < 0$
12. a. No, $\Delta G° > 0$ b. $T = 7320$ K
13. $\Delta G = -17,800$ J. Spontaneous in the forward direction.

SELF-TEST B1

1. $\Delta S_{vap} = 92.6$ J/K · mol
2. $\Delta G° = 68.6$ kJ
3. 79.8 kJ
4. 4400 K
5. $\Delta G = 112.9$ kJ

SELF-TEST B2

1. 630 K
2. $\Delta S° = 127$ J/K
3. $K = 1.0 \times 10^{-14}$
4. Positive
5. 1.5×10^{-23} atm

Chapter Twenty
ELECTROCHEMISTRY

- Galvanic Cells and Standard Electrode Potentials
- Spontaneity of Redox Reactions
- Effect of Concentration on Cell emf
- Electrolysis: Faraday's Law

GALVANIC CELLS AND STANDARD ELECTRODE POTENTIALS

STUDY OBJECTIVES

You should be able to:
1. Diagram a galvanic cell, labeling the anode, the cathode, the charges on the electrodes, and the directions of electron and ion flows.
2. Calculate the standard emf of a galvanic cell.
3. Arrange given redox reagents in order of increasing strength as oxidizing or reducing agents.

Electrochemical Cells. In Chapter 3 we discussed oxidation-reduction reactions. In many of these reactions, electrons are transferred from the reducing agent to the oxidizing agent.

$$\underset{\substack{\text{reducing}\\\text{agent}}}{Cu(s)} + \underset{\substack{\text{oxidizing}\\\text{agent}}}{2Ag^+(aq)} \rightarrow Cu^{2+}(aq) + 2Ag(s)$$

In the above reaction, copper metal is immersed in a solution containing Ag^+ ions. As silver ions come in contact with the metal surface, electrons are transferred spontaneously from Cu atoms to the Ag^+ ions, forming Ag atoms that plate out on the copper surface. The newly formed Cu^{2+} ions go into solution. The half-reactions are

oxidation $\quad Cu \rightarrow Cu^{2+} + 2e^-$

reduction $\quad 2e^- + 2Ag^+ \rightarrow 2Ag$

Copper metal is the *reducing agent* because it supplies electrons to the silver ion. The silver ion is the *oxidizing agent* because it removes electrons from copper.

It is possible to build a device in which electrons must move through an external electrical circuit rather than transferring directly from an atom of the reducing agent to an atom of the oxidizing agent. Such a device, which utilizes a spontaneous redox reaction, is called a *galvanic* or *voltaic cell*. The design of a galvanic cell is such that the reactants are prevented from direct contact with each other. The oxidation half-reaction occurs at an electrode called the *anode*, and the reduction half-reaction occurs at an electrode called the *cathode*.

Figure 20.1 shows a copper-silver galvanic cell. The anode is a bar of copper metal that is partially immersed in a solution of $CuSO_4$. The cathode is a small bar of silver that is partially immersed in a solution of $AgNO_3$. The reducing agent, Cu metal, does not come into direct contact with the oxidizing agent, Ag^+ ion. The electrons given up by Cu atoms must travel through the outer circuit to the Ag

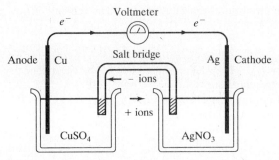

Figure 20.1. A galvanic cell consisting of a copper electrode (anode) and a silver electrode (cathode).

electrode, where Ag^+ ions from the solution are reduced. As the reaction proceeds, the copper electrode loses mass and the silver electrode gains mass.

While electrons travel through the outer circuit from Cu to Ag (from the anode to the cathode), negative ions move through a porous barrier or a "salt bridge" from the $AgNO_3$ solution into the $CuSO_4$ solution. This motion resupplies negative charge to the anode compartment and maintains electric neutrality in the solutions surrounding the electrodes. An electric current will flow until either the Cu metal or the Ag^+ ions are used up.

The fact that electrons flow from the anode to the cathode means that a voltage difference exists between the two electrodes. This voltage difference depends on the nature of the species involved in the half-reactions. The voltage measured across the two electrodes is called the *electromotive force*, or *cell emf*. The cell emf is represented by the symbol E and is measured in units of volts.

Standard Electrode Potentials. The cell emf has been found to be dependent upon the nature of the half-reactions, the concentrations of the ions, and the temperature. The *standard cell emf is the voltage generated under standard state conditions*. The standard cell emf has the symbol $E°$. The standard state of all solute ions and molecules is a concentration of 1 mol/L, and for all gases it is a partial pressure of 1 atm. Unless stated otherwise, the temperature is assumed to be 25°C. When the concentration of Cu^{2+} and Ag^+ ions in the above cell (Figure 20.1) are both 1.0 M, the cell emf is 0.46 V at 25°C.

Rather than construct a cell and measure its emf every time a standard cell emf is needed, a method has been devised that allows the use of a table of half-reactions and standard electrode potentials. It is impossible to measure the absolute value of the potential of a single electrode, so a relative scale has been established by measuring the standard cell emf's of a number of half-cell reactions with respect to a reference electrode. The standard hydrogen electrode (SHE), shown in Figure 20.2 of the textbook, is the reference electrode, and its potential is assigned a value of exactly zero:

$$2H^+(aq, 1\,M) + 2e^- \rightarrow H_2(g, 1\,atm) \qquad E°_{SHE} = 0.0\,V$$

The superscript degree denotes the standard state conditions.

A number of standard reduction potentials are listed in Table 20.1 of the textbook. The values range from $+2.87$ V for the F_2 half-reaction to -3.05 V for the Li^+ half-reaction. Note the 0.00 potential assigned to the hydrogen ion. The greater the value of $E°_{red}$, the greater the tendency for the reduction reaction to occur as written. A positive reduction potential means that the oxidizing agent (such as F_2) has a greater tendency to be reduced than the hydrogen ion. A negative reduction potential for an oxidizing agent indicates that the hydrogen ion is more readily reduced than that substance.

Calculation of the Standard Cell emf. In electrochemistry, it is convenient to think of an overall cell reaction as the sum of two half-reactions. The cell emf can also be thought of as the sum of two half-cell potentials. One half-cell potential is that due to the loss of electrons at the anode, E_{ox}, and the other is due to the gain of electrons at the cathode, E_{red}. The cell emf E_{cell} is given by the equation

$$E_{cell} = E_{ox} + E_{red}$$

For a standard cell, $E°_{cell} = E°_{ox} + E°_{red}$. The standard emf of the silver-copper cell discussed above is calculated as follows. The reaction is

$$Cu + 2Ag^+ \rightarrow Cu^{2+} + 2Ag$$

The half-reactions are

oxidation $Cu \rightarrow Cu^{2+} + 2e^-$

reduction $2Ag^+ + 2e^- \rightarrow 2Ag$

Therefore, the standard cell emf is given by

$$E°_{cell} = E°_{ox} + E°_{red} = E°_{Cu/Cu^{2+}} + E°_{Ag^+/Ag}$$

Notice that the symbols here represent the direction of the reaction,

$$E°_{reactant/product}$$

Therefore,

$$E°_{ox} = E°_{Cu/Cu^{2+}} \quad \text{and} \quad E°_{red} = E°_{Ag^+/Ag}$$

Table 20.1 in the text lists the reduction potentials of a number of half-reactions. The standard oxidation potential for an anode half-reaction is equal in magnitude but of *opposite sign* to that of the reduction potential for the reverse reaction:

$$E°_{red} = E°_{Cu^{2+}/Cu} = 0.34 \text{ V}$$

Therefore

$$E°_{ox} = E°_{Cu/Cu^{2+}} = -0.34 \text{ V}$$

For the cell

$$E°_{ox} = E°_{Cu/Cu^{2+}} = -0.34 \text{ V} \quad \text{and} \quad E°_{red} = E°_{Ag^+/Ag} = 0.80 \text{ V}$$

$$E°_{cell} = E°_{ox} + E°_{red} = -0.34 \text{ V} + 0.80 \text{ V} = 0.46 \text{ V}$$

The following list summarizes the information contained in the standard reduction potentials given in Table 20.1 in the text:

1. The $E°$ values are reduction potentials and apply to the reduction half-reactions. The more positive the value of $E°$, the greater the tendency for the substance shown on the left to be reduced. The species listed on the left-hand side of the reactions shown in the table are all capable of acting as oxidizing agents (thus they are reduced). The species shown on the right-hand side are reducing agents. When predicting the direction of reaction under standard state conditions, you can use the *diagonal rule*. This rule states that any species on the left of a given half-cell reaction will react spontaneously with a species that is shown on the right of any half-reaction located *above* it in the table. This rule works because the sum of any two half-cell potentials, such as those just described, will always give a positive $E°_{cell}$.
2. All half-cell reactions are reversible. Depending on the reducing strength of the other electrode that is chosen, a given electrode may act as an anode or as a cathode.

3. The standard cell emf ($E°_{cell}$) changes sign whenever a half-cell reaction is reversed. The standard oxidation potential for an anode half-reaction is equal in magnitude but opposite in sign to that of the reduction potential for the same half-reaction.
4. The standard reduction potential is not affected when stoichiometric coefficients of a half-cell reaction are changed. For example,

$$Ag^+ + e^- \rightarrow Ag(s) \qquad E°_{red} = 0.80 \text{ V}$$

$$2Ag^+ + 2e^- \rightarrow 2Ag(s) \qquad E°_{red} = 0.80 \text{ V}$$

The electrode potentials are the same. The charge on the electrode is related to the number of electrons per surface area of silver. The ratio of e^- to Ag atoms in both cases is the same:

$$\frac{e^-}{\text{Ag atom}} = \frac{2e^-}{2 \text{ Ag atoms}}$$

Therefore, electrode potentials are *intensive properties* and do not depend on the size of the electrode or the manner in which the half-reaction is balanced.

Cell Diagrams. Rather than always representing a cell by a sketch, we can use a *cell diagram*. For instance, the copper-silver cell discussed above and shown in Figure 20.1 is represented by

$$Cu(s)|CuSO_4(aq)\|AgNO_3(aq)|Ag(s)$$

In the diagram the anode (where oxidation occurs) is on the left, and the cathode (where reduction occurs) is on the right. The single vertical lines separate the solid electrode from the liquid solution with which it is in contact. The double vertical line indicates a salt bridge or porous barrier between the two solutions.

EXAMPLE 20.1 Galvanic Cell Reactions

Consider a galvanic cell constructed from the following half-cells linked by a KNO_3 salt bridge: a Cu electrode immersed in 1.0 M $Cu(NO_3)_2$ and a Sn electrode in 1.0 M $Sn(NO_3)_2$. As the reaction proceeds the Cu electrode gains mass and the Sn electrode loses mass.
a. Which electrode is the anode and which is the cathode?
b. Write a balanced chemical equation for the overall cell reaction.
c. Sketch the half-cells and show the direction of flow of the electrons in the external circuit.
d. Which electrodes do the positive and the negative ions diffuse toward? Include a salt bridge in your sketch.

METHOD OF SOLUTION

a. Reduction occurs at the cathode and oxidation occurs at the anode. Since the Cu electrode gains mass, its half-reaction must involve the plating out of Cu^{2+} ions:

$$2e^- + Cu^{2+}(aq) \rightarrow Cu(s)$$

Reduction occurs at the copper electrode; therefore, it is the *cathode*. Since the Sn electrode loses mass, Sn metal must be dissolving as it is oxidized, and so Sn is the *anode*. The Sn^{2+} ions go into solution:

$$Sn(s) \rightarrow Sn^{2+}(aq) + 2e^-$$

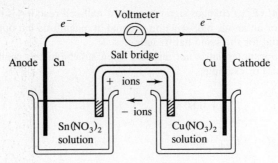

Figure 20.2. A galvanic cell consisting of a tin electrode (anode) and a copper electrode (cathode).

b. Add the two half-reactions:

$$Sn \rightarrow Sn^{2+} + 2e^-$$
$$2e^- + Cu^{2+} \rightarrow Cu$$

cell reaction $Sn + Cu^{2+} \rightarrow Sn^{2+} + Cu$

c. Electrons flow from the anode to the cathode. See Figure 20.2.
d. The positive ions (cations) diffuse toward the Cu electrode (the cathode) and the negative ions (anions) diffuse toward the Sn electrode (the anode).

EXAMPLE 20.2 Standard Cell Voltage

Calculate E°_{cell} for the following galvanic cell:

$$Sn(s) + Pb^{2+}(aq) \rightarrow Sn^{2+}(aq) + Pb(s)$$

METHOD OF SOLUTION

The standard cell emf is the sum of the standard oxidation and reduction potentials of the appropriate half-reactions; $E^\circ_{cell} = E^\circ_{ox} + E^\circ_{red}$.

Separate the reaction into half-reactions, and obtain the oxidation and reduction potentials from Table 20.1 of the text.

CALCULATION

reduction	$2e^- + Pb^{2+}(aq) \rightarrow Pb(s)$	$E^\circ_{Pb^{2+}/Pb}$	$= -0.13$ V
oxidation	$Sn(s) \rightarrow Sn^{2+}(aq) + 2e^-$	$E^\circ_{Sn/Sn^{2+}}$	$= 0.14$ V
	$Sn(s) + Pb^{2+}(aq) \rightarrow Sn^{2+}(aq) + Pb(s)$	E°_{cell}	$= 0.01$ V

COMMENT

The positive value of the standard cell potential indicates that the reaction is spontaneous under standard conditions.

EXAMPLE 20.3 Relative Strength of Oxidizing and Reducing Agents

Consider the following species: Mg, MnO_4^-, Cl^-, Zn^{2+}, and Co^{2+}. Using Table 20.1 of the text, separate these into oxidizing and reducing agents. Then arrange the reducing agents according to increasing strength. Also arrange the oxidizing agents according to strength.

METHOD OF SOLUTION

The oxidizing agents appear on the left-hand side of the half-reactions in Table 20.1, and the reducing agents appear on the right-hand side. Therefore, the oxidizing agents are MnO_4^-, Co^{2+}, and Zn^{2+}, and the reducing agents are Co^{2+}, Cl^-, and Mg.

Recall that oxidizing agents are reduced in redox reactions. The more positive the value of the reduction potential, the greater the tendency to undergo reduction, and the greater the strength as an oxidizing agent.

MnO_4^- 1.51 V (in acid solution)

Co^{2+} −0.28 V

Zn^{2+} −0.76 V

The order of increasing oxidizing strength is

$$Zn^{2+} < Co^{2+} < MnO_4^-$$

For a substance on the right-hand side of the half-reactions in Table 20.1 to act as a reducing agent, it must react in the reverse direction, whereby it is oxidized. Write the oxidation potentials for each:

Mg   2.37 V

Cl^- −1.36 V

Co^{2+} −1.82 V

The greater the value of the oxidation potential, the greater the tendency to be oxidized. Substances that are easily oxidized are good reducing agents. The order of increasing reducing strength is

$$Co^{2+} < Cl^- < Mg$$

COMMENT

The Co^{2+} ion can be both a reducing agent and an oxidizing agent, because Co^{2+} participates in two different half-reactions:

$$Co^{2+} + 2e^- \rightarrow Co \quad \text{and} \quad Co^{3+} + 2e^- \rightarrow Co^{2+}$$

SPONTANEITY OF REDOX REACTIONS

STUDY OBJECTIVES

You should be able to:
1. Predict whether a redox reaction will be spontaneous or nonspontaneous.
2. Calculate $\Delta G°$ and K for a redox reaction, given $E°_{cell}$.

Criterion for Spontaneity. For a process carried out at constant temperature and pressure, the Gibbs free energy change is equal to the maximum amount of work (w_{max}) that can be done by the process:

$$\Delta G = w_{max}$$

Any reaction that occurs spontaneously can be utilized to perform useful work. Combustion of fuels and the oxidation of food components both provide energy to do work. The difference in free energy between the reactants and the products is the energy available to do work.

Electrical work is equal to the product of the cell emf (E_{cell}) and the total charge in coulombs carried through the circuit. The total charge in coulombs is nF, where *n is the number of moles of electrons transferred from the reducing agent to the oxidizing agent according to the balanced equation* and F is the Faraday constant, 96,500 C/mol of electrons. One faraday equals the charge carried by one mole of electrons. The electrical work is given by

$$w_{ele} = -nFE_{cell}$$

Work has units of joules, the same as those for energy. This means that Faraday's constant should be expressed in joules. Since 1 J = 1 coulomb volt (C · V), then 1 F = 96,500 C/mol = 96,500 J/V · mol. The maximum work from an electrochemical process is

$$w_{max} = w_{ele} = -nFE_{cell}$$

The negative sign is in accord with the sign convention that when work is done on the surroundings, the system loses energy.

The Gibbs free energy change for a redox reaction is

$$\Delta G = -nFE$$

For a spontaneous redox reaction, ΔG will be negative, and E must be positive. For reactions in which reactants and products are in their standard states:

$$\Delta G° = -nFE°$$

The Equilibrium Constant. In Chapter 18 you learned that $\Delta G°$ is related to the equilibrium constant K for the reaction

$$\Delta G° = -RT \ln K$$

In electrochemistry it is still conventional to use common logarithms. Since $\ln K = 2.303 \log K$, then $\Delta G° = -2.303 RT \log K$, where 2.303 is a factor relating $\ln K$ to $\log K$. Therefore, the equilibrium constant of a redox reaction is related to the standard cell emf ($E°$) by the equation

$$-nFE° = -2.303RT \log K$$

and

$$E° = \frac{2.303RT}{nF} \log K$$

At 25°C, the term $2.303RT/F$ equals 0.0591 V, since 2.303(8.31 J/K · mol)(298 K)/96,500 J/V · mol = 0.0591 V:

$$E° = \frac{0.0591 \text{ V}}{n} \log K$$

Thus reactions with large equilibrium constants generate higher standard cell emf's. Rearranging yields

$$\log K = \frac{nE°}{0.0591 \text{ V}}$$

If any one of the three quantities $\Delta G°$, K, or $E°$ is known, both of the others can be calculated. Table 20.1 summarizes the criteria for spontaneous redox reactions.

Table 20.1 Criteria for Spontaneous Redox Reactions

$\Delta G°$	K	$E°_{cell}$	Reaction under Standard State Conditions
Negative	> 1	Positive	Spontaneous
0	= 1	0	At equilibrium
Positive	< 1	Negative	Nonspontaneous. Reaction is spontaneous in the reverse direction.

EXAMPLE 20.4 Predicting Spontaneous Redox Reactions

Predict whether a spontaneous reaction will occur when the following reactants and products are in their standard states:
a. $2Fe^{3+}(aq) + 2I^-(aq) \rightarrow 2Fe^{2+}(aq) + I_2(s)$
b. $Cu(s) + 2H^+(aq) \rightarrow Cu^{2+}(aq) + H_2(g)$

METHOD OF SOLUTION

For a redox reaction to be spontaneous, it must have a positive cell emf.
a. To calculate the cell emf for this reaction, first separate it into half reactions:

reduction $\quad 2Fe^{3+} + 2e^- \rightarrow 2Fe^{2+}$

oxidation $\quad 2I^- \rightarrow I_2 + 2e^-$

$E°_{cell} = E°_{ox} + E°_{red}$

From Table 20.1

$E°_{cell} = E°_{I^-/I_2} + E°_{Fe^{3+}/Fe^{2+}} = -0.53 + 0.77 = 0.24$ V

The positive value of the standard cell emf indicates that the reaction is spontaneous under standard conditions.

b. Separating into half-reactions gives

reduction $\quad 2H^+ + 2e^- \rightarrow H_2(g)$

oxidation $\quad Cu(s) \rightarrow Cu^{2+}(aq) + 2e^-$

$E°_{cell} = E°_{ox} + E°_{red}$

$E°_{cell} = -0.34$ V $+ 0.0 = -0.34$ V

The negative standard cell emf indicates that this reaction is not spontaneous at standard conditions. Copper will not dissolve in 1 M HCl.

COMMENT

Since the cell emf changes sign when the reaction is reversed, the reaction $Cu^{2+} + H_2 \rightarrow Cu + 2H^+$ will have a positive cell emf, and the reduction of Cu^{2+} by H_2 would be spontaneous under standard conditions. It is also very important to notice that we could have answered this problem by application of the *diagonal rule*. In part a, I^- appears *above* and to the right of Fe^{3+} in Table 20.1 in the textbook. Thus Fe^{3+} will oxidize I^-. In part b, Cu appears to the right but *below* H^+. Thus H^+ cannot oxidize Cu.

EXAMPLE 20.5   Free Energy Change and the Standard Cell emf

Calculate $\Delta G°$ and the equilibrium constant at 25°C for the reaction

$$2Br^-(aq) + I_2(s) \rightarrow Br_2(l) + 2I^-(aq)$$

METHOD OF SOLUTION

First find the standard cell emf as we have described previously. Then the standard Gibbs free energy change is

$$\Delta G° = -nFE°$$

The equilibrium constant at 25°C is related to $E°_{cell}$ by

$$\log K = \frac{nE°}{0.0591 \text{ V}}$$

CALCULATION

The half-reactions are

reduction	$2e^- + I_2(s) \rightarrow 2I^-(aq)$	$E°_{red} =$	0.53 V
oxidation	$2Br^-(aq) \rightarrow Br_2(l) + 2e^-$	$E°_{ox} =$	-1.07 V
	$2Br^-(aq) + I_2(s) \rightarrow Br_2(l) + 2I^-(aq)$	$E°_{cell} =$	-0.54 V

Substituting into $\Delta G° = -nFE°$ yields

$$\Delta G° = (-2 \text{ mol})(96{,}500 \text{ J/V} \cdot \text{mol})(-0.54 \text{ V})$$

where $n = 2$ mol of electrons transferred in the balanced equation. The units of the Faraday constant must be expressed in J/V · mol in order to cancel the units of volts from $E°_{cell}$. Remember that 1 J = 1 C · V. Continuing with our calculation gives

$$\Delta G°_{rxn} = 1.04 \times 10^5 \text{ J} = 104 \text{ kJ}$$

The positive value of $\Delta G°$ indicates the reaction is not spontaneous under standard conditions. Next we calculate the equilibrium constant:

$$\log K = \frac{nE°}{0.0591 \text{ V}}$$

$$= \frac{2(-0.54 \text{ V})}{0.0591 \text{ V}}$$

$$= -18.3$$

Taking the antilog of both sides, we get

$$K = 10^{-18.3}$$

$$= 5 \times 10^{-19}$$

EFFECT OF CONCENTRATION ON CELL emf

STUDY OBJECTIVES

You should be able to:
1. Calculate the emf of a galvanic cell in which the reactants and products are present at nonstandard concentrations.
2. From a knowledge of E and $E°$, calculate the concentrations of a given ion in a galvanic cell.

The Nernst Equation. We mentioned earlier that the cell emf depends on the nature of the reactants and products and on their concentrations. Since the cell emf is a measure of the spontaneity of the cell reaction, we might reasonably expect the voltage to fall as reactants are consumed and products accumulate.

The equation that relates the cell emf to the concentrations of reactants and products is named after Walter Nernst. At 298 K, for a redox reaction of the type

$$a\text{A} + b\text{B} \rightarrow c\text{C} + d\text{D}$$

The Nernst equation is

$$E = E° - \frac{0.0591 \text{ V}}{n} \log Q$$

where $E°$ is the standard cell emf, E is the nonstandard cell emf, 0.0591 V is a constant at 298 K, and n is the number of moles of electrons transferred according to the balanced equation. The reaction quotient Q contains the concentrations of reactants and products:

$$Q = \frac{[\text{C}]^c[\text{D}]^d}{[\text{A}]^a[\text{B}]^b}$$

The Nernst equation predicts that E will decrease as reactant concentrations decrease and as product concentrations increase. As Q increases, log Q increases, and thus an increasingly large number is *subtracted* from $E°$. Therefore, *E is not a constant* but decreases as a reaction proceeds, whereas $E°$ is a constant for a reaction and is characteristic of the reaction.

At equilibrium, this equation reduces to one we have seen before. When the reaction is at equilibrium, no net reaction occurs and no net transfer of electrons occurs, and $E = 0$. Also, at equilibrium $Q = K$, and the Nernst equation becomes

$$0 = E° - \frac{0.0591 \text{ V}}{n} \log K$$

$$E° = \frac{0.0591 \text{ V}}{n} \log K$$

Therefore, the standard cell emf is related to the equilibrium constant.

When both E_{cell} and $E°_{\text{cell}}$ are known, the Nernst equation can be used to calculate an unknown concentration. This will be illustrated in Example 20.7.

EXAMPLE 20.6 emf for a Nonstandard Cell

Calculate the voltage of a cell at 25°C in which the following reaction occurs at the concentrations given:

$$\text{Zn}(s) + 2\text{H}^+(aq, 1 \times 10^{-4} \text{ M}) \rightarrow \text{Zn}^{2+}(aq, 1.5 \text{ M}) + \text{H}_2(g, 1 \text{ atm})$$

METHOD OF SOLUTION

The cell emf can be calculated by use of the Nernst equation:

$$E = E° - \frac{0.0591 \text{ V}}{n} \log \frac{[Zn^{2+}]P_{H_2}}{[H^+]^2}$$

where

$E°_{cell} = E°_{ox} + E°_{red} = 0.76 \text{ V} + 0.00 \text{ V} = 0.76 \text{ V}$

$n = 2$ (number of moles of electrons transferred according to balanced equation)

$P_{H_2} = 1$, because hydrogen is in its standard state, 1 atm

$[Zn^{2+}] = 1.5 \; M$

$[H^+] = 1 \times 10^{-4} \; M$

CALCULATION

Substitution into the Nernst equation above yields

$$E = 0.76 \text{ V} - \frac{0.0591 \text{ V}}{2} \log \frac{1.5}{(1 \times 10^{-4})^2}$$

$$= 0.76 \text{ V} - 0.24 \text{ V}$$

$$= 0.52 \text{ V}$$

COMMENT

The low concentration of H^+ ($1 \times 10^{-4} \; M$) compared to its standard state value ($1 \; M$) means that the driving force for reaction will be less than in the standard state, and as we see, $E < E°$.

EXAMPLE 20.7 Determining Ion Concentrations

An electrochemical cell is constructed from a silver half-cell and a copper half-cell. The copper half-cell contains $0.10 \; M$ $Cu(NO_3)_2$ and the concentration of silver ions in the other half-cell is unknown. If the Ag electrode is the cathode and the cell emf is measured and found to be 0.10 V at 25°C, what is the Ag^+ ion concentration?

METHOD OF SOLUTION

Since both E_{cell} and $E°_{cell}$ are known, the Nernst equation can be used to calculate an unknown concentration. To write the reaction quotient Q, we need the cell reaction. If Ag is the cathode, the reduction half-reaction is

cathode $Ag^+ + e^- \rightarrow Ag$

Then Cu must be oxidized:

anode $Cu \rightarrow Cu^{2+} + 2e^-$

The net reaction is

$Cu + 2Ag^+ \rightarrow Cu^{2+} + 2Ag$

where $n = 2$. The Nernst equation is

$$E = E° - \frac{0.0591 \text{ V}}{2} \log \frac{[\text{Cu}^{2+}]}{[\text{Ag}^+]^2}$$

Substitution gives

$$0.10 \text{ V} = 0.46 \text{ V} - \frac{0.0591 \text{ V}}{2} \log \frac{0.10}{[\text{Ag}^+]^2}$$

This equation must be solved for the unknown concentration $[\text{Ag}^+]$:

$$\frac{2(0.10 \text{ V} - 0.46 \text{ V})}{-0.0591 \text{ V}} = \log \frac{0.10}{[\text{Ag}^+]^2}$$

$$12.2 = \log \frac{0.10}{[\text{Ag}^+]^2}$$

Taking the antilog of both sides yields

$$10^{12.2} = \frac{0.10}{[\text{Ag}^+]^2}$$

$$1.5 \times 10^{12} = \frac{0.10}{[\text{Ag}^+]^2}$$

$$[\text{Ag}^+] = \sqrt{\frac{0.10}{1.5 \times 10^{12}}}$$

$$= 2.6 \times 10^{-7} \, M$$

ELECTROLYSIS: FARADAY'S LAW

STUDY OBJECTIVES

You should be able to:
1. Diagram an electrolytic cell, showing the reactions that occur at the anode and cathode.
2. Choose the most likely oxidation and reduction processes to be involved in the electrolysis of a given aqueous solution.
3. Apply Faraday's law to calculate the amount of a substance produced by the flow of a measured electrical current for a given time, or calculate the time required for a given current to produce a given amount of product.

Electrolysis of Molten Salts. Nonspontaneous redox reactions can be forced to occur by the application of an electrical current in an electrolytic cell. For instance, molten sodium chloride can be decomposed into the elements sodium and chlorine in an electrolytic cell. The process in which electrical energy is converted into chemical energy is called *electrolysis*. A schematic diagram of an electrolytic cell is shown in Figure 20.3. A battery or other DC power supply serves as an "electron pump" that supplies electrons to the cathode, where chemical species are reduced. Electrons resulting from the oxidation of chemical species are withdrawn from the anode and return to the battery. In the

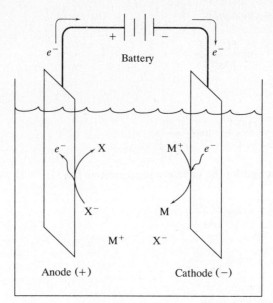

Figure 20.3. Schematic diagram of an electrolytic cell. During electrolysis the cations (M^+) migrate to the cathode, where they are reduced. Anions (X^-) migrate to the anode, where they are oxidized.

electrolytic cell the cathode is negative and the anode is positive. This is the opposite of the galvanic cell. As always, the electrons flow through the external circuit from the anode to the cathode.

Within the cell, current is carried by cations and anions. No current can flow through the cell unless it is filled with a liquid because ions in a solid crystal do not move from place to place. Molten salts such as liquid KCl and NaCl are often used for the study of electrolysis.

Voltage Requirement. The minimum voltage required to bring about electrolysis can be estimated. Consider the following spontaneous reaction:

$$2Ag^+ + Cu \rightarrow 2Ag + Cu^{2+} \qquad E° = 0.46 \text{ V}$$

This reaction can be reversed by connecting the electrodes to an external source of electrical energy such as a battery, as in the electrolytic cell. The electrons could then be made to flow in the opposite direction, *but only if the voltage of the battery is greater than 0.46 V.* Reversing this reaction would cause Cu and Ag^+ ions to be formed in the electrolysis cell.

Competing Electrode Reactions. A complicating factor in the electrolysis of species in aqueous solution is that water molecules may be oxidized or reduced in preference to the solute species. For example, in the electrolysis of KI solution, $H_2(g)$ is formed at the *cathode* and I_2 is formed at the *anode*. Water has a higher (more positive) standard reduction potential than K^+, which indicates that it will be reduced more readily,

$$K^+(aq) + e^- \rightarrow K(s) \qquad E°_{red} = -2.93 \text{ V}$$

$$H_2O(l) + e^- \rightarrow \tfrac{1}{2}H_2(g) + OH^-(aq) \qquad E°_{red} = -0.83 \text{ V}$$

and releases hydrogen at the cathode. We can usually predict the outcome of the electrolysis reaction by consulting the table of standard reduction potentials. Remember that the more positive the $E°$ value, the greater is the tendency for a certain reduction half-reaction to occur.

During KI(aq) electrolysis, I_2 is formed at the anode. This means the anode reaction is

$$2I^-(aq) \rightarrow I_2 + 2e^-$$

rather than

$$H_2O(l) \rightarrow \tfrac{1}{2}O_2(g) + 2H^+(aq) + 2e^-$$

This preference is consistent with the respective oxidation potentials:

$$2I^- \rightarrow I_2(s) + 2e^- \qquad E^\circ_{ox} = -0.53 \text{ V}$$

$$H_2O \rightarrow \tfrac{1}{2}O_2(g) + 2H^+ + 2e^- \qquad E^\circ_{ox} = -1.23 \text{ V}$$

The higher oxidation potential of I^- indicates that it should be more readily oxidized than H_2O. However, keep in mind that *standard oxidation potentials do not always make the correct prediction of which species will be oxidized in an electrolysis reaction*.

The existence of an overvoltage is a complicating factor. *Overvoltage* is the additional voltage over and above the standard potential required to cause electrolysis.

Faraday's Law. Our knowledge of the quantitative relationships in electrolysis is due mostly to the work of Michael Faraday. He observed that the quantity of a substance undergoing chemical change during electrolysis is proportional to the quantity of electrical charge that passes through the cell. The quantity of electrical charge is expressed in coulombs. The electrical charge of 1 mole of electrons is called 1 faraday, which is equal to 96,500 coulombs: 1 mol e^- = 1 F = 96,500 C.

The number of electrons shown in the half-reaction is the link between the number of moles of substance reacted and the number of faradays of electricity required. We obtain the following conversion factors for the two half-reactions shown below:

$$K^+ + e^- \rightarrow K \qquad\qquad Ca^{2+} + 2e^- \rightarrow Ca$$

$$\frac{1 \text{ mol K}}{1 \text{ mol } e} = \frac{1 \text{ mol K}}{1 \text{ F}} \qquad \frac{1 \text{ mol Ca}}{2 \text{ mol } e} = \frac{1 \text{ mol Ca}}{2 \text{ F}}$$

The number of coulombs passing through an electrolytic cell in a given period of time is related to the current in amperes (A):

$$1 \text{ A} = 1 \text{ C/s}$$

In an electrolysis experiment, the current is the quantity measured along with the period of time for which it flows. Therefore, the number of coulombs passing through the cell is

total charge in coulombs (C) = current(A) × time(t)

charge in coulombs = A × t = $\dfrac{C}{s}$ × s = C

When an electrical current of so many amperes flows for a certain time, the amount of chemical change can be calculated according to the following road map:

current × time → coulombs → faradays → moles of reactant → grams of reactant

The number of faradays is the key term. It links the experimental variables to the theoretical yield. Reversing the road map allows one to calculate the time required to produce a specific amount of chemical change with a given electrical current.

EXAMPLE 20.8 Electrolysis of Aqueous Solutions

Predict the products of the electrolysis of aqueous $MgCl_2$ solution.

METHOD OF SOLUTION

In aqueous solution the metal ion is not the only species that can be reduced. H_2O can be reduced as well. The two possible reduction reactions are

$$Mg^{2+} + 2e^- \rightarrow Mg \qquad E°_{red} = -2.37 \text{ V}$$

$$2H_2O + 2e^- \rightarrow H_2 + 2OH^- \qquad E°_{red} = -0.83 \text{ V}$$

Of the two, water has a higher standard reduction potential and therefore a greater tendency to be reduced than Mg^{2+}. The two possible oxidation reactions and their oxidation potentials are

$$2Cl^- \rightarrow Cl_2 + 2e^- \qquad E°_{ox} = -1.36 \text{ V}$$

$$H_2O \rightarrow \tfrac{1}{2}O_2 + 2H^+ + 2e^- \qquad E°_{ox} = -1.23 \text{ V}$$

The oxidation potentials indicate that H_2O is more readily oxidized than the chloride ion. However, in this particular case, as discussed in Section 20.7 of the text, Cl^- is actually oxidized. The large overvoltage for O_2 formation prevents its production when the Cl^- ion is there to compete. The overall reaction is

$$2H_2O + 2Cl^- \rightarrow H_2 + 2OH^- + Cl_2$$

COMMENT

Keep in mind that when two standard oxidation potentials have similar values, as in this case, they are not a reliable predictive tool.

EXAMPLE 20.9 Minimum Voltage Requirement

What is the minimum voltage that must be applied to bring about the electrolysis of a 1.0 M KI solution to give H_2, I_2, and OH^-?

METHOD OF SOLUTION

In the discussion of electrolysis of KI solution, we saw that the half-reactions were

$$\begin{array}{ll} \text{cathode} & 2H_2O(l) + 2e^- \rightarrow H_2(g) + 2OH^-(aq) \\ \text{anode} & 2I^- \rightarrow I_2 + 2e^- \\ \hline \text{overall} & 2H_2O + 2I^- \rightarrow H_2 + I_2 + 2OH^- \end{array}$$

The standard cell emf is

$$E°_{cell} = E°_{ox} + E°_{red}$$

$$= E°_{I^-/I_2} + E°_{H_2O/H_2} = -0.53 \text{ V} + (-0.83 \text{ V}) = -1.36 \text{ V}$$

As expected, this reaction is nonspontaneous.
Answer: To force this reaction to go, at least 1.36 V must be applied to an electrolysis cell.

EXAMPLE 20.10 Faraday's Law

How many faradays are transferred in an electrolytic cell when a current of 12 amps flows for 16 hours?

METHOD OF SOLUTION

First find the number of coulombs passing through the cell in 16 h and convert to faradays:

current × time → coulombs → faradays
12 A = 12 C/s

The number of coulombs in 16 h is given by multiplying the current times the time (in seconds):

$$\frac{12\ C}{1\ s} \times 16\ h \times \frac{3600\ s}{1\ h} = 691{,}000\ C$$

One faraday is the charge of 1 mol of electrons:

$$1\ F = 96{,}500\ C = 9.65 \times 10^4\ C$$

$$691{,}000\ C \times \frac{1\ F}{9.65 \times 10^4\ C} = 7.2\ F$$

EXAMPLE 20.11 Faraday's Law

How many grams of copper metal would be deposited from a solution of $CuSO_4$ by the passage of 3.0 A of electrical current through an electrolytic cell for 2.0 h?

METHOD OF SOLUTION

Cu^{2+} is reduced according to the half-equation

$$Cu^{2+} + 2e^- \rightarrow Cu$$

This half-equation tells us that 2 mol of electrons are required to produce 1 mol of Cu. In terms of faradays this is

$$\frac{2\ mol\ e}{1\ mol\ Cu} = \frac{2\ F}{1\ mol\ Cu}$$

The number of coulombs passing through a cell is given by the current times the time:

$$30\ A \times 2.0\ h \times \frac{1\ C/s}{1\ A} \times \frac{3600\ s}{1\ h} = 2.16 \times 10^4\ C$$

From this number of coulombs we can find the number of faradays of charge passing through the cell. The road map will be

current × time → coulombs → [faradays] → mol Cu → g Cu

The faraday is the link from the current measurements to the molar amount of Cu formed. The number of faradays is

$$2.16 \times 10^4\ C \times \frac{1\ F}{9.65 \times 10^4\ C} = 0.224\ F$$

The number of moles of Cu is

$$0.224 \text{ F} \times \frac{1 \text{ mol Cu}}{2 \text{ F}} = 0.112 \text{ mol Cu}$$

The number of grams of Cu is

$$0.112 \text{ mol Cu} \times \frac{63.5 \text{ g}}{1 \text{ mol Cu}} = 7.11 \text{ g Cu}$$

Of course, this calculation could be carried out using the factor-label method:

$$3.0 \text{ A} \times \frac{1 \text{ C/s}}{1 \text{ A}} \times 2.0 \text{ h} \times \frac{3600 \text{ s}}{1 \text{ h}} \times \frac{1 \text{ F}}{9.65 \times 10^4 \text{ C}} \times \frac{1 \text{ mol Cu}}{2 \text{ F}} \times \frac{63.5 \text{ g}}{1 \text{ mol Cu}} = 7.11 \text{ g Cu}$$

This predicted amount of Cu is based on a process that is 100% efficient. Any side reactions or any oxidation and reduction of impurities will cause the actual yield to be less than the theoretical yield.

EXAMPLE 20.12 Faraday's Law

How long would it take to electrodeposit (plate out) 1.0 g of Ni from a $NiSO_4$ solution using a current of 2.5 A?

METHOD OF SOLUTION

First find how many faradays are required to electrodeposit 1.0 g of Ni from solution. The half-equation is

$$2e^- + Ni^{2+} \rightarrow Ni(s)$$

$$?\text{faradays} = 1.0 \text{ g Ni} \times \frac{1 \text{ mol Ni}}{58.7 \text{ g Ni}} \times \frac{2 \text{ F}}{1 \text{ mol Ni}} = 0.034 \text{ F}$$

Then find how long it will take for 0.034 F of electrical charge to flow through a cell when the current is 2.5 C/s:

$$?\text{seconds} = 0.034 \text{ F} \times \frac{9.65 \times 10^4 \text{ C}}{1 \text{ F}} \times \frac{1 \text{ s}}{2.5 \text{ C}} = 1300 \text{ s } (22 \text{ min})$$

TRUE-FALSE QUESTIONS

1. In a galvanic cell chemical energy is converted into electrical energy.

2. Spontaneous redox reactions are carried out in an electrolytic cell.

3. The anode is the electrode at which the oxidation half-reaction takes place.

4. The standard hydrogen electrode is the reference for the establishment of standard electrode potentials, and its potential is assigned to be 10 V.

5. The weakest oxidizing agent in Table 20.1 in the text is Li^+.

6. Under standard conditions H^+ will oxidize Pb, Zn, and Al, but not Fe.

7. H^+ will not oxidize copper under standard conditions.

8. When $E°_{cell}$ is negative, $\Delta G°$ is positive.

9. The value of $E°$ for an electrode is an intensive property and is independent of the area of the electrode.

10. According to the Nernst equation, if the concentrations of products are increased, E_{cell} will increase (become more positive).

11. In electrochemical calculations, n is the number of moles of electrons transferred from the reducing agent to the oxidizing agent, consistent with the balanced overall equation.

12. Nonspontaneous redox reactions can be made to occur by the application of an external voltage.

13. One ampere equals 1 faraday per second.

14. The reduction of 1 mol of Al^{3+} to Al requires 3 F of electricity.

SELF-TEST A

1. Draw the cell diagrams for each of the redox reactions given below. You may use platinum as an inert electrode:
   a. $2Al(s) + 3H_2SO_4(aq) \rightarrow Al_2(SO_4)_3(aq) + 3H_2(g)$
   b. $Fe_2(SO_4)_3 + 3Pb(s) \rightarrow 3PbSO_4(s) + 2Fe(s)$
   c. $CuSO_4(aq) + H_2(g) \rightarrow Cu(s) + H_2SO_4(aq)$
   d. $2NaBr(aq) + I_2(g) \rightarrow Br_2(l) + 2NaI(aq)$
   e. $2FeCl_2(aq) + SnCl_2(aq) \rightarrow 2FeCl_3(aq) + Sn(s)$
   f. $2FeCl_2(aq) + SnCl_4(aq) \rightarrow 2FeCl_3(aq) + SnCl_2(aq)$

2. Consider a galvanic cell constructed from the following half-cells that are linked by a porous membrane: (1) an Au electrode dipped into 1.0 M $Au(NO_3)_3$; (2) an Fe electrode dipped into 1.0 M $FeSO_4$. Answer the following questions:
   a. Which electrode is the cathode?
   b. Write a balanced equation for the reaction occurring while the cell is discharging.
   c. What emf should the cell generate?
   d. In what direction will electrons flow in the outer circuit?
   e. Toward which electrode will positive ions migrate?

3. Arrange the following species in order of increasing strength as oxidizing agents: Ce^{4+}, O_2, H_2O_2, SO_4^{2-}.

4. Arrange the following species in order of increasing strength as reducing agents: Zn, Ni, H_2, F_2.

5. a. Will $O_2(g)$ oxidize an I^- ion in acid solution under standard conditions?
   b. Will $O_2(g)$ oxidize Br^-?
   c. Will $O_2(g)$ oxidize Cl_2?

6. Calculate $\Delta G°$ and the equilibrium constant (K) for the following reaction:

   $Fe^{3+}(aq) + Ag(s) \rightarrow Fe^{2+}(aq) + Ag^+(aq)$

7. At what ratio of $[Fe^{2+}]/[Fe^{3+}]$ would the reaction given in Problem 5 become spontaneous if $[Ag^+] = 1.0\ M$?

8. Calculate the cell emf for the following reaction:

   $2Ag^+(0.10\ M) + H_2(1\ atm) \rightarrow 2Ag(s) + 2H^+(pH = 8)$

9. What is the minimum voltage required to bring about electrolysis of a solution of Cu^{2+} and Br^- at standard concentrations?

10. What change in the emf of a hydrogen-copper cell will occur when $NaOH(aq)$ is added to the solution in the hydrogen half-cell?

11. Consider a uranium-bromine galvanic cell in which U is oxidized and Br_2 is reduced. The half-reactions are

$$U^{3+}(aq) + 3e^- \rightarrow U(s) \qquad E°_{U^{3+}/U} = ?$$

$$Br_2(l) + 2e^- \rightarrow 2Br^-(aq) \qquad E°_{Br_2/Br^-} = 1.07 \text{ V}$$

If the standard cell emf is 2.91 V, what is the standard reduction potential for uranium?

12. When the concentration of Zn^{2+} is 0.15 M, the measured voltage of the Zn-Cu galvanic cell is 0.40 V. What is the Cu^{2+} ion concentration?

13. A hydrogen electrode is immersed in an acetic acid solution. This electrode is connected to another consisting of an iron nail dipping into 0.10 M $FeCl_2$. If E_{cell} is found to be 0.24 V, what is the pH of the acetic acid solution?

14. How many faradays are transferred in an electrolytic cell when a current of 2.0 amps flows for 6 hours?

15. How many faradays are required to electroplate 6.0 g of chromium from a solution containing Cr^{3+}?

16. a. How many grams of nickel can be electroplated by passing a constant current of 5.2 A through a solution of $NiSO_4$ for 60.0 min?
    b. How many grams of cobalt can be electroplated by passing a constant current of 5.2 A through a solution of $CoCl_3$ for 60.0 min?

17. How long will it take to produce 54 kg of Al metal by the reduction of Al^{3+} in an electrolytic cell using a current of 500 amps?

GENERAL PROBLEMS

18. Given the reduction potential of the SHE and the half-equation

$$2H_2O + 2e^- \rightarrow H_2(g) + 2OH^-(aq) \qquad E° = -0.83 \text{ V}$$

calculate the value of the ionization constant for water:

$$H_2O \rightarrow H^+(aq) + OH^-(aq)$$

19. A cell was constructed using the standard hydrogen electrode as one half-cell and a lead electrode in a 0.10 M K_2CrO_4 solution in contact with undissolved $PbCrO_4$ as the other half-cell. The potential of the cell was measured to be 0.50 V with the Pb electrode as the anode. From these data determine the K_{sp} of $PbCrO_4$.

20. A 40-watt light bulb is powered by a lead storage battery that has 20.0 g of Pb available for the anode. For how many hours will the bulb provide illumination? Assume the voltage is constant at 1.5 V. Recall: 1 watt = 1 J/s.

SELF-TEST B1

1. How many faradays are transferred in an electrolytic cell when a current of 5.0 A flows for 10.0 hours?

2. How many faradays are required to electroplate 29.4 g of nickel from a solution containing the Ni^{2+} ion?

3. What mass of copper metal can be produced by the electrolysis of a copper(II) sulfate solution at a constant current of 20.0 A for 1.00 h?

4. Find the standard cell potential for an electrochemical cell in which the following reaction takes place:

$$Cl_2(g) + 2Br^-(aq) \rightarrow Br_2(l) + 2Cl^-(aq)$$

5. Calculate the equilibrium constant for the following redox reaction at 25°C:

$$2Fe^{2+}(aq) + Ni^{2+}(aq) \rightarrow 2Fe^{3+}(aq) + Ni(s)$$

6. Determine the cell voltage for the following reaction where the concentrations are as shown:

$$2Ag^+(1\ M) + H_2(1\ atm) \rightarrow 2Ag(s) + 2H^+(pH = 7.12)$$

7. Calculate the standard free energy change $\Delta G°$ for the following redox reaction at 25°C:

$$3Sn(s) + 2NO_3^-(aq) + 8H^+(aq) \rightarrow 3Sn^{2+}(aq) + 2NO(g) + 4H_2O$$

SELF-TEST B2

1. How many hours are required for 1.87 F of charge to flow through a cell if the current is constant at 5.0 A?

2. What mass of nickel can be electroplated from a solution of $NiCl_2$ by 1.00 F of electricity?

3. How many minutes are required to electroplate 23.7 g of metallic Cu using a constant current of 20.0 A?

4. The standard cell voltage for the following reaction is 0.29 V.

$$Cl_2(g) + 2Br^-(aq) \rightarrow 2Cl^-(aq) + Br_2(l)$$

The reduction half reaction is

$$2e^- + Cl_2(g) \rightarrow 2Cl^-(aq) \quad E° = 1.36\ V$$

Determine the standard reduction potential of $Br_2(l)$?

5. Consider the following redox reaction:

$$2Fe^{2+}(aq) + Ni^{2+}(aq) \rightleftharpoons 2Fe^{3+}(aq) + Ni(s)$$

The equilibrium constant $K = 3.0 \times 10^{-35}$ at 25°C. Calculate the value of the standard cell voltage.

6. The cell voltage for the following reaction is 1.23 V:

$$2Ag^+(aq, 1\ M) + H_2(g, 1\ atm) \rightarrow 2Ag(s) + 2H^+(aq, ?M)$$

Calculate the concentration of hydrogen ions.

7. Consider the following redox reaction:

$$3Sn(s) + 2NO_3^-(aq) + 8H^+(aq) \rightarrow 3Sn^{2+}(aq) + 2NO(g) + 4H_2O$$

The standard free energy change $\Delta G° = -637$ kJ at 25°C. Calculate the standard cell voltage for this reaction.

ANSWERS

TRUE-FALSE QUESTIONS

1. True.
2. False. Spontaneous redox reactions are carried out in a galvanic cell.
3. True.
4. False. It is assigned 0.0 V.
5. True.
6. False. H^+ will oxidize Fe along with the others.
7. True.
8. True.
9. True.
10. False. E_{cell} decreases as product concentrations increase.
11. True.
12. True.
13. False. 1 A = 1 C/s.
14. True.

SELF-TEST A

1. a. $Al(s)|Al^{3+}(aq)\|H^+(aq)|H_2(g)|Pt(s)$
   b. $Pb(s)|Pb^{2+}(aq)\|Fe^{3+}(aq)|Fe(s)$
   c. $Pt(s)|H_2(g)|H^+(aq)\|Cu^{2+}(aq)|Cu(s)$
   d. $Pt(s)|Br^-(aq)|Br_2(l)\|I_2(s)|I^-(aq)|Pt(s)$
   e. $Pt(s)|Fe^{2+}(aq), Fe^{3+}(aq)\|Sn^{2+}(aq)|Sn(s)$
   f. $Pt(s)|Fe^{2+}(aq), Fe^{3+}(aq)\|Sn^{4+}(aq), Sn^{2+}(aq)|Pt(s)$
2. a. Au   b. $2Au^{3+}(aq) + 3Fe(s) \rightarrow 2Au(s) + 3Fe^{2+}(aq)$ c. 1.94 V   d. from Fe to Au
   e. Au
3. $SO_4^{2-} > O_2 > Ce^{4+} > H_2O_2$
4. $F_2 < H_2 < Ni < Zn$
5. a. Yes   b. Yes   c. No
6. $\Delta G° = 2.9$ kJ; $K = 0.31$
7. $[Fe^{2+}]/[Fe^{3+}] = 0.31$
8. $E = 1.21$ V
9. 0.73 V

10. The cell emf will increase due to a decrease in [H^+]. The cell reaction is $Cu^{2+} + H_2 \rightarrow Cu + 2H^+$.
11. -1.84 V
12. [Cu^{2+}] = 3.1×10^{-25} M
13. pH = 3.88
14. 0.45 F
15. 0.35 F
16. a. 5.69 g Ni   b. 3.81 g Co
17. 322 h
18. $K_w = 9.0 \times 10^{-15}$
19. [Pb^{2+}] = 3.0×10^{-13} M; $K_{sp} = 3.0 \times 10^{-14}$
20. 0.19 h

SELF-TEST B1

1. 1.87 F
2. 1.00 F
3. 23.7 g
4. $E°_{cell} = 0.29$ V
5. $K = 3.0 \times 10^{-35}$
6. 1.23 V
7. $\Delta G° = -637$ kJ

SELF-TEST B2

1. 10.0 h
2. 29.4 g
3. 60 min
4. 1.07 V
5. $E°_{cell} = -1.02$ V
6. 7.6×10^{-8} M
7. $E°_{cell} = 1.10$ V

Chapter Twenty-One
METALLURGY AND THE CHEMISTRY OF METALS

- Metallurgical Processes
- Bonding in Metals: The Band Theory
- The Alkali Metals
- The Alkaline Earth Metals
- Aluminum

METALLURGICAL PROCESSES

STUDY OBJECTIVES

You should be able to:
1. List the procedures used to separate minerals from gangue.
2. Describe briefly the means used to produce and purify a metal.
3. Describe the steps in the production of iron and copper.

Preliminary Treatment. Metallurgy is the science and technology of extracting metals from their ores and preparing these metals for use. The steps necessary to prepare a metal for use are outlined in this section. Most metals are found in nature in the combined state. Ores generally must be cleaned before the metal is extracted. The waste materials, called gangue, are removed by *flotation, ferromagnetic separation*, or *distillation of mercury amalgams*.

Reduction of Metals. The process of producing a free metal is always one of reduction. *Roasting* is a preliminary operation used to convert a sulfide or a carbonate into an oxide. Reduction of oxides is usually less complex than reduction of sulfides, for instance.

Two types of reductions are ordinarily employed:

1. *Chemical reduction* is the use of a reducing agent to prepare the elemental form of a metal from a compound. Calcium, magnesium, aluminum, hydrogen, and carbon are often used in chemical reductions.
2. *Electrolytic reduction* is necessary to produce the very electropositive metals such as sodium, magnesium, and aluminum.

Iron. The raw materials for making iron are (1) iron ore, either hematite, Fe_2O_3, or magnetite, Fe_3O_4; (2) coke; and (3) limestone. These three materials are mixed and fed into a huge blast furnace (Figure 21.3 in the text). A strong blast of air preheated to 1500°C is blown in at the bottom, where oxygen in the air reacts with the coke. This reaction supplies heat for the furnace and carbon monoxide, which is the reducing agent in the reaction of the iron oxides:

$$2C + O_2 \rightarrow 2CO$$

$$Fe_2O_3(s) + 3CO(g) \rightarrow 2Fe(l) + 3CO_2(g)$$

A temperature gradient is set up in the furnace. The temperature at the bottom is 1500°C, and it decreases with height, falling to about 250°C at the top. Much of the resulting carbon dioxide is reduced by reaction with excess coke,

$$CO_2 + C \rightarrow 2CO$$

thereby forming more of the principal reducing agent. The gas that escapes at the top of the furnace is mostly nitrogen and carbon monoxide. The molten iron runs to the bottom where it is withdrawn periodically.

Limestone serves as a flux in the removal of impurities such as silica (SiO_2) and alumina (Al_2O_3). The limestone decomposes at temperatures above 900°C, forming calcium oxide and carbon dioxide:

$$CaCO_3 \rightarrow CaO + CO_2$$

Calcium oxide unites with silica and alumina to form a glassy, molten substance called *slag*, which is composed mainly of $CaSiO_3$ and some calcium aluminate. Slag is less dense than iron, and so it collects as a pool on top of the metal. It is drawn off, leaving the molten iron. Iron prepared in this manner contains many impurities and is called pig iron or cast iron. Cast iron is made into steel by further treatment in a basic oxygen furnace.

Purification of Metals. Once a metal is prepared, it may need to be further purified. Distillation, electrolysis, and zone refining are three common procedures.

Metals with low boiling points, such as mercury (357°C), cadmium (767°C), and zinc (907°C), are referred to as volatile. They can be purified by *fractional distillation*. On heating, these metals vaporize, leaving behind any nonvolatile impurities. Condensation of the vapor yields the purified metal.

Metals can also be purified by *electrolysis*. The metals that plate out on a cathode can be controlled by the voltage at which the electrolysis is carried out. In the purification of copper, Cu^{2+} is reduced much more readily than iron or zinc ions, which are common impurities. Thus the more electropositive metal impurities are removed in this way.

Zone refining takes advantage of the fact that when liquids begin to freeze, the impurities tend to remain in the liquid phase. See Figure 21.8 of the textbook. Extremely pure metals can be obtained by repeating this process a number of times.

EXAMPLE 21.1 Ores and Minerals

As we will see in the next chapter, beryl is the only important beryllium ore, yet there are other minerals richer in this metal. These minerals, however, are too costly and scarce to be used for its preparation. Distinguish between an ore and a mineral.

METHOD OF SOLUTION

A mineral is a naturally occurring substance with a characteristic range of chemical composition. An ore is a mineral, or a mixture of minerals, from which a particular metal can be *profitably* extracted.

EXAMPLE 21.2 Chemical Reduction

Why are calcium, magnesium, aluminum, hydrogen, and carbon often used in chemical reductions?

392 / Metallurgy and the Chemistry of Metals

METHOD OF SOLUTION

These substances are good reducing agents. They all have high oxidation potentials, which means they have a strong tendency to lose electrons and thereby reduce other metal cations to the pure metal:

$$Ca \rightarrow Ca^{2+} + 2e^- \quad E°_{ox} = 2.87 \text{ V}$$

$$Mg \rightarrow Mg^{2+} + 2e^- \quad E°_{ox} = 2.37 \text{ V}$$

$$Al \rightarrow Al^{3+} + 3e^- \quad E°_{ox} = 1.66 \text{ V}$$

Ca, Mg, Al, and C as well are used to reduce most of the active metals. H_2 is used to reduce the more noble metals such as copper.

EXAMPLE 21.3 Electrolytic Reduction

Why must electrolytic reduction be used in the production of some metals such as lithium and sodium?

METHOD OF SOLUTION

The electropositive metals can be reduced by a more electropositive metal such as Li, but there is no metal electropositive enough to reduce Li^+ to Li. Electrolytic reduction is the only way to prepare Li.

It is possible to reduce other metals with Li. Rather than prepare Li by electrolysis and then use it to reduce aluminum, magnesium, or sodium, it is more convenient simply to prepare all these metals in one step by electrolytic reduction.

EXAMPLE 21.4 Steel Production

How are the impurities present in iron ore, such as silica and sulfur, removed during steel production?

METHOD OF SOLUTION

Acidic impurities such as silica and sulfides combine with a base such as CaO to form a molten slag:

$$SiO_2 + CaO \rightarrow CaSiO_3$$

$$FeS + CaO \rightarrow CaS + FeO$$

The slag has a lower density than molten iron. Therefore it collects as a pool on top of the metal. This slag is drawn off and used as a component in making cement.

BONDING IN METALS: THE BAND THEORY

STUDY OBJECTIVES

You should be able to:
1. Compare the energies of the conduction bands and valence bands in metals, insulators, and semiconductors.
2. Use the band theory to describe conduction in metals.

The Conduction Band. The band theory is the result of the application of molecular orbital theory to metals. In metals, the atoms lie in a three-dimensional array and take part in bonding that

Figure 21.1. The formation of an energy band by the successive overlap of atomic orbitals. (*a*) When the orbitals of two atoms overlap, one bonding orbital and one antibonding orbital of significantly different energies are formed. (*b*) When the orbitals of four atoms overlap, four molecular orbitals are formed. (*c*) When the orbitals of N atoms overlap and N is very large, as in Avogadro's number, N orbitals are formed, but the orbital energies differ only infinitesimally from each other and form a virtually continuous band.

spreads over the entire crystal. Take, for example, an alkali metal atom that carries a single valance electron in an s orbital. Recall that the atomic orbitals of two atoms will overlap when the atoms are close together. This results in the formation of two molecular orbitals: a bonding orbital and an antibonding orbital. The total number of molecular orbitals produced always equals the number of atomic orbitals that overlap to produce them. The overlap of four atomic orbitals from four atoms, for instance, will form four molecular orbitals: two bonding and two antibonding orbitals. A crystal made up of N atoms with overlapping atomic orbitals will have $\frac{1}{2}N$ bonding and $\frac{1}{2}N$ antibonding molecular orbitals. These bonding and antibonding orbitals are so closely spaced in terms of energy that they are called a "band." The formation of a band is shown in Figure 21.1.

The band containing the valence shell electrons ($3s$ for Na) is called the *valance band*. For alkali metals there are N valence electrons and N orbitals in the band. These N orbitals have a capacity for $2N$ electrons. Any band that is either vacant or partially filled is called a *conduction band*.

Electrical Conduction. The band theory explains conduction in the alkali metals in the following way. When a voltage is applied across a piece of sodium metal, conduction occurs. The current is the result of electrons in the $3s$ band being fee to jump from atom to atom. The free movement of electrons is possible for two reasons. In alkali metals the conduction band and the valence band are the same. The orbitals within the band are so similar in energy that an electron does not need to gain appreciable energy to reach the conduction band. Also, as discussed above, the conduction band is only half filled to capacity, which means that an electron is free to move through the entire metal.

In an insulator, such as glass or plastic, the valence band is filled. Thus the next vacant higher energy band becomes the conduction band. An energy gap exists between the valence band and the conduction band, as shown in Figure 21.10 of the text. This large separation prevents electrons in insulators from entering the conduction band.

A *semiconductor*, such as Si or Ge, has a filled valence band and an empty conduction band, but in contrast to an insulator, a relatively small gap exists between these bands. A relatively small amount of thermal energy will promote an electron into the conduction band. Thus as temperature increases, the conductivity of semiconductors increases correspondingly.

EXAMPLE 21.5 Band Theory

Show that for magnesium the valence band is completely filled.

METHOD OF SOLUTION

Magnesium atoms have filled $1s$, $2s$, and $2p$ orbitals; therefore the corresponding bands in the solid are filled. The valence band for Mg is the $3s$. In the metal the $3s$ orbitals of N magnesium atoms overlap to form a total of N molecular orbitals that make up the valence band of Mg. The capacity of this band is $2N$ electrons. Since each Mg atom has two $3s$ electrons, the N magnesium atoms have $2N$ electrons in the valence band. This means that the valence band is filled. The $3p$ band of Mg must be the conduction band. The band structure of Mg is

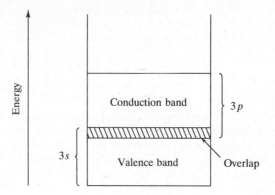

COMMENT

At first glance it appears that Mg should be an insulator, since the valence band is filled. However, Mg is metallic and does conduct electricity, as our experience suggests it should. The reason Mg conducts is that the empty $3p$ orbitals form an energy band that, because of the closeness in energy of the $3s$ and $3p$ orbitals, overlaps with the $3s$ band. Therefore, electrons from the valence band (the $3s$) can easily enter the conduction band (the $3p$).

THE ALKALI METALS

STUDY OBJECTIVES
1. Describe the chemical properties of the alkali metal elements.
2. Describe the sources and preparation of sodium and potassium metals.

As a group, the alkali metals are extremely reactive and never occur naturally in elemental form. They are the least electronegative group of elements and exist as $+1$ cations combined with halides, sulfate, carbonate, and silicate ions. A number of properties of these elements are listed in Table 21.4 of the text. Compared with other metals, they are very soft and have low melting points and low densities. Lithium, sodium, and potassium will float on water. The low density is a result of their large atomic volume and the fact that they all possess a body-centered crystal structure that has a low packing efficiency.

Sources and Preparations. Lithium is obtained principally from spodumene ($LiAlSi_2O_6$) and from refining of brine. The element was discovered in 1817, but it was not isolated as the free metal until 1855.

The preparation of Group 1A metals requires large amounts of energy because their positive ions are difficult to reduce. Lithium, being the most active metal, is prepared by electrolysis of molten LiCl.

Sodium and potassium occur in a wide variety of minerals. Sodium compounds are so abundant and widespread that it is difficult to find matter free of this element. Sodium chloride makes up about two-thirds of the solid matter dissolved in sea water. Sodium is the sixth most abundant element on Earth and is the most abundant of the alkali metals.

Potassium is the seventh most abundant element on Earth. The minerals polyhalite, langbeinite ($K_2SO_4 \cdot 2MgSO_4$), and sylvite (KCl) are found in ancient lake and seabeds and serve as commercial sources of potassium and its compounds.

Both metals, sodium and potassium, were first prepared in 1807 by the English chemist Humphry Davy. He used the electrolysis of the corresponding moist hydroxides. As their discoverer, Davy was allowed to name the elements, giving them the above names. However, you will notice that neither has an element symbol consistent with its name. At the time both potash and soda were recognized as carbonates of these unisolated metal elements. Potassium obtained its name because it was the metallic element in potash, potassium carbonate. The metallic element of soda (sodium carbonate) Davy called sodium.

The symbols for the elements sodium and potassium are, of course, Na and K, reflecting the strong German influence in chemistry in the early nineteenth century. The German chemists used the Latin names for these elements, which were natron and kalium. The symbols Na and K are derived from these names.

Sodium and Potassium.

Metallic sodium is prepared commercially by the electrolysis of molten sodium chloride. Most of the sodium made in the United States is produced in the Downs cell by the electrolysis of a mixture of sodium chloride and calcium chloride. This electrolyte mixture melts at 505°C, whereas pure NaCl melts at 801°C. The lower temperature reduces the cost of production.

A Downs cell consists of a carbon anode and an iron cathode. The chloride ions are oxidized at the anode, and the sodium ions are reduced at the cathode. The half-reactions are

$$\text{anode} \quad 2Cl^- \rightarrow Cl_2(g) + 2e^-$$

$$\text{cathode} \quad 2Na^+ + 2e^- \rightarrow 2Na(l)$$

The liquid sodium is drawn off and kept from contact with chlorine and oxygen. Sodium can also be obtained by the electrolysis of molten sodium hydroxide. Potassium is made by reaction of potassium chloride with sodium vapor in the absence of air:

$$KCl(l) + Na(g) \xrightarrow{\Delta} K(g) + NaCl(l)$$

This reaction should occur to only a small extent. However, removal of potassium vapor as it is formed drives the reaction to completion.

Reactions of Alkali Metals.

The alkali metals are extremely reactive. The reactivity increases with atomic mass. Lithium metal combines directly with nitrogen to form lithium nitride:

$$6Li(s) + N_2(g) \rightarrow 2Li_3N(s)$$

It also combines with oxygen, the halogens, and hydrogen. Lithium reacts with water to yield the hydroxide in a manner similar to that of sodium.

Sodium forms two oxides on reaction with oxygen: sodium oxide (Na_2O) and sodium peroxide (Na_2O_2),

$$2Na(s) + \tfrac{1}{2}O_2(g) \rightarrow Na_2O(s)$$

$$2Na(s) + O_2(g) \rightarrow Na_2O_2(s)$$

Potassium burns in air to form potassium superoxide:

$$K(s) + O_2(g) \rightarrow KO_2(s)$$

THE ALKALINE EARTH METALS

STUDY OBJECTIVES

You should be able to:
1. List the mineral ores that serve as sources of the alkaline earth metals, and write chemical equations to show the preparations of these elements.
2. Describe the chemical properties and cite the major uses of these elements and their compounds.

Sources and Preparations. The important ores of the Group 2A metals and their formulas are given in Table 21.1.

Table 21.1 Ores of the Alkaline Earth Metals

Element	Rank (Earth's Crust)	Ore	Formula
Beryllium	32nd	Beryl	$Be_3Al_2Si_6O_{18}$
Magnesium	8th	Brucite	$Mg(OH)_2$
		Dolomite	$MgCO_3 \cdot CaCO_3$
		Epsomite	$MgSO_4 \cdot 7H_2O$
Calcium	5th	Limestone	$CaCO_3$
		Gypsum	$CaSO_4 \cdot 2H_2O$
		Fluorite	CaF_2
Strontium	—	Strontianite	$SrCO_3$
		Celestite	$SrSO_4$
Barium	—	Witherite	$BaCO_3$
		Barytes	$BaSO_4$

Pure beryllium is obtained by first converting the ore (beryl) to the oxide (BeO). The oxide is then converted to the fluoride or chloride. Then the salt is reduced with magnesium by heating to about 1000°C to give beryllium:

$$BeF_2(s) + Mg(l) \rightarrow Be(s) + MgF_2(s)$$

Seawater is an important source of the element magnesium. It contains 0.13% Mg^{2+} by weight. $Mg(OH)_2$ is precipitated from seawater by adding $Ca(OH)_2$ until a high enough concentration of OH^- is reached:

$$Mg^{2+}(aq) + Ca^{2+}(aq) + 2OH^-(aq) \rightarrow Mg(OH)_2(s) + Ca^{2+}(aq)$$

Neutralization of magnesium hydroxide with hydrochloric acid yields magnesium chloride. Metallic magnesium is obtained by electrolysis of molten magnesium chloride.

$$MgCl_2(l) \xrightarrow{\text{electrolysis}} Mg(l) + Cl_2(g)$$

Calcium metal is prepared by a thermal reduction process similar to that for preparation of beryllium. Calcium oxide, obtained from heating limestone, is reduced with aluminum at high temperature (1200°C):

$$CaCO_3(s) \rightarrow CaO(s) + CO_2(g)$$

$$6CaO(s) + 2Al(l) \rightarrow 3Ca(l) + Ca_3Al_2O_6(s)$$

Continuing down the column of alkaline earth elements in the periodic table, we come to strontium and barium. Strontium is prepared by electrolysis of molten $SrCl_2$, while barium can be prepared by electrolysis of the chloride, or by thermal reduction of barium oxide with aluminum.

Properties of Beryllium. In proceeding down the column of Group 2A elements, we note increasing metallic properties. Beryllium, the first element in the group, is the most nonmetallic. Several of the differences between Be and the rest of the Group 2A members are summarized as follows:

1. Be is quite *unreactive* toward oxygen and water at room temperature.
2. BeF_2 and $BeCl_2$ are covalent compounds, in contrast to the salts of the rest of the alkaline earth metals, which are ionic.
3. Be^{2+} is extensively hydrolyzed in water and resembles Al^{3+} in its ability to produce acidic solutions:

$$Be(H_2O)_4^{2+} \rightleftharpoons Be(H_2O)_3(OH)^+ + H^+$$

$$Al(H_2O)_6^{3+} \rightleftharpoons Al(H_2O)_5(OH)^{2+} + H^+$$

Such a resemblance is called a *diagonal relationship*. Other well-known examples are (1) Li and Mg and (2) B and Si.

4. The other alkaline earth oxides are basic,

$$MO(s) + H_2O(l) \rightarrow M(OH)_2(s)$$

but beryllium oxide is amphoteric and resembles aluminum oxide. It reacts with acids, as does aluminum oxide:

$$BeO + 2H^+ + 3H_2O \rightarrow Be(H_2O)_4^{2+}$$

$$Al_2O_3 + 6H^+ + 9H_2O \rightarrow 2Al(H_2O)_6$$

And it reacts with bases:

$$BeO + 2OH^- + H_2O \rightarrow Be(OH)_4^{2-}$$

$$Al_2O_3 + 6OH^- + 3H_2O \rightarrow 2Al(OH)_6^{3-}$$

Properties of Magnesium and Calcium. The chemistry of magnesium is intermediate between that of beryllium and calcium. Magnesium does not react with cold water, but it reacts slowly with steam. Calcium, on the other hand, reacts with cold water:

$$Mg(s) + H_2O(g) \rightarrow MgO(s) + H_2(g)$$

$$Ca(s) + 2H_2O(l) \rightarrow Ca(OH)_2(aq) + H_2(g)$$

Both MgO and CaO react with water to give hydroxides:

$$MgO(s) + H_2O(g) \rightarrow Mg(OH)_2(s) \quad \text{(reacts with steam)}$$

$$CaO(s) + H_2O(l) \rightarrow Ca(OH)_2(s) \quad \text{(reacts with cold water)}$$

Both Mg and Ca react with hydrochloric and sulfuric acids to produce hydrogen gas:

$$Mg(s) + 2H^+(aq) \rightarrow Mg^{2+}(aq) + H_2(g)$$

$$Ca(s) + 2H^+(aq) \rightarrow Ca^{2+}(aq) + H_2(g)$$

On heating, Mg and Ca react with a number of nonmetals such as Cl_2, S, N_2, and O_2. For example, when magnesium burns in air, considerable nitride is formed along with the oxide:

$$2Mg(s) + O_2(g) \rightarrow 2MgO(s)$$

$$3Mg(s) + N_2(g) \rightarrow Mg_3N_2(s)$$

Properties of Strontium and Barium. The properties of strontium and barium resemble those of calcium. As with the other families, the oxides and salts of the heavier, larger sized members are more ionic, and more reactive, than those of the lighter members. This trend of increasing metallic character as you look down a group of elements is consistent with BeO being amphoteric, while BaO is very basic.

Uses of the Alkaline Earth Metals. Beryllium has excellent alloying qualities. It is used in tools to provide a nonsparking and nonmagnetic alloy. Beryllium alloys resist corrosion. Beryllium is used as a "window" in X-ray tubes.

The major uses of magnesium are also in alloys. Because of its low atomic mass, it is a good lightweight structural metal. It is also used in flash bulbs, in batteries, and for cathodic protection of buried metal pipelines and storage tanks.

Metallic calcium finds use mainly in alloys. Calcium serves to harden lead that is a component of cables. Calcium salts are used as dehydrating agents; anhydrous calcium chloride, for example, has a strong affinity for water.

Strontium nitrate and carbonate are used in fireworks and highway flares to provide a brilliant red color. Barium metal is the most active of the alkaline earth metals so it has very few uses. Its reactivity is employed in high vacuum work to remove the last traces of oxygen and nitrogen gases.

EXAMPLE 21.6 Properties of Alkaline Earth Metals

Account for the trend in solubilities of alkaline earth hydroxides given in Table 21.6 of the text.

METHOD OF SOLUTION

The table shows that the solubilities of the hydroxides increase with increasing atomic radius of the alkaline earth atoms. As the degree of ionic character of the M — OH bond increases, so will the solubility of the metal hydroxide. The electronegativity of Be is quite high for a metal (1.5), and for the other Group 2A metals it decreases as we read down the group. Thus the electronegativity difference between M and O increases as the atomic radius of the metal atom increases, and the bonds become increasingly ionic, $M^{\delta+}$— $OH^{\delta-}$. The greater the ionic character of the M — O bond, the more readily the hydroxides dissociate in aqueous solution. $Be(OH)_2$ is much more covalent than $Ba(OH)_2$.

EXAMPLE 21.7 Properties of Alkaline Earth Metals

Write chemical equations describing the reaction of H_2O with the alkaline earth metals.

METHOD OF SOLUTION

In the discussion we saw that barium, strontium, and calcium react with cold water to give the corresponding hydroxide. Magnesium reacts slowly with boiling water, but beryllium is inert. Mg, Ca, Sr, and Ba react according to the equation

$$M(s) + 2H_2O(l) \rightarrow M(OH)_2(aq) + H_2(g)$$

EXAMPLE 21.8 Electrolytic Reduction of Calcium

Starting with Ca^{2+} ions in limestone, write equations to show how you would obtain pure calcium using electrolysis rather than thermal reduction.

METHOD OF SOLUTION

Electrolysis of molten metal chlorides is often used to prepare pure metals. First limestone, chalk, or sea shells, all of which contain $CaCO_3$, can be calcined above 900°C. Calcination is the conversion of a metal carbonate to its oxide by heating:

$$CaCO_3(s) \xrightarrow{\Delta} CaO(s) + CO_2(g)$$

The $CaO(s)$ is slaked to yield calcium hydroxide:

$$CaO(s) + H_2O(l) \rightarrow Ca(OH)_2(s)$$

This base is then neutralized with hydrochloric acid to give the desired $CaCl_2$:

$$Ca(OH)_2(aq) + HCl(aq) \rightarrow CaCl_2(aq) + 2H_2O(l)$$

After drying, the $CaCl_2$ can be melted and then electrolyzed:

$$CaCl_2(l) \xrightarrow{electrolysis} Ca(l) + Cl_2(g)$$

ALUMINUM

STUDY OBJECTIVES

You should be able to:
1. Write chemical equations to show the reactions of aluminum with acids, bases, oxygen, and other metal oxides.
2. Write chemical equations that show the amphoteric nature of $Al(OH)_3$.
3. Mention several uses of aluminum and its compounds.

Source and Preparation. Aluminum is the third most abundant element in Earth's crust, making up 7.5% by mass. It is too active chemically to occur free in nature, but it is found in compounds in over 200 different minerals. The most important ore of Al is bauxite, which contains hydrated aluminum oxide ($Al_2O_3 \cdot 2H_2O$), along with silica and hydrated iron oxide ($Fe_2O_3 \cdot 2H_2O$). Another important mineral is cryolite (Na_3AlF_6), which is used in the metallurgy of aluminum. The Hall process for the production of aluminum metal is described in Figure 21.26 of the text.

Aluminum is obtained from bauxite ($Al_2O_3 \cdot 2H_2O$). The first step in its preparation is to separate pure aluminum oxide from the silica and iron impurities. The bauxite is pulverized and digested with sodium hydroxide solution. This converts the silica into soluble silicates, and the aluminum oxide is converted to the aluminate ion $AlO_2^-(aq)$, which remains in solution. However, this digestion treatment has no effect on the iron that remains as insoluble $Fe_2O_3(s)$ and is removed by filtration. Aluminum hydroxide is then precipitated by acidification (with carbonic acid) to about pH 6. The precipitate is heated strongly to produce pure Al_2O_3. The chemical changes involving aluminum are

$$Al_2O_3(s) + 2OH^-(aq) \rightarrow 2AlO_2^-(aq) + H_2O(l)$$

$$AlO_2^-(aq) + H^+(aq) + H_2O(l) \rightarrow Al(OH)_3(s)$$

$$2Al(OH)_3(s) \rightarrow Al_2O_3(s) + 3H_2O(g)$$

The aluminum ions in Al_2O_3 can be reduced to metallic aluminum efficiently only by electrolysis. The melting point of Al_2O_3 is 2050°C, which makes electrolysis of pure molten Al_2O_3 extremely expensive owing to the need to maintain the high temperature. In the Hall process, Al_2O_3 is dissolved in molten cryolite (Na_3AlF_6), which melts at 1000°C. The use of cryolite makes it possible to lower the temperature of electrolysis by 1050°C! The mixture is electrolyzed to produce aluminum and oxygen:

$$\begin{array}{ll} \text{cathode} & 2Al^{3+} + 6e^- \rightarrow 2Al(l) \\ \text{anode} & 3O^{2-} \rightarrow \tfrac{3}{2}O_2 + 6e^- \\ \hline \text{overall} & 2Al^{3+} + 3O^{2-} \rightarrow 2Al(l) + \tfrac{3}{2}O_2 \end{array}$$

Properties. Pure aluminum is a silvery-white metal with low density (2.7 g/cm^3) and high tensile strength. It is malleable and can be rolled into thin foils. Its electrical conductivity is about 65% that of copper.

Aluminum is an amphoteric element, reacting with both acids and bases:

$$2Al(s) + 6HCl(aq) \rightarrow AlCl_3(aq) + 3H_2(g)$$

$$2Al(s) + 6NaOH(aq) \rightarrow 2Na_3AlO_3(aq) + 2H_2(g)$$

Aluminum has a strong affinity for oxygen; the metal is usually covered by an oxide film:

$$4Al(s) + 3O_2(g) \rightarrow 2Al_2O_3(s)$$

This layer of Al_2O_3 forms a compact, adherent, protective, surface coating on the metal and is responsible for preventing further oxidation and corrosion. Because of this surface oxide layer, Al is practically insoluble in weak or dilute acids.

The large enthalpy of formation of aluminum oxide ($\Delta H_f^\circ = -1670$ kJ/mol) makes the metal an excellent reducing agent. Thus a variety of metals can be produced in a series of similar reactions involving Al powder with the corresponding metal oxides. So much heat is liberated in these reactions, called *aluminothermic reactions*, that the metal is usually obtained in the molten state. The *thermite reaction*, used in welding steel and iron, is one example:

$$2Al(s) + Fe_2O_3(s) \rightarrow Al_2O_3(l) + 2Fe(l)$$

Aluminum hydroxide is an amphoteric hydroxide, dissolving in both acid and base:

$$Al(OH)_3(s) + 3H^+(aq) \rightarrow Al^{3+}(aq) + 3H_2O(l)$$

$$Al(OH)_3(s) + OH^-(aq) \rightarrow Al(OH)_4^-(aq)$$

Uses. Aluminum metal is used in high voltage transmission lines because it is cheaper and lighter than copper. Its chief use is in aircraft construction. Aluminum is employed also as a solid propellant in rockets. This is another example of the great affinity that aluminum has for oxygen. Ammonium perchlorate, NH_4ClO_4, is the oxidizer in the reaction.

The formation of aluminum hydroxide precipitate is used in water treatment plants. The process requires large amounts of aluminum sulfate:

$$Al_2(SO_4)_3(aq) + 3Ca(OH)_2(aq) \rightarrow 2Al(OH)_3(s) + 3CaSO_4(aq)$$

Aluminum hydroxide is a gelatinous substance. As it settles in treatment pools, it coprecipitates suspended matter such as bacteria and colloidal sized particles. This process clarifies drinking water.

Alums are compounds that have the general formula

$$M^+M^{3+}(SO_4)_2 \cdot 12H_2O$$

where $M^+ = K^+, Na^+, NH_4^+$ and $M^{3+} = Al^{3+}, Cr^{3+}, Fe^{3+}$. They are used in the dying industry, where the formation of the gelatinous aluminum hydroxide plays a part in "fixing" the dye to the cloth.

EXAMPLE 21.9 Production of Aluminum

What is the role of cryolite, Na_3AlF_6, in aluminum production?
Answer: Cryolite is not an aluminum ore; rather, molten cryolite is used as a solvent for alumina, Al_2O_3, in aluminum production. Cryolite melts at 1000°C, as compared to 2050°C for alumina.

EXAMPLE 21.10 Amphoterism

Explain the term *amphoterism*. The oxides of what elements described thus far in this chapter are amphoteric?

METHOD OF SOLUTION

An amphoteric substance may react either as an acid or as a base. The oxides of aluminum and beryllium exhibit amphoteric behavior. They can react as bases,

$$BeO + 2H^+ \rightarrow Be^{2+} + H_2O$$

$$Al_2O_3 + 6H^+ \rightarrow 2Al^{3+} + 2H_2O$$

and they can react as acids,

$$BeO + 2OH^- + H_2O \rightarrow Be(OH)_4^{2-}$$

$$Al_2O_3 + 2OH^- + 3H_2O \rightarrow 2Al(OH)_4^- \quad \text{(aluminate ion)}$$

COMMENT

The similarity of beryllium to aluminum was referred to as a diagonal relationship previously in this chapter. Diagonal relationships occur only in periods 2 and 3. Pairs of elements on the diagonal exhibit many similarities:

```
1A   2A   3A   4A
Li   Be   B    C
Na   Mg   Al   Si
```

EXAMPLE 21.11 Oxidation of Aluminum

Aluminum is a good reducing agent, and its reduction potential is quite negative ($E° = -1.67$ V), which means that aluminum should react with water and liberate hydrogen. But we know that airplanes do not dissolve in thunderstorms. Explain.

METHOD OF SOLUTION

Aluminum readily forms the oxide Al_2O_3 when exposed to air. The oxide is a tenacious surface film that protects the aluminum metal from further corrosion due to water. *Answer:* Owing to the presence of this layer, aluminum is practically insoluble even in very dilute acids and in concentrated HNO_3.

TRUE-FALSE QUESTIONS

1. The process for producing most free metals requires reduction of the metal ion.

2. Electrolysis and zone refining are two types of reduction.

3. Metals can be purified by distillation, zone refining, and electrolysis.

4. Metallic sodium is obtained by the electrolysis of aqueous sodium chloride.

5. Metallic aluminum is efficiently prepared by chemical reduction with magnesium metal.

6. Ferromagnetic separation is used to remove iron impurities from bauxite.

7. Limestone is used in iron production.

8. Copper is usually found as native copper, Cu.

9. A metal crystal made up of N atoms will have N bonding and N antibonding molecular orbitals.

10. The conduction band of a conductor will be either vacant or only partially filled.

11. Solids with a large energy gap between the valence band and the conduction band are semiconductors.

12. Beryllium, being the first element in the group, is the most metallic of the alkaline earth metals.

13. The oxides of magnesium and calcium react with water to give acidic solutions.

14. Barium metal is the most active of the alkaline earth metals, and so it has the most uses in construction.

15. The formula of magnesium nitride is Mg_2N_3.

16. Aluminum is an amphoteric element, which means that it can act as an oxidizing agent and as a reducing agent.

17. Aluminum has a strong affinity for oxygen.

18. Aluminothermic reactions are endothermic.

19. Cryolite (Na_3AlF_6) is an important ore of aluminum.

SELF-TEST A

1. Distinguish between
   a. An ore and a mineral
   b. Leaching and flotation
   c. Roasting and reduction

2. Can carbon be used to reduce Al_2O_3 to $Al(s)$? *Hint*: Is the following reaction spontaneous?

   $$2Al_2O_3(s) + 2C(s) \rightarrow 4Al(s) + 3CO_2(g)$$

3. When a solution of NaOH is added dropwise to a test tube containing a solution of Al^{3+}, a white gelatinous precipitate is formed. Upon continued addition of NaOH, the precipitate disappears. Write the chemical equations to explain this observation.

4. In the production of sodium, the metal is kept from contact with Cl_2 and O_2. Write chemical equations for the reactions of these elements with sodium.

5. What is the agent that actually reduces Fe in the blast furnace? Write a chemical equation to show the reaction.

6. Write equations for the electrode reactions when copper is purified electrochemically.

7. Distinguish between an alloy and an amalgam.

8. According to the band theory, why is copper a conductor?

9. Considering that electrons in metals have random thermal motion like molecules of gas, explain why electrical conductivity of conductors *decreases* with increasing temperature?

10. Why does the electrical conductivity of semiconductors *increase* with increasing temperature?

11. Write two equations that illustrate the amphoteric behavior of metallic beryllium.

12. How do the solubilities of the alkaline earth hydroxides vary from lightest to heaviest in the group?

13. In what ways does the chemistry of beryllium differ from that of the other members of the group?

14. What are the sources of calcium and magnesium?

15. What is quicklime? How is it made? Define *calcining*.

16. Aluminum metal was first prepared in 1825 by the action of potassium metal on aluminum chloride. Write a balance equation for this reaction.

17. What is an alum? Which alum is used in baking powder? *Hint*: Read a label. What is its function in baking powder?

18. What is the role of aluminum sulfate in water purification?

19. What is the pH of a 0.10 M aluminum chloride solution if it is hydrolyzed to the extent of 10% at room temperature:

$$[Al(H_2O)_6]^{3+} \rightarrow H^+ + [Al(H_2O)_5OH]^{2+}$$

ANSWERS

TRUE-FALSE QUESTIONS

1. True.
2. False. Zone refining is not reduction; it is a purification technique.
3. True.
4. False. The electrolysis must be of molten NaCl.
5. False. Aluminum can be prepared that way, but not efficiently.
6. False. Iron is removed from bauxite by precipitation of $Fe(OH)_3$.
7. True.
8. False. A chief source of copper is the sulfide ore, $CuFeS_2$.
9. False. It will have $0.5N$ bonding and $0.5N$ antibonding molecular orbitals.
10. True.
11. False. Insulators have the large energy gap.
12. False. Metallic character increases in going down a group.
13. False. The pH's are greater than 7.0. MgO and CaO are basic.
14. False. Being active, the metal would tend to corrode.
15. False. It is Mg_3N_2.
16. False. Amphoteric substances act as acids and bases.
17. True.
18. False. Exothermic.
19. False. Bauxite is the ore for aluminum production.

SELF-TEST A

1. a. A *mineral* is a naturally occurring substance with a characteristic chemical composition. Although an *ore* is a mineral, it may be a single mineral or a mixture of minerals from which a particular metal can be profitably extracted.
   b. *Leaching* is the selective dissolution of a metal from an ore. *Flotation* is a technique used to separate mineral particles from waste clays and silicates called gangue.
   c. *Roasting* involves heating an ore in the presence of air. The idea is to convert metal carbonates and sulfides to oxides, which can be more conveniently reduced to yield pure metals. *Reduction* is the gain of electrons by a cation to yield the pure metal.
2. No. Use Appendix 2 in the text to calculate $\Delta G°$. The sign of $\Delta G°$ indicates whether the reaction will go or not.
3. $Al^{3+} + 3OH^- \rightarrow Al(OH)_3(s)$

   $Al(OH)_3(s) + OH^- \rightarrow Al(OH)_4^- \ (aq)$

4. $2Na + Cl_2 \rightarrow 2NaCl$

   $2Na + O_2 \rightarrow Na_2O_2$

   $2Na + \frac{1}{2}O_2 \rightarrow Na_2O$

5. Carbon monoxide. $Fe_2O_3 + 3CO \rightarrow 2Fe + 3CO_2$
6. Anode: Cu (impure) $\rightarrow Cu^{2+} + 2e^-$. Cathode: $Cu^{2+} + 2e^- \rightarrow$ Cu (pure).
7. An *alloy* is a mixture of two or more metals or of metals and nonmetal elements that have metallic properties. An *amalgam* is an alloy containing mercury.
8. Copper atoms have a half-filled $4s$ orbital. This means that copper metal has a half-filled valence band. A partially filled valence is a conduction band.
9. The increasingly random motion of electrons due to a temperature rise opposes motion in one unified direction.
10. As temperature increases, more electrons acquire enough energy to "jump" the energy gap between the valence band and the conduction band.
11. $Be + 2OH^- + 2H_2O \rightarrow [Be(OH)_4]^{2-} + H_2$

    $Be + 2H^+ \rightarrow Be^{2+} + H_2$

12. Solubility increases from the lightest to the heaviest member.
13. In contrast to the other members of Group 2A, Be is unreactive toward O_2 and water, BeF_2 and $BeCl_2$ are covalent, Be^{2+} is extensively hydrolyzed, and BeO is amphoteric.
14. The source of Ca is limestone; of Mg, seawater.
15. Quicklime is CaO. It is made by calcining limestone. Calcination is the conversion of a substance to an oxide by heating.
16. $3K + AlCl_3 \xrightarrow{\Delta} Al + 3KCl$
17. Alums are sulfates with the general formula $M^+M^{3+}(SO_4)_2 \cdot 12H_2O$, where $M^+ = K^+, Na^+, NH_4^+$ and $M^{3+} = Al^{3+}, Cr^{3+}, Fe^{3+}$. Sodium aluminum sulfate is used in baking powder as an anticaking agent.
18. $Al_2(SO_4)_3$ is used to clarify water through its reaction with $Ca(OH)_2$. The precipitation of $Al(OH)_3$ traps dirt and dust particles.
19. pH = 2

Chapter Twenty-Two
NONMETALLIC ELEMENTS AND THEIR COMPOUNDS

- Hydrogen
- Carbon
- Nitrogen and Phosphorus
- Oxygen and Sulfur
- The Halogens

HYDROGEN

STUDY OBJECTIVES

You should be able to:
1. Write chemical equations to show the reactions used to prepare hydrogen.
2. Describe the chemical properties and major uses of hydrogen.

In this chapter we continue our survey of the chemistry of some of the elements. The elements discussed in this chapter are all nonmetals, or metalloids. For the more familiar of these we will summarize (1) their natural occurrences and preparations, (2) their characteristic chemical and physical properties, and (3) their uses in society.

Source and Preparation. Commercially, the most important large-scale preparation is the reaction between propane and steam in the presence of a catalyst at 900°C:

$$C_3H_8(g) + 3H_2O(g) \rightarrow 3CO(g) + 7H_2(g)$$

In another process steam is passed over a bed of red-hot coke (the water gas reaction):

$$C(s) + H_2O(g) \xrightarrow{1000°C} CO(g) + H_2(g)$$

Small quantities of hydrogen are prepared in the laboratory by the reaction of zinc or aluminum with hydrochloric acid:

$$Zn(s) + 2HCl(aq) \rightarrow ZnCl_2(aq) + H_2(g)$$

$$Al(s) + 3HCl(aq) \rightarrow AlCl_3(aq) + \tfrac{3}{2}H_2(g)$$

Properties of Hydrogen. Hydrogen can be oxidized to the H^+ ion and can be reduced to the H^- ion (hydride ion) as well. Thus, it resembles the alkali metals and the halogens, respectively. Like the halogens, hydrogen is a gas and forms diatomic molecules. Like the alkali metals, hydrogen forms ions with a +1 charge. As a result of these properties, there is no totally satisfactory position for hydrogen in the periodic table.

Hydrogen forms binary compounds called hydrides with a large number of elements; it reacts directly with most active metals to form ionic hydrides:

$$2Na(s) + H_2(g) \xrightarrow{\Delta} 2NaH(s)$$

$$2Al(s) + 3H_2(g) \xrightarrow{\Delta} 2AlH_3(s)$$

The ionic character of these metal hydrides increases as we move down a group of the periodic table. Metal hydrides react with water to form hydrogen gas and the metal hydroxide:

$$NaH(s) + H_2O(l) \rightarrow NaOH(aq) + H_2(g)$$

Uses. Hydrogen is used in the synthesis of ammonia, which is an important fertilizer. The Haber process for the synthesis of ammonia was discussed in Chapter 14. Ammonia that results from the reaction of hydrogen with nitrogen is an example of a covalent hydride:

$$N_2(g) + 3H_2(g) \rightarrow 2NH_3(g)$$

Hydrogen is used in the production of margarine and cooking oil to convert polyunsaturated fats (they contain C = C bonds) to partially hydrogenated products.

Hydrogen may play an important role in our energy future in terms of the "hydrogen economy." As fossil fuel supplies dwindle, hydrogen gas could take on more of a role as a fuel. Atomic energy could be used to generate electricity, which in turn could be used to electrolyze water to obtain hydrogen:

$$\text{electrical energy} + H_2O(l) \rightarrow H_2(g) + \tfrac{1}{2}O_2(g)$$

The combustion of hydrogen as a fuel would liberate energy:

$$H_2(g) + \tfrac{1}{2}O_2(g) \rightarrow H_2O(l) \qquad \Delta H° = -285.8 \text{ kJ}$$

EXAMPLE 22.1   Hydrogen Compounds

Give an example of an ionic hydride and an example of a covalent hydride. What is the oxidation number of hydrogen in each?

METHOD OF SOLUTION

In ionic compounds, hydrogen has an oxidation number of -1. The hydrides of the alkali metals and some of the alkaline earth metals are ionic hydrides. CaH_2 is an example of an ionic hydride. Hydrogen has an electronegativity of 2.1 and that of calcium is 1.0. Thus hydrogen has an oxidation number of -1 in CaH_2.

H_2S is a covalent hydride. The electronegativity of sulfur is 2.5. The lower polarity of the H — S bond is indicated by a Δ value of only 0.4. The oxidation number of hydrogen is $+1$ in H_2S.

EXAMPLE 22.2   Comparison of Ionic and Covalent Hydrids

Compare the reactions of water with covalent hydrides and ionic hydrides.

METHOD OF SOLUTION

Covalent hydrides either do not react with water at all, which is the case with CH_4, or are acids in water. One of them, ammonia, is a weak base:

$$H_2S(aq) + H_2O(l) \rightleftharpoons HS^-(aq) + H_3O^+(aq)$$

Ionic hydrides are *strong* Brønsted bases; they readily accept a proton from water:

$$H^-(aq) + H_2O(l) \rightarrow H_2(g) + OH^-(aq)$$

Lithium hydride is an ionic hydride:

$$LiH(s) + H_2O(l) \rightarrow LiOH(aq) + H_2(g)$$

CARBON

STUDY OBJECTIVES

You should be able to:
1. Describe the chemical and physical properties of carbon.
2. Describe some of the inorganic compounds of carbon.

Properties of Carbon. Carbon occurs in nature as the free element in graphite and diamond. Carbon occurs as CO_2 in the atmosphere and as the carbonate in limestone and chalk. Fossil fuels contain a large percentage of carbon. Carbon is the essential element in living matter.

Pure carbon in the form of diamond is the hardest substance known. The microscopic structure of diamond is such that each carbon atom is covalently linked to four other carbon atoms (see Figure 11.29a in the text). Thus a diamond is a single giant molecule. The melting point of a diamond is extremely high, 3550°C.

On heating carbon and silicon at 1500°C a compound called silicon carbide or carborundum (SiC) is formed. Silicon carbide is almost as hard as diamond and has the diamond structure. Its melting point is about 2700°C. The hardness of diamond and carborundum is put to important uses. Synthetic diamonds are not of gem quality but are used as abrasives in cutting concrete and many other hard substances. Carborundum is used for cutting, grinding, and polishing metals and glasses.

Carbon forms two important classes of compounds called carbides and cyanides. CaC_2 and Be_2C contain carbon in the form of a C_2^{2-} and a C^{4-} ion, respectively. The C_2^{2-} ion contains a triple bond, $-C \equiv C-$. Cyanides contain the anion, $C \equiv H^-$. Hydrogen cyanide, sodium cyanide, and potassium cyanide are well-known cyanide compounds.

Oxides of Carbon. Carbon forms two important oxides, CO and CO_2. Carbon monoxide is a colorless, odorless, and poisonous gas. Carbon monoxide burns readily in air and is an important component of synthesis gas. Carbon dioxide is a colorless and odorless gas. Its concentration in air is 350 ppm. At this concentration, it is not toxic. In enclosed low-lying places such as abandoned mines, CO_2 concentrations can build up to suffocating levels.

Carbon dioxide is formed by the complete combustion of carbon-containing compounds and fuels. If oxygen is restricted from reaching the combustion zone, carbon monoxide is formed instead. Since CO_2 cannot be oxidized further, it finds use in fire extinguishers. Dry ice (solid carbon dioxide) is used as a refrigerant.

NITROGEN AND PHOSPHORUS

STUDY OBJECTIVES

You should be able to:
1. Describe the chemical properties and uses of nitrogen and some of its compounds.
2. Describe the allotropic forms of phosphorus and list the main phosphorus compounds.

Nitrogen (N_2). The elemental form of nitrogen is extremely stable owing to its triple bond. Molecular nitrogen is obtained from air, of which nitrogen makes up 78% by volume. When air under a pressure of 30 atm is cooled to $-190°C$, it liquefies. Fractional distillation of liquid air yields pure N_2. Nitrogen is used in the laboratory to create atmospheres devoid of oxygen. This is desirable when working with chemicals that are attacked and degraded by oxygen or that burn spontaneously in it. Liquid N_2 is used as a coolant, providing a temperature of $-196°C$.

Nitrides **(N^{3-}).** Most nitrogen compounds are covalent. However, nitrogen forms ionic compounds with certain metals on heating. Li_3N, Mg_3N_2, and Ca_3N_2 are examples. The nitride ion is a very strong Brønsted base and readily accepts protons from water, forming ammonia:

$$Mg_3N_2(s) + 6H_2O \rightarrow 3Mg(OH)_2 + 2NH_3$$

Ammonia **(NH_3).** The quantity of ammonia produced each year is second in tonnage only to sulfuric acid production. Ammonia is synthesized by the Haber process, which uses nitrogen gas and hydrogen gas (from hydrocarbon cracking):

$$3H_2 + N_2 \rightarrow 2NH_3$$

The main use of ammonia is in the production of ammonium and nitrate fertilizers. Ammonia is an important ingredient in such diverse substances as explosives and nylon.

Liquid ammonia (b.p. $-33.4°C$) is a solvent resembling water. It undergoes autoionization:

$$2NH_3(l) \rightleftharpoons NH_4^+ + NH_2^-$$

The value of the equilibrium constant at $-50°C$ is $K = [NH_4^+][NH_2^-] = 1 \times 10^{-33}$. Alkali metals dissolve in liquid ammonia, producing beautiful blue solutions containing a positive ion and an electron:

$$Li(s) \rightarrow Li^+ + e^-$$

Both the cation and the electron are highly solvated by ammonia. The solvated electron e^- is responsible for the characteristic color of these solutions.

Nitrogen Oxides **(N_2O, NO, NO_2).** "Fixed nitrogen" refers to nitrogen present in chemical compounds. Nitrogen is fixed, that is, incorporated into compounds, by the action of lightning in the atmosphere, by high temperature, by the Haber process, and by certain types of soil bacteria. An appreciable amount of nitric oxide (NO) is formed in electrical storms and in internal combustion engines:

$$N_2 + O_2 \rightarrow 2NO$$

Nitric oxide reacts very rapidly with oxygen to yield nitrogen dioxide:

$$2NO + O_2 \rightarrow 2NO_2$$

Nitrogen dioxide is a key molecule in the formation of photochemical smog. NO_2 is prepared in the laboratory by the action of concentrated nitric acid on copper:

$$Cu(s) + 4HNO_3(aq) \rightarrow Cu(NO_3)_2(aq) + 2H_2O(l) + 2NO_2(g)$$

Nitric acid is a major inorganic acid. Over 8.0 million tons were produced in the United States in 1985. Nitric acid is synthesized by the Ostwald process, which uses ammonia and oxygen as starting materials. Reaction occurs at $1000°C$ in the presence of a platinum-rhodium catalyst:

$$4NH_3(g) + 5O_2(g) \rightarrow 4NO(g) + 6H_2O(g)$$

Then oxygen is introduced:

$$2NO(g) + O_2(s) \rightarrow 2NO_2(s)$$

The resulting NO_2 readily dissolves in water, yielding a mixture of nitrous acid and nitric acid:

$$2NO_2(g) + H_2O(l) \rightarrow HNO_2(aq) + HNO_3(aq)$$

The nitrous acid is eliminated by heating:

$$3HNO_2(aq) \rightarrow HNO_3(aq) + H_2O(l) + 2NO(g)$$

Nitric acid is a strong acid and an oxidizing acid. *Aqua regia*, the only acid that can dissolve gold, is one part nitric acid and three parts hydrochloric acid. Nitric acid is used to manufacture fertilizers and explosives such as nitroglycerin and TNT.

Nitrous oxide (N_2O) is a colorless gas with a pleasing odor and a sweet taste. It is used as an anesthetic in dental and minor surgery.

Phosphorus. There are several allotropic forms of phosphorus; the most important ones are white and red phosphorus. White phosphorus consists of P_4 molecules. It is extremely reactive and bursts into flame when exposed to air. Red phosphorus has a polymeric structure and is more stable than white phosphorus. It is obtained by heating white phosphorus to 300°C in the absence of air.

The two important oxides of phosphorus are tetraphosphorus hexaoxide (P_4O_6) and tetraphosphorus decaoxide (P_4O_{10}) (Figure 22.28 in the text). These can also be called tetraphosphorus hexoxide and tetraphosphorus decoxide. Both oxides are acidic:

$$P_4O_6(s) + 6H_2O(l) \rightarrow 4H_3PO_3(aq)$$

$$P_4O_{10}(s) + 6H_2O(l) \rightarrow 4H_3PO_4(aq)$$

Phosphoric Acid. Industrially, phosphoric acid is prepared by the reaction of sulfuric acid with calcium phosphate:

$$Ca_3(PO_4)_2(s) + 3H_2SO_4(aq) \rightarrow 2H_3PO_4(aq) + 3CaSO_4(aq)$$

Phosphoric acid is not a particularly strong acid, as indicated by its first ionization constant:

$$H_3PO_4 \rightleftharpoons H^+ + H_2PO_4^- \qquad K_{a_1} = 7.5 \times 10^{-3}$$

In 1985, 11 million tons of phosphoric acid (measured as its anhydride P_2O_5) were produced in the United States. Its main uses are in the production of phosphate fertilizers, cattle feed additives, and water softeners (builders) in household detergents.

EXAMPLE 22.3 Oxidation Numbers of Nitrogen

List the known oxidation numbers of nitrogen and write the formula of a compound or ion that exemplifies each oxidation number.

METHOD OF SOLUTION

Oxidation Number	Compound	Formula
+5 (maximum)	Nitrate ion	NO_3^-
+4	Nitrogen dioxide	NO_2
+3	Nitrite ion	NO_2^-
+2	Nitric oxide	NO
+1	Nitrous oxide	N_2O
0	Nitrogen gas (not a compound)	N_2
−1	Chloramine	NH_2Cl
−2	Hydrazine	N_2H_4
−3	Ammonia and nitride	NH_3 and N^{3-}

EXAMPLE 22.4 Allotropic Forms

What is meant by the term *allotropy*? Describe the allotropic forms of phosphorus.

METHOD OF SOLUTION

The term *allotropy* refers to the existence of different chemical forms of an element; examples are white phosphorus and red phosphorus. White phosphorus consists of discrete tetrahedral P_4 molecules, as shown in Figure 22.28 in the text. The white solid melts at 44°C and is insoluble in water. It bursts into flame spontaneously when exposed to air. White phosphorus is usually stored under water.

At room temperature white phosphorus slowly converts to red phosphorus. The structure of this polymeric form is also depicted in Figure 22.28. Red phosphorus differs in reactivity from white phosphorus. It melts at 600°C and is stable toward oxygen at room temperature. Red phosphorus must be heated to about 400°C before it will ignite in air.

OXYGEN AND SULFUR

STUDY OBJECTIVES

You should be able to:
1. Describe the allotropic forms of oxygen and sulfur.
2. List some of the compounds of oxygen and sulfur and describe their chemical properties.

Allotropic Forms of Oxygen. On Earth the element oxygen exists in two allotropic forms, diatomic oxygen (O_2) and ozone (O_3). Diatomic oxygen makes up 21% of the atmosphere by volume. Industrially, oxygen is obtained by liquefaction of air followed by fractional distillation. In the laboratory, oxygen is obtained by heating potassium chlorate in the presence of MnO_2 catalyst:

$$KClO_3(s) \xrightarrow{\Delta} KCl(s) + \tfrac{3}{2}O_2(g)$$

Ozone is a less stable form of oxygen. It is produced by the action of an electrical discharge or ultraviolet (UV) light on diatomic oxygen:

$$O_2(g) + \text{UV light} \rightarrow O(g) + O(g)$$

$$O(g) + O_2(g) \rightarrow O_3(g)$$

Ozone's presence in the atmosphere prevents the sun's UV light from reaching Earth's surface. Ozone absorbs the light in the following reaction:

$$O_3(g) + \text{UV light} \rightarrow O_2(g) + O(g)$$

Both ozone and diatomic oxygen are strong oxidizing agents:

$$O_2(g) + 4H^+ + 4e^- \rightarrow 2H_2O(l) \qquad E° = 1.23 \text{ V}$$

$$O_3(g) + 2H^+ + 2e^- \rightarrow O_2(g) + H_2O(l) \qquad E° = 2.02 \text{ V}$$

Ozone is used mainly to purify drinking water, to deodorize air and sewage gases, and to bleach waxes, oils, and textiles. About 65% of all the diatomic oxygen produced in the United States is used in the *basic oxygen process* for making steel. Pure oxygen gas is blown over the surface of molten iron to lower the carbon content by burning it to CO_2. At the same time silicon and phosphorus impurities are oxidized and later removed as slag.

Hydrogen Peroxide. Besides forming water, oxygen forms another important binary compound with hydrogen: hydrogen peroxide (H_2O_2). Hydrogen peroxide is never used as a pure substance because of its tendency to explode. In aqueous solutions, H_2O_2 can act both as an oxidizing agent and as a reducing agent. For example, H_2O_2 reduces silver oxide to metallic silver,

$$H_2O_2(aq) + Ag_2O(s) \rightarrow 2Ag(s) + H_2O(l) + O_2(g)$$

and oxidizes iron(II) to iron(III),

$$H_2O_2(aq) + 2Fe^{2+}(aq) + 2H^+(aq) \rightarrow 2Fe^{3+}(aq) + 2H_2O(l)$$

Metal peroxides contain the peroxide ion, O_2^{2-}.

These ions are strong Brønsted bases. All the metal peroxides hydrolyze in acidic solutions to form hydrogen peroxide:

$$Na_2O_2(s) + 2H_2O(l) \rightarrow H_2O_2(aq) + 2NaOH(aq)$$

Sulfur. Although it is not very abundant, sulfur is readily available because of its occurrence in large deposits of the free element. Sulfur is extracted from these underground deposits by the *Frasch process*. There are several allotropic forms of sulfur. The most important are rhombic sulfur and monoclinic sulfur. Rhombic sulfur is the more stable form at room temperature. In this form, sulfur atoms are linked to each other as puckered, eight-membered rings, S_8. Above 96°C, monoclinic sulfur is more stable. This form melts at 119°C. Above 150°C, the pale yellow liquid becomes more viscous as the rings break open and the newly formed molecular chains entangle. Sulfur has a wide range of oxidation numbers in its compounds, as shown in Table 22.3 of the text. The major end uses of sulfur are in fertilizers (60%) and in chemicals (20%).

Hydrogen Sulfide. H_2S (oxidation number -2) is a colorless, highly toxic gas with an offensive odor. When dissolved in water, the resulting solution is called hydrosulfuric acid. $H_2S(aq)$ is a weak acid.

In the laboratory, $H_2S(aq)$ is used to separate groups of metal ions for their qualitative analysis. It is often prepared in the laboratory by the hydrolysis of thioacetamide:

$$CH_3\overset{\overset{S}{\|}}{C}-NH_2(aq) + 2H_2O + H^+(aq) \rightarrow CH_3\overset{\overset{O}{\|}}{C}OH(aq) + NH_4^+(aq) + H_2S(aq)$$

Oxides of Sulfur. The major oxides of sulfur are sulfur dioxide and sulfur trioxide, SO_2 and SO_3. Sulfur dioxide is a colorless gas with a pungent odor and is quite toxic. It dissolves in water, forming sulfurous acid:

$$SO_2(g) + H_2O(l) \rightarrow H_2SO_3(aq)$$

Sulfur dioxide is slowly oxidized in air to sulfur trioxide:

$$2SO_2(g) + O_2(g) \rightarrow 2SO_3(g)$$

Sulfur trioxide dissolves in water to form sulfuric acid:

$$SO_3(g) + H_2O(l) \rightarrow H_2SO_4(aq)$$

Sulfur dioxide is a major air pollutant. It is formed during combustion of sulfur-containing oil and coal. Smelters and coal-fired electric power plants are the primary sources of sulfur oxides. In smelters, SO_2

is also formed during roasting of sulfide ores, as in the case of lead sulfide:

$$PbS(s) + \tfrac{3}{2}O_2(g) \xrightarrow{\Delta} PbO(s) + SO_2(g)$$

Sulfuric acid, carbon disulfide (CS_2), and sulfur hexafluoride (SF_6) are important sulfur compounds.

EXAMPLE 22.5   Oxygen Compounds

Distinguish between the three types of oxides: the normal oxide, the peroxide, and the superoxide.

METHOD OF SOLUTION

Normal oxides contain the O^{2-} ion in which oxygen has an oxidation state of -2. *Peroxides* contain an oxygen-oxygen bond. Metal peroxides are ionic and contain the O_2^{2-} ion. Hydrogen peroxide is a covalent peroxide (H — O — O — H) with oxygen in the -1 oxidation state. The *superoxide* ion O_2^- has oxygen in the $-\tfrac{1}{2}$ oxidation state.

EXAMPLE 22.6   Oxidation Numbers of Sulfur

List the known oxidation numbers of sulfur and write the formula of a compound or ion that exemplifies each oxidation number.

METHOD OF SOLUTION

Oxidation Number	Compound	Formula
+6	sulfate ion	H_2SO_4
+4	sulfite ion	$CaSO_3$
+2	sulfur dichloride	SCl_2
+1	disulfur dichloride	S_2Cl_2
0	sulfur (the element)	S_8
−2	sulfide ion	Na_2S

THE HALOGENS

STUDY OBJECTIVES

You should be able to:
1. Write chemical equations to show the preparations of the halogens.
2. Describe the chemical properties and cite the major uses of these elements and their compounds.

Occurrence and Preparation. The elements in Group 7 are called the halogens. Because of their high reactivity, the halogens are always found combined with other elements. Chlorine, bromine, and iodine occur as halides in seawater, and fluorine occurs in the minerals fluorspar (CaF_2) and cryolite (Na_3AlF_6).

Fluorine is prepared by the electrolysis of liquid HF at 70°C.

$$2HF(l) \xrightarrow{\text{electrolysis}} H_2(g) + F_2(g)$$

Chlorine is prepared by the electrolysis of a concentrated aqueous NaCl solution (brine) as described in Chapter 20:

$$2NaCl(aq) + 2H_2O(l) \xrightarrow{\text{electrolysis}} 2NaOH(aq) + H_2(g) + Cl_2(g)$$

Because chlorine is a stronger oxidizing agent than bromine or iodine, free Br_2 and I_2 can be generated by bubbling Cl_2 gas through solutions containing bromide and iodide ions:

$$Cl_2(g) + Br^-(aq) \rightarrow 2Cl^-(aq) + Br_2(l)$$

$$Cl_2(g) + 2I^-(aq) \rightarrow 2Cl^-(aq) + I_2(s)$$

Chlorine, bromine, and iodine can be prepared in the laboratory by heating the corresponding alkali halides in concentrated sulfuric acid in the presence of MnO_2:

$$2NaCl(aq) + 2H_2SO_4(aq) + MnO_2(s) \rightarrow Cl_2(g) + Na_2SO_4(aq) + MnSO_4(aq) + 2H_2O(l)$$

Properties of the Halogens. Most of the halides (halogens with an oxidation number of -1, as in chloride) can be classified into two categories: ionic and covalent. The fluorides and chlorides of many metallic elements, especially those belonging to the Group 1A and 2A metals, are ionic compounds. The halides of nonmetals are covalent compounds.

Because fluorine is the most electronegative element and because of its unusually weak F — F bond, fluorine has a chemistry somewhat different from the rest of the halogens. Five differences are pointed out in Section 22.7 of the textbook.

The halogens form a series of oxoacids with the following general formulas:

HOX	HXO_2	HXO_3	HXO_4
hypohalous acid	halous acid	halic acid	perhalic acid

Table 22.5 in the text lists the oxoacids of the halogens.

Hypochlorous, *hypobromous*, and *hypoiodous* acids can be prepared by the reaction of the elemental halogen with water:

$$X_2 + H_2O \rightleftharpoons \underset{\text{hydrohalic acid}}{HX} + \underset{\text{hypohalous acid}}{HOX}$$

The smaller the radius of the halogen atom, the farther the equilibrium lies to the right. Mixtures of these halogen acids react with bases to produce *hypohalite salts* (NaOX):

$$HX + HOX + 2NaOH \rightarrow NaX + NaOX + 2H_2O$$

When the elemental form of a halogen, X_2, is dissolved in a base, the net reaction is

$$X_2 + 2NaOH \rightarrow NaX + NaOX + H_2O$$

This type of reaction is called a *disproportionation* reaction because the same element (the halogen) is both oxidized and reduced in the reaction. One halogen atom in X — X is the reducing agent and the other is the oxidizing agent.

The compounds formed when molecules of two different halogens react are interhalogens. The interhalogen molecules consist of an atom of the larger halogen bound to an odd number of atoms of the smaller halogen. They have the following formulas: XX', XX'_3, XX'_5, and XX'_7, where X and X' are different halogens and X is the larger atom of the two.

Uses of the Halogens. The halogens and their compounds find many uses in industry, the home, and the field of health. Fluorine is used in fluoride toothpaste, in uranium isotope separation as UF_6, and in the manufacture of Teflon for electrical insulators, plastics, and cooking utensils.

Chlorine is widely used as a bleaching agent and as a disinfectant for water purification. Household bleach is 5.25% NaOCl by mass. The major end uses of chlorine are in manufacturing polymers such as polyvinyl chloride, in chemical solvents, in the pulp and paper industry, and in water treatment.

Bromine is used to prepare dibromoethane ($BrCH_2CH_2Br$), which is a gasoline additive, and silver bromide (AgBr), used in photographic film. Iodine finds uses as a medicinal antiseptic.

EXAMPLE 22.7 Preparation of F_2

The chemistry of fluorine differs in many ways from that of the rest of the halogens. For instance, it cannot be prepared by the electrolysis of an aqueous solution of its sodium salt, as Cl_2 is prepared. Write chemical equations to show the steps in the industrial process of making F_2. Give the source of the fluorine, and mention why it cannot be prepared in aqueous solution.

METHOD OF SOLUTION

The industrial process starts with the mineral fluorspar, CaF_2. This is treated with sulfuric acid to make hydrogen fluoride:

$$CaF_2(s) + H_2SO_4(l) \rightarrow 2HF(l) + CaSO_4(s)$$

The hydrogen fluoride is mixed with KF and electrolyzed:

$$2HF(l) \rightarrow H_2(g) + F_2(g)$$

The F_2 molecule has such a strong attraction for electrons that it reacts violently with water, hot glass, and most metals. If F_2 were formed in an aqueous solution, it would immediately oxidize water to oxygen gas.

EXAMPLE 22.8 Reaction of Chlorine with Sodium Hydroxide

When chlorine is dissolved in a base, a solution of aqueous sodium hypochlorite is formed. A 5.25% solution is sold as liquid laundry bleach. Write the chemical equation for the reaction. Assign oxidation numbers and identify the oxidizing agent and the reducing agent.

METHOD OF SOLUTION

$$Cl_2(g) + 2NaOH(aq) \rightarrow NaCl(aq) + NaOCl(aq) + H_2O(l)$$

If we assign oxidation numbers to all the chlorine atoms, we can see that chlorine is both oxidized and reduced. None of the other elements in the equation changes oxidation state. Thus chlorine is the oxidizing agent and the reducing agent:

```
              reduction
         ┌───────────────┐
         0   0          -1    +1
         Cl—Cl + 2NaOH → NaCl + NaOCl + H2O
             └──────────────────────┘
                     oxidation
```

This is an example of a disproportionation reaction.

TRUE-FALSE QUESTIONS

1. The oxidation number of hydrogen in AlH_3 is +1.

2. Covalent hydrides are usually acids in water.

3. A carbide is an ionic form of carbon.

4. The "lead" in pencils is not Pb, but rather another element in the same chemical group.

5. Nitrogen gas is quite inert and is referred to as fixed nitrogen.

6. The raw materials for making ammonia come from the atmosphere and from petroleum.

7. N_2 is used in the Ostwald process to make nitric acid.

8. P_4O_6 and P_4O_{10} are allotropes of phosphorus.

9. Phosphoric acid is a strong acid.

10. Ozone is a stronger oxidizing agent than diatomic oxygen.

11. The main use of O_2 in the United States is in steelmaking.

12. O_2^{2-} is called the superoxide ion.

13. One of the steps in the production of sulfuric acid involves burning elemental sulfur.

14. The major pollutant acids in acid rain are nitrous acid and sulfurous acid.

15. All the hydrogen halides are strong acids in water.

16. Fluorine is the strongest oxidizing agent among the elements.

17. Br_2 is prepared by bubbling fluorine gas through seawater containing bromide ions.

18. The least volatile of the hydrogen halides is HF.

19. A halide is a binary compound of halogen atoms formed by the union of two different halogens.

20. The shape of IF_5 is trigonal bipyramidal.

SELF-TEST A

1. Write chemical equations to show the two commercial methods of preparation of $H_2(g)$.

2. What are two important uses of hydrogen?

3. Write a chemical equation to show the synthesis of boric acid from kernite.

4. Coke is used to reduce Fe_2O_3 in iron production. What is it? How can it be converted to graphite?

5. List the major uses of
   a. N_2 b. NH_3 c. HNO_3

6. What is *aqua regia*?

7. What are the acid anhydrides of HNO_3, HNO_2, H_3PO_4, and H_3PO_3?

8. Write a balanced overall equation that shows the synthesis of nitric acid from ammonia.

9. Why is calcium phosphate not used directly as a fertilizer?

10. What chemical compounds contain phosphate in commercial fertilizers?

11. Give the formulas and names of the allotropic forms of oxygen and sulfur.

12. List several uses of O_2.

13. Distinguish between the peroxide ion and the superoxide ion.

14. Write chemical equations to show the preparations of the elemental forms of the halogens.

15. Write the chemical formulas of the halic acids.

16. Describe the colors and the standard physical states at 25°C of the diatomic halogens.

17. Write equations for the preparation of three interhalogens: ClF_3, BrF_5, and IF_7.

ANSWERS

TRUE-FALSE QUESTIONS

1. False. The oxidation number is -1.
2. True.
3. True.
4. True.
5. False. Nitrogen gas is quite inert, but fixed nitrogen is "combined" nitrogen.
6. True.
7. False. NH_3 is used.
8. False. *Allotropes* refers to elements, not compounds.
9. False. It is a weak acid.
10. True.
11. True.
12. False. Peroxide ion.
13. True.
14. False. Nitric acid and sulfuric acid.
15. False. HF is a weak acid.
16. True.
17. False. Br_2 is prepared by bubbling Cl_2 gas through seawater.
18. True.
19. False. The sentence describes an interhalogen.
20. False. The shape is a square pyramid.

SELF-TEST A

1. $C_3H_8(g) + 3H_2O(g) \rightarrow 3CO(g) + 7H_2(g)$

   $C(s) + H_2O(g) \rightarrow CO(g) + H_2(g)$

2. To make ammonia and partially hydrogenated food products
3. $Na_2B_4O_7 \cdot 4H_2O + H_2SO_4 + H_2O \rightarrow 4H_3BO_3 + Na_2SO_4$
4. When coal is distilled in the absence of oxygen to release volatile hydrocarbons, the high carbon residue is called coke. Coke is made into graphite by the Acheson process, in which an electric current is passed through coke for several days.
5. a. Create atmospheres devoid of O_2; a coolant
   b. Ammonia fertilizer
   c. Nitrate fertilizer and explosives
6. A mixture of 3 volumes of concentrated HCl to 1 volume of concentrated HNO_3
7. NO_2, NO, P_4O_{10}, P_4O_6
8. $12NH_3 + 21O_2 \rightarrow 8HNO_3 + 14H_2O + 4NO$
9. It is insoluble.
10. Diammonium phosphate, $(NH_4)_2HPO_4$; superphosphate $Ca(HPO_4)_2 \cdot H_2O$

11. O, O_2, O_3; rhombic S_8, monoclinic S_8
12. In the blast furnace for steelmaking, medical uses, welding.
13. Peroxide ion O_2^{2-}; superoxide O_2^-
14. $CaF_2 + H_2SO_4 \rightarrow 2HF + CaSO_4$

    $2HF \rightarrow H_2 + F_2$

    $2NaCl + 2H_2O \rightarrow 2NaOH + H_2 + Cl_2$

    $Cl_2 + 2I^- \rightarrow 2Cl^- + I_2$

15. HIO_3, $HBrO_3$, $HClO_3$. The comparable fluorine acid does not exist.
16. $I_2(s)$, black crystals; $Br_2(l)$, dark red liquid; $Cl_2(g)$, a yellow-green gas; $F_2(g)$, pale yellow gas
17. $Cl_2 + 3F_2 \rightarrow 2ClF_3$

    $KBr + 3F_2 \rightarrow KF + BrF_5$

    $KI + 4F_2 \rightarrow KF + IF_7$

Chapter Twenty-Three
TRANSITION METAL CHEMISTRY AND COORDINATION COMPOUNDS

- Properties of the Transition Metals
- Coordination Compounds
- Stereochemistry of Coordination Compounds
- Bonding in Coordination Compounds

PROPERTIES OF THE TRANSITION METALS

STUDY OBJECTIVES

You should be able to:
1. Describe several general characteristics of the transition metal elements.
2. Explain why there is only a gradual decrease in atomic radius when moving across a series of transition elements.

Electron Configurations of Atoms and Ions. In this chapter, we will focus on the first-row transition metals from scandium to copper. Recall that the transition metal elements have atoms or ions with incompletely filled d subshells.

The electron configurations of the first-row transition elements were discussed in Section 7.10 of the text. For the representative elements potassium and calcium, the $4s$ orbital is lower in energy than the $3d$ orbital. The $3d$ orbitals begin to fill only after the $4s$ orbital is complete. Thus, for elements after Ca[Ar]$4s^2$, electrons are added one at a time to the $3d$ sublevel beginning with scandium, the first transition element.

The electron configurations are given in Table 23.1 below. Two exceptions occur, one at chromium [Ar]$4s^1 3d^5$ and the other at copper [Ar]$4s^1 3d^{10}$. The basis for these exceptions is that the energies of a $3d$ orbital and a $4s$ orbital are similar for these elements. For chromium, the $4s$ and $3d$ orbitals are almost identical in energy, and so the electrons enter all *six* of the available orbitals singly in accordance with Hund's rule. For chromium, a configuration of [Ar]$4s^2 3d^4$ would be achieved only by forcing two

Table 23.1 Electron Configurations of Atoms and Ions

	Atom	+2 ion	+3 ion
Sc	$3d^1 4s^2$	$3d^1$	[Ne]
Ti	$3d^2 4s^2$	$3d^2$	$3d^1$
V	$3d^3 4s^2$	$3d^3$	$3d^2$
Cr	$3d^5 4s^1$	$3d^4$	$3d^3$
Mn	$3d^5 4s^2$	$3d^5$	$3d^4$
Fe	$3d^6 4s^2$	$3d^6$	$3d^5$
Co	$3d^7 4s^2$	$3d^7$	$3d^6$
Ni	$3d^8 4s^2$	$3d^8$	
Cu	$3d^{10} 4s^1$	$3d^9$	

electrons into the same orbital (the $4s$) where they are closer together and of higher potential energy due to greater repulsion. In the case of copper, the exception shows the importance of extra stability associated with a completely filled $3d$ subshell.

To understand the electron configurations of the ions shown in Table 23.1, it helps to recall that electrons are removed from the $4s$ orbital before they are taken out of the $3d$. The energies of the $3d$ and $4s$ orbitals are not as close together in the *ions* of the transition metals as they are in the neutral *atoms*. In fact, for the ions of transition elements the $3d$ orbitals are lower in energy than the $4s$ orbitals. This means that when electrons come out of an atom, the electrons most easily lost are those in the outermost principal energy level, the ns. Additional electrons may then be lost from the $(n-1)d$ orbital.

Periodic Properties. For the representative elements, properties such as the atomic radius, ionization energy, and electronegativity vary markedly from element to element as the atomic number increases across any period (Chapter 8). In contrast, the chemical and physical properties of the transition metal elements vary only slightly as we read across a period. Table 22.1 in the text gives a number of properties of the transition elements for comparison.

The decrease in radii is consistent with an increase in effective nuclear charge as atomic number increases. All the first-row transition elements have $4s$ orbitals as the outermost occupied orbitals. The atomic radius will depend on the strength of the nuclear charge felt by the $4s$ electrons. The greater the charge, the smaller the atom. The effective nuclear charge experienced by the outermost electrons depends on the shielding provided by inner electrons.

Recall that due to its lesser extent of penetration, an electron in a d orbital does not shield the outer electrons from the nuclear charge as effectively as an electron in an s or p orbital. They provide poor shielding of the outer $4s$ electrons of these elements. Consequently the $4s$ electrons of the first-row transition elements feel a greater effective nuclear charge as the atomic number increases and the atomic radii decrease.

The more dramatic decrease in atomic radius among the representative elements (compare K and Ca) results because the last electrons added to those elements go into outermost orbitals. Electrons in the same orbital are not very effective in shielding each other. Thus the effective nuclear charge felt by the $4s$ electrons in Ca is much greater than in K. Since the last electrons added to transition metal atoms enter inner d electrons, the decrease in atomic radius of the first-row transition elements is not as dramatic.

Properties of Transition Elements. The characteristics of transition metals are summarized below.

1. *General physical properties*. Transition metals have relatively high densities, high melting and boiling points, and high heats of fusion and vaporization. See Table 23.2 in the text.
2. *Variable oxidation states*. The common oxidation states for the transition elements from Sc to Cu are $+2$ and $+3$ (Table 23.1 in the text). Some of the elements exhibit the $+4$ state, and Mn even shows $+5$, $+6$, and $+7$. Transition metals usually exhibit their highest oxidation states in compounds with oxygen, fluorine, or chlorine. $KMnO_4$ and $K_2Cr_2O_7$ are examples. The variability of oxidation states for transition metal ions results from the fact that the $4s$ and $3d$ subshells are similar in energy. Therefore, an atom can form ions of roughly the same stability by losing different numbers of electrons. The $+3$ oxidation states are more stable at the beginning of the series, but toward the end the $+2$ oxidation states are more stable.
3. *Color*. Transition metal compounds are often highly colored, a fact that has attracted many chemists to the study of these compounds.
4. *Magnetic properties*. Paramagnetism is a property of many transition metal compounds.
5. *Complex ion formation*. Transition metals form a wide variety of complex ions.
6. *Catalytic properties*. Many of the transition metals and their compounds are used as catalysts for chemical reactions.

EXAMPLE 23.1 Transition Metal Elements

What distinguishes a transition element from a representative element?

METHOD OF SOLUTION

The *representative elements* are those in which all the inner subshells are filled and the last electron enters an outer s or p subshell. The *transition elements* are those in which an inner d subshell is partially filled. Among the *inner transition elements* (actinides), an inner f subshell is partially filled.

EXAMPLE 23.2 Properties of Transition Elements

Why do the transition elements have relatively high densities?

METHOD OF SOLUTION

Densities of solids are related to atomic and ionic radii. As the atoms get smaller, the solids tend to get more dense. The effective nuclear charge experienced by the outermost electrons in transition metal atoms is greater than that experienced in Group 1A and 2A metal atoms.
Answer: The atomic radii of the transition elements, in general, are smaller than those of the representative elements of the same period.

COORDINATION COMPOUNDS

STUDY OBJECTIVES

You should be able to:
1. Define the terms *complex ion*, *ligand*, *donor atom*, *coordination number*, and *coordination compound*.
2. Determine the oxidation number and coordination number of metal atoms in coordination compounds.
3. Name coordination compounds when given their formulas, and write their formulas when given their names.

Terminology. A *complex ion* consists of a central metal cation to which several anions or molecules are bonded. The complex ion may be positively or negatively charged. The metal cation is called the *central atom*, and the attached anions and molecules are called *ligands*. The free ligands each have at least one unshared pair of electrons which can be donated to the electron-deficient metal ions. The *donor atom* is the atom in the ligand that is directly bonded to the metal. Some typical ligands are Cl^-, CN^-, NH_3, H_2O, and $H_2NCH_2CH_2NH_2$.

A neutral species containing a complex ion is called a *coordination compound*. These compounds usually have complicated formulas. Two examples are

$$[Ag(NH_3)_2]Cl \quad \text{and} \quad K_3[Fe(CN)_6]$$

The complex ion is shown enclosed in brackets. In the silver compound, Cl^- is a free (meaning uncomplexed) chloride ion, and in the iron compound each K^+ is a free potassium ion. The K^+ and Cl^- ions in the above formulas are examples of counterions. They serve to balance or neutralize the charge of the complex ion. The *coordination number* is defined as the number of donor atoms surrounding the central metal atom. Thus the coordination number of Pt^{2+} in $[Pt(NH_3)_4]^{2+}$ is 4, and that of Co^{2+} in $[Co(NH_3)_6]^{2+}$ is 6.

Some ligands contain more than one donor atom. Ligands that coordinate through two bonds are called *bidentate ligands*. Those with more than two donor atoms are referred to as *polydentate ligands*.

Examples are

oxalate ion
(bidentate)

ethylenediaminetetraacetic acid

Bidentate and polydentate ligands are also called *chelating agents* (pronounced key-lating) because these ligands form rings of atoms that include the central metal ion:

The coordination number of the Cu^{2+} ion in this complex ion is 4, as there are four donor atoms. Another important property characterizing coordination compounds is the oxidation number of the central metal atom. Oxidation numbers were an important part of our discussion in Chapter 3. The determination of the oxidation number of the metal atom in coordination compounds is explained below.

Oxidation Number. Each complex ion carries a net charge that is the sum of the charges on the central atom (or ion) and on each of the ligands. In $Fe(CN)_6^{3-}$, for example, each cyanide (CN^-) contributes a -1 charge and the iron ion a $+3$ charge. The charge of the complex ion then is $(+3) + 6(-1) = -3$. The *oxidation number* of iron is $+3$, which is the same as its ionic charge. See Example 23.3.

Naming Coordination Compounds. Thousands of coordination compounds are known. The rules that have been developed for naming them are summarized below:

1. The cation is named before the anion.
2. Within the complex ion, the ligands are named first, followed by the metal ion. Ligands are listed in alphabetical order.
3. The names of anionic ligands end with the letter *o*, whereas neutral ligands are usually called by the names of the molecules. The exceptions are H_2O (aquo), CO (carbonyl), and NH_3 (ammine). Table 23.3 in the textbook lists the names of some common ligands.
4. When several ligands of a particular kind are present, use the Greek prefixes *di*, *tri*, *tetra*, *penta*, and *hexa*. Thus the ligands in $[Co(NH_3)_4Cl_2]^+$ are tetraamminedichloro. If the ligand itself contains a Greek prefix, use the prefixes *bis* (2), *tris* (3), and *tetrakis* (4) to indicate the number of ligands present. The ligand ethylenediamine already contains the term *di*; therefore bis(ethylenediamine) is used to indicate two ethylenediamine ligands.
5. If the complex is an anion, attach the ending *-ate* to the name of the metal. Thus $Fe(CN)_6^{3-}$ is named hexacyanoferrate(III) ion.
6. The oxidation number of the metal is written in Roman numerals following the name of the metal.

See Examples 23.3, 23.4, and 23.5 for the application of these nomenclature rules.

EXAMPLE 23.3 Terminology of Coordination Compounds

A certain coordination compound has the formula $[Co(NH_3)_4Cl_2]Cl$.
a. Which atom is the central atom?
b. Name the ligands and point out the donor atoms.

c. What is the charge on the complex ion?
d. What is the oxidation number (O.N.) of the central atom?

METHOD OF SOLUTION

First sketch the structure of the complex ion:

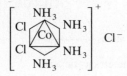

Answer:
a. Cobalt, the transition metal, is the central atom.
b. The ligands are ammonia and chloride, of which the donor atoms are N and Cl, respectively.
c. Since only one chloride ion is needed to balance the charge of the complex ion, the complex ion must have a charge of +1.
d. Ammonia is neutral and two chlorides are −2; thus in order for the complex ion to have a +1 charge, Co must be a +3 ion. That is, the central ion charge + sum of charges of ligands = charge of the complex ion.

O.N.(Co) + 2(−1) = +1

O.N.(Co) = +3

EXAMPLE 23.4 Naming Coordination Compounds

Give a systematic name for each of the following compounds:
a. [Ag(NH$_3$)$_2$]Br
b. Ni(CO)$_4$
c. K$_2$[Cd(CN)$_4$]

METHOD OF SOLUTION

a. The two ammonia ligands are represented by the term *diammine*. The oxidation state of the Ag is +1. The bromide ion is not complexed with silver. It is a counter ion. *Answer:* diamminesilver(I) bromide.
b. The term *tetracarbonyl* signifies that there are four carbon monoxide ligands. The oxidation state of the nickel atom is zero. *Answer:* tetracarbonylnickel(0).
c. In this case the complex ion is an anion; thus cadmium will be referred to as cadmate. Since K is +1 and each cyanide ligand is −1, Cd must have a +2 charge. *Answer:* The name is potassium tetracyanocadmate(II).

EXAMPLE 23.5 Writing Formulas from the Names

Write formulas for the following coordination compounds:
a. diaquodicyanocopper(II)
b. potassium hexachloropalladate(IV)
c. dioxalatocuprate(II) ion

METHOD OF SOLUTION

a. *Diaquo* refers to two water molecules and *dicyano* to two cyanide ions. Since the copper ion has a +2 charge, this coordination compound is neutral. *Answer:* [Cu(H$_2$O)$_2$(CN)$_2$].
b. The ending *−ate* indicates that the complex ion must be an anion. *Answer:* K$_2$[Pd(Cl)$_6$].
c. The complex ion is an anion. *Answer:* [Cu(C$_2$O$_4$)$_2$]$^{2-}$.

STEREOCHEMISTRY OF COORDINATION COMPOUNDS

STUDY OBJECTIVES

You should be able to:
1. Describe geometric and optical isomers.
2. Define the terms *chiral* and *enantiomers*.

Geometric Isomers. Stereoisomers of complex ions or of neutral coordination compounds have the same chemical bonds but different spatial arrangements. There are two types of stereoisomers: geometric isomers and optical isomers.

Geometric, or *cis-trans*, isomers are distinguished by the position of like ligands or groups. The isomer with the like groups close to each other is called a *cis* isomers, and the one with like groups farther apart is called the *trans* isomers. *Cis* and *trans* isomers of coordination compounds generally have different physical and chemical properties.

Two forms of the complex $Pt(NH_3)_2Cl_2$ have been prepared. Both have square planar structures. The *trans* form has no dipole moment, whereas the *cis* form has an appreciable dipole moment:

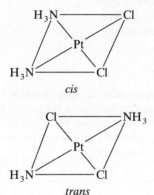

The simplest type of geometric isomerism of octahedral complexes occurs in cases where four of the six ligands of the complex are the same. Then the *cis* and *trans* forms correspond to those shown below:

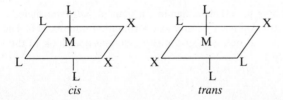

Optical Isomers. Optical isomers are nonsuperimposable mirror images of one another. They are also called *enantiomers*. Enantiomers have the same relationship to one another as do your right and left hands. If you place your left hand parallel to your right hand facing each other, you get the same effect that you would get if you placed one hand in front of a mirror. Your left hand is the mirror image of your right hand. However, your left hand is not superimposable upon your right hand. Your hands are enantiomers, that is, nonsuperimposable mirror images of one another. The enantiomers of the *cis* bis(ethylenediamine)dichlorocobalt (III) ion are show in Figure 23.1.

Not all objects are enantiomers. An ordinary chair, for example, looks the same in the mirror as when viewed directly. A chair is superimposable on its mirror image. An object that is not identical with its mirror image is said to be *chiral*, from the Greek word for hand (see Figure 23.20 in the text).

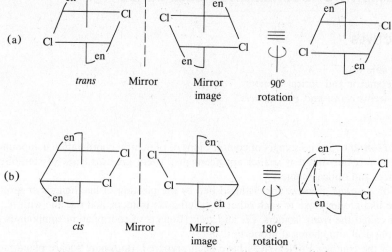

(a)

trans | Mirror | Mirror image | 90° rotation

(b)

cis | Mirror | Mirror image | 180° rotation

Figure 23.1. The *cis* and *trans* isomers of bis(ethylenediamine)dichlorocobalt (III) ion. The *cis* isomer exists as enantiomers, but there is only one form of the *trans* isomer. The "en" stands for ethylenediamine. (*a*) The mirror image of the *trans* isomer can be superimposed on the original after rotating it by 90°. (*b*) The mirror image of the *cis* isomer *cannot be superimposed* on the original no matter how it is rotated.

Enantiomers are called optical isomers because they differ with respect to the direction in which they rotate the plane of polarization of plane-polarized light when light is passed through the substance. The isomer that rotates the plane of polarized light to the right is said to be *dextrorotatory*. Its mirror image will rotate the plane to the left, *levorotatory*. A *racemic* mixture is an equimolar mixture of the two optical isomers. Such a mixture produces no net rotation of plane-polarized light.

EXAMPLE 23.6 Geometric Isomers

Sketch the geometric isomers of $[Cr(en)_2Br_2]^+$.

METHOD OF SOLUTION

First identify the ligands. Recall that "en" stands for ethylenediamine, a neutral *bidentate* ligand. Bromide ion is the other ligand. Thus, the central ion is Cr^{3+}, and its coordination number is 6. The complex will be octahedral. The following steps will help you give a three-dimensional look to the octahedron.

Draw a hexagon:

Draw a triangle:

For an octahedral complex, ligands at adjacent corners are said to be *cis* to each other. The ligands at opposite corners are *trans* to each other:

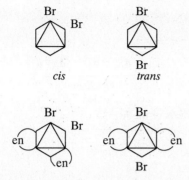

EXAMPLE 23.7 Optical Isomers

For the complex ion in the preceding example, indicate if any of the isomers exhibit optical isomerism.

METHOD OF SOLUTION

Sketch the mirror images of the *cis* and *trans* isomers, and look for nonsuperimposable mirror images, the enantiomers.

The *cis* isomer is chiral. The *trans* isomer is not.

BONDING IN COORDINATION COMPOUNDS

STUDY OBJECTIVES

You should be able to:
1. Explain how the interaction between the ligands and the d orbitals of the central metal atom in a coordination compound produces crystal field splitting.
2. Explain the relationship of the color of a complex ion to the wavelength of light absorbed.
3. Predict the number of unpaired electrons in a given complex ion.

Crystal Field Theory. In the case of coordination compounds a satisfactory theory must account for properties such as color, magnetism, and stereochemistry. Although the crystal field theory is not the only theory of coordination compounds, it explains satisfactorily their colors and magnetic properties.

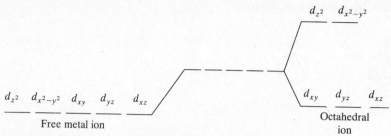

Figure 23.2. Crystal field splitting. On the left are the energy levels of the five $3d$ orbitals in a free Fe^{3+} ion. On the right are the same d orbitals in the same ion when it is a part of the octahedral complex ion $[Fe(CN)_6]^{3-}$.

The crystal field model considers the ligand-to-metal bonding in a complex ion to be primarily electrostatic rather than covalent. In an isolated transition metal ion all five $3d$ orbitals have the same energy. The effect of the ligands is to change the relative energies of these orbitals through electrostatic interactions. There are two types of electrostatic interactions. One is the force that holds the ligands to the metal ion. This is the attraction between the positive charge of the metal ion and the lone-electron pairs of the ligands. Second, there is the repulsion between the lone pairs of the ligand donor atoms and the electrons in the $3d$ orbitals of the metal. It is this latter interaction that gives rise to crystal field splitting and its effect on the color and magnetic properties of the complex ion.

Consider the $Fe(CN)_6^{3-}$ complex ion, for example. An isolated Fe^{2+} ion has five $3d$ electrons, all with the same energy:

Fe^{3+} ↑ ↑ ↑ ↑ ↑

When the six cyanide ligands become positioned in an octahedral arrangement, all the $3d$ orbitals are raised in energy due to the repulsion we have mentioned. The repulsion energy is not the same for all $3d$ orbitals. Rather, those orbitals that are directed straight toward a ligand experience greater repulsion than those directed between ligands. Thus, in an octahedral complex the five $3d$ orbitals are split into two groups in terms of energy. One group, comprising the d_{xy}, d_{yz}, and d_{xz} orbitals, is not raised in energy as much as the other group, consisting of the d_{z^2} and $d_{x^2y^2}$ orbitals. The situation is shown here schematically in Figure 23.2. The energy difference between these two sets of d orbitals is called the *crystal field splitting energy* and is given the symbol Δ.

Color. The color of a coordination compound results from electron transitions from the lower-energy set of orbitals to the higher-energy orbitals. Absorption of light occurs when the energy of an incoming photon is equal to the difference in energy Δ, and an electron transition occurs. The requirement for light absorption is

$$\Delta = E_{photon}$$

Recall from Chapter 7 that the energy of a photon is $E_{photon} = h\nu$, where h is Planck's constant and ν is the frequency of the radiation. Thus $\Delta = h\nu$.

A substance has a color because it absorbs light at certain wavelengths in the visible part of the electromagnetic spectrum (from 400 to 700 nm) and it reflects the other wavelengths. Figure 7.4 shows the relationship of wavelength to color. Light that is a combination of all wavelengths appears white. When white light impinges on a coordination compound and light of a certain wavelength is absorbed, the reflected light is missing that component and no longer appears white to the eye.

Table 23.2 shows the relationship of *wavelength absorbed* to *color observed*. It allows you to estimate the color of reflected light when light of a known color is absorbed. For instance, the hexacyanoferrate(II) ion absorbs light in the visible region at about 410 nm. The absorbed light is violet. Thus, according to Table 23.2, the reflected light with violet missing is yellow.

Table 23.2 Relationship of Wavelength Absorbed to Color

Wavelength Absorbed (nm)	Color Observed
400 (violet)	Greenish yellow
450 (blue)	Yellow
490 (blue green)	Red
570 (yellow green)	Violet
580 (yellow)	Dark blue
600 (orange)	Blue
650 (red)	Green

The extent of crystal field splitting, that is, the magnitude of Δ, determines the wavelength of light absorbed and thus the color of the complex. Each ligand has a different strength of interaction with the d orbitals of the metal ion. Thus each splits the orbital energies by a different amount. The ligands can be arranged according to increasing values of Δ. This establishes a *spectrochemical series*:

$$I^- < Br^- < Cl^- < OH^- < F^- < H_2O < NH_3 < en < CN^- < CO$$

CO and CN^- are called *strong-field* ligands because they produce a relatively large splitting, whereas I^- and Br^- are *weak-field* ligands because they split d orbital energies to a lesser extent.

Magnetic Properties. In Chapter 7 you learned that paramagnetic substances have at least one unpaired electron and diamagnetic substances have all their electrons paired. The Fe^{3+} ion has a d^5 configuration. Two of its octahedral complexes, $[FeF_6]^{3-}$ and $[Fe(CN)_6]^{3-}$, are both paramagnetic, and yet they have different magnetic properties. $[FeF_6]^{3-}$ has five unpaired electrons, but $[Fe(CN)_6]^{3-}$ has only one. The former is called a *high-spin complex* and the latter a *low-spin complex*.

The reason for this difference lies in the spectrochemical series and the value of Δ. Remember that according to Hund's rule, electrons in orbitals with similar energies prefer to be unpaired. Energy is required to make electrons pair up. Therefore, if Δ is small, as in the case of the weak-field ligand F^-, the electrons enter all five orbitals one at a time without pairing in accord with Hund's rule. For small Δ values, pairing of electrons would not occur until the sixth electron is added.

However, when Δ is large enough, Hund's rule no longer applies to the entire set of d orbitals. The lowest-energy configuration corresponds to placing all of the electrons in the three lowest-energy $3d$ orbitals first, before any enter the two higher-energy orbitals. This is the case for $[Fe(CN)_6]^{3-}$ because CN^- is a small-field ligand in the spectrochemical series and produces large Δ values. Figure 23.3 shows the energy-level diagrams for hexafluoroferrate(III) ion and hexacyanoferrate(III) ion.

So far we have focused on octahedral complexes. Crystal field splitting occurs for tetrahedral and square planar complexes as well. See Figures 23.23 and 23.24 in the text for the splitting diagrams of these complexes.

EXAMPLE 23.8 Crystal Field Theory

The $[Ti(H_2O)_6]^{3+}$ ion absorbs light at a wavelength of 498 nm.
a. Calculate the crystal field splitting energy Δ.
b. Determine the color of the complex.

METHOD OF SOLUTION FOR a

Absorption of light occurs when the energy of an incoming photon is equal to the difference in energy Δ between the lower-energy $3d$ orbitals and the higher-energy $3d$ orbitals of the titanium ion. The requirement for light absorption is

$$\Delta = E_{photon}$$

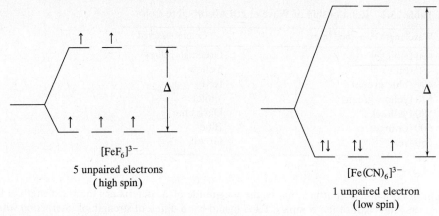

Figure 23.3. Effect of crystal field splitting on pairing of d electrons in two Fe(III) complex ions. When Δ is small, Hund's rule applies to all five orbitals. When Δ is large, the three lower-energy orbitals fill before the two higher energy orbitals.

or

$$\Delta = h\nu = hc/\lambda$$

where h is Planck's constant, c is the velocity of light, and λ is the wavelength of electromagnetic radiation (light), 498 nm.

CALCULATION FOR a

$$\Delta = \frac{(6.63 \times 10^{-34} \text{ J s})(3.00 \times 10^8 \text{ m/s})}{498 \text{ nm} \times \left(\frac{10^{-9} \text{ m}}{1 \text{ nm}}\right)}$$

$$= 3.99 \times 10^{-19} \text{ J}$$

METHOD OF SOLUTION FOR b

The complex absorbs light with a wavelength of 498 nm. According to Table 23.2, this is green light. A solution that absorbs green light will reflect all the colors except green. According to the color wheel (or Table 23.6 in the text), the solution will be red.

EXAMPLE 23.9 Crystal Field Theory

$[CoF_6]^{4-}$ is a high-spin paramagnetic complex, and $[Co(en)_3]^{2+}$ is a low-spin paramagnetic complex. Draw the crystal field splitting diagrams for these two complex ions and show the proper relationship of the two Δ values.

METHOD OF SOLUTION

The crystal field splitting diagram for a Co^{2+} ion ($3d^7$) must show seven electrons in the $3d$ orbitals. In a high-spin complex Δ is small enough so the electrons can enter $3d$ orbitals one at a time according to Hund's rule. Pairing of electrons will not occur until after the fifth electron has been added. In the low-spin complex, Δ is much larger and the electrons enter the lower-energy orbitals first. Only after six electrons are positioned in the lower-energy set of orbitals can electrons then be placed in the upper two orbitals.

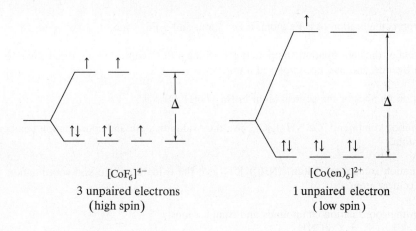

TRUE-FALSE QUESTIONS

1. The oxidation number of Fe in $[Fe(CN)_6]^{4-}$ is $+6$.

2. In transition metal complexes, the ligands act as Lewis acids.

3. The charge on the complex ion in $K_2[Pt(Cl)_4]$ is $+2$.

4. The coordination number of Hg in $[Hg(en)_2]^{2-}$ is 4.

5. The best name for $[CoCl_3Br_3]^{4-}$ is trichlorotribromocobalt(II) ion.

6. *Cis*-diamminedichloroplatinum(II) is a chiral species.

7. Enantiomers will rotate the plane of polarized light in opposite directions.

8. The following figure represents a square planar complex:

$$\left[\begin{array}{c} \text{CN} \\ \text{NC} \diagup\!\!\!\diagdown \text{CN} \\ \text{Fe} \\ \text{NC} \diagdown\!\!\!\diagup \text{CN} \\ \text{CN} \end{array} \right]^{4-}$$

9. The color of a coordination compound is the same as the color corresponding to the wavelength absorbed by the complex ion.

10. Ethylenediamine is a stronger-field ligand than a bromide ion.

11. High-spin complexes have more unpaired electrons than low-spin complexes.

12. The greater the value of Δ, the greater the likelihood of having a high-spin complex.

SELF-TEST A

1. Write an electron configuration of a Fe atom, a Fe^{2+} ion, and a Fe^{3+} ion.

2. Write the formula of the coordination compound containing a Pt^{4+} central atom, three chlorides, three ammonia ligands, and one uncomplexed nitrate ion.

3. Find the oxidation number of the central atom in $Na_2[Co(H_2O)_2I_4]$.

4. For the coordination compound $[Co(NH_3)_6]Cl_3$ give the oxidation state and coordination number of the central atom.

5. For the coordination compound $[Ni(en)_2(NH_3)_2]Cl_2$ give the oxidation state and coordination number of the central atom.

6. Name the following coordination compounds and complex ions:
   a. $[Pt(NH_3)_4Cl_2]Cl_2$
   b. $[Ni(NH_3)_6]^{2+}$
   c. $[Cr(OH)_4]^-$
   d. $Co(CN)_6^{3-}$
   e. $K_2[Cu(CN)_4]$

7. Write the formula for each of the following coordination compounds and ions:
   a. tetrahydroxoaluminate(III) ion
   b. tetraiodomercurate(II) ion
   c. potassium dichlorobis(oxalato)cobaltate(III)
   d. tris(ethylenediamine)nickel(II) sulfate

8. Sketch two geometrical isomers for $[Ni(NH_3)_2(CN)_4]^{2-}$.

9. Sketch the structures of the square planar complex trans-$[Pt(Cl)_2(NH_3)_2]$ and its mirror image. Are the two structures enantiomers?

10. Why are chiral substances said to be optically active?

11. Illustrate each of the following by giving formulas, geometry, or energy diagrams:
    a. octahedral complex
    b. cis-trans isomers
    c. $[Fe(H_2O)_6]^{3-}$ is a high-spin complex
    d. bidentate ligand
    e. coordination number 2

12. If green light is absorbed by a complex in solution, what is the color of the solution?

13. Sketch the crystal field splitting diagram for the low-spin $[Fe(CN)_6]^{4-}$ ion.

14. Sketch the crystal field splitting diagram for the tetrahedral complex $[FeCl_4]^-$.

15. Distinguish between inert and labile complexes.

16. Complexes of zinc are never paramagnetic. Explain.

17. In the complex ion $[ML_6]^{n+}$, M^{n+} has four d electrons, and L is a weak-field ligand. How many unpaired electrons are there in this complex?

18. $[Ni(H_2O)_6]^{2+}$ is green, and $[Ni(en)_3]^{2+}$ is violet. Which has the larger value of Δ?

ANSWERS

TRUE-FALSE QUESTIONS

1. False. The oxidation number of iron is +2.
2. False. The ligands are Lewis bases.
3. False. The charge is −2.
4. True.
5. False. Tribromotrichlorocobaltate(II) ion.
6. False. There are no optical isomers of this species.
7. True.
8. False. It is an octahedral complex.
9. False. The color of a compound is the same as the reflected light, not the absorbed light.
10. True.
11. True.
12. False. The greater likelihood is a low-spin complex.

SELF-TEST A

1. Fe[Ar]$4s^2 3d^6$ Fe[Ar]$3d^6$ Fe[Ar]$3d^5$
2. [Pt(NH$_3$)$_3$Cl$_3$]NO$_3$
3. +2
4. The coordination number is 6; the oxidation state is +3.
5. The coordination number is 6; the oxidation state is +2.
6. a. tetraamminedichloroplatinum(IV) chloride
   b. hexaamminenickel(II) ion
   c. tetrahydroxochromate(III) ion
   d. hexacyanocobaltate(III) ion
   e. potassium tetracyanocuprate(II)
7. a. [Al(OH)$_4$]$^-$ b. [HgI$_4$]$^{2-}$ c. K$_3$[Co(C$_2$O$_4$)$_2$Cl$_2$]   d. [Ni(en)$_3$]SO$_4$
8.

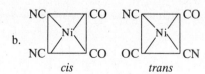

cis   trans

9. The two structures are not enantiomers.
10. Because when plane-polarized light passes through a solution of a chiral substance, the plane of the polarized light is rotated.
11.

a.

b. (cis and trans square planar Ni(CN)$_2$(CO)$_2$ structures)

c. (orbital diagram with arrows)

d. $^-$O−C(=O)−C(=O)−O$^-$

e. [H$_3$N:Ag:NH$_3$]$^+$

12. Red

13.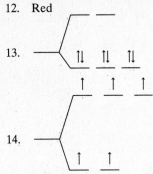

14.

15. The terms *inert* and *labile* are kinetic terms; that is, they refer to relative rates of ligand substitution reaction. Complexes that undergo rapid ligand exchange are said to be labile complexes. In inert complexes the bonds between the central atom and the ligands are broken and re-formed relatively infrequently.
16. Zinc has ten d electrons that completely fill the five d orbitals.
17. Four
18. $[Ni(en)_3]^{2+}$

Chapter Twenty-Four
NUCLEAR CHEMISTRY

- The Nature of Nuclear Reactions
- The Stability of Nuclei
- Natural and Artificial Radioactivity
- Nuclear Fission and Fusion
- Biological Effects of Radiation

THE NATURE OF NUCLEAR REACTIONS

STUDY OBJECTIVES

You should be able to:
1. Write balanced nuclear equations for radioactive decay and nuclear transmutation processes.
2. Compare chemical reactions and nuclear reactions.

Nuclear Reactions and Nuclear Equations. The textbook discusses two types of nuclear reactions: radioactive decay and nuclear transmutation. *Radioactivity* was discovered in 1896 and can be described as the spontaneous emission of radiation by unstable atomic nuclei. In 1902, E. Rutherford and F. Soddy established that radioactivity results from the spontaneous decomposition of an atom of an element, resulting in the formation of a new element. The changes are often accompanied by the emission of particles and rays. The kinds of particles ejected from various nuclei are shown in Table 24.1.

A nucleus can also undergo change by nuclear *transmutation*. In this process a nucleus reacts with another nucleus, an elementary particle, or a photon (gamma particle) to produce one or more new nuclei.

Radioactive decay and nuclear transmutation processes can be described by nuclear equations. These equations uses isotopic and elementary particle symbols to represent the reactants and products of nuclear reactions. For instance, in the first nuclear transmutation ever observed (in 1919), alpha particles were used to bombard nitrogen-14 nuclei. The observed products were atoms of oxygen-17 and protons. The nuclear equation is

$$^{14}_{7}N + ^{4}_{2}He \rightarrow ^{17}_{8}O + ^{1}_{1}H$$

Table 24.1 Particles from Radioactive Decay

Particle	Mass (amu)	Charge	Symbol
Alpha	4.0	+2	$^{4}_{2}\alpha$ or $^{4}_{2}He$
Beta	0.0005	−1	$^{0}_{-1}\beta$ or $^{0}_{-1}e$
Positron	0.0005	+1	$^{0}_{+1}\beta$ or $^{0}_{+1}e$
Gamma	0	0	$^{0}_{0}\gamma$

Note that the balancing rules given in the textbook are followed here:

1. The sum of the mass numbers of the reactants must equal the sum of the mass numbers of the products.
2. The sum of the nuclear charges of the reactants must equal the sum of the nuclear charges of the products.

These rules will be employed further in the example problems.

Comparison of Chemical and Nuclear Reactions. Table 24.1 of the textbook lists a number of comparisons between chemical and nuclear processes. Keep in mind that in chemical reactions the number of atoms of each element is conserved. Only changes in chemical bonding occur. However, in nuclear reactions the composition of the atomic nucleus is altered, and so elements are converted from one to another. Since a nucleus is so extremely small and in some cases contains large numbers of positively charged protons, the energy associated with a nucleus is tremendous compared with that of chemical bonds.

EXAMPLE 24.1 Nuclear Equations

Complete the following nuclear equations:
a. $^{14}_{7}\text{N} + ^{1}_{0}\text{n} \rightarrow ^{14}_{6}\text{C} + \underline{\qquad}$
b. $^{226}\text{Ra} \rightarrow ^{4}_{2}\alpha + \underline{\qquad}$

METHOD OF SOLUTION

a. According to rule 1, the sum of the mass numbers of the reactants must equal the sum of the mass numbers of the products. Thus the unknown product will have a mass number of 1. According to rule 2, the sums of the nuclear charges on both sides of the equation must be the same. Thus the nuclear charge of the unknown product must be 1, making it a proton:

$$^{14}_{7}\text{N} + ^{1}_{0}\text{n} \rightarrow ^{14}_{6}\text{C} + ^{1}_{1}\text{p}$$

b. Note that the atomic number of radium is missing. The periodic table indicates that Ra is element number 88. Balancing the mass numbers first, we find that the unknown product must have a mass of 222. Balancing the nuclear charges next, we find that the nuclear charge of the unknown must be 86. Element number 86 is radon:

$$^{226}_{88}\text{Ra} \rightarrow ^{4}_{2}\alpha + ^{222}_{86}\text{Rn}$$

COMMENT

The product isotope radon-222 is called the *daughter product*, or *progeny*, of radium-226.

EXAMPLE 24.2 Nuclear Transformations

Alpha particles from the decay of Po-214 were used to produce the first nuclear transmutation:

$$^{4}_{2}\text{He} + ^{14}_{7}\text{N} \rightarrow ^{17}_{8}\text{O} + ^{1}_{1}\text{H}$$

What forces must be overcome in order for this process to occur?

METHOD OF SOLUTION

The alpha particle and the nitrogen nucleus must first combine. Therefore, the kinetic energy of the alpha particle must be sufficient to overcome the force of coulombic repulsion between the alpha particle with its +2 charge and the nitrogen nucleus with its +7 charge.

COMMENT

Since the energy of alpha particles is fixed, depending on the decaying nucleus that serves as their source, there is an upper limit to the atomic number of the target nuclei for a successful transmutation. For target elements beyond chlorine (atomic number 17), alpha particles from all sources are repelled by the highly charged nucleus. This repulsion can be overcome by boosting the energy of the alpha particles in an accelerator.

THE STABILITY OF NUCLEI

STUDY OBJECTIVES

You should be able to:
1. Compare relative stabilities of given nuclei by applying stability rules.
2. Calculate the nuclear binding energies of given nuclei.

Nuclear Stability. Little is known about the forces that hold the nucleus together. However, some interesting facts emerge if we examine the numbers of protons and neutrons found in those nuclei that are stable. Nuclei can be classified according to whether they contain even or odd numbers of protons and neutrons. The number of stable isotopes of each of the four types of nuclei classified in this way is shown in Table 24.2.

Table 24.2 Number of Stable Isotopes

| Number of Protons | Even | Even | Odd | Odd |
Number of Neutrons	Even	Odd	Even	Odd
Number of stable isotopes	157	52	50	8

The following rules are useful in predicting nuclear stability:

1. Nuclei with even numbers of both protons and neutrons are generally more stable than those with odd numbers of these particles. Nuclei that contain certain specific numbers of protons and neutrons within a nucleus ensure an extra degree of stability. These so-called *magic numbers* for protons and for neutrons are 2, 8, 20, 28, 50, 82, and 126.
2. Nuclei with even numbers of protons or neutrons are generally more stable than those with odd numbers of these particles.
3. All isotopes of elements after bismuth ($Z = 83$) are radioactive.

The Belt of Stability. The principal factor for determining whether a nucleus is stable is the neutron-to-proton ratio. Figure 24.1 shows a plot of the number of neutrons versus the number of protons in various isotopes. Each dot represents a *stable* isotope. The stable nuclei are located in an area of the graph known as the *belt of stability*. In this figure we see that at low atomic numbers stable nuclei possess a neutron-to-proton ratio of about 1.0. Above $Z = 20$, the number of neutrons always exceeds the number of protons in stable isotopes. The n:p ratio increases to about 1.5 at the upper end of the belt of stability.

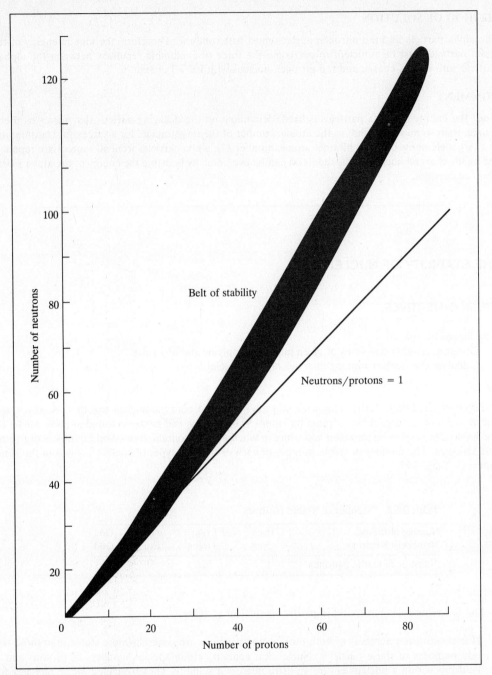

Figure 24.1. A graph showing all the stable isotopes when plotted as neutron number versus proton number. In order for "light-weight" isotopes to be stable, the ratio of $n:p$ must be about 1.0. In order for isotopes of high-atomic-number elements to be stable, the ratio of $n:p$ must be about $1.5:1.0$.

Nuclear Binding Energy. One of the important consequences of Einstein's theory of relativity was the discovery of the equivalence of mass and energy. The total energy content (E) of a system of mass m is given by the relation

$$E = mc^2$$

where c is the velocity of light (3.0×10^8 m/s). Therefore, the *mass of a nucleus is a direct measure of its energy content*. It was discovered in the 1930s that the measured mass of a nucleus is always *smaller* than the sum of the separate masses of its constituent nucleons. This difference in mass is called the *mass defect*. When the mass defect is expressed as energy by applying Einstein's equation, it is called the binding energy of the nucleus. The *binding energy* is the energy required to break up a nucleus into its component protons and neutrons. The binding energy provides a quantitative measure of nuclear stability. The greater the binding energy, the more stable the nucleus is toward decomposition.

In terms of the $^{17}_{8}O$ nucleus, for instance, the binding energy (BE) is the energy required for the process

$$^{17}_{8}O + BE \rightarrow 8^{1}_{1}p + 9^{1}_{0}n$$

The mass defect Δm is equal to the total mass of the products minus the total mass of the reactants:

$$\Delta m = 8(\text{proton mass}) + 9(\text{neutron mass}) - \left(^{17}_{8}O \text{ nuclear mass}\right)$$

Here we can substitute atomic masses, which include the electrons, for the nuclear masses:

$$\Delta m = 8\left(^{1}_{1}H \text{ atomic mass}\right) + 9(\text{neutron mass}) - \left(^{17}_{8}O \text{ atomic mass}\right)$$

This works because the mass of eight electrons in the eight hydrogen atoms is canceled by the mass of the eight electrons in the oxygen atom:

$$\Delta m = 8(\text{proton mass}) + 8(\text{electron mass}) + 9(\text{neutron mass})$$
$$- \left[\left(^{17}_{8}O \text{ nuclear mass}\right) + 8(\text{electron mass})\right]$$

We continue by using atomic masses given in Chapter 24 and Table 24.3.

$$\Delta m = 8\left(^{1}_{1}H \text{ atomic mass}\right) + 9(\text{neutron mass}) - \left(^{17}_{8}O \text{ atomic mass}\right)$$

$$= 8(1.007825 \text{ amu}) + 9(1.008665 \text{ amu}) - (16.999131 \text{ amu}) = 0.141454 \text{ amu}$$

The eight protons and nine neutrons have more mass than the oxygen-17 nucleus. The binding energy is

$$\Delta E = \Delta mc^2$$

The calculation of binding energy will be illustrated in Example 24.4.

In comparing the stability of any two different nuclei, we must account for the different numbers of nucleons per nucleus. A satisfactory comparison of nuclear stabilities can be made by using the binding energy per nucleon, that is, *the binding energy of each nucleus divided by the total number of nucleons* in the nucleus. This is one of the most important features of a nucleus. This quantity is plotted as a function of the atomic mass in Figure 24.2 of the text. Note that first it rises rapidly with increasing atomic mass and then it hits a maximum at mass 56. After mass 56, the binding energy drops slowly as atomic mass increases. Table 24.3 compares the total binding energy and the binding energy per nucleon for several isotopes.

Table 24.3 Binding Energies of Selected Isotopes

	Mass (amu)	Binding Energy (J)	
		Total	Per Nucleon
$^{2}_{1}H$	2.01410	3.57×10^{-13}	1.78×10^{-13}
$^{3}_{2}He$	3.01603	1.24×10^{-12}	4.13×10^{-13}
$^{4}_{2}He$	4.00260	4.52×10^{-12}	1.13×10^{-12}
$^{16}_{8}O$	15.99491	2.04×10^{-11}	1.28×10^{-12}
$^{17}_{8}O$	16.999131	2.10×10^{-11}	1.24×10^{-12}
$^{56}_{26}Fe$	55.934939	7.90×10^{-11}	1.41×10^{-12}
$^{238}_{92}U$	238.0508	2.89×10^{-10}	1.22×10^{-12}

EXAMPLE 24.3 Nuclear Stability

Using the stability rules, rank the following isotopes in order of increasing nuclear stability:

$$^{39}_{20}\text{Ca} \quad ^{222}_{86}\text{Rn} \quad ^{98}_{43}\text{Tc}$$

METHOD OF SOLUTION

Technetium-98 should be unstable because it has odd numbers of both protons and neutrons. Radon-222 will perhaps be the least stable because all elements beyond bismuth are radioactive. Calcium-39 has 20 protons—a "magic number." Of the three isotopes, it should be most stable:

$$^{222}_{86}\text{Rn} < ^{98}_{43}\text{Tc} < ^{39}_{20}\text{Ca}$$

EXAMPLE 24.4 Nuclear Binding Energy

Calculate the nuclear binding energy of the light isotope of helium, helium-3. The atomic mass of $^{3}_{2}\text{He}$ is 3.01603 amu.

METHOD OF SOLUTION

The binding energy is the energy required for the process

$$^{3}_{2}\text{He} \rightarrow 2^{1}_{1}\text{p} + ^{1}_{0}\text{n}$$

where $\Delta m = 2(\text{proton mass}) + (\text{neutron mass}) - ^{3}_{2}\text{He}$ (nuclear mass). The mass difference is calculated using atomic masses:

$$\Delta m = 2(^{1}_{1}\text{H atomic mass}) + (\text{neutron mass}) - (^{3}_{2}\text{He atomic mass})$$

$$= 2(1.007825 \text{ amu}) + 1.008665 \text{ amu} - 3.01603 \text{ amu}$$

$$= 8.29 \times 10^{-3} \text{ amu}$$

Using Einstein's equation, we get

$$\Delta E = (\Delta m)c^2 = (8.29 \times 10^{-3} \text{ amu}) \times (3.00 \times 10^8 \text{ m s}^{-1})^2$$

$$= 7.46 \times 10^{14} \text{ amu m}^2 \text{ s}^{-2}$$

$$= 7.46 \times 10^{14} \text{ amu m}^2 \text{ s}^{-2} \times \frac{1.00 \text{ kg}}{6.022 \times 10^{26} \text{ amu}} \times \frac{1 \text{ J}}{1 \text{ kg m}^2 \text{ s}^{-2}}$$

$$= 1.24 \times 10^{-12} \text{ J/atom}$$

Each $^{3}_{2}\text{He}$ atom contains three nucleons.

$$= \frac{1.24 \times 10^{-12} \text{ J/atom}}{3 \text{ nucleons/atom}}$$

The binding energy per nucleon is

$$\Delta E = 4.13 \times 10^{-13} \text{ J/nucleon}$$

COMMENT

By combining the above conversion factors into one constant, the number of steps in future calculations can be lessened:

$$? \text{ J/amu} = (3.00 \times 10^8 \text{ m s}^{-1})^2 \times \frac{1.00 \text{ kg}}{6.022 \times 10^{26} \text{ amu}} \times \frac{1 \text{ J}}{1 \text{ kg m}^2 \text{ s}^{-2}}$$

$$= 1.49 \times 10^{-10} \text{ J/amu}$$

This constant is a useful factor relating energy to mass in amu. Applying this to the mass defect (0.14145 amu) calculated earlier in the discussion for $^{17}_{8}\text{O}$, the binding energy energy is 2.10×10^{-11} J/atom. See Table 24.3.

NATURAL AND ARTIFICIAL RADIOACTIVITY

STUDY OBJECTIVES

You should be able to:
1. Predict, using the n:p ratio, whether a given isotope will decay by beta decay or positron decay.
2. Predict the product resulting from neutron capture by a given isotope, which is then followed by beta decay.
3. Calculate the age of a rock sample, given information about the amount of a particular radioisotope present in the rock.

Natural Radioactivity. A number of isotopes exist in nature that have an n:p ratio that places them outside the belt of stability. Radioisotopes that occur naturally (those not produced by human activities) give rise to *natural radioactivity*. The uranium decay series is shown in Table 24.3 of the textbook. All the radioisotopes in this series decay by alpha or beta emission. The uranium decay series includes all the elements between lead and uranium.

If you were given the symbol of a radioisotope, without any experience, it would be impossible to tell its mode of decay. But with knowledge of the belt of stability (Figure 24.1), you can make accurate predictions of the expected mode of decay.

Isotopes with too many neutrons lie above the belt of stability. The nuclei of these isotopes decay in such a way as to lower their n:p ratio. Thus one of the neutrons may decay into a proton and a beta particle:

$$^{0}_{1}\text{n} \rightarrow {}^{1}_{1}\text{p} + {}^{0}_{-1}\beta$$

The proton remains in the nucleus, and the beta particle is ejected from the atom. The loss of a neutron and the gain of a proton produce a new isotope with two important properties. It has a *lower n:p ratio* and thus is more likely to be stable. Also, the daughter product has an atomic number that is *one greater* than the decaying isotope due to the additional proton. Consider the decay of carbon-14, for example. Carbon-14 is continually produced in the upper atmosphere by the interaction of cosmic rays with nitrogen. Carbon-14 has a higher n:p ratio than either of carbon's stable isotopes (C-12 and C-13), and decays by beta decay:

$$^{14}_{6}\text{C} \rightarrow {}^{14}_{7}\text{N} + {}^{0}_{-1}\beta$$

Note that the product isotope $^{14}_{7}\text{N}$ is one atomic number greater than carbon. Also, it is stable. Its n:p ratio is 1.0.

Isotopes with too many protons have a low n:p ratio and lie below the belt of stability. These isotopes tend to decay by positron emission because this process produces a new isotope with a higher n:p ratio.

During *positron emission* a proton ejects a positron ($_{+1}^{0}\beta$) and becomes a neutron:

$$_{1}^{1}p \rightarrow\, _{0}^{1}n +\, _{+1}^{0}\beta$$

The neutron remains in the nucleus, and the positron is ejected from the atom. Thus, a product nucleus will contain one less proton and one more neutron than the parent nucleus. The n:p ratio increases due to positron decay.

Electron capture accomplishes the same end, that is, a higher n:p ratio. Some nuclei decay by capturing an orbital electron of the atom:

$$_{1}^{1}p +\, _{+1}^{0}e \rightarrow\, _{0}^{1}n$$

Lanthanum-138, a naturally occurring isotope with an abundance of 0.089%, decays by electron capture:

$$_{57}^{138}La +\, _{-1}^{0}e \rightarrow\, _{58}^{138}Ba$$

Electron capture is accompanied by X-ray emission.

Artificial Radioactivity

Artificial radioactivity results when an unstable nucleus is produced by transmutation. Irene Curie and Frederic Joliot discovered this phenomenon in 1933 while bombarding light elements with alpha particles. For example, in the reaction between aluminum and alpha particles the product is ^{30}P, which decays by positron emission. Phosphorus-30 has a low n:p ratio:

$$_{13}^{27}Al +\, _{2}^{4}He \rightarrow\, _{15}^{30}P +\, _{0}^{1}n$$

$$_{15}^{30}P \rightarrow\, _{14}^{30}Si +\, _{+1}^{0}\beta$$

Neutrons readily produce "artificial" radioactivity because they are easily captured by stable nuclei, with the result that the new nucleus has a higher n:p ratio. Thus, the product decays by beta decay:

$$_{0}^{1}n +\, _{17}^{37}Cl \rightarrow\, _{17}^{38}Cl$$

$$_{17}^{38}Cl \rightarrow\, _{18}^{38}Ar +\, _{-1}^{0}\beta$$

Chlorine-38 has a short half-life and is not found naturally on Earth. Note that capture of a neutron, followed by beta decay, yields a new element (Ar) with an atomic number one greater than the original element. This technique has been used to synthesize elements that were "missing" from the periodic table, such as Tc and Pm, elements number 43 and 61, respectively. All isotopes of these elements are radioactive. Note that these elements have *odd* numbers of protons.

Rates of Decay

Radioactive decay rates obey first-order kinetics:

$$\text{decay rate} = \text{number of atoms disintegrating per unit time} = kN$$

where k is the rate constant, or decay constant as it is called, and N is the number of atoms of the particular radioisotope present in the sample. Recall that the half-life is related to the rate constant by

$$t_{1/2} = 0.693/k$$

The integrated first-order equation is

$$\ln \frac{N_0}{N} = kt$$

where N is the number of atoms of the radioisotope present in the sample after time t has elapsed, and

N_0 is the number of atoms of the radioisotope present initially. Normally, N_0, N, and t are known, and this information is used to calculate the rate constant, k.

However, the object of radioactive dating is to determine the age of geological and archaeological samples and specimens. The age (t in the calculation) of certain rocks, for instance, can be estimated from analysis of the number of atoms of a particular radioisotope present now (N), as compared with the number present originally when the rock was formed (N_0).

When the above equation is rearranged, the age t is given by

$$t = \frac{1}{k} \ln \frac{N_0}{N}$$

The value of the initial number of atoms, N_0, is the sum $N + D$, where D is the number of daughter nuclei resulting from the decay of atoms of the radioisotope. The original number of atoms of a radioisotope present in a rock sample is equal to the number N remaining at time t plus the number of daughter atoms (D).

EXAMPLE 24.5 Types of Radioactive Decay

The only stable isotope of fluorine is F-19. What type of radioactivity would you expect from the isotope F-21?

METHOD OF SOLUTION

Fluorine-19 has 9 protons and 10 neutrons. Fluorine-21 must have two more neutrons, and so has a higher n:p ratio than the stable isotope. Fluorine-21 will decay by $_{-1}^{0}\beta$ emission.

EXAMPLE 24.6 Synthesizing a Transuranium Element

The first transuranium element to be synthesized by scientists was neptunium, atomic number 93. Devise a means to produce neptunium starting with U-238 and neutrons.

METHOD OF SOLUTION

Neutron capture by a nucleus followed by beta decay produces a new nucleus of one atomic number higher than the original. Neptunium is one atomic number beyond uranium. Therefore,

$$_{92}^{238}U + _{0}^{1}n \rightarrow _{92}^{239}U$$

$$_{92}^{239}U \rightarrow _{93}^{239}Np + _{-1}^{0}\beta$$

COMMENT

Uranium's other naturally occurring isotope, U-235, will not produce neptunium in the same way because it undergoes neutron-induced fission.

EXAMPLE 24.7 Radioactive Dating

The Rb-87 and Sr-87 method of dating rocks was used to analyze lunar samples:

$$_{37}^{87}Rb \rightarrow _{38}^{87}Sr + _{-1}^{0}\beta \qquad t_{1/2} = 4.9 \times 10^{10} \text{ yr}$$

Estimate the age of a rock sample in which the present mole ratio of Rb-87 to Sr-87 is 40.

METHOD OF SOLUTION

The age of the rock can be calculated from

$$t = \frac{1}{k} \ln \frac{N_0}{N}$$

Since the decay constant k is not given, we must use the equation relating k and $t_{1/2}$:

$$k = \frac{0.693}{t_{1/2}}$$

This can be substituted for k, yielding

$$t = \frac{t_{1/2}}{0.693} \ln \frac{N_0}{N}$$

Since Rb decays to Sr, the initial number of Rb atoms is equal to the sum of the Rb atoms and Sr atoms present:

$$N_0^{Rb} = N^{Rb} + N^{Sr}$$

Given that $N^{Rb}/N^{Sr} = 40$, then

$$N^{Sr} = \frac{1}{40} N^{Rb}$$

Therefore, after substitution for N^{Sr}, we get

$$N_0^{Rb} = N^{Rb} + \tfrac{1}{40} N^{Rb} = 1.025 N^{Rb}$$

Substituting into the rate equation, we get

$$t = \frac{4.9 \times 10^{10} \text{ yr}}{0.693} \ln \frac{1.025 N^{Rb}}{N^{Rb}} = (7.07 \times 10^{10} \text{ yr}) \ln 1.025$$

$$= 1.7 \times 10^9 \text{ yr}$$

NUCLEAR FISSION AND FUSION

STUDY OBJECTIVES

You should be able to:
1. Describe both fission and fusion, and relate these processes to the curve of binding energy per nucleon versus mass number.
2. Compare the energy released by fission and fusion with that of an ordinary chemical reaction.

Nuclear Fission. The curve of binding energy per nucleon versus mass number, Figure 24.2 in the text, shows that the most stable nuclei are those with masses close to iron-56, which is the most stable nucleus. During *fission* a heavy nucleus of mass greater than about 230 amu splits into two lighter nuclei whose masses are usually between 80 and 160 amu. Since the two smaller nuclei are more stable than the larger nucleus, energy is released in the process.

Although many nuclei heavier than uranium can undergo fission, the most important ones are U-233, U-235, and Pu-239. These isotopes undergo fission upon capture of a neutron. It is important to realize that naturally occurring uranium consists of two isotopes, U-235 and U-238, but that only U-235 is fissionable with thermal neutrons. The term *thermal neutrons* refers to those existing at temperatures around 25°C. Since the nucleus does not repel a neutron, as it does an alpha particle, for instance, neutron-induced fission will occur at ordinary temperatures.

The reaction is quite complex because the same two products are not formed by all fissioning nuclei. The two reactions below show just two out of many possibilities:

$$^{235}_{92}U + ^{1}_{0}n \longrightarrow \begin{cases} ^{141}_{56}Ba + ^{92}_{36}Kr + 3^{1}_{0}n \\ ^{137}_{52}Te + ^{96}_{40}Zr + 2^{1}_{0}n \end{cases}$$

The actual distribution of product yields is shown in Figure 24.7 of the text.

A significant feature of fission is that, on the average, two to three neutrons are released per fission event. Since neutrons are required to initiate fission, and because neutrons are also products of fission, a nuclear chain reaction is possible.

The energy released during nuclear fission depends somewhat on just what products are formed. The energy released from the fission of 1 mol of U-235 atoms can be calculated from the equation $\Delta E = \Delta mc^2$. The calculation shows that about 2.0×10^{10} kJ is released per mole of uranium. This amount of energy is *70 million times* the amount released in the exothermic chemical reaction in which 1 mol of H_2 reacts with $\frac{1}{2}$ mol of O_2 to form 1 mol of water!

Nuclear Fission. Radioactivity and nuclear fission are processes in which matter "comes apart." The energy-releasing processes that occur on the sun are ones in which matter is fused. *Nuclear fusion* is the combining of small nuclei, such as hydrogen, to form a larger, more stable nucleus. Such a nucleus will have a higher average binding energy per nucleon, and fusion reactions will be exothermic. Because all nuclei are positively charged, they must collide with enormous force in order to combine (fusion). This means that the atoms that will undergo fusion must be heated to millions of degrees. Fusion reactions are called *thermonuclear* reactions because they occur only at very high temperatures, such as those in the sun.

The reaction that accounts for the tremendous release of energy by the sun is believed to be the stepwise fusion of four hydrogen nuclei to produce one helium nucleus. The net process is

$$4^{1}_{1}H \rightarrow ^{4}_{2}He + 2^{0}_{+1}e \quad \Delta E = -4.3 \times 10^{-12} \text{ J}$$

Upon fusion, 1 g of hydrogen releases the energy equivalent to the combustion of 20 tons of coal. The fusion of 4 mol of H atoms by the preceding equation releases 2.6×10^9 kJ of energy.

Our sun is made up of mostly hydrogen (0%) and helium (9%). In its interior the temperatures are estimated to reach 15 million °C. At these temperatures, hydrogen will fuse to form helium; but helium, with its greater nuclear charge, will not fuse to form the heavier elements. The heavier elements up to iron are formed by fusion reactions that occur in exploding stars, called nova and supernova, where the temperatures can reach 2 billion °C.

EXAMPLE 24.8   Relationship of Mass and Energy

Calculate the mass of hydrogen that must undergo nuclear fusion each day in order to provide just the fraction of the daily energy output of the sun that reaches Earth, which is 1.5×10^{22} J.

METHOD OF SOLUTION

Using the equation

$$4^{1}_{1}H \rightarrow ^{4}_{2}He + 2^{0}_{+1}e \quad \Delta E = -4.3 \times 10^{-12} \text{ J}$$

we find that fusion of 4 H atoms = 4.3×10^{-12} J. We can set up the calculation:

$$? \text{ g H atoms} = \frac{1.5 \times 10^{22} \text{ J}}{1 \text{ day}}$$

$$= \frac{1.5 \times 10^{22} \text{ J}}{1 \text{ day}} \times \frac{4 \text{ H atoms}}{4.3 \times 10^{-12} \text{ J}} \times \frac{1.66 \times 10^{-24} \text{ g}}{\text{H atom}}$$

$$= 2.3 \times 10^{10} \text{ g}$$

Expressing the answer in tons, we get

$$2.3 \times 10^{10} \text{ g} = 25{,}500 \text{ tons of hydrogen}$$

EXAMPLE 24.9 Comparing Fission and Fusion

Compare fission and fusion with respect to the temperatures required and the nature of the by-products of these processes.

METHOD OF SOLUTION

Neutron-induced fission occurs at ordinary temperatures, whereas nuclear fusion requires temperatures in the millions of degrees. Fission of heavy elements yields hundreds of isotopes of elements with intermediate atomic numbers. These isotopes, by and large, have an excess of neutrons and are beta emitters. On the other hand, fusion of "light" nuclei yields stable isotopes of low-to-medium atomic mass.

BIOLOGICAL EFFECTS OF RADIATION

STUDY OBJECTIVES

You should be able to:
1. Describe how radiation interacts with matter.
2. Define radiation dose in units of rads and rems.

Interactions of Radiation with Matter. In passing through matter, alpha, beta, and gamma rays lose energy chiefly by interaction with electrons, which make up most of the volume of matter. Alpha and beta particles, on colliding with electrons, forcefully eject these electrons from atoms and molecules and thereby produce ions. These particles lose only a small fraction of their energy in a single collision with an electron. Because alpha and beta particles are extremely energetic, thousands of collisions are required to bring them to rest. These particles produce "tracks" of ionization. Alpha, beta, and gamma rays are known as ionizing radiation.

Most devices for detecting radioactivity depend on the formation of ions. The best known instrument for detecting radiation is the Geiger counter, in which ions produced by a particle trigger a pulse of electricity that is counted electronically. Darkening of photographic plates, discharging of electroscopes, and damage to biological tissue all involve ionization.

Units of Radiation Dose. Two units used to measure radiation dose are the rad and the rem. The *rad* (*r*adiation *a*bsorbed *d*ose) is defined as 0.01 J of energy absorbed per kilogram of any absorbing material. The millirad is one-thousandth of a rad. Because beams of different radiations cause very different biological damage even when the body absorbs the same amount of energy from each type, it is necessary to define a unit specifically for biological tissue. The unit of biologically effective dose is the *rem* (*r*adiation *e*quivalent in *m*an), which is the absorbed dose in rads multiplied by the *r*elative *b*iological *e*ffectiveness factor, *RBE*. The millirem is one-thousandth of a rem. For beta and gamma rays, $RBE = 1.0$; for fast neutrons and alpha particles, $RBE = 10$. Thus, a dose of 1 rad of alpha radiation is equivalent to 10 rem.

$$\text{dose (rem)} = RBE \times \text{dose (rad)}$$

TRUE-FALSE QUESTIONS

1. The process represented by the following nuclear equation is an example of radioactive decay:

   $^{9}_{4}Be + ^{4}_{2}He \rightarrow ^{12}_{6}C + ^{1}_{0}n$

2. $^{3}_{2}He$ should have greater nuclear stability than $^{4}_{2}He$.

3. The following nuclear equation is balanced:

   $^{238}_{92}U + ^{14}_{7}N \rightarrow ^{247}_{99}Es + 5^{1}_{0}n$

4. Beta particles are distinguished from orbital electrons because they are ejected from the atomic nucleus. However, once formed, β particles and electrons are indistinguishable.

5. Alpha particles are atoms of helium-4.

6. The measured mass of a nucleus is always greater than the combined masses of its protons and neutrons.

7. The mass defect of a nucleus is directly related to its binding energy.

8. Isotopes with a high n:p ratio decay by ejecting a neutron from the nucleus.

9. Positron emission by a nucleus produces a new atom that is one atomic number higher than the original.

10. Artificial radioactivity is produced by nuclear transmutation.

11. The rates of radioactive decay obey second-order rate equations.

12. Fission occurs among the heaviest nuclides, whereas nuclear fusion occurs more readily for nuclides of low mass.

13. Nuclear fusion reactions require temperatures in the millions of degrees.

14. When U-235 nuclei undergo neutron-induced fission, they all split into the same two nuclei plus three neutrons.

SELF-TEST A

1. What similarities and differences exist between beta particles and positrons?

2. Complete and balance the following nuclear equations:
   a. $^{239}_{94}Pu \rightarrow ^{4}_{2}He +$ _____
   b. _____ $+ ^{6}_{3}Li \rightarrow 2^{4}_{2}He$
   c. $^{90}_{38}Sr \rightarrow$ _____ $+ ^{0}_{-1}\beta$
   d. $^{10}_{5}B + ^{4}_{2}He \rightarrow$ _____ $+ ^{1}_{0}n$
   e. $^{56}_{26}Fe + ^{1}_{0}n \rightarrow$ _____

3. Rank the following nuclides in order of increasing nuclear stability:

   $^{40}_{20}Ca$ $^{39}_{20}Ca$ $^{11}_{5}B$

4. With reference to the belt of stability, state the modes of decay you would expect for the following:
   a. $^{13}_{7}N$ b. $^{26}_{13}Al$ c. $^{28}_{13}Al$

5. a. Calculate the binding energy and the binding energy per nucleon of $^{27}_{13}Al$. The atomic mass of $^{27}_{13}Al$ is 26.98154 amu.
   b. Compare this result to the binding energy of $^{28}_{14}Si$, which has an even number of protons and neutrons. The atomic mass of $^{28}_{14}Si$ is 27.976928 amu.

6. Cobalt-60 is used in radiation therapy. It has a half-life of 5.26 yr.
   a. Calculate the rate constant for radioactive decay.
   b. What fraction of a certain sample will remain after 12 yr?

7. Radioactive decay follows first-order kinetics. If 20% of a certain radioisotope decays in 4.0 yr, what is the half-life of this isotope?

8. Estimate the age of a bottled wine that has a tritium ($^{3}_{1}H$) content three-fourths that of environmental water obtained from the area where the grapes were grown. $t_{1/2}$ = 12.3 yr.

9. Analysis of a sample of uranite ore shows that the ratio of U-238 atoms to Pb-206 atoms is 3.8. Assuming there was no Pb-206 present initially, how old is the rock? $t_{1/2}$ (U-238) = 4.5 × 10^9 yr.

10. Consider the following fusion reactions:
    a. $^{1}_{1}H + 2H \rightarrow ^{3}_{2}He$
    b. $^{3}_{2}He + ^{3}_{2}He \rightarrow ^{4}_{2}He + 2^{1}_{1}H$
    c. $^{2}_{1}H + ^{3}_{1}H \rightarrow ^{4}_{2}He + ^{1}_{0}n$
    Given the atomic masses $^{3}_{2}He$ = 3.016029 amu, $^{4}_{2}He$ = 4.002603 amu, and $^{3}_{1}H$ = 3.017005 amu, which of the above has the largest change in energy as indicated by its change in mass, Δm?

11. One atom of element 109, the heaviest element yet made, was prepared by bombardment of a target of bismuth-209 with accelerated nuclei of iron-58. Write a balanced nuclear equation to show the formation of the isotope of element 109 with a mass number of 266.

12. How does the fusion-type reaction discussed in Problem 11 differ from fusion reactions discussed in the textbook?

13. Write nuclear equations that show how Pu-239 is formed in a "breeder" reactor.

14. $^{137}_{55}Cs$ is a fission product of $^{235}_{92}U$. If it is formed along with three neutrons, what is the other isotope formed?

15. What is the mode of decay expected for "light" nuclei which are unstable because of a low n:p ratio?

16. One natural radioactive series begins with $^{238}_{92}U$ and ends with $^{206}_{82}Pb$. All steps in the series are either α or β decay. How many α particles and β particles are emitted?

17. The β particles emitted by carbon-14 atoms have a maximum energy of 2.5 × 10^{-14} J per particle. What is the dose in rads when 8.0 × 10^{10} carbon-14 atoms decay and all the energy from their decay is absorbed by 2.0 kg of matter?

18. A sample of biological tissue absorbs a dose of 1.0 rad of alpha radiation. How many rems is this?

GENERAL PROBLEM

19. Radium-226 is an α emitter with a half-life of 1600 yr. If 2.0 g of radium were allowed to undergo decay for 10 yr, and all the α particles were collected over that time as helium gas, what would be the mass and the volume of the He at STP?

ANSWERS

TRUE-FALSE QUESTIONS

1. False. Transmutation.
2. False. Nuclei with even numbers of protons and neutrons are more stable than those with an even number of protons and an *odd* number of neutrons.
3. True.
4. True.
5. False. Alpha particles are the nuclei of He-4 atoms.
6. False. The measured mass is always less.
7. True.
8. False. Such isotopes decay by ejecting an electron from the nucleus.
9. False. Positron emission yields an atom one atomic number lower.
10. True.
11. False. First-order rate equations.
12. True.
13. True.
14. False. Hundreds of different isotopes are produced.

SELF-TEST A

1. Beta particles and positrons have the same mass, 0.00055 amu. Beta particles have one unit of negative charge, while positrons have one unit of positive charge.
2. a. $^{235}_{92}U$ b. $^{2}_{1}H$ c. $^{90}_{39}Y$ d. $^{13}_{7}N$ e. $^{57}_{26}Fe$
3. $^{11}_{5}B > ^{39}_{20}Ca < ^{40}_{20}Ca$. Boron-11 has an odd number of protons and an odd number of neutrons (rule 2). Calcium-39 has an even number of protons and an odd number of neutrons (rule 1). Calcium-40 has an even number of protons and an even number of neutrons. It also has a "magic number" of protons and neutrons (rule 2).
4. a. Positron emission   b. Positron emission   c. Beta decay
5. a. 3.6×10^{-11} J/atoms; 1.3×10^{-12} J/nucleon
   b. 3.8×10^{-11} J/atoms; 1.4×10^{-12} J/nucleon
6. a. $k = 0.132$/yr   b. $N/N_0 = 0.205$
7. $t_{1/2} = 12.4$ yr
8. $t = 5.1$ yr
9. $t = 1.5 \times 10^9$ yr
10. c
11. $^{209}_{98}Bi + ^{58}_{26}Fe \rightarrow ^{266}_{109}X + ^{1}_{0}n$
12. The fusion reactions discussed in the textbook are those of the "light" elements, which are exothermic. In the synthesis of element-109, energy is not evolved; rather, energy is absorbed.
13. $^{238}_{92}U + ^{1}_{0}n \rightarrow ^{239}_{92}U$
    $^{239}_{92}U \rightarrow ^{239}_{93}Np + ^{0}_{-1}\beta$
    $^{239}_{93}Np \rightarrow ^{239}_{94}Pu + ^{0}_{-1}\beta$
14. $^{239}_{92}U + ^{1}_{0}n \rightarrow ^{137}_{55}Cs + ^{96}_{37}Rb + 3^{1}_{0}n$
15. Positron emission
16. There are eight α particles, which account for the mass change; eight α and six β particles account for the change in positive charge.
17. 0.10 rad
18. 10 rem
19. 2.3×10^{19} Ra atoms decayed in 10 yr; 2.3×10^{19} He atoms have a mass of 1.6×10^{-4} g and occupy a volume of 0.87 cm^3 at STP.

Chapter Twenty-Five
ORGANIC CHEMISTRY

- Hydrocarbons
- Functional Groups

HYDROCARBONS

STUDY OBJECTIVES

You should be able to:
1. Name four different families of hydrocarbons and draw a structural formula of a typical example of each.
2. Name hydrocarbons of these four groups, given a structural formula. Given an IUPAC name, draw a structural formula.
3. Write equations for addition reactions to alkenes and alkynes.
4. Apply Markovnikov's rule concerning the addition of unsymmetrical reagents to alkenes.

Organic Chemistry. The heart of organic chemistry is the carbon atom. Carbon is a key ingredient of about 6 million chemical compounds primarily because of its ability to form long chains of self-linked atoms. The existence of structural and geometric isomers contributes strongly to the number of organic or carbon-containing compounds.

The *hydrocarbons* are an important class of organic compound that consists only of the elements carbon and hydrogen. Four groups of families of hydrocarbons are known. These are alkanes, alkenes, alkynes, and aromatics.

The general formula for an alkane is C_nH_{2n+2}, where n is the number of carbon atoms in the molecule; $n = 1, 2, 3, \ldots$. When $n = 1$, we have the simplest member of the alkane family, methane (CH_4). The alkanes make up a *homologous series*: a series of compounds differing in the number of carbon atoms. As n increases one at a time, we can generate the formulas of the entire series of alkanes. The names and formulas of the first 10 straight-chain *alkanes* are given in Table 25.1. The first part of

Table 25.1 Names of the First 10 Alkanes

Formula	Name
CH_4	Methane
C_2H_6	Ethane
C_3H_8	Propane
C_4H_{10}	Butane
C_5H_{12}	Pentane
C_6H_{14}	Hexane
C_7H_{16}	Heptane
C_8H_{18}	Octane
C_9H_{20}	Nonane
$C_{10}H_{22}$	Decane

each name represents the number of C atoms in the molecule. The ending *-ane* is common to all alkanes. If you learn these names, they will prove to be very useful in naming other organic compounds.

Alkenes contain C = C double bonds, and members of this homologous series have the general formula C_nH_{2n}. Alkenes are named with the same root word as alkanes to indicate the number of carbon atoms, but all names end in *ene*:

C_2H_4 $CH_2 = CH_2$ ethene (ethylene)

C_3H_6 $CH_3CH = CH_2$ propene

C_4H_8 $CH_3CH_2 = CH_2$ butene

Alkynes contain C ≡ C triple bonds and have the general formula C_nH_{2n-2}. The names of alkynes end with *yne*:

C_2H_2 $CH \equiv CH$ ethyne (acetylene)

C_3H_4 $CH_3C \equiv CH$ propyne

C_4H_6 $CH_3CH_2C \equiv CH$ butyne

Aromatics contain benzene rings and do not have a simple general formula. Benzene is the simplest compound in this group:

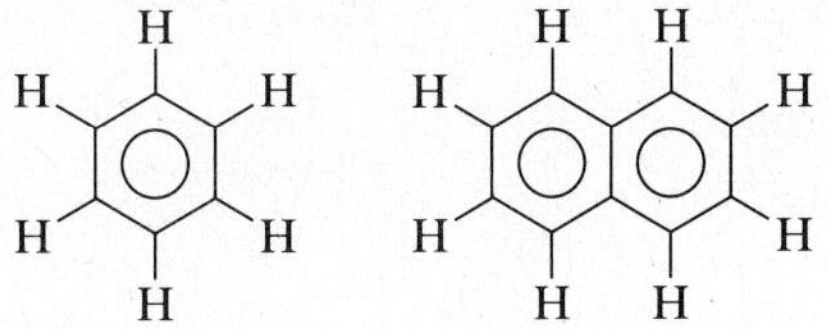

benzene (C_6H_6)   naphthalene ($C_{10}H_8$)

There is also a type of *alkane* that has atoms bonded into ring configurations. These are the *cycloalkanes*. They have the same general formula as the alkenes, C_nH_{2n+2}, but do not have double bonds. Neither are these aromatics. Two cycloalkanes are shown below:

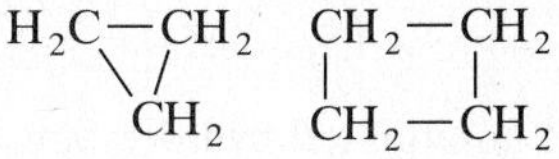

cyclopropane   cyclobutane

Structural Isomers. Hydrocarbons that do not contain the benzene ring (the alkanes, cycloalkanes, alkenes, and alkynes) are known as *aliphatic hydrocarbons*. Until now, except for cycloalkanes, we have considered only aliphatic hydrocarbons that have straight chains of carbon atoms. Branching of hydrocarbon chains is very common. Butane (C_4H_{10}) can be straight-chained, but it can also be branched-chained:

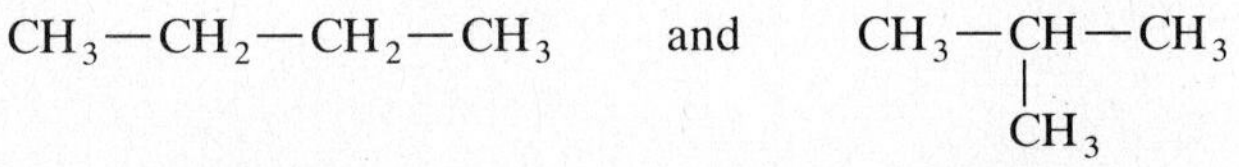

$CH_3—CH_2—CH_2—CH_3$ and $CH_3—CH—CH_3$
                                                                                                  CH_3

n-butane   2-methylpropane

Note that both molecules have the same molecular formula, but they have different arrangements of atoms (different structures). These are actually two distinguishable compounds, with their different structures producing slightly different chemical and physical properties. Molecules that have the same molecular formula but different structures are called *structural isomers*. Straight-chain hydrocarbons are called normal and use the symbol *n*. The straight-chain form of C_4H_{10} has the name *n*-butane. You will learn how to name branched-chain hydrocarbons in the next section. In the alkane series, as the number of C atoms increases, the number of possible structural isomers increases dramatically. For example, C_4H_{10} has 2 isomers, C_6H_{14} has 5, and $C_{10}H_{22}$ has 75.

Nomenclature. The rules for naming hydrocarbons according to the IUPAC system are briefly summarized as follows:

1. Decide whether the compound is an alkane, alkene, or alkyne. For alkanes, give the longest carbon chain a root name corresponding to the alkane with the same number of C atoms. For alkenes and alkynes, name the longest chain that contains the double or triple bond. Any group branching from the chain is called a side group.
2. Number the carbon atoms in the longest chain. For alkanes, start numbering at the end of the chain such that the side groups will have the smallest numbers. For alkenes and alkynes, number the C atoms in the main chain starting at the end nearer to the double or triple bond.
3. Identify and number side groups.
4. Combine the names of side groups and the main chain into the hydrocarbon name. Arrange the side groups alphabetically followed by the root name of the main chain.

These rules are applied in Examples 25.1 and 25.2

Addition Reactions. Alkenes, alkynes, and aromatics are called *unsaturated hydrocarbons*. This means that they can acquire more hydrogen atoms in an addition reaction called *hydrogenation*:

$$CH_2 = CH_2 + H_2 \rightarrow CH_3 - CH_3$$

$$CH \equiv CH + 2H_2 \rightarrow CH_3 - CH_3$$

$$\text{C}_6\text{H}_6 + 3H_2 \rightarrow \text{C}_6\text{H}_{12}$$

Alkanes are called *saturated hydrocarbons* because they cannot acquire additional hydrogen atoms. The carbon atoms in a saturated hydrocarbon are already bonded to the maximum number of H atoms:

$$CH_3 - CH_3 + H_2 \rightarrow \text{no reaction}$$

In an *addition reaction* a small molecule such as H_2 is added to an unsaturated hydrocarbon. The addition reaction occurs at the $C = C$ double bond. One atom of the small molecule links to one of the carbon atoms of the double bond, while the other atom attaches to the other carbon atom. Other examples are

$$CH_2 = CH_2 + Cl_2 \rightarrow CH_2Cl - CH_2Cl$$

$$CH_2 = CH_2 + H_2O \rightarrow CH_3 - CH_2OH$$

$$CH_2 = CH_2 + HCl \rightarrow CH_3 - CH_2Cl$$

Markovnikov's Rule. The two carbons of a double bond suffer different fates during addition reactions with unsymmetrical reagents such as HCl, HBr, and H—OH (water). For example, two different compounds might possibly be formed by reaction of 1-butene with HCl:

$$CH_3CH_2CH = CH_2 + HCl \rightarrow CH_3CH_2CHCl-CH_3 \quad \text{or} \quad CH_3CH_2CH_2-CH_2Cl$$

$$\qquad\qquad\qquad\qquad\qquad\qquad\qquad \text{observed product} \qquad\qquad\qquad \text{not found}$$

The rule that predicts which of the two products is formed is called *Markovnikov's rule*. This rule states: *In the addition of unsymmetrical reagents to alkenes, the positive group of the reagent adds to the carbon that already has the most hydrogen atoms*. In HCl, HBr, and H₂O the hydrogen atoms have a partially positive charge because they are bonded to more electronegative atoms. Thus, in the above reaction with $CH_3CH_2CH = CH_2$, the H atom from HCl bonds to the terminal C atom. The Cl atom bonds to the CH group.

Geometric Isomers of Alkenes. $C = C$ double bonds are completely rigid. Consider the structures of *cis*-2-butene and *trans*-2-butene:

$$\begin{array}{cc} CH_3 \quad H & CH_3 \quad CH_3 \\ \diagdown \;\; \diagup & \diagdown \;\; \diagup \\ C=C & C=C \\ \diagup \;\; \diagdown & \diagup \;\; \diagdown \\ H \quad CH_3 & H \quad H \\ \textit{trans-2-butene} & \textit{cis-2-butene} \end{array}$$

Neither structure can rotate around the $C = C$ bond to become the other. These two molecules have different physical properties and can be separated from one another. Therefore, these two structures represent two different compounds and are isomers. Many pairs of such isomers have been isolated, and they have no tendency to undergo interconversion.

The rigidity of the double bond gives rise to a new kind of isomerism, *geometric isomerism*, which is sometimes called *cis-trans* isomerism. The relative positions of atoms in space are known as a *configuration*. The isomers are named *cis* or *trans* depending on their configuration. The *cis* configuration has like groups on the same side of the double bond:

$$\begin{array}{c} H_3C \quad\quad CH_3 \\ \diagdown \quad\quad \diagup \\ ——C=C—— \\ \diagup \quad\quad \diagdown \\ H \quad\quad H \end{array}$$

The *trans* configuration has like groups on the opposite sides of the double bond. *Trans* means "across":

$$\begin{array}{c} H_3C \quad\quad H \\ \diagdown \quad\quad \diagup \\ ——C=C—— \\ \diagup \quad\quad \diagdown \\ H \quad\quad CH_3 \end{array}$$

EXAMPLE 25.1 Naming Hydrocarbons

Name the molecule that has the following structure:

$$\begin{array}{c} \quad\quad\quad\quad H_3C \quad CH_2—CH_3 \\ \quad\quad\quad\quad\;\; | \quad\quad\;\; | \\ CH_3—CH_2—C—CH—CH_2—CH_3 \\ \quad\quad\quad\quad 4 \;\; | \quad 3 \quad\;\; 2 \quad\quad 1 \\ \quad\quad\quad\quad 5 \;\; CH_2 \\ \quad\quad\quad\quad\quad\;\; | \\ \quad\quad\quad\quad 6 \;\; CH_2 \\ \quad\quad\quad\quad\quad\;\; | \\ \quad\quad\quad\quad 7 \;\; CH_3 \end{array}$$

METHOD OF SOLUTION

First identify the longest continuous carbon chain. Two equivalent chains containing seven carbon atoms are noticeable. Number the carbon atoms as shown. Two ethyl side groups appear at carbons 3 and 4, and one methyl group appears at carbon 4. The root name is heptane. Placing the names of the side groups (alphabetically, ethyl is before methyl) and their position numbers in front of the root name, we get

    3,4-diethyl-4-methylheptane

COMMENT

If the chain had been numbered in reverse order, the name would be

    4,5-diethyl-5-methylheptane

Since 3,4 is smaller than 4,5, the first name is the correct one.

EXAMPLE 25.2   Naming Hydrocarbons

Name the following hydrocarbon:

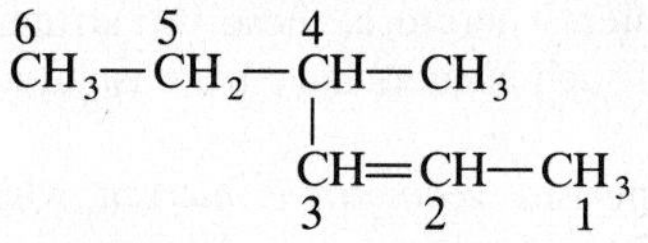

$$\overset{6}{C}H_3-\overset{5}{C}H_2-\overset{4}{C}H-CH_3$$
$$|$$
$$\underset{3}{C}H=\underset{2}{C}H-\underset{1}{C}H_3$$

METHOD OF SOLUTION

Locate the longest chain that contains the double bond. This chain contains six C atoms, so the root name will be hexene. Number the chain so that the double bond has the smallest number, as shown. Note the methyl group on carbon 4. The name is

    4-methyl-2-hexene

COMMENT

The 2 indicates the position of the C = C double bond, which connects carbons numbered 2 and 3.

EXAMPLE 25.3   Addition Reactions

Give the structure of the product of the following reaction:

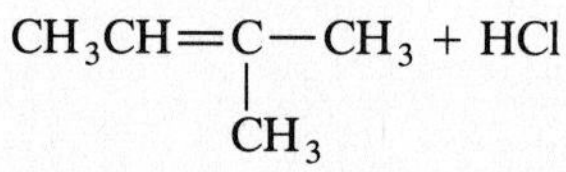

$$CH_3CH=C-CH_3 + HCl$$
$$|$$
$$CH_3$$

METHOD OF SOLUTION

This solution involves the addition of an unsymmetrical reagent to a double bond. According to Markovnikov's rule, the positive portion of the reagent (in this case an H atom) adds to the carbon atom in the double bond that already has the most hydrogen atoms. The Cl atom adds to the other C atom. The product will be

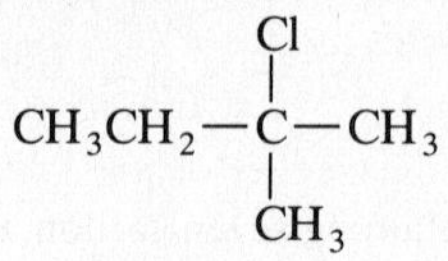

$$\overset{\quad\;\; Cl}{\underset{\quad\;\; CH_3}{CH_3CH_2-\overset{|}{\underset{|}{C}}-CH_3}}$$

FUNCTIONAL GROUPS

STUDY OBJECTIVES

You should be able to:
1. Define the term *functional group*. Sketch the general structural formulas of alcohol, ether, ketone, aldehyde, carboxyl, ester, and amine groups.
2. Write structural formulas for products of the oxidation of alcohols and aldehydes, for products of the reactions of carboxylic acids with alcohols, and for products of the reactions of esters with water and sodium hydroxide.

Alcohols. Alkanes are quite inert toward most substances, the main exception being oxygen. You have seen many examples of combustion reactions throughout the text. Certain groups of atoms within molecules always react in the same way regardless of the chain length of the alkane portion. The group of atoms that is largely responsible for the chemical behavior of a molecule is called a *functional group*. For instance, *alcohols* contain an alkane group and a hydroxyl group, OH. Methanol (CH_3OH), ethanol (C_2H_5OH), and *n*-propanol (C_3H_7OH) are the first three alcohols of a homologous series. In each of the three, chemical reactions occur at the OH group, rather than at the "inert" alkane group. The hydroxyl group is the functional group in each, and the three alcohols react toward other reagents in the same way.

Reactions of alcohols discussed in the textbook are oxidation by oxidizing agents, displacement of hydrogen by alkali metals, and esterification. An oxidizing agent removes two H atoms, one from the hydroxyl group and one from the adjacent carbon atom, forming an aldehyde (shown below):

$$CH_3CH_2CH_2-OH \xrightarrow{MnO_4^-} CH_3CH_2CHO$$

Quite often, in organic reactions, we do not write completely balanced equations. Here the emphasis is on the chemical change of the alcohol functional group.

Alkali metals displace $H_2(g)$ from alcohols:

$$2CH_3CH_2OH + 2K \rightarrow 2CH_3CH_2OK + H_2$$

Esterification will be discussed under carboxylic acids.

Aldehydes and Ketones. The functional group in these compounds is the carbonyl group, $C=O$. In aldehydes the carbonyl group is at the end of the alkane chain. In ketones it is not a terminal group:

$$\underset{\text{butanal (an aldehyde)}}{CH_3CH_2CH_2\overset{\overset{O}{\|}}{C}H \quad \text{or} \quad CH_3CH_2CH_2CHO} \qquad \underset{\text{2-butanone (a ketone)}}{CH_2CH_2\overset{\overset{O}{\|}}{C}CH_3 \quad \text{or} \quad CH_3CH_2COCH_3}$$

Aldehydes are easily oxidized to carboxylic acids:

$$CH_3CH_2\overset{\overset{O}{\|}}{C}H \xrightarrow{MnO_4^-} CH_3CH_2\overset{\overset{O}{\|}}{C}-OH$$

Ketones generally are less reactive than aldehydes:

$$CH_3CH_2\overset{\overset{O}{\|}}{C}CH_3 \xrightarrow{MnO_4^-} \text{no reaction}$$

Carboxylic Acids. The functional group in organic acids is the carboxyl group:

$$-\overset{\overset{\displaystyle O}{\|}}{C}-OH \qquad \text{abbreviated as } -COOH$$

This group is weakly acidic, and a homologous series of organic acids exists:

HCOOH	methanoic acid (formic acid)
CH_3COOH	ethanoic acid (acetic acid)
CH_3CH_2COOH	propanoic acid
$CH_3CH_2CH_2COOH$	butanoic acid

Fatty acids are carboxylic acids that contain more than four carbon atoms. Two fatty acids are

$CH_3(CH_2)_{14}COOH$	palmitic acid
$CH_3(CH_2)_{16}COOH$	stearic acid

The ionizable hydrogen atom accounts for the acidic properties of carboxylic acids:

$$CH_3COOH(aq) \rightleftharpoons CH_3COO^-(aq) + H^+(aq)$$

Esters. The ester functional group is

$$-\overset{\overset{\displaystyle O}{\|}}{C}-OR \qquad \text{or} \qquad -COOR$$

where R stands for any alkyl group such as those listed in Table 25.2 of the textbook. This functional group resembles the carboxylic acid group except that the R group replaces the H atom. Carboxylic acids react with alcohols to form the compounds called esters:

$$CH_3\overset{\overset{\displaystyle O}{\|}}{C}-\boxed{OH} + \boxed{H}-OCH_3 \rightarrow CH_3\overset{\overset{\displaystyle O}{\|}}{-}OCH_3 + \boxed{H_2O}$$

acetic acid   methanol   methyl acetate

Esterification reactions are reversible, and at equilibrium a mixture is formed that contains all four substances.

Esterification is an example of a *condensation* reaction. A "condensation" is a reaction in which parts of two molecules become joined by formation of a new covalent bond to make a new, larger molecule.

Saponification is the alkaline hydrolysis of an ester. In this reaction the base reacts directly with an ester to split it into *a salt of an organic acid* and *an alcohol*:

$$C_2H_5COOCH_3 + NaOH \rightarrow C_2H_5COONa + CH_3OH$$

Amines. Organic bases contain the $-NH_2$ functional group, which is called an amino group. Molecules containing an amino group are called amines. The amines may be considered as derivatives of ammonia (NH_3). Amines have the general formula R_3N, where R may be H, an alkyl group, or an aromatic group. Molecules of the type RNH_2 are called primary amines. Those with R_2NH are secondary amines, and R_3N is a tertiary amine:

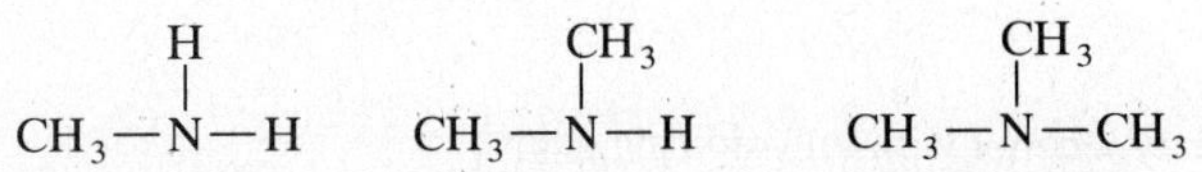

$$\begin{array}{ccc} \text{H} & \text{CH}_3 & \text{CH}_3 \\ | & | & | \\ \text{CH}_3-\text{N}-\text{H} & \text{CH}_3-\text{N}-\text{H} & \text{CH}_3-\text{N}-\text{CH}_3 \\ \text{methylamine,} & \text{dimethylamine,} & \text{trimethylamine,} \\ \text{a primary amine} & \text{a secondary amine} & \text{a tertiary amine} \end{array}$$

Ethers. The functional group in ethers is C — O — C or, more generally, R — O — R. Ethers and alcohols are structural isomers, as illustrated in the following diagram:

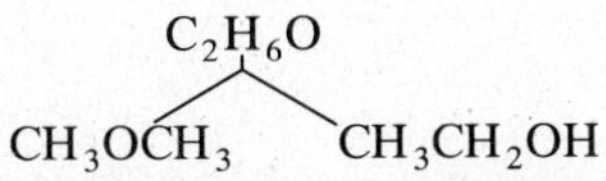

$$\begin{array}{cc} & C_2H_6O \\ \diagup & \diagdown \\ CH_3OCH_3 & CH_3CH_2OH \\ \text{dimethyl ether} & \text{ethanol} \\ T_{\text{b.p.}} \; -25°C & T_{\text{b.p.}} \; 78°C \end{array}$$

The boiling points given show that ethers are much more volatile than the corresponding alcohols. This is because of the absence of hydrogen bonding in ethers. Alcohols contain the polar $O^{\delta-} - H^{\delta+}$ group which can participate in hydrogen bonding to neighboring alcohol molecules. Diethyl ether, $CH_3CH_2OCH_2CH_3$, was used as an anesthetic for many years.

EXAMPLE 25.4 Identifying Functional Groups

Indicate the functional groups in the following formulas:
a. $C_5H_{11}OH$
b. CH_3CHO
c. $C_3H_7OCH_3$
d. $CH_3COC_2H_5$
e. CH_3COOCH_3

METHOD OF SOLUTION

Learning to recognize functional groups requires memorization of their structural formulas. Table 25.5 in the textbook shows a number of the important functional groups.

a. $C_5H_{11}OH$ contains a *hydroxyl* group. It is an alcohol.

b. CH_3CHO is a way to represent $CH_3\overset{\overset{\displaystyle O}{\|}}{CH}$ on a single line of type. C = O is a carbonyl group. CH_3CHO is an aldehyde.

c. $C_3H_7OCH_3$ contains a C — O — C group and is thus an ether.

d. $CH_3COC_2H_5$ is a way to represent a ketone on a single line. $\diagdown_{\diagup}C=O$ is a carbonyl group.

e. CH_3COOCH_3 is a condensed structural formula for an ester:

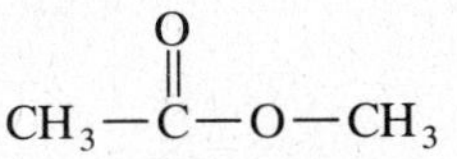

$$CH_3-\overset{\overset{\displaystyle O}{\|}}{C}-O-CH_3$$
ester functional group

EXAMPLE 25.5 Predicting Reaction Products

Predict the products of the following reactions:
a. $CH_3CH(OH)CH_3 + Cr_2O_7^{2-} \rightarrow$
b. $CH_3CHO + MnO_4^- \rightarrow$

c. HCOOH + CH$_3$OH →
d. CH$_3$CH$_2$COOCH$_3$ + NaOH →

METHOD OF SOLUTION

a. This is the reaction of a secondary alcohol with an oxidizing agent. The product is a ketone: CH$_3$$\overset{\overset{O}{\|}}{C}CH_3$.

b. This is the reaction of an aldehyde with an oxidizing agent. The product is a carboxylic acid: CH$_3$COOH.

c. The reaction of an alcohol with an acid yields an ester and water:

$\overset{\overset{O}{\|}}{H C}OCH_3$ + H$_2$O

d. This is a saponification of an ester to form the sodium salt of a carboxylic acid and an alcohol:

CH$_3$CH$_2$$\overset{\overset{O}{\|}}{C}$ONa + CH$_3$OH

TRUE-FALSE QUESTIONS

1. The general formula of an alkane is C$_n$H$_{2n-2}$.

2. Alkenes contain double bonds.

3. The best name for the compound with the following structure is 4-methylhexane:

$$CH_3-CH_2-CH_2-\underset{\underset{CH_3}{|}}{CH}-CH_2-CH_3$$

4. The product of the addition reaction Cl$_2$ + CH$_3$CH = CH$_2$ is CH$_3$CHCl — CH$_2$Cl.

5. CH$_3$CH = CHCH$_3$ is the formula of an unsaturated hydrocarbon.

6. The product of the addition reaction CH$_3$CH = CH$_2$ + H$_2$O is CH$_3$CH$_2$CH$_2$OH.

7. The aldehyde functional group is —$\overset{\overset{O}{\|}}{C}$—OH.

8. Alcohols are easily oxidized.

9. The reaction of an alcohol and a carboxylic acid yields a ketone and water.

10. Amines contain the element nitrogen.

11. Alcohols exhibit hydrogen bonding, but ethers do not.

12. Saponification of C$_6$H$_{13}$COOC$_2$H$_5$ with NaOH yields C$_6$H$_{13}$COONa and C$_2$H$_5$OH.

SELF-TEST A

1. Give the systematic name for each of the following structural formulas:

   a. $CH_3(CH_2)_7CH_3$

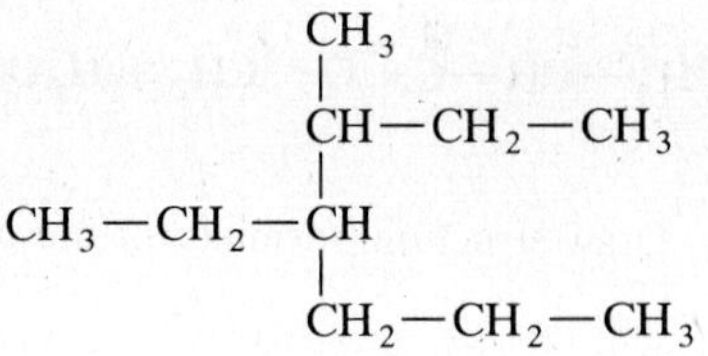

   b. 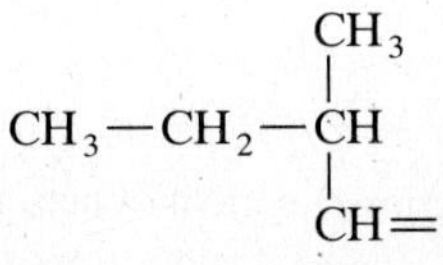 $CH_3-CH_2-\underset{\underset{CH=CH_2}{|}}{\overset{\overset{CH_3}{|}}{CH}}$

   c. 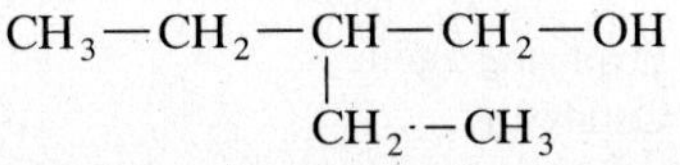 $CH_3-CH_2-\underset{\underset{CH_2-CH_3}{|}}{CH}-CH_2-OH$

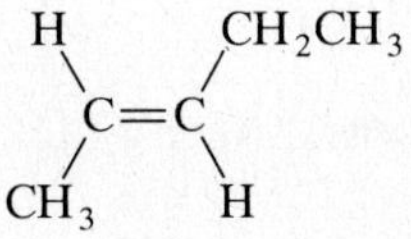

   d. 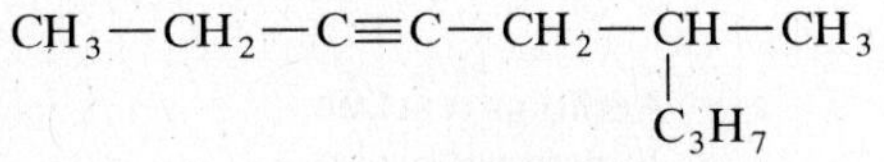 $CH_3-CH_2-C{\equiv}C-CH_2-\underset{\underset{C_3H_7}{|}}{CH}-CH_3$

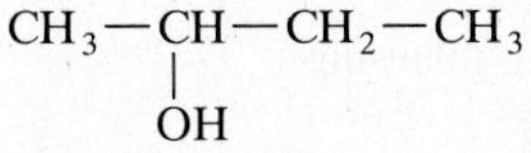

   e. 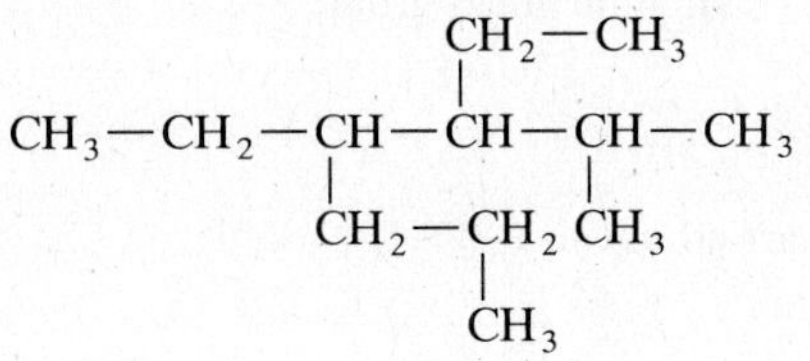 $CH_3-CH_2-\underset{\underset{CH_2-CH_2}{|}}{CH}-\underset{\underset{\underset{CH_3}{|}}{CH_3}}{\overset{\overset{CH_2-CH_3}{|}}{CH}}-\underset{\underset{CH_3}{|}}{CH}-CH_3$

   f. $CH_3-CH_2-\underset{\underset{CH_2-CH_2-CH_3}{|}}{\overset{\overset{\overset{CH_3}{|}}{CH-CH_2-CH_3}}{CH}}$

   g. $\underset{\underset{CH_3}{}}{\overset{\overset{H}{}}{C}}{=}\underset{\underset{H}{}}{\overset{\overset{CH_2CH_3}{}}{C}}$

   h. $CH_3-\underset{\underset{OH}{|}}{CH}-CH_2-CH_3$

   i. CH_3OH

2. Draw structural formulas for
   a. 3-ethyl-4-methyl-4-isopropylheptane
   b. 3-bromo-2,5-dimethyl-*trans*-3-hexene
   c. 3-methyl-3-hexanol
   d. Trichloroacetic acid

3. Draw structural formulas of all isomers of C_5H_{10}. Include both structural and geometric isomers.

4. List the functional groups by name that are shown in the following molecules:

   a. 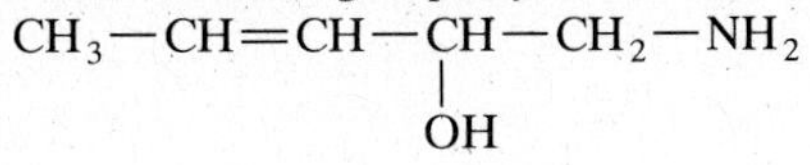 $CH_3-CH{=}CH-\underset{\underset{OH}{|}}{CH}-CH_2-NH_2$

   b. 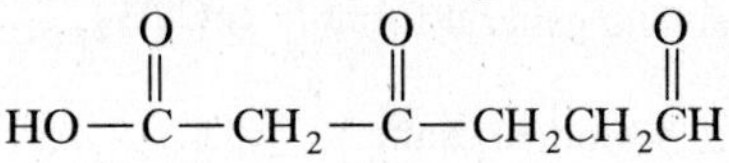 $HO-\overset{\overset{O}{\|}}{C}-CH_2-\overset{\overset{O}{\|}}{C}-CH_2CH_2\overset{\overset{O}{\|}}{CH}$

5. Give the structure of the organic product of each of the following reactions:

   a. 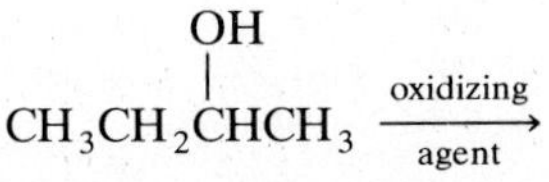 $CH_3CH_2\underset{\underset{}{}}{\overset{\overset{OH}{|}}{CH}}CH_3 \xrightarrow{\text{oxidizing agent}}$

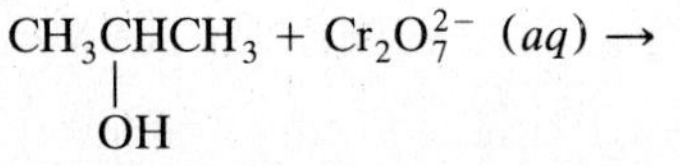

   b. 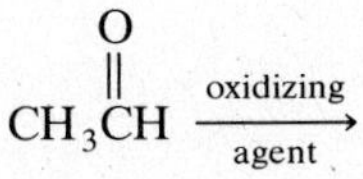 $CH_3\overset{\overset{O}{\|}}{CH} \xrightarrow{\text{oxidizing agent}}$

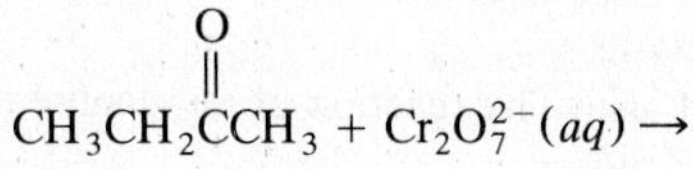

   c. $CH_3C{\equiv}CH + Cl_2 \rightarrow$

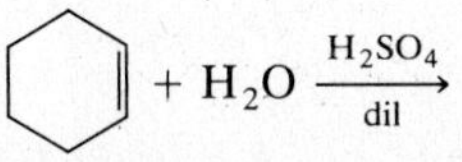

   d. $C_3H_8 + Cl_2 \xrightarrow{\text{light}}$

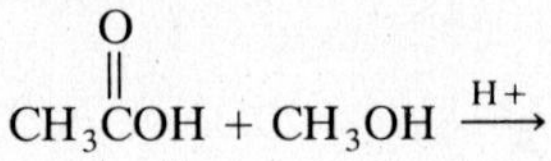

   e. 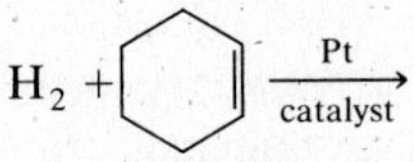 $H_2 + \bigcirc \xrightarrow{\text{Pt catalyst}}$

   f. $CH_3CH_2CH{=}CHCH_2CH_3 + H_2O \xrightarrow[\text{dil}]{H_2SO_4}$

   g. $CH_3\underset{\underset{OH}{|}}{CH}CH_3 + Cr_2O_7^{2-} (aq) \rightarrow$

   h. $CH_3CH_2\overset{\overset{O}{\|}}{C}CH_3 + Cr_2O_7^{2-} (aq) \rightarrow$

   i. $\bigcirc + H_2O \xrightarrow[\text{dil}]{H_2SO_4}$

   j. $CH_3\overset{\overset{O}{\|}}{C}OH + CH_3OH \xrightarrow{H+}$

   k. $CH_3CH_2CH{=}CHCH_3 + H_2 \xrightarrow{Pt}$

6. a. Draw structures of the hydrolysis products of the following ester:

$$CH_3-\underset{\underset{CH_3}{|}}{CH}-\underset{\underset{O}{\|}}{C}-O-CH_3 + H_2O \xrightarrow{H^+}$$

b. Draw structural formulas of the products of the following saponification reaction:

$$CH_3(CH_2)_{12}COO(CH_2)_4CH_3 + NaOH \rightarrow$$

7. A certain reagent converts methanol into methanoic acid. In this reaction the alcohol acts as which of the following?
   a. Lewis base
   b. Reducing agent
   c. Lewis acid
   d. Oxidizing agent
   e. Catalyst

8. Define the following:
   a. Alkene
   b. Homologous series
   c. Addition reaction
   d. Functional group

9. How many structural isomers are there of $C_4H_{10}O$?

10. Show how the following chemical change can be carried out in two steps:

$$CH_3CH_2CH_2OH \rightarrow CH_3\underset{\underset{OH}{|}}{CH}CH_3$$

ANSWERS

TRUE-FALSE QUESTIONS

1. False. The general formula is C_nH_{2n+2}.
2. True.
3. False. 3-Methylhexane.
4. True.
5. True.
6. False. $CH_3CH(OH)CH_3$. See Markovnikov's rule.
7. False.
8. True.
9. False. The reaction of an alcohol and an acid yields an ester.
10. True.
11. True.
12. True.

SELF-TEST A

1. a. *n*-nonane
   b. 3-methyl-1-pentene
   c. 2-ethyl-1-butanol
   d. 6-methyl-3-nonyne
   e. 3,4-diethyl-2-methylheptane
   f. 4-ethyl-3-methylheptane
   g. *trans*-2-pentene
   h. 2-butanol
   i. methanol

2. a.
$$CH_3-CH_2-\underset{\underset{\underset{CH_3}{|}}{\overset{|}{CH_3-CH}}}{\overset{\overset{C_2H_5}{|}}{CH}}-\underset{\underset{CH_3}{|}}{\overset{\overset{CH_3}{|}}{C}}-CH_2-CH_2-CH_3$$

c.
$$CH_3-CH_2-\underset{\underset{OH}{|}}{\overset{\overset{CH_3}{|}}{C}}-CH_2-CH_2-CH_2$$

b.

d. CCl_3-COOH

3. There are five alkenes: 1-pentene, 3-methyl-1-butene, 2-methyl-1-butene, and *cis*- and *trans*-2-pentene. There are three cycloalkanes: cyclopentane, methylcyclobutane, and dimethylcyclopropane.
4. a. Carbon-carbon double bond, hydroxyl, amine
   b. Carboxyl, carbonyl (ketone), carbonyl (aldehyde)
5.
   a. $CH_3CH_2\overset{\overset{O}{\|}}{C}CH_3$
   b. CH_3COOH
   c. $CH_3CCl_2CHCl_2$
   d. C_3H_7Cl, $C_3H_6Cl_2$, $C_3H_5Cl_3$, etc.
   e. (cyclohexane ring)
   f. $CH_3CH_2CH_2\underset{\underset{OH}{|}}{CH}CH_2CH_3$
   g. $CH_3\overset{\overset{O}{\|}}{C}CH_3$
   h. No reaction
   i. (cyclohexanol)
   j. $CH_3\overset{\overset{O}{\|}}{C}OCH_3$
   k. $CH_3CH_2CH_2CH_2CH_3$

6. a.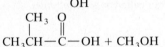
   b. $CH_3(CH_2)_{12}COONa + CH_3(CH_2)_4OH$

7. b
8. a. A straight-chain or branch-chain hydrocarbon with one or more double bonds
   b. Organic compounds with the same functional group that differ only in the number of carbon atoms
   c. A reaction in which a small molecule is added to an organic compound. The organic molecule must contain a $C=C$ bond or $C=O$ (carbonyl) group.
   d. A characteristic grouping of atoms that imparts certain chemical and physical properties when incorporated into organic molecules
9. Seven (four alcohols and three ethers)
10. $CH_3CH_2CH_2OH \xrightarrow[\text{conc}]{H_2SO_4} CH_3CH=CH_2 + H_2O$

   $CH_3CH=CH_2 + H_2O \xrightarrow[\text{dil}]{H_2SO_4} CH_3\underset{\underset{OH}{|}}{CH}CH_3$

Chapter Twenty-Six
ORGANIC POLYMERS: SYNTHETIC AND NATURAL

- Synthetic Organic Polymers
- Proteins
- Nucleic Acids

SYNTHETIC ORGANIC POLYMERS

STUDY OBJECTIVES

You should be able to:
1. Define *monomer* and give several examples of addition polymers (polyaddition).
2. Define *copolymer* and give several examples of condensation polymers (polycondensation).

Monomers. The word *polymer* means "many parts." A *polymer* is a compound with an unusually high molecular mass, consisting of a large number of small molecular units that are linked together. The small unit that is repeated many times is called a *monomer*. A typical polymer molecular contains a chain of monomers several thousand units long. Polymers are often called *macromolecules*.

Addition Polymers. Addition polymers are made by adding monomer to monomer until a long chain is produced. Ethylene and its derivatives are excellent monomers for addition polymers. In an addition reaction, the polymerization process is initiated by a radical or an ion. When ethylene is heated to 250°C under high pressure (1000–3000 atm) in the presence of a little oxygen or benzoyl peroxide (the initiator), addition polymers with molecular masses of about 30,000 amu are obtained. This reaction is represented by

$$n \underset{\text{ethylene}}{\overset{H}{\underset{H}{C}}=\overset{H}{\underset{H}{C}}} \rightarrow \underset{\text{a segment of polyethylene}}{-\overset{H}{\underset{H}{C}}-\overset{H}{\underset{H}{C}}-\overset{H}{\underset{H}{C}}-\overset{H}{\underset{H}{C}}-\overset{H}{\underset{H}{C}}-\overset{H}{\underset{H}{C}}-\overset{H}{\underset{H}{C}}-\overset{H}{\underset{H}{C}}-}$$

The general equation for addition polymerization is

$$n \underset{\text{monomer}}{\overset{H}{\underset{H}{C}}=\overset{H}{\underset{H}{C}}} \rightarrow \underset{\text{repeating unit}}{\left[\overset{H}{\underset{H}{C}}-\overset{H}{\underset{H}{C}} \right]_n}$$

Substitution of one or more hydrogen atoms in ethylene with Cl atoms, phenyl, acetate, CN, or F

460

provides a wide selection of monomers from which to make addition polymers with various properties. For instance, substitution of a Cl atom for a H atom in ethylene gives the monomer called vinyl chloride, $CH_2 = CHCl$. Polymerization of vinyl chloride yields the polymer polyvinyl chloride:

$$n \begin{array}{c} H \quad H \\ | \quad \; | \\ C=C \\ | \quad \; | \\ H \quad Cl \end{array} \rightarrow \left[\begin{array}{c} H \quad H \\ | \quad \; | \\ -C-C- \\ | \quad \; | \\ H \quad Cl \end{array} \right]_n$$

vinyl chloride monomer   polyvinyl chloride repeating unit

Table 26.1 in the text gives the names, structures, and uses of a number of monomers and addition polymers. Polymers that are made from one monomer such as polyvinyl chloride are called *homopolymers*.

Addition Polymerization. The addition polymerization process occurs by several identifiable steps. The first step is called initiation. This is followed by a chain growth process and finally a termination reaction.

1. *Initiation.* Polymerization is initiated by a radical or an ion. A radical (also called a free radical) is a species that has an unpaired electron. The symbol for a radical contains the usual letters for the elements followed by a dot to represent the unpaired electron. $CH_3 \cdot$ is the symbol for a methyl radical. The letter R is the general symbol for an alkyl group and $R\cdot$ for an alkyl radical. Peroxides with their $-O-O-$ bond are frequently chosen as sources of radicals.

   In a polymerization process, first the peroxide initiator ($RO-OR$) is dissociated by heating to yield two radicals. Then the peroxide radical adds to an ethylene molecule, which generates a new radical:

   $$RO-OR \xrightarrow{\Delta} 2RO\cdot$$

   $$RO\cdot + CH_2=CH_2 \rightarrow RO-CH_2-CH_2\cdot$$

2. *Chain growth.* Next, the free radical formed above can add to another molecule of ethylene:

   $$RO-CH_2-CH_2\cdot + CH_2=CH_2 \rightarrow RO-CH_2-CH_2-CH_2-CH_2\cdot$$

   The length of the carbon chain grows as the last free radical formed reacts with yet another ethylene molecule, and so on:

   $$RO-(CH_2-CH_2)_n-CH_2-CH_2\cdot$$

3. *Termination.* The polymerization process continues until a termination reaction occurs. When the radical ends of two chains meet, they may combine. When this happens, there is no new radical formed and the chain-lengthening process ceases. Of course, the result of this termination is the formation of a molecule of polyethylene, which may contain up to 800 carbon atoms:

   $$RO-(CH_2-CH_2)_n-CH_2-CH_2\cdot + \cdot CH_2-CH_2-(CH_2-CH_2)_n-OR \rightarrow$$
   $$RO-(CH_2-CH_2)_n-CH_2-CH_2-CH_2-CH_2-(CH_2-CH_2)_n-OR$$

Condensation Polymers. *Copolymers* are polymers that contain two or more different monomers. If monomers A and B are mixed and then polymerized, the following copolymers might be obtained:

   $-ABABAB$ (I)

   $-AABABB$ (II)

   $-BABBAA$ (III)

When A and B are linked by condensation reactions, the uniform copolymer I can be made. All the copolymers discussed in the text are of type I.

Polyesters, such as the well-known Dacron, are copolymers. When one monomer is an alcohol and the other is a carboxylic acid, they can be joined by an esterification reaction. The alcohol and the acid both must contain two functional groups. The monomers in polyester are the dicarboxylic acid called phthalic acid and the dialcohol called ethylene glycol:

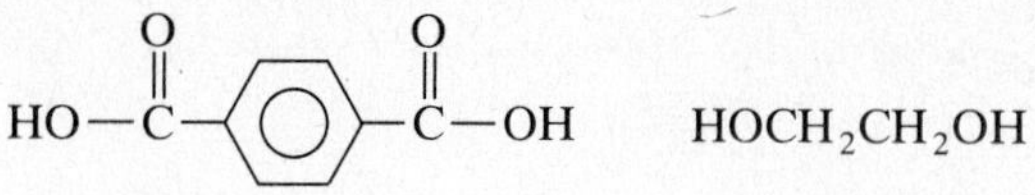

$$\text{HO—C(=O)—C}_6\text{H}_4\text{—C(=O)—OH} \qquad \text{HOCH}_2\text{CH}_2\text{OH}$$

phthalic acid   ethylene glycol

Condensation reactions differ from addition reactions in that the former always result in the formation of a small molecule such as water. Polyesters are produced from the esterification reaction between an alcohol and an acid. When phthalic acid and ethylene glycol react to form an ester, the first products are

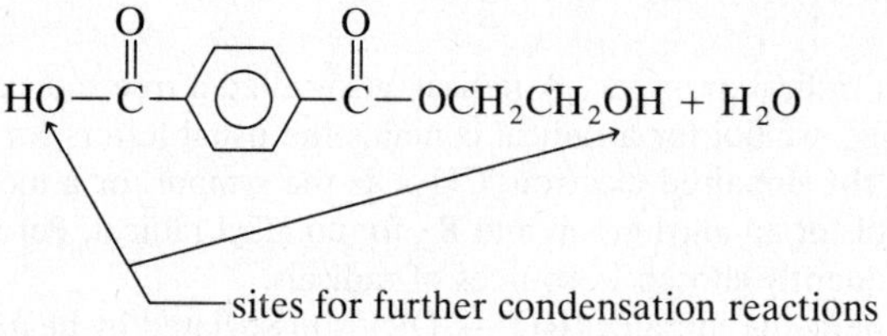

$$\text{HO—C(=O)—C}_6\text{H}_4\text{—C(=O)—OCH}_2\text{CH}_2\text{OH} + \text{H}_2\text{O}$$

—sites for further condensation reactions

When this product reacts with another molecule of the diacid, the polymer chain grows longer:

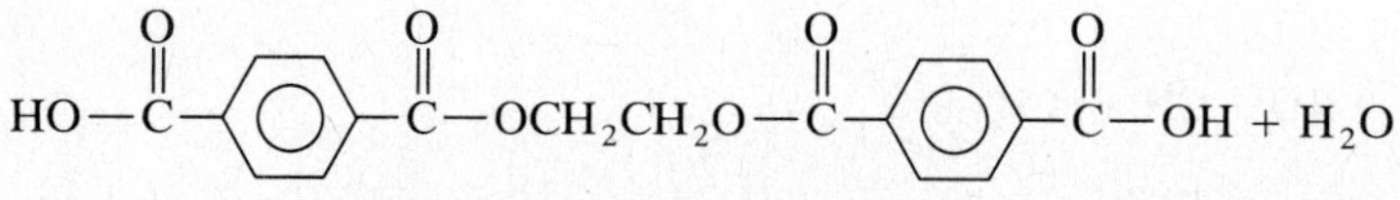

$$\text{HO—C(=O)—C}_6\text{H}_4\text{—C(=O)—OCH}_2\text{CH}_2\text{O—C(=O)—C}_6\text{H}_4\text{—C(=O)—OH} + \text{H}_2\text{O}$$

segment of a condensation polymer chain

The general formula for this polyester is

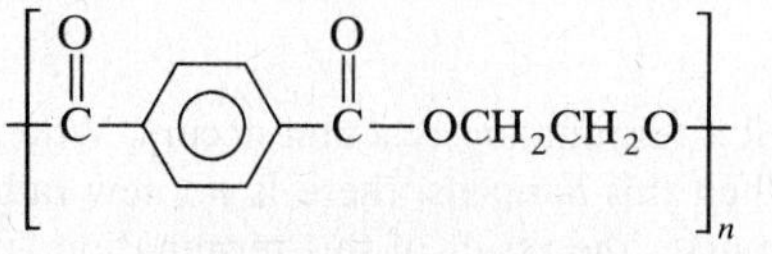

$$\left[\text{C(=O)—C}_6\text{H}_4\text{—C(=O)—OCH}_2\text{CH}_2\text{O} \right]_n$$

EXAMPLE 26.1 Monomers and Polymers

Write the formulas of the monomers used to prepare the following polymers:
a. Teflon
b. Polystyrene
c. PVC

METHOD OF SOLUTION

Refer to Table 26.1 of the text.
a. Teflon is an addition polymer with the formula $+\text{CF}_2-\text{CF}_2+_n$. It is prepared from the monomer tetrafluoroethylene ($\text{CF}_2 = \text{CF}_2$).

b. The monomer used to prepare polystyrene is styrene:

$-(CH-CH_2)_n-$ $CH=CH_2$
   | |
   (phenyl)   (phenyl)

polystyrene   styrene

c. Polyvinylchloride is prepared by the successive addition of vinyl chloride molecules ($CH_2 = CHCl$).

EXAMPLE 26.2 Condensation Polymers

List three examples of condensation polymers mentioned in the textbook.
Answer: Nylon 66 (first prepared in 1931), polyester, and polyurethane.

PROTEINS

STUDY OBJECTIVES

You should be able to:
1. Write a general formula for an α-amino acid.
2. Describe the condensation reaction between two amino acids that results in the formation of a dipeptide.
3. Discuss the terms used to describe the four levels of protein structure.

Proteins. Proteins are truly giant molecules having molecular masses that range from about 10,000 to several million amu. Proteins play many roles in living organisms, where they function as catalysts (enzymes), transport molecules (hemoglobin), contractile fibers (muscle), protective agents (blood clots), hormones (chemical messengers), and structural members (feathers, horn, nails). The word *protein* comes from the Greek word *proteios*, meaning "first." From this partial list of functions it is easy to see why proteins occupy "first place" among biomolecules in their importance to life.

Amino Acids. Even though each protein is unique, all proteins are built from the same set of amino acids. An amino acid consists of an amino group, a carboxylic acid group, a hydrogen atom, and a distinctive R group, all bonded to the same carbon atom:

$$R - \underset{\underset{NH_2}{|}}{\overset{\overset{H}{|}}{C}} - \overset{O}{\underset{}{\overset{\|}{C}}} - OH$$

← common to all amino acids

side chain (R), α-carbon atom

All amino acids in proteins have a common structural feature. This is the attachment of the amino group to the carbon atom adjacent to the carboxylic acid group. This carbon atom is called the α carbon, meaning the first carbon from the carboxylic acid functional group. Thus, these acids are sometimes called α-amino acids. The R group is different from each amino acid, and some 20 different R groups, or *side chains* as they are called, are found in the proteins from natural sources. Indeed, all proteins in all species, from bacteria to humans, are constructed from the same set of 20 amino acids. The structural formulas of the amino acids are shown in Table 26.2 of the text.

In proteins the amino acid units are hooked together to form a polypeptide chain. The carboxyl group of one amino acid is joined to the amino group of another amino acid by the formation of a peptide bond:

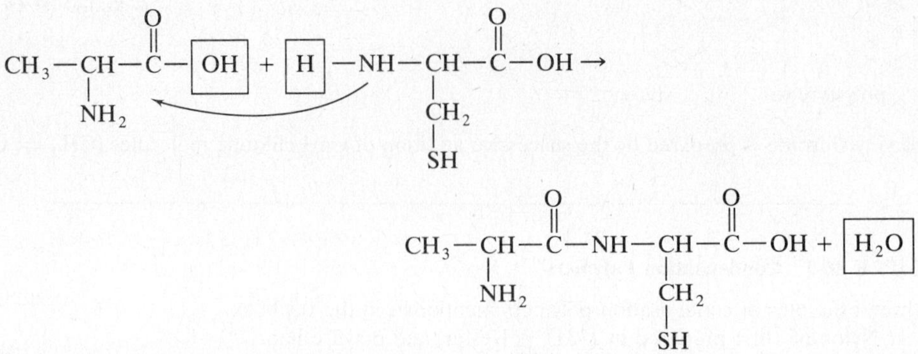

$$CH_3-CH(NH_2)-C(=O)-\boxed{OH} + \boxed{H}-NH-CH(CH_2SH)-C(=O)-OH \rightarrow$$

$$CH_3-CH(NH_2)-C(=O)-NH-CH(CH_2SH)-C(=O)-OH + \boxed{H_2O}$$

This type of reaction is another example of a condensation reaction. The new C — N covalent bond is called a *peptide bond* or an *amide bond*. The amide functional group present in all proteins is

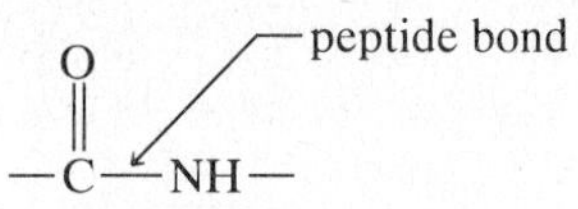

$$-C(=O)-NH- \quad \text{(peptide bond)}$$

The molecule above, in which two amino acids are joined, is called a *dipeptide*. Peptides are structures intermediate in size between amino acids and proteins. The term *polypeptide* refers to long molecular chains containing many amino acid units. An amino acid unit in a polypeptide chain is called a residue.

Disulfide Bonds. In some proteins, the polypeptide chains are cross-linked by disulfide bonds. Cysteine is a sulfur-containing amino acid. These cross-links are formed when two cysteine residues on the chain, or on two different chains, are oxidized (loss of H atoms). The disulfide bond is a covalent bond that is much stronger than a hydrogen bond:

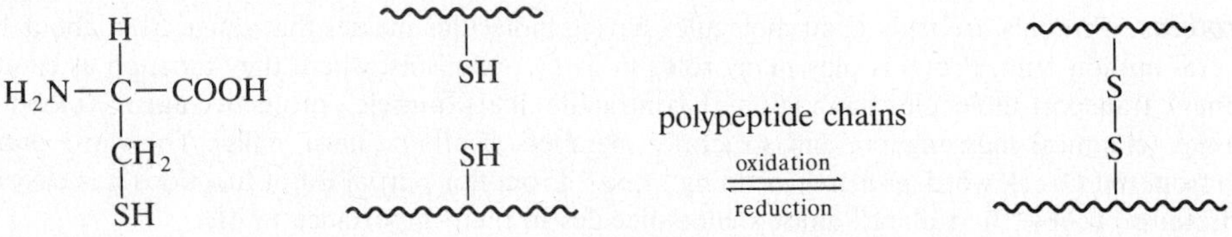

$$H_2N-CH(CH_2SH)-COOH \quad \text{cysteine}$$

polypeptide chains with SH groups ⇌ (oxidation/reduction) a disulfide bond (S–S)

Structure of Proteins. Proteins are so complex that four levels of structural features have been identified. The structure of proteins is extremely important in determining just how efficiently and effectively a protein will function. A stretched-out or unfolded polypeptide chain does not exhibit biological activity and is said to be *denatured*. The four levels of protein structure are summarized below.

1. *The Primary Structure.* Each protein has a unique amino acid sequence of its polypeptide chain. It is the amino acid sequence that distinguishes one protein from another. Proteins differ in the numbers and kinds of amino acids, but especially in the sequence of amino acid residues.
2. *The Secondary Structure.* This refers to the spatial relationship of amino acid units that are close to one another in sequence. The configuration that appears in many proteins is the α helix, shown in Figure 26.13 in the textbook. In this configuration the polypeptide is coiled much like the arrangement of stairs in a spiral staircase. The tightly coiled polypeptide chain forms the inner part of the spiral stairway, and the side chains extend outward forming a helical pattern. The figure shows that the α helix is stabilized by the presence of hydrogen bonds (dashed lines) between the C = O and NH groups in the peptide chains. The CO group of each amino acid is hydrogen-bonded to the NH group of the amino acid that is located four amino acids ahead in the sequence. Thus

amino acids spaced four apart in the sequence are actually close to one another due to the coiled or spiral arrangement.

The α helix is the main structural feature of the oxygen-storage protein, myoglobin, and of many other proteins. The presence of this feature in a polypeptide chain is shown in Figure 26.14 in the text. The β-pleated sheet is another common protein structure. In this structure a polypeptide chain interacts strongly with adjacent chains by forming many hydrogen bonds (Figure 26.15 in the text).

3. *The Tertiary Structure*. Myoglobin, like many other proteins, is a globular protein. The polypeptide chain is folded into a compact globular shape. The folding of the chain results in some amino acid units being in very close proximity to each other even though they are widely separated in the amino acid sequence. The term *tertiary structure* refers to the spatial relationship of amino acid units that are far apart in the sequence, or more simply, to just its three-dimensional structure. In myoglobin, for example, the tertiary structure is a unique three-dimensional shape resulting from the folding of the polypeptide chain. The folding takes place quite naturally in aqueous medium due to the interactions of hydrophobic and hydrophilic side chains with water. The tertiary structure is stabilized by hydrogen bonding, dispersion forces, and ionic forces.

4. *The Quaternary Structure*. Proteins that consist of more than one polypeptide chain exhibit an additional level of structural organization called the *quaternary structure*. This structural feature involves the way in which separate chains fit together. Hemoglobin, the oxygen-carrying molecule of the blood, exhibits a quaternary structure (Figure 26.19 in the textbook). It consists of four separate polypeptide chains or subunits (labeled $\alpha_1, \alpha_2, \beta_1, \beta_2$). The quaternary structure results from interaction between chains. Ionic forces and hydrogen bonds are important in holding the subunits together.

Prosthetic Groups. Many proteins require a tightly bound, non–amino acid unit for their biological activity. Such a unit is called a *prosthetic group*. The combination of a polypeptide chain and a prosthetic group is called a *conjugated protein*. In addition to their polypeptide chains, myoglobin and hemoglobin each contain a prosthetic group called a *heme group*. Heme groups serve to bind molecular oxygen. The heme group has the same structure in myoglobin and in hemoglobin and is responsible for the red color of blood and of muscle tissue. Heme contains an iron ion enclosed in a planar grouping called a *porphyrin*. Iron is the atom in the heme group to which the O_2 molecule is attached.

EXAMPLE 26.3 Amino Acids

Find and name five amino acids in Table 26.2 of the textbook that have polar side chains.

METHOD OF SOLUTION

Recall that polar groups are those that have relatively large electronegativity differences between bonded atoms. Thus, — COOH, — OH, and — SH are polar side chains. *Answer*: The side chains in *aspartic acid* and *glutamic acid* contain carboxylic acid groups. The side chains in *serine* and *threonine* contain hydroxyl groups. And the side chain in cysteine contains a polar — SH group.

EXAMPLE 26.4 Proteins

What are the five chemical elements found in proteins?

METHOD OF SOLUTION

Proteins are made up of the same elements as the amino acids. Therefore, proteins contain C, H, O, and N. Two amino acids—methionine and cysteine—also contain S.

EXAMPLE 26.5 A Polypeptide Chain

Sketch a portion of a polypeptide chain consisting of the amino acids cysteine, glycine, valine, and phenylalanine. Point out the peptide bonds and amide groups.

METHOD OF SOLUTION

The main backbone of a polypeptide chain is made up of the α-carbon atoms and the amide group repeating alternately along the chain:

$$-\underset{\alpha\text{ carbons}}{CH}\underset{R}{|}-\underset{\|}{\overset{O}{C}}-NH-\underset{R}{\overset{|}{CH}}-\underset{\|}{\overset{O}{C}}-NH-$$

(amide groups are the boxed C(=O)—NH segments)

Table 26.2 in the text gives the structures of the amino acids. For each R group shown above, substitute the distinctive side groups of the four amino acids. The chain structure will be

$$-\underset{\text{cysteine}}{CH(CH_2SH)-C(=O)-NH}-\underset{\text{glycine}}{CH(H)-C(=O)-NH}-\underset{\text{valine}}{CH(CH(CH_3)_2)-C(=O)-NH}-\underset{\text{phenylalanine}}{CH(CH_2C_6H_5)-C(=O)-NH}-$$

Arrows mark the peptide bonds.

EXAMPLE 26.6 Tertiary Structure of Proteins

What kinds of forces stabilize the tertiary structures of proteins?

METHOD OF SOLUTION

The term *tertiary structure* is given to the folded three-dimensional structure. The polypeptide chain folds into a specific three-dimensional structure under the influence of hydrophobic and hydrophilic interactions. The hydrophilic side groups (polar or ionic groups are attracted to water) are spread uniformly over the outside surface of the folded polypeptide chain where they can interact with water via hydrogen bonding and ion-dipolar interactions. The chain folds in such a way that the hydrophobic side groups (nonpolar groups are repelled by water) appear on the "inside" of the molecule where water molecules cannot reach.

NUCLEIC ACIDS

STUDY OBJECTIVES

You should be able to:
1. Describe the composition of nucleic acids.
2. Distinguish chemical and structural differences between DNA and RNA.
3. Draw structures of nucleotides and indicate where they link together to form strands of nucleic acids.

The chemical composition of the cell nucleus was first studied in the 1860s by Friedrich Miescher. He found the major components to be protein and a new material not previously isolated. This material was found to be acidic, and so it was referred to as *nucleic acid*.

Nucleic acids are now known to be giant molecules with molecular masses in the range 1 billion to 10 billion amu. These molecules carry information in the form of the genetic code. Enough information is stored in nucleic acid molecules to allow the complete assembly of an entire organism. The nucleic acid molecule called *DNA* carries this information from generation to generation.

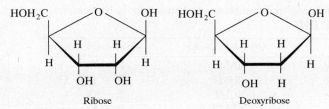

Figure 26.1. The two sugars present in nucleic acids. Ribonucleic acid contains ribose, whereas deoxyribonucleic acid contains deoxyribose.

Hydrolysis of nucleic acid shows that it is composed of 1 part phosphoric acid, 1 part sugar, and 1 part nitrogen base. One of two sugars is present—ribose or deoxyribose (Figure 26.1). Five different nitrogen bases are found in nucleic acids. Their names are adenine, thymine, guanine, cytosine, and uracil (Figure 26.23 in the text).

Two types of nucleic acids are recognized. These are called DNA and RNA. The compositions of DNA and RNA are quite similar, but they differ in two significant ways. DNA contains the sugar deoxyribose and is called deoxyribonucleic acid. RNA contains the sugar ribose, hence the name ribonucleic acid. The second difference is that DNA contains the four bases adenine, thymine, guanine, and cytosine, while RNA contains adenine, uracil, guanine, and cytosine. DNA contains thymine but no uracil, whereas RNA contains uracil but no thymine.

Table 26.1 summarizes the building blocks of DNA and RNA and gives useful abbreviations.

Table 26.1 Component Parts of DNA and RNA

	DNA	RNA
Acid	Phosphoric acid (P)	Phosphoric acid (P)
Sugar	Deoxyribose (D)	Ribose (R)
N bases	Adenine (A)	Adenine (A)
	Thymine (T)	Uracil (U)
	Guanine (G)	Guanine (G)
	Cytosine (C)	Cytosine (C)

Nucleotides. The repeating unit in nucleic acids is called a nucleotide. It is a combination of a phosphate group, a five-carbon sugar, and a nitrogen base. See Figure 26.22 of the text for the full structure of the nucleotide deoxyadenosine monophosphate (dAMP), which contains the sugar deoxyribose and the base adenine. Figure 26.2 shows the structure of the nucleotide containing the base

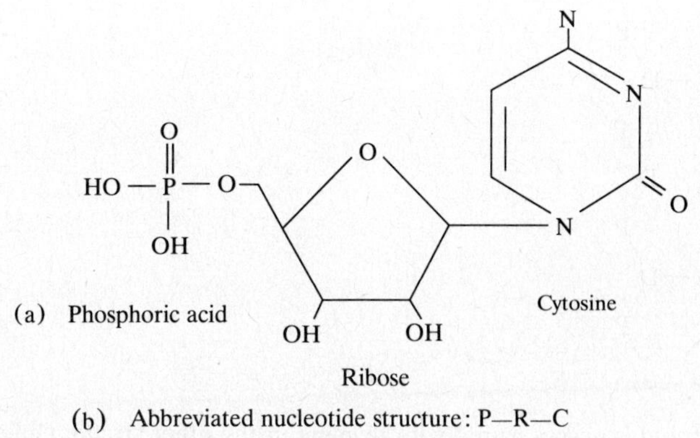

Figure 26.2. (*a*) The molecular structure of the nucleotide cytosine monophosphate (CMP). (*b*) The abbreviated structure of CMP.

Table 26.2 Abbreviated Names and Structures of Nucleotides Found in DNA

Base	Abbreviated Name	Abbreviated Structure
Adenine	dAMP	P — D — A
Guanine	dGMP	P — D — G
Cytosine	dCMP	P — D — C
Thymine	dTMP	P — D — T

cytosine and the sugar ribose, called cytosine monophosphate (CMP). To simplify DNA and RNA structures, we will use the abbreviated nucleotide structures shown in Table 26.2 in our discussion.

A strand of DNA or RNA is constructed by linking the nucleotides together with a bond from the sugar unit of one nucleotide to the phosphate unit of another nucleotide. Such a strand of DNA is represented in abbreviated form as follows:

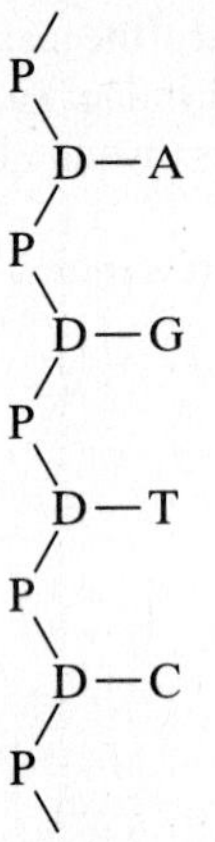

```
   /
  P
   \
    D—A
   /
  P
   \
    D—G
   /
  P
   \
    D—T
   /
  P
   \
    D—C
   /
  P
   \
```

In the 1940s, Edwin Chargaff studied the composition of DNA. His analysis of the base composition of DNA showed that the amount of adenine (A) always equaled that of thymine (T) and that the amount of guanine (G) equaled that of cytosine (C). These relations became known as Chargaff's rules. Watson and Crick in 1953 proposed a two-stranded structure for DNA that provided an explanation of Chargaff's rules. The two-stranded structure can be shown as follows:

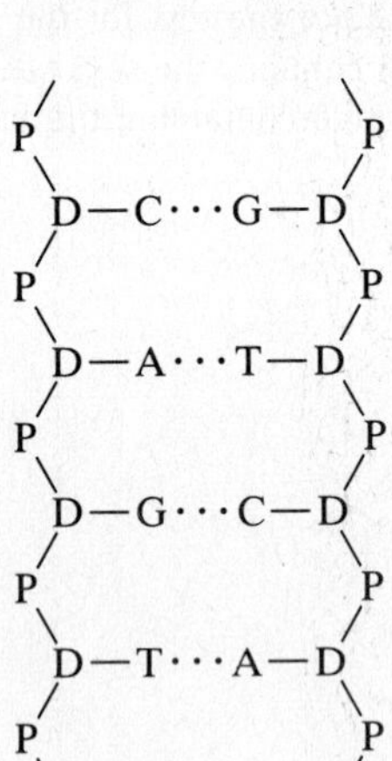

```
   /         \
  P           P
   \         /
    D—C···G—D
   /         \
  P           P
   \         /
    D—A···T—D
   /         \
  P           P
   \         /
    D—G···C—D
   /         \
  P           P
   \         /
    D—T···A—D
   /         \
  P           P
   \         /
```

Adenine in one strand is always paired with thymine in the other strand. Guanine is always paired with cytosine. The two strands are not identical; rather they are *complementary*. Base pairing and the resulting association of the two strands are the result of hydrogen bonding. Hydrogen atoms in a base in one strand are attracted to unshared electron pairs on oxygen and nitrogen atoms of a base attached to the other strand [see Figure 26.23(a) in textbook]. X-ray data suggested that DNA had the helical structure shown in Figure 26.23(b) of the text.

RNA, on the other hand, does not follow the base-pairing rules. X-ray data and other evidence ruled out a double-helical structure for RNA. RNA is single-stranded.

EXAMPLE 26.7   Nucleotide Structures

Draw abbreviated structures of the four different nucleotides that appear in RNA.

METHOD OF SOLUTION

Nucleotides contain three main groups: phosphate (P), a five-carbon sugar, and a nitrogen base. In RNA the sugar is ribose, symbolized by R. The four bases in RNA are symbolized by A, U, C, and G. Therefore, the nucleotides are

P — R — A

P — R — U

P — R — C

P — R — G

EXAMPLE 26.8   Base Pairing

What types of forces cause base pairing in the double-stranded helical DNA molecule?

METHOD OF SOLUTION

The purine bases (thymine and cytosine) in one strand of DNA form hydrogen bonds to the pyrimidine bases (adenine and guanine) in the other DNA strand. Hydrogen atoms covalently bonded to nitrogen carry a partial positive charge. These H atoms are attracted to lone-electron pairs on oxygen and nitrogen atoms of another base. Since two complementary bases are attached to different strands, the hydrogen bonds hold the strands together.

TRUE-FALSE QUESTIONS

1. Polythylene contains double bonds.

2. The monomer used to make Teflon is C_2F_4.

3. Condensation polymers are copolymers.

4. Polypeptide chains consist of amino acid residues linked by peptide bonds.

5. A denatured protein could have the same primary structure as the active (undenatured) protein.

6. The α helix is an example of a primary structure of a protein.

7. Heme is an example of a prosthetic group.

8. Hemoglobin is a conjugated protein.

9. Nucleic acids are giant molecules constructed from five different nucleotide monomers.

10. A nucleotide monomer in RNA contains a phosphate group, a deoxyribose unit (a sugar), and four nitrogen bases (A, U, C, and G).

11. The nitrogen bases in one strand of DNA pair with bases in the other strand by forming peptide bonds.

12. Adenine pairs with guanine.

SELF-TEST A

1. Sketch the structures of the monomers from which these polymers are formed:

    a.
    $$-\underset{\underset{H}{|}}{\overset{\overset{Cl}{|}}{C}}-\underset{\underset{CH_3}{|}}{\overset{\overset{H}{|}}{C}}-\underset{\underset{H}{|}}{\overset{\overset{Cl}{|}}{C}}-\underset{\underset{CH_3}{|}}{\overset{\overset{H}{|}}{C}}-\underset{\underset{H}{|}}{\overset{\overset{Cl}{|}}{C}}-\underset{\underset{CH_3}{|}}{\overset{\overset{H}{|}}{C}}-$$

    b. $-CH_2-CCl_2-CH_2-CCl_2-CH_2-CCl_2-$

2. Draw structures for the monomers used to make the following polyester:

    $$\left[-\overset{\overset{O}{\|}}{C}-(CH_2)_4-\overset{\overset{O}{\|}}{C}-O-(CH_2)_3-O- \right]_n$$

3. Sketch the structure of the monomer of natural rubber.

4. What chemical elements are found in proteins?

5. List several functions of proteins in living systems.

6. What is the role of enzymes in biochemical systems?

7. Describe the structure of an α-amino acid.

8. Describe a dipeptide and polypeptide.

9. Draw the structural formula of the tripeptide formed from the sequence of amino acids below (see Table 26.2 of the textbook):

    Ser—Cys—Lys

    Label the peptide bonds and side chains.

10. Consider two polypeptide chains: (1) Lue — Phe — Pro — Gly — Ala and
    (2) Gly — Ser — Lys — Asp — Tyr.
    a. Which would be more soluble in water?
    b. Which would be more soluble in a nonpolar solvent?

11. Explain how the primary structure of a protein differs from the tertiary structure.

12. Distinguish between the following:
    a. A purine and a pyrimidine base
    b. Ribose and deoxyribose

13. List four differences between DNA and RNA.

14. Sketch the molecular structure of the nucleotide consisting of phosphate (P), deoxyribose (D), and adenine (A).

15. What is the biological role of DNA?

16. If the base sequence in one strand of DNA is A, T, G, C, T, then the base sequence in the complementary strand is _____, _____, _____, _____, _____.

ANSWERS

TRUE-FALSE QUESTIONS

1. False. The structure is $-CH_2-CH_2-CH_2-CH_2-$.
2. True.
3. True.
4. True.
5. True.
6. False. The α helix is an example of a secondary structure.
7. True.
8. True.
9. True.
10. False. RNA contains ribose, not deoxyribose.
11. False. The complementary bases in DNA are attracted by hydrogen bonds.
12. False. A pairs with C.

SELF-TEST A

1. a. $CH_3CH = CHCl$ b. $CH_2 = CCl_2$
2. $$HOC\overset{O}{\overset{\|}{-}}(CH_2)_4-\overset{O}{\overset{\|}{C}}OH \quad \text{and} \quad HO-(CH_2)_3-OH$$
3. $$CH_2=\underset{\underset{CH_3}{|}}{C}-CH=CH_2$$
4. C, H, O, N, S
5. Catalysts, transport, contractile, protection, hormones, structural elements
6. Enzymes are catalysts.
7. Amino acids consist of an amino group, a carboxylic acid group, a hydrogen atom, and side chain all bonded to the same carbon atom.
8. A dipeptide is two amino acids linked by a peptide bond. A polypeptide consists of many amino acid residues each linked to the next by peptide bonds.
9.

$$H_2N-CH-\overset{O}{\overset{\|}{C}}-NH-CH-\overset{O}{\overset{\|}{C}}-NH-CH-\overset{O}{\overset{\|}{C}}-OH$$

with side chains: CH_2-OH, CH_2-SH, and $CH_2-CH_2-CH_2-NH_2$

10. a. 2   b. 1

11. The primary structure of a protein refers to the number and the sequence of amino acids within the polypeptide chain. The chain twists and folds so that some amino acids far apart in sequence come into proximity with each other. The tertiary structure is the three-dimensional structure that shows the folding of the polypeptide chain.
12. a. Purines are nitrogen bases that consist of two-ring systems, such as in adenine and guanine. Pyrimidines have only one ring, as in thymine and cytosine.
    b. Ribose, $C_5H_{10}O_5$; deoxyribose, $C_5H_{10}O_4$. See also the structures given in the section titled Nucleic Acids.
13. (1) RNA contains uracil but no thymine; DNA contains thymine but no uracil. (2) RNA contains ribose as the 5-carbon sugar; DNA contains deoxyribose. (3) RNA is single-stranded; DNA is double helical. (4) DNA is found only in the nucleus; RNA is found outside the nucleus.

15. DNA is the molecule of heredity; it is responsible for passing genetic information from one generation to the next.
16. T, A, C, G, A